Volkswagen

Beetle
Super Beetle
Karmann Ghia

Official Service Manual Type 1

1970, 1971, 1972, 1973, 1974, 1975, 1976, 1977, 1978, 1979

Volkswagen Service Manuals from Robert Bentley

Volkswagen Fox Service Manual: 1987-1993, including GL, GL Sport and Wagon. Robert Bentley
ISBN 0-8376-0363-3
Volkswagen Part No. LPV 800 504

Cabriolet, Scirocco Service Manual: 1985-1993, including 16V. Robert Bentley. ISBN 0-8376-0362-5
Volkswagen Part No. LPV 800 113

Corrado Official Factory Repair Manual: 1990-1993
Volkswagen United States. ISBN 0-8376-0387-0
Volkswagen Part No. LPV 800 300

GTI, Golf, Jetta Service Manual: 1985-1992 Gasoline, Diesel, and Turbo Diesel, including 16V.
Robert Bentley. ISBN 0-8376-0342-0
Volkswagen Part No. LPV 800 112

Passat Service Manual: 1990-1992 including Wagon.
Robert Bentley. ISBN 0-8376-0377-3
Volkswagen Part No. LPV 800 204

Passat Electrical Troubleshooting Manual: 1990-1991
Robert Bentley. ISBN 0-8376-0381-1
Volkswagen Part No. LPV 800 150

Vanagon Official Factory Repair Manual: 1980-1991
Including Diesel, Syncro and Camper.
Volkswagen United States. ISBN 0-8376-0336-6
Volkswagen Part No. LPV 800 148

Quantum Official Factory Repair Manual: 1982-1988
Gasoline and Turbo Diesel, including Wagon and Syncro.
Volkswagen United States. ISBN 0-8376-0341-2
Volkswagen Part No. LPV 800 202

Rabbit, Scirocco, Jetta Service Manual: 1980-1984
Gasoline Models, including Pickup Truck, Convertible, and GTI. Robert Bentley. ISBN 0-8376-0183-5
Volkswagen Part No. LPV 800 104

Rabbit, Jetta Service Manual: 1977-1984 Diesel Models, including Pickup Truck and Turbo Diesel.
Robert Bentley. ISBN 0-8376-0184-3
Volkswagen Part No. LPV 800 122

Rabbit, Scirocco Service Manual: 1975-1979 Gasoline Models. Robert Bentley. ISBN 0-8376-0107-X
Volkswagen Part No. LPV 997 174

Dasher Service Manual: 1974-1981 including Diesel.
Robert Bentley. ISBN 0-8376-0083-9
Volkswagen Part No. LPV 997 335

Super Beetle, Beetle and Karmann Ghia Official Service Manual Type 1: 1970-1979
Volkswagen United States. ISBN 0-8376-0096-0
Volkswagen Part No. LPV 997 109

Beetle and Karmann Ghia Official Service Manual Type 1: 1966-1969 Volkswagen United States.
ISBN 0-8376-0416-8
Volkswagen Part No. LPV 997 169

Station Wagon/Bus Official Service Manual Type 2: 1968-1979 Volkswagen United States.
ISBN 0-8376-0094-4
Volkswagen Part No. LPV 997 288

Fastback and Squareback Official Service Manual Type 3: 1968-1973 Volkswagen United States.
ISBN 0-8376-0057-X
Volkswagen Part No. LPV 997 383

Volkswagen 1200 Workshop Manual: 1961-1965, Types 11, 14 and 15 Volkswagen United States.
ISBN 0-8376-0390-0
Volkswagen Part No. LPV 800 121

Volkswagen Transporter Workshop Manual: 1963-1967, Type 2 all models including Kombi, Micro Bus, Micro Bus De Luxe, Pick-up, Delivery Van and Ambulance.
Volkswagen United States. ISBN 0-8376-0391-9
Volkswagen Part No. LPV 800 135

Audi Service Manuals from Robert Bentley

Audi 100, 200 Official Factory Repair Manual: 1989-1991 Audi of America. ISBN 0-8376-0372-2
Audi Part No. LPV 800 701

Audi 80, 90, Coupe Quattro Official Factory Repair Manual: 1988-1991 including 80 quattro, 90 quattro and 20-valve models. Audi of America. ISBN 0-8376-0367-6
Audi Part No. LPV 800 601

Audi 80, 90, Coupe Quattro Electrical Troubleshooting Manual: 1988-1992
Robert Bentley. ISBN 0-8376-0375-7
Audi Part No. LPV 800 603

Audi 5000S, 5000CS Official Factory Repair Manual: 1984-1988 Gasoline, Turbo, and Turbo Diesel, including Wagon and Quattro. Audi of America. ISBN 0-8376-0370-6
Audi Part No. LPV 800 445

Audi 5000, 5000S Official Factory Repair Manual: 1977-1983 Gasoline and Turbo Gasoline, Diesel and Turbo Diesel. Audi of America. ISBN 0-8376-0352-8
Audi Part No. LPV 800 443

Audi 4000S, 4000CS, and Coupe GT Official Factory Repair Manual: 1984-1987 including Quattro and Quattro Turbo. Audi of America. ISBN 0-8376-0373-0
Audi Part No. LPV 800 424

Audi 4000, Coupe Official Factory Repair Manual: 1980-1983 Gasoline, Diesel, and Turbo Diesel.
Audi of America. ISBN 0-8376-0349-8
Audi Part No. LPV 800 422

Audi Fox Service Manual: 1973-1979
Robert Bentley. ISBN 0-8376-0097-9
Audi Part No. LPA 997 082

Volkswagen

Beetle
Super Beetle
Karmann Ghia

Official Service Manual Type 1

1970, 1971, 1972, 1973, 1974,
1975, 1976, 1977, 1978, 1979

Volkswagen of America, Inc.

Published and Distributed by:

Robert Bentley, Publishers
1000 Massachusetts Avenue
Cambridge, Massachusetts 02138

CAUTION—Important Safety Notice

Do not use this manual unless you are familiar with basic automotive repair procedures and safe workshop practices. This manual illustrates the workshop procedures required for most service work; it is not a substitute for full and up-to-date information from the vehicle manufacturer or for proper training as an automotive technician. Note that it is not possible for us to anticipate all of the ways or conditions under which vehicles may be serviced or to provide cautions as to all of the possible hazards that may result.

The vehicle manufacturer will continue to issue service information updates and parts retrofits after the editorial closing of this manual. Some of these updates and retrofits will apply to procedures and specifications in this manual. We regret that we cannot supply updates to purchasers of this manual.

We have endeavored to ensure the accuracy of the information in this manual. Please note, however, that considering the vast quantity and the complexity of the service information involved, we cannot warrant the accuracy or completeness of the information contained in this manual.

FOR THESE REASONS, NEITHER THE PUBLISHER NOR THE AUTHOR MAKES ANY WARRANTIES, EXPRESS OR IMPLIED, THAT THE INFORMATION IN THIS BOOK IS FREE OF ERRORS OR OMISSIONS AND WE EXPRESSLY DISCLAIM THE IMPLIED WARRANTIES OF MERCHANTABILITY AND OF FITNESS FOR A PARTICULAR PURPOSE, EVEN IF THE PUBLISHER OR AUTHOR HAVE BEEN ADVISED OF A PARTICULAR PURPOSE, AND EVEN IF A PARTICULAR PURPOSE IS INDICATED IN THE MANUAL. THE PUBLISHER AND AUTHOR ALSO DISCLAIM ALL LIABILITY FOR DIRECT, INDIRECT, INCIDENTAL OR CONSEQUENTIAL DAMAGES WHICH RESULT FROM ANY USE OF THE EXAMPLES, INSTRUCTIONS OR OTHER INFORMATION IN THIS BOOK. IN NO EVENT SHALL OUR LIABILITY WHETHER IN TORT, CONTRACT OR OTHERWISE EXCEED THE COST OF THIS MANUAL.

Your common sense and good judgment are crucial to safe and successful service work. Read procedures through before starting them. Think about whether the condition of your car, your level of mechanical skill, or your level of reading comprehension might result in or contribute in some way to an occurrence which might cause you injury, damage your car, or result in an unsafe repair. If you have doubts for these or other reasons about your ability to perform safe repair work on your car, have the work done at an authorized Volkswagen dealer or other qualified shop.

Part numbers listed in this manual are for identification purposes only, not for ordering. Always check with your authorized Volkswagen dealer to verify part numbers and availability before beginning service work that may require new parts.

Before attempting any work on your Volkswagen, read the warnings and cautions on page ix and any warning or caution that accompanies a procedure in the service manual. Review the warnings and cautions on page ix each time you prepare to work on your Volkswagen.

Special tools required to perform certain service operations are identified in the manual and are recommended for use. Use of tools other than those recommended in this service manual may be detrimental to the car's safe operation as well as the safety of the person servicing the car.

Copies of this manual may be purchased from authorized Volkswagen dealers, most automotive accessories and parts dealers specializing in Volkswagens, from selected booksellers, or directly from the publisher by mail.

The publisher encourages comments from the reader of this manual. These communications have been and will be carefully considered in the preparation of this and other manuals. Please write to Robert Bentley, Inc. at the address listed on the top of this page.

Library of Congress Catalog Card No. 78-75039
ISBN 0-8376-0096-0
VWoA Part No. LPV 997 109
Bentley Stock No. V179

94 93 92   13 12 11

The paper used in this publication is acid free and meets the requirements of the National Standard for Information Sciences-Permanence of Paper for Printed Library Materials.∞

© Copyright 1979 Volkswagen of America, Inc.

All rights reserved. All information contained in this manual is based on the information available to the publisher at the time of editorial closing. The right is reserved to make changes at any time without notice. No part of this publication may be reproduced, stored in a retrieval system, or transmitted in any form or by any means, electronic, mechanical, photocopying, recording, or otherwise, without the prior written consent of the publisher. This includes text, figures, and tables. All rights reserved under Berne and Pan-American Copyright conventions.

Manufactured in the United States of America

Engine 1

Fuel System 2

Transmission and Rear Axle 3

Body and Frame 4

Brakes and Wheels 5

Front Axle 6

Electrical System 7

Strut Front Suspension 8

Automatic Stick Shift 9

Lubrication and Maintenance 10

Fuel Injection 11

FOREWORD

Service to VW owners is of top priority to the Volkswagen organization and has always included the continuing development and introduction of new and expanded services. In line with this purpose, Volkswagen of America, Inc. has introduced this Volkswagen Official Service Manual.

This Type I Manual covers the Beetles, Super Beetles, VW Convertibles, and Karmann Ghias of Model Years 1970, 1971, 1972, 1973, 1974, 1975, 1976, 1977, 1978, and 1979. (Cars of each Model Year are usually introduced in August of the preceding year.) In 1970 the VW Convertible was a Beetle. From 1971, it was a Super Beetle. Though much of the contents can be applied to VW cars sold elsewhere, this Manual is specifically written to cover U.S.A. and Canadian Type I VWs only.

The chassis numbers assigned to the Beetles, Super Beetles, VW Convertibles, and Karmann Ghias for the Model Years covered in this Manual are:

Beetle
1970: 110 2000 001 to 110 3100 000
1971: 111 2000 001 to 111 3200 000
1972: 112 2000 001 to 112 3200 000
1973: 113 2000 001 to 113 3200 000
1974: 114 2000 001 to 114 3200 000
1975: 115 2000 001 to 115 3200 000
1976: 116 2000 001 to 116 3200 000
1977: 117 2000 001 to 117 3200 000
1978: None
1979: None

Karmann Ghia/Karmann Ghia Convertible
1970 140 2000 001 to 140 3100 000
1971: 141 2000 001 to 141 3200 000
1972 142 2000 001 to 142 3200 000
1973 143 2000 001 to 143 3200 000
1974: 144 2000 001 to 144 3200 000
1975: None
1976: None
1977: None
1978: None
1979: None

Super Beetle/La Grande Bug (1975)
1970: None
1971: 131 2000 001 to 131 3200 000
1972: 132 2000 001 to 132 3200 000
1973: 133 2000 001 to 133 3200 000
1974: 134 2000 001 to 134 3200 000
1975: 135 2000 001 to 135 3200 000
1976: None
1977: None
1978: None
1979: None

VW Convertible
1970: 150 2000 001 to 150 3100 000
1971: 151 2000 001 to 151 3200 000
1972: 152 2000 001 tp 152 3200 000
1973: 153 2000 001 to 153 3200 000
1974: 154 2000 001 to 154 3200 000
1975: 155 2000 001 to 155 3200 000
1976: 156 2000 001 to 156 3200 000
1977: 157 2000 001 to 157 3200 000
1978: 158 2000 001 to 158 3200 000
1979: 159 2000 001 to 159 3200 000

The chassis number of your VW is found in three places: on the frame tunnel beneath the back seat, under the hood, and on the dashboard. This Manual is organized so that changes from model year to model year are noted, and if a change within one model year is made, the chassis number of the first VW with the change is given.

© 1976 VWoA

FOREWORD

For the Volkswagen owner with mechanical skills and for independent garages, this Manual gives Volkswagen-approved specifications and procedures. In addition, a Volkswagen owner who has no intention of working on his car will find that reading and owning this Manual will enable him or her to discuss repairs intelligently with a professional mechanic.

The aim throughout has been simplicity, clarity, and completeness with step-by-step procedures and accurate specifications. We have endeavored to ensure the highest degree of accuracy possible. When the vast array of data presented in this Manual is taken into account, however, no claim to infallibility can be made.

The Volkswagen owner intending to do maintenance and repairs should have a set of metric wrenches, a torque wrench, screwdrivers, and feeler gauges, since these basic hand tools will be used in accomplishing a majority of the repairs described in this Manual. Usually, there will be a caution in the text when a repair requires special tools or special skills.

If you are a professional mechanic already working on imported cars, you may have some Volkswagen special tools that are shown in some of the illustrations in this Manual. If you have previously worked only on American-manufactured cars, you will not have to replace your expensive micrometers, vernier calipers, and other precision tools because specifications are given both in millimeters and in inches, except when special Volkswagen metric tools are indispensable (such measurements are given only in millimeters).

Volkswagens are constantly being improved and sometimes such changes—both in parts and specifications—are made applicable to older Volkswagens. Thus, a replacement part to be used on an older Volkswagen may not be the same as the part used in the original installation. Such changes are noted in this Manual. If a specification given in this Manual differs from one in an earlier source, disregard the earlier specification. The specifications in this Volkswagen Official Service Manual are accurate as of the publication date of this Manual.

Volkswagen offers an extensive warranty. Therefore, before deciding to repair a Volkswagen that is covered by the new-car warranty, consult your Authorized Volkswagen Dealer. You may find that he can make the repair either free or at minimum cost. Regardless of its age, or whether it is under warranty, your Volkswagen is both an easy car to service and an easy car to get serviced. So if at any time a repair is needed that you feel is too difficult to do yourself, a trained Volkswagen mechanic is ready to do the job for you.

Volkswagen of America, Inc.

Please read these warnings and cautions before proceeding with maintenance and repair work.

WARNING—

- Do not re-use any fasteners that are worn or deformed in normal use. Many fasteners are designed to be used only once and become unreliable and may fail when used a second time. This includes, but is not limited to, nuts, bolts, washers, self-locking nuts or bolts, circlips, cotter pins. Always replace these fasteners with new parts.

- Never work under a lifted car unless it is solidly supported on stands designed for the purpose. Do not support a car on cinder blocks, hollow tiles, or other props that may crumble under continuous load. Never work under a car that is supported solely by a jack. Never work under the car while the engine is running.

- If you are going to work under a car on the ground, make sure that the ground is level. Block the wheels to keep the car from rolling. Disconnect the battery negative (–) terminal (ground strap) to prevent others from starting the car while you are under it.

- Never run the engine unless the work area is well ventilated. Carbon monoxide kills.

- Do not attempt to work on your car if you do not feel well. You increase the danger of injury to yourself and others if you are tired, upset or have taken medicine or any other substance that may impair you from being fully alert.

- Friction materials such as brake or clutch discs may contain asbestos fibers. Do not create dust by grinding, sanding, or by cleaning with compressed air. Avoid breathing asbestos fibers and asbestos dust. Breathing asbestos can cause serious diseases such as asbestosis or cancer, and may result in death.

- Tie long hair behind your head. Do not wear a necktie, a scarf, loose clothing, or a necklace when you work near machine tools or running engines. If your hair, clothing, or jewelry were to get caught in the machinery, severe injury could result.

- Disconnect the battery negative (–) terminal (ground strap) whenever you work on the fuel system or the electrical system. When you work around fuel, do not smoke or work near heaters or other fire hazards. Keep an approved fire extinguisher handy.

- Illuminate your work area adequately but safely. Use a portable safety light for working inside or under the car. Make sure the bulb is enclosed by a wire cage. The hot filament of an accidentally broken bulb can ignite spilled fuel or oil.

- Catch draining fuel, oil, or brake fluid in suitable containers. Do not use food or beverage containers that might mislead someone into drinking from them. Store flammable fluids away from fire hazards. Wipe up spills at once, but do not store the oily rags, which can ignite and burn spontaneously.

- Finger rings should be removed so that they cannot cause electrical shorts, get caught in running machinery, or be crushed by heavy parts.

- Batteries produce explosive hydrogen gas. Keep sparks, lighted matches, and open flame away from the top of the battery. If hydrogen gas escaping from the cap vents is ignited, it will ignite gas trapped in the cells and cause the battery to explode.

- Always observe good workshop practices. Wear goggles when you operate machine tools or work with battery acid. Gloves or other protective clothing should be worn whenever the job requires working with harmful substances.

- Some aerosol tire inflators are highly flammable. Be extremely cautious when repairing a tire that may have been inflated using an aerosol tire inflator. Keep sparks, open flame or other sources of ignition away from the tire repair area. Inflate and deflate the tire at least four times before breaking the bead from the rim. Completely remove the tire from the rim before attempting any repair.

CAUTION—

- If you lack the skills, tools and equipment, or a suitable workshop for any procedure described in this manual, we suggest you leave such repairs to an authorized Volkswagen dealer or other qualified shop. We especially urge you to consult an authorized Volkswagen dealer before beginning repairs on any car that may still be covered wholly or in part by any of the extensive warranties issued by Volkswagen.

- Volkswagen is constantly improving its cars and sometimes these changes, both in parts and specifications, are made applicable to earlier models. Therefore, part numbers listed in this manual are for reference only. Always check with your authorized Volkswagen dealer Parts department for the latest information.

- Before starting a job, make certain that you have all the necessary tools and parts on hand. Read all the instructions thoroughly, do not attempt shortcuts. Use tools appropriate to the work and use only replacement parts meeting Volkswagen specifications. Makeshift tools, parts and procedures will not make good repairs.

- Use pneumatic and electric tools only to loosen threaded parts and fasteners. Never use these tools to tighten fasteners, especially on light alloy parts. Always use a torque wrench to tighten fasteners to the tightening torque specification listed.

- Be mindful of the environment and ecology. Before you drain the crankcase, find out the proper way to dispose of the oil. Do not pour oil onto the ground, down a drain, or into a stream, pond, or lake. Consult local ordinances that govern the disposal of wastes.

Directions for using torque wrenches calibrated in newton meters

In adopting the SI *(Systeme International)* units of measure, which constitute the Modernized Metric System, tool manufacturers are beginning to introduce torque wrenches that are calibrated in newton meters. As metrication proceeds, torque specifications given in foot pounds (ft. lb.) and meter kilograms (mkg) will eventually be replaced by torque specifications given in newton meters (N·m or Nm).

At present, there are in use too few torque wrenches calibrated in newton meters to justify the inclusion of newton meter torque specifications in this Manual. Nevertheless, if you purchase a new torque wrench, we recommend that you try to obtain one that is calibrated in newton meters. Such a tool can easily be used with this Manual by converting the meter kilogram specifications to newton meters.

To convert meter kilograms (mkg) to newton meters, simply disregard the decimal point. For example, 3.5 mkg would become 35 Nm. To convert centimeter kilograms (cmkg) to newton meters, point-off the one place with a decimal. For example, 50 cmkg would become 5.0 Nm. These conversions are not mathematically precise (3.5 mkg actually equals 34.3 Nm) but they are adequate for normal workshop purposes.

Section 1

ENGINE

Contents

Introduction 3	6.2 Removing and Installing Crankshaft Pulley 14
1. General Description 4	**7. Distributor and Drive** 14
Engine Mounting 5	7.1 Removing and Installing Distributor 14
Crankcase 5	7.2 Removing and Installing Distributor Drive Shaft 15
Crankshaft and Bearings 5	**8. Oil Cooler** 15
Valve Train 5	8.1 Removing and Installing Oil Cover 16
Cylinder Heads 5	8.2 Testing and Replacing Oil Pressure Switch 16
Connecting Rods 5	**9. Valve Gear and Cylinder Heads** 17
Pistons and Cylinders 5	9.1 Removing and Installing Rocker Arm Shaft 18
Flywheel and Drive Plate 5	9.2 Disassembling and Assembling Rocker Arm Shaft Assembly 18
Cooling System 5	9.3 Removing and Installing Cylinder Head and Pushrod Tubes 19
Emission Controls 5	
Ignition System 5	**10. Valves** 20
Lubrication System 5	10.1 Removing and Installing Valves 21
Clutch 5	10.2 Removing and Installing Valve Springs (head installed) 21
2. Maintenance 6	10.3 Checking and Replacing Valve Guides .. 22
3. Removing and Installing Engine 6	10.4 Replacement of Valve Seats 22
4. Replacing Outlet Pipes and Heater Control Flaps 8	10.5 Refacing Valve Seats 23
	10.6 Refacing Valves 24
5. Exhaust System 9	10.7 Lapping Valves 25
5.1 Removing and Installing Muffler 9	10.8 Valve Clearance 25
Carburetor Engine Muffler 9	10.9 Adjusting Valves 25
Fuel Injection Muffler 10	**11. Cylinders** 26
5.2 Removing and Installing Heat Exchangers 11	11.1 Removing and Installing Cylinders 26
6. Cooling Air System and Intake Manifold 12	**12. Pistons and Rings** 27
Partial Removal of Fan Housing 12	
Removing and Installing Cooling Air System 12	
6.1 Adjusting Thermostat 14	

© 1976 VWoA

2 ENGINE

- 12.1 Removing and Installing Pistons 27
- 12.2 Piston Rings 28
- 12.3 Piston Clearance 29

13. Torque Converter 30
- 13.1 Removing and Installing Drive Plate 30

14. Clutch and Flywheel 30
- 14.1 Removing and Installing Clutch 30
- 14.2 Removing and Installing Flywheel 32
- 14.3 Removing and Installing Crankshaft Oil Seal 33
- 14.4 Replacing Clutch Release Bearing 34
- 14.5 Adjusting Clutch Pedal Freeplay 35
  - Replacing Clutch Cable 35
- 14.6 Troubleshooting Clutch 36
  - Troubleshooting Table 36

15. Oil Pump 37
- 15.1 Removing and Installing Oil Pump 37
- 15.2 Checking Oil Pump 38
- 15.3 Removing and Installing Oil Pressure Valves 38

16. Crankcase 40
- 16.1 Removing and Installing Oil Strainer ... 40
- 16.2 Disassembling and Assembling Crankcase 41
- 16.3 Removing and Installing Oil Filler and Breather 42
- 16.4 Repairing Tapped Holes 43

17. Camshaft and Crankshaft 43
- 17.1 Checking and Installing Camshaft and Bearings 43
- 17.2 Removing and Installing Crankshaft 45
- 17.3 Disassembling and Assembling Crankshaft 46
- 17.4 Crankshaft End Play 47
- 17.5 Crankshaft Reconditioning 47

18. Connecting Rods 48
- 18.1 Removing Connecting Rods 48
- 18.2 Checking and Installing Connecting Rods 48
- 18.3 Reconditioning Connecting Rods 50

19. Ignition System 51
- 19.1 Coil 51
- 19.2 Distributor 51
  - Point Adjustments 52
  - Adjusting Timing 53
  - Checking Spark Advance Mechanism 54
- 19.3 Disassembling and Assembling Distributor 55
- 19.4 Secondary Circuit 55
- 19.5 Spark Plugs 56

20. Engine Technical Data 57
- I. Tightening Torques 57
- II. General Engine Data 57
- III. Tolerances, Wear Limits, and Settings ... 58
- IV. Basic Tune-up Specifications 60

TABLES
- a. Valve Stem Diameters 21
- b. Clutch Troubleshooting 36
- c. Spring Specifications 39
- d. Bearing Clearance 45
- e. Ignition Timing 53
- f. Ignition Timing with Distributor Installed 54

©1978 VWoA

Engine

The VW "1600" engine, which is installed in the cars covered by this Manual, is a four-cycle powerplant, rear mounted, and air cooled. It displaces 1584 cc (96.6 cu. in.) and has pushrod-operated overhead valves. As on other similar VW engines, the four cylinders are divided into two horizontally-opposed pairs. The 1970 through 1974 model engines are equipped with a single carburetor. On the 1975 and later models, the carburetor and the mechanical fuel pump were eliminated and an electronic fuel injection system was introduced. All engines, whether equipped with carburetors or fuel injection, have conventional coil and battery ignition systems.

Many VWs are equipped with a network of test wiring for the VW Computer Analysis system. On the engine, a sensor for obtaining spark timing data is located in the high tension cable leading to the No. 1 cylinder and a battery-testing wire is attached to the starter solenoid. On many 1974 and later models, another timing sensor is pressed into the right-hand side of the crankcase, near the flywheel. All sensor wires must remain properly attached if the VW Computer Analysis system is to work as designed. Never connect any device other than the test plug of the VW Computer Analysis system to the test network central socket in the engine compartment. Incorrect equipment could damage the plug connectors, the test sensors, or the components containing them.

The information in this section of the Manual is intended to serve as a guide to both car owners and professional mechanics. Some of the operations, however, may require special equipment and experience that only a trained VW mechanic will normally have. If you lack the skills, tools, or a suitable workshop for servicing the engine, we suggest you leave such repairs to an Authorized VW Dealer or other qualified shop. We especially urge you to consult your Authorized VW Dealer before attempting any repairs on a car still covered by the new-car warranty.

However limited your experience or equipment may be, we hope that all the information included here will be of interest. We direct your special attention to **1. General Description** on the following pages. The material presented here can be informative to all VW owners and even to those experienced mechanics who may be unfamiliar with the VW engine.

4 Engine

1. General Description

A cutaway view of the air-cooled engine appears in Fig. 1-1. Note the oil cooler, the deeply finned heat exchangers, cylinders, and cylinder heads.

Fig. 1-1. Cutaway view of the "1600" engine (with transmission and clutch) used in the cars covered by this Manual.

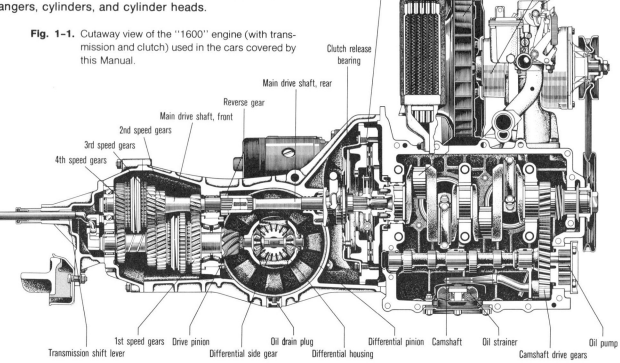

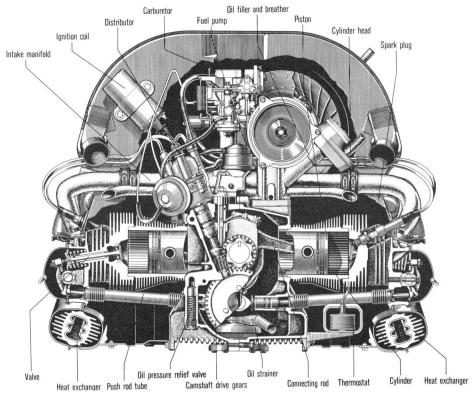

Engine 5

Engine Mounting

Four threaded fasteners join the engine and transmission. Since the engine is totally supported by the transmission, there are no rear engine mounts.

Crankcase

The two-piece crankcase is cast in lighweight alloy and so precisely machined that the halves fit together without a gasket. If either half becomes defective, both must be replaced to ensure an exact fit.

Crankshaft and Bearings

The crankshaft rotates in four main bearings. Only one (No. 2) is of the split-shell type. The No. 1 bearing ring is flanged to take crankshaft end thrust. A seal at the clutch end of the crankshaft and a slinger at the pulley end prevent oil leakage.

Valve Train

The camshaft is gear driven off the crankshaft and runs in three split-shell bearings. A Woodruff key positions the crankshaft gear while the camshaft gear is riveted in place. Solid lifters, pushrods, and adjustable rocker arms make up the valve-operating linkage.

Cylinder Heads

Each pair of cylinders is covered by a deep-finned cast aluminum alloy cylinder head. The valve guides and seats are pressed in but they can be replaced only with special machinery. No cylinder head gaskets are used. However, oil seals are fitted at both ends of each pushrod tube.

Connecting Rods

The connecting rods are steel forgings. Split-shell bearings are used at the crankshaft end and lead-bronze coated steel bushings at the piston pin end.

Pistons and Cylinders

The pistons are of the three-ring type with full-floating pins secured by circlips. The cylinders are detachable and, with their matched pistons, fully interchangeable. The cylinders are cast iron and finned for cooling.

Flywheel and Drive Plate

The flywheel, used on cars with manual transmissions, is mounted on the crankshaft with four steel dowels and a gland nut. The nut contains the needle bearings that serve as a pilot bearing for the transmission's rear driveshaft. Instead of a flywheel, engines used with Automatic Stick Shift transmissions have a drive plate to which the torque converter is bolted.

© 1976 VWoA

Cooling System

A fan, mounted on one end of the generator's armature shaft, forces cooling air over the finned cylinders and cylinder heads. Thermostat-operated flaps regulate the volume of this cooling air according to engine heat. Some air can be diverted through the heat exchangers. This warmed air is used to heat the car's interior. A control lever between the front seats operates the flaps that control the volume of heated air reaching the car's interior.

Emission Controls

Emission control servicing is covered in **FUEL SYSTEM** but you must remove and install some of the emission control devices when you make engine repairs. All cars covered by this Manual have the following emission controls: (1) intake air preheating, (2) engine modifications, (3) evaporative emission control (EEC), and (4) closed positive crankcase ventilation (PCV). The PCV system has no PCV valve that requires periodic cleaning or replacement.

In addition to the above, a throttle valve positioner is used on 1970 and 1971 cars with manual transmissions and on 1972 California models with manual transmissions. 1971 through 1974 cars with manual transmissions have a throttle valve dashpot. Exhaust gas recirculation (EGR) is used on 1972 and later cars sold in California with Automatic Stick Shift, on all 1973 and 1974 cars with Automatic Stick Shift, and on all 1975 and later cars regardless of transmission type. Beginning in 1975 all models are equipped with an electronic fuel injection system and, in California, a catalytic converter.

Ignition System

The ignition is a conventional coil and battery system. Unlike the ignition distributor on most other engines, the ignition distributor on the VW engine is driven by a gear on the crankshaft, not by the camshaft.

Lubrication System

A gear-type oil pump, driven by the camshaft, draws oil through a strainer in the bottom of the crankcase sump. The pump then forces the oil through an oil cooler and into the engine's oil passages. The system is described and illustrated in **15.3 Removing and Installing Oil Pressure Valves**.

Clutch

The engines used with manual transmissions have a dry, single-plate clutch fitted to the flywheel. The pressure plate assembly, which has a diaphragm spring, and the driven plate remain on the flywheel when the engine is removed from the car. The clutch release bearing remains with the clutch release shaft inside the transmission bellhousing. The friction clutch for the Automatic Stick Shift is built into the transmission itself. Its servicing and repair are covered in **AUTOMATIC STICK SHIFT**.

6 ENGINE

2. Maintenance

The following maintenance steps are covered in **LUBRICATION AND MAINTENANCE** or under the listed headings.

1. Changing engine oil, cleaning the oil strainer, and checking the oil level
2. Servicing and replacing the spark plugs. **19.5**
3. Checking distributor and ignition timing. **19.2**
4. Checking valve clearance. **10.9**
5. Checking compression
6. Servicing the air cleaner
7. Checking exhaust system and EGR system
8. Checking clutch pedal freeplay. **14.5**
9. Checking fan belt adjustment
10. Checking and adjusting idle
11. Replacing catalytic converter. **5.1**.

3. Removing and Installing Engine

All major engine repairs, including valve grinding, require that you remove the engine from the car.

To remove:

1. Remove the air cleaner as described in **FUEL SYSTEM** or **FUEL INJECTION**. Drain the engine oil. Remove the air hoses that connect the fan housing with the heat exchanger connections.
2. Disconnect the battery ground strap. Disconnect the wires from the coil, the generator or alternator, the oil pressure switch, and—on 1970 through 1974 cars—the carburetor.
3. On cars with fuel injection, disconnect the injection system wiring harness from the coil, the injectors, the crankcase, the temperature sensors, and the intake air distributor and related components. For correct installation mark each wire as you disconnect it.
4. Working under the car, disconnect the wires and the battery cable from the starter solenoid. On cars with carburetors, pull the fuel hose off the metal tube on the frame, then plug the tube. On cars with fuel injection, disconnect the fuel hoses from the pressure regulator, then plug the hoses.

   WARNING—
   Do not smoke or work near heaters or other fire hazards. Have a fire extinguisher handy.

5. Disconnect the back-up light wires at the right-hand side of the engine. Disconnect the vacuum hoses for the throttle valve positioner on cars that are so equipped.
6. Disconnect the accelerator cable from the throttle arm. Remove the spring retainer, the accelerator cable return spring, and the housing from the cable. Then pull the cable guide tube out of the fan housing as far as it will go. Reaching behind the fan housing, pull the exposed cable out of the guide tube.
7. Remove the two small cover plates over the intake manifold preheater pipe connections.
8. Except on fuel injection engines, remove the crankshaft pulley cover plate. Then remove the rear cover plate and the air hose connecting pipes.
9. Raise the car on a hydraulic lift. Alternatively, you can raise the car with a floor jack and support it on safety stands about 1 m (3 ft.) off the ground.
10. On cars with Automatic Stick Shift, disconnect the wire and the vacuum hoses from the control valve. Disconnect the ATF connections from the engine (Fig. 3-1) and seal them with M 16 × 1.5 union nuts that have been soldered shut.

Fig. 3-1. Connection for ATF line from ATF tank. (The ATF line to the torque converter is at the opposite side of the engine.)

11. Disconnect the heater cables from the arms on the heat control box flaps.
12. Detach the flexible heater ducts from the fronts of the heat exchangers or the heat control boxes. Move the ducts aside so that they will not interfere with lowering the engine.
13. On cars with Automatic Stick Shift transmissions, remove the 12-point bolts that hold the torque converter to the drive plate. These bolts are accessible through openings in the bellhousing (Fig. 3-2). Hand-turn the crankshaft to position each of

the four bolts in the opening. If the engine is seized, you can remove the engine with the converter attached but the converter seal will be ruined and must then be replaced.

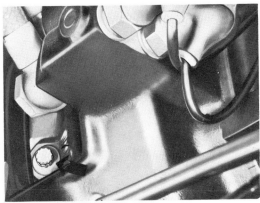

Fig. 3-2. TOP: Bolt in access opening on carburetor engine. BOTTOM: Bolt access opening on fuel injection engine. On fuel injection engine, you must pry a cap out of the opening.

14. Remove the nuts from the two lower engine mounting studs as shown in Fig. 3-3.

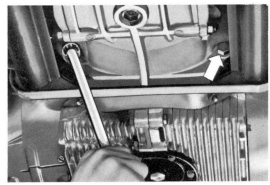

Fig. 3-3. Lower engine mounting nuts being removed. The second nut is indicated by the arrow.

15. Lower the car until it is about 1 m (3 ft.) above the ground. Then position a wheeled floor jack with an engine adapter under the engine. Raise the jack until it contacts the crankcase firmly.

16. Using the procedure appropriate to the car model, remove the two upper engine mounting nuts, bolts, or combination (Fig. 3-4).

    NOTE —
    1970 cars with Automatic Stick Shift have two nuts on studs; 1970 cars with manual transmissions have two nuts on bolts. The bolts must be held from under the car while the nuts are being removed. All 1971 and later cars have a bolt only at the left side that must be removed from the transmission side. The fastener at the right side is a nut and bolt as on earlier models.

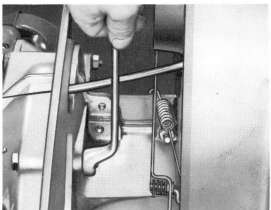

Fig. 3-4. Upper engine mounting fasteners (early cars). It may be necessary to have a helper hold the bolts from under the car.

17. Using the floor jack, pull the engine to the rear until the transmission rear driveshaft clears the clutch pressure plate assembly or until the drive plate is clear of the torque converter. Make sure that the converter is not pulled off the transmission. If you must rock the engine to free it from a manual transmission, do so gently. Otherwise, you could damage the transmission's rear driveshaft or the clutch.

18. With the floor jack, carefully lower the engine and, at the same time, pull the engine to the rear and out from under the car. To prevent damage to the connections, make sure that no wires or hoses remain attached to the engine. If necessary, disconnect the overlooked wires or hoses.

19. On cars with Automatic Stick Shift, immediately install a bracket on one of the bellhousing studs to prevent the torque converter from falling off while the engine is out. Such a bracket is illustrated in **AUTOMATIC STICK SHIFT**.

20. To prevent the entry of abrasive dirt, thoroughly clean the engine before you begin to disassemble it.

8 ENGINE

Installation is the reverse of removal, with the addition of the following steps:

1. Check the clutch release bearing for wear. (See **14.4 Replacing Clutch Release Bearing**.) On both new and used release bearings, roughen the plastic facing with emery cloth. Then rub in molybdenum grease.

2. Wipe off the mating surfaces of the engine and transmission. Lightly lubricate the manual transmission rear driveshaft splines with molybdenum disulfide powder. On engines equipped with Bosch starters, lubricate the starter drive bushing in the transmission case with multipurpose grease.

3. When installing the engine on cars with manual transmissions, be careful not to damage the transmission rear driveshaft or the clutch driven plate. Engage 1st gear, set the parking brake, then hand-turn the crankshaft until the splines mesh.

4. Start the lower engine mounting studs in their holes, then press the engine into firm contact with the transmission flange. Install the upper mounting nuts and bolts. Torque all four mounting fasteners to 3.0 mkg (22 ft. lb.).

5. Check the position of the rubber seal that is between the engine cover plates and the body.

6. Fill the crankcase with oil.

7. Adjust the clutch pedal freeplay as described in **14.5 Adjusting Clutch Pedal Freeplay**; adjust the accelerator cable as described in conjunction with carburetor installation in **FUEL SYSTEM** (or in conjunction with throttle valve housing installation in **FUEL INJECTION**); adjust ignition timing as described in **19.2 Distributor**; and adjust the idle as described in **FUEL SYSTEM** or **FUEL INJECTION**.

4. REPLACING OUTLET PIPES AND HEATER CONTROL FLAPS

Except on fuel injection engines, you can gain access to the heater control flaps as shown in Fig. 4-1.

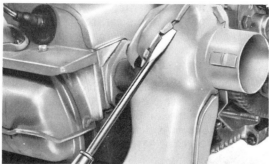

Fig. 4-1. Screwdriver being used to pry cover loose.

Check the heater flaps for free movement. Replace them if they are badly stuck or corroded. Install the cover plate, then crimp the tabs to hold it in place.

To replace the outlet pipes, use a 6-mm (¼-in.) drill to drill out the spot welds as shown in Fig. 4-2.

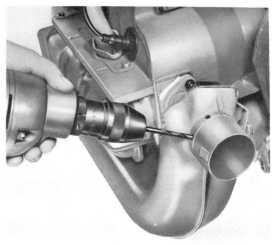

Fig. 4-2. Spot welds being drilled out. Centerpunch the welds first.

After drilling, pry up the lip that holds the heat exchanger casing together (Fig. 4-3). Then bend the casing halves slightly apart and remove the outlet pipe.

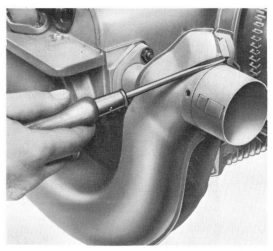

Fig. 4-3. Heat exchanger casing halves being separated.

After removing the old outlet pipe, inspect the metal in the heat exchanger casing to make sure that it is sound. Keeping the heater flaps open, tack-weld the new outlet pipe to the heat exchanger. Then coat the pipe with cold zinc paint or a heat-resistant equivalent.

ENGINE 1

5. Exhaust System

If you intend to disassemble the engine following its removal, the exhaust system is the first item to take off. However, you can also remove and install the muffler and the heat exchangers with the engine still in the car.

5.1 Removing and Installing Muffler

The exhaust system used on fuel injection engines is completely different from the system used on engines with carburetors. The two systems will be covered separately under this heading.

Carburetor Engine Muffler

The components of the carburetor engine's exhaust system are shown in Fig. 5-1. The main heat exchangers below the cylinder head covers seldom require replacement—you need not remove them when replacing the muffler. The small rear heat exchangers are a unit with the muffler and must be replaced with the muffler. You can replace the tailpipes separately and without replacing the muffler.

To remove muffler:

1. If the engine is in the car, loosen the two hose clamps. Then remove the air hoses and the air hose grommets. Remove the cover plates over the intake manifold preheater pipe connections, and unbolt the preheater pipes from the muffler. Then remove the crankshaft pulley cover plate, the rear cover plate, and the air hose connecting pipes.

2. Remove the screws and nuts from the flat clamps on the heat exchanger connections. On cars with exhaust gas recirculation, disconnect the pipe fitting on the exhaust pipe flange for the No. 4 cylinder.

3. Remove the M 6 nuts and bolts, then remove the clamps that join the heat exchanger outlet pipes to the flanged pipes at the ends of the muffler.

4. Remove the four M 8 self-locking nuts that hold the muffler on the cylinder heads. Then, grasping the tailpipes, pull the muffler to the rear and off the car.

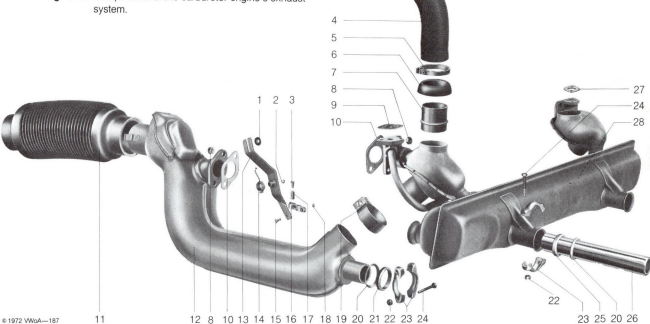

Fig. 5-1. Components of the carburetor engine's exhaust system.

1. Clamp washer
2. E-clip
3. M 5 bolt
4. Air hose from fan housing
5. Hose clamp
6. Air hose grommet
7. Air hose connecting pipe
8. M 8 self-locking nut (2)
9. Left preheating pipe gasket
10. Exhaust pipe flange gasket (2)
11. Flexible heated air duct
12. Heat exchanger
13. Heater flap lever
14. Flap lever return spring
15. Pin
16. Cable connecting link
17. Cable retaining pin
18. E-clip
19. Flat clamp
20. Tailpipe retaining ring (2)
21. Heat exchanger seal
22. M 6 nut
23. Clamp
24. M 6 bolt (2)
25. Tailpipe seal
26. Tailpipe
27. Right preheating pipe gasket
28. Muffler

10 ENGINE

To install:

1. Inspect the muffler for leaks and damage and make sure that the exhaust pipe flanges that bolt to the cylinder heads are not warped. If you install a new muffler, also install new tailpipes.

2. Measure new tailpipes; three lengths are used. Depending on the car model, these lengths are 276 mm (10 7/8 in.), 249 mm (9 13/16 in.), and 226 mm (8 29/32 in.). The last length is on 1973 and 1974 cars only. (However, the 1974 pipes have an inside diameter of 23 mm (29/32 in.), not 20 mm (51/64 in.) as in 1973.)

3. On 1974 cars only, install the tailpipes so that they project 155 mm (6 7/64 in.) from the muffler.

4. On 1970 through 1973 cars, insert the left tailpipe. Then insert a steel tape measure through the center of the tailpipe until it contacts the preheating pipe inside the muffler.

5. Move the tailpipe in and out until its outer end aligns with the correct graduation on the tape. Then install the tailpipe retaining ring, the tailpipe seal, and the clamp. Install the right tailpipe so that it projects the same distance from the muffler as the left tailpipe.

The correct graduations are 270 mm (10 5/8 in.) for 276-mm (10 7/8-in.) tailpipes, 243 mm (9 9/16 in.) for 249-mm (9 13/16-in.) tailpipes, and 220 mm (8 21/32 in.) for 1973-model 226-mm (8 29/32-in.) tailpipes. The remainder of installation is the reverse of removal. Use new gaskets at all points. Torque the M 8 self-locking nuts to 2.0 mkg (14 ft. lb.) and the M 6 nuts and bolts to 1.0 mkg (7 ft. lb.).

Fuel Injection Muffler

The exhaust system of the fuel injection engine is shown installed in Fig. 5-2. (The two black steel straps shown passing beneath the engine and muffler are part of a trailer hitch.) A car with a catalytic converter is shown in Fig. 5-3. On such cars, the muffler is shorter and all exhaust gases enter the muffler from one end—after passing through the catalytic converter.

Fig. 5-3. Location of catalytic converter (arrow).

To replace muffler:

1. On cars with catalytic converters, simply remove the clamp that holds the muffler to the converter, then remove the muffler. If necessary, remove the tailpipe clamp and pull the tail pipe off the muffler.

2. If you must remove the catalytic converter, do so by removing the clamp that holds the converter to the exhaust pipe.

3. On cars without catalytic converters, remove the four nuts and bolts that hold the EGR pipe (Fig. 5-4). Then remove the EGR pipe.

Fig. 5-2. Exhaust system used with fuel injection.

Fig. 5-4. Bolts (arrows) that hold EGR pipe.

ENGINE

4. On cars without catalytic converters, unbolt the muffler inlet pipes from the heat exchangers, then take off the muffler. If necessary, remove the tailpipe clamp and pull the tailpipe off the muffler.

Installation is the reverse of removal. Always use new gaskets on all exhaust connections. Replace any clamps and clamp bolts that are no longer serviceable.

5.2 Removing and Installing Heat Exchangers

The exhaust system used on fuel injection engines is completely different from the system used on engines with carburetors. The two systems will be covered separately under this heading. On either engine, you should remove the muffler before you remove the heat exchanger.

To remove carburetor heat exchangers:

1. Remove the two fillister head screws that hold the heat exchanger casings to the lower warm air duct plates under each pair of cylinders.

2. Disconnect the heater control cables from the heater flap levers.

3. Pull the flexible heated air ducts off the front of the heat exchangers. Then remove the four M 8 self-locking nuts that hold the heat exchangers on the cylinder heads.

4. Take a good grip on the heat exchanger (it is heavy); then push it forward and remove it.

Installation is the reverse of removal. Use new gaskets and torque the M 8 self-locking nuts to 2.0 mkg (14 ft. lb.).

To remove fuel injection engine heat exchangers:

1. Loosen the flat clamp indicated in Fig. 5-5.

Fig. 5-5. Air inlet duct clamp (arrow).

2. Disconnect the heater control cables from the arms of the heat control box flaps.

3. Detach the flexible heater ducts from the fronts of the heat control boxes. Move the ducts aside so they will not interfere with lowering the control box.

4. Remove the fillister head screw and the bolt that hold the heat exchanger casings to the lower warm air duct plates under each pair of cylinders. See Fig. 5-6. Remove the nuts and bolts that hold the heat exchangers to the exhaust pipe. (The bolts and nuts at the exhaust outlets of the exchangers will have been removed during muffler removal.)

5. Carefully lower each heat exchanger and its attached heat control box from the car. If necessary, you can loosen the flat clamp and separate the heat control box from the exchanger after you have removed both components from the car.

Fig. 5-6. Fasteners (arrows) that hold heat exchanger to exhaust pipe and air ducts.

Installation is the reverse of removal. Always use new gaskets at all exhaust connections. Replace any clamps, clamp bolts, and flange bolts that are no longer serviceable.

To replace elbow and head pipe:

1. You can replace the elbow and head pipe (the exhaust pipes bolted to the cylinder heads) without first removing the muffler or the heat exchanger. To remove the elbow and the head pipe, first remove the nuts that hold the front elbow to the studs on the cylinder head.

2. Remove the nuts and bolts that hold the front elbow to the head pipe. Remove the elbow and its gaskets.

3. Remove the nuts that hold the rear of the head pipe to the studs on the cylinder head. Unbolt the front of the head pipe from the heat exchanger, then remove the head pipe.

Installation is the reverse of removal. Use new gaskets and torque the M 8 self-locking nuts to 2.0 mkg (14 ft. lb.).

12 ENGINE

6. COOLING AIR SYSTEM AND INTAKE MANIFOLD

With the engine in the car, you must partially remove the fan housing before you can remove the intake manifold, or before you can remove the generator or alternator as described in **ELECTRICAL SYSTEM**.

Partial Removal of Fan Housing

To partially remove the fan housing, remove the V-belt. Remove the air cleaner and the carburetor or intake air sensor. See **FUEL SYSTEM** or **FUEL INJECTION**. Pull out the accelerator cable guide tube.

Remove the strap that holds the generator or alternator to its support. Disconnect the ignition cables and any hoses, air ducts, and wires that might prevent the fan housing from being lifted upward. Except on 1970 cars, use a 10-mm wrench to remove the bolt for the oil cooler cover and the bolt that holds the fan housing to the oil cooler flange. Remove the cover and the air duct.

Remove the screws that are at each end of the fan housing. Unbolt the thermostat as shown later in Fig. 6-3, then unscrew the thermostat from its rod. Except on 1970 cars, pry off the spring clip and disconnect the connecting link from the left control flaps (Fig. 6-1). Raise up the fan housing so that you can remove the intake manifold or the generator or alternator. Installation is the reverse of removal.

Removing and Installing Cooling Air System

This procedure is easiest with the engine removed. However, after removing the generator or alternator and the intake manifold using the procedure just given, you can fully remove the fan housing from the installed engine if you first remove the rear hood and its hinges as described in **BODY AND FRAME**. Then remove the remaining cooling air system parts as described here.

To remove air ducts, fan housing, and control flaps:

1. If not removed previously, remove the air hoses that carry fresh air to the heat exchangers. Disconnect the evaporative emission control hose from the fan housing.

2. Remove the distributor cap and pull the spark plug connectors off the spark plugs. Disconnect any other wires or hoses that might prevent you from lifting the fan housing straight up.

3. If not removed previously, remove the crankshaft pulley cover plate, the two small cover plates over the intake manifold preheating pipes, and the rear cover plate. Fuel injection engines have only the rear cover plate.

4. Disconnect the vacuum hose(s) that go to the distributor. On carburetor engines, disconnect the fuel line(s) from the fuel pump and carburetor. (Fuel pump and carburetor removal are covered separately in **FUEL SYSTEM**. Removal of fuel injection components is covered in **FUEL INJECTION**.)

5. Remove the generator pulley outer half as shown in Fig. 6-2 and then remove the fan belt.

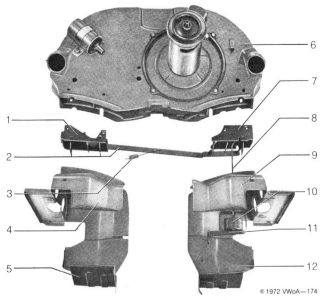

1. Left control flaps
2. Connecting link
3. Link return spring
4. Left cylinder cover plate
5. Left air duct lower part
6. Fan housing with fan
7. Right control flaps
8. Thermostat rod
9. Right cylinder cover plate
10. Thermostat
11. Thermostat bracket
12. Right air duct lower part

Fig. 6-1. Components of the cooling air system.

Fig. 6-2. Fan belt being removed. Screwdriver is held in a notch on the back of the pulley and braced against a generator through bolt. Use a 21-mm ($^{13}/_{16}$-in.) wrench to remove the nut.

6. Take out the fillister head screws that hold the air duct lower parts on the crankcase, on the cylinder cover plates, and on the heat exchangers (if the latter have not already been removed).

7. Unbolt the thermostat from its bracket (Fig. 6-3). Unscrew the thermostat from its rod. If the thermostat bracket is to be removed from the side of the crankcase, scribe its location. Doing so will simplify thermostat adjustment.

8. Except on 1970 cars, remove the bolt for the oil cooler cover and the bolt that holds the fan housing to the oil cooler flange. Disconnect the connecting link from the left control flaps.

Fig. 6-3. Thermostat being unbolted from its bracket.

9. Loosen the bolt in the generator or alternator mounting strap. Then slide the strap forward and off the generator (or alternator) support stand.

10. Remove the fillister head screw at each side of the fan housing. Then lift the fan housing with the attached generator or alternator straight up until it is clear of the oil cooler.

11. If necessary, you can remove the right and left control flaps and the connecting link by unhooking the return spring and taking out the eight fillister head screws that hold the flap assembly in the lower part of the fan housing.

NOTE
When installing the control flaps, make sure that the rubber stop is inserted in the right-hand flap housing.

Installation of the fan housing is the reverse of removal. Following installation, check and adjust the fan belt tension as described in **LUBRICATION AND MAINTENANCE**. Check the thermostat adjustment as described in **6.1 Adjusting Thermostat**.

To remove intake manifold and cylinder cover plates:

1. First remove the fan housing. If the muffler has not already been removed (carburetor engines only), unbolt the intake manifold preheating pipes at their flanges.

2. On fuel injection engines, remove the injectors. Also remove the intake air sensor and unbolt the intake air distributor from the crankcase. All of these jobs are described in **FUEL INJECTION**.

3. Remove the nuts that hold the intake manifold or the two cast aluminum intake pipes to the cylinder heads. (It is unnecessary to separate the intake manifold or the intake air distributor from the intake pipes by removing the hose clamps and the rubber sleeves.)

4. If the engine is in the car, disconnect the throttle valve positioner vacuum hose from the intake manifold on cars that are so equipped.

5. On fuel injection engines, disconnect all air and vacuum hoses from the intake air distributor, marking each one for correct installation.

6. On cars with exhaust gas recirculation, disconnect the exhaust feed pipe from the EGR valve. See Fig. 6-4.

7. Lift off the intake manifold/carburetor or intake manifold/intake air distributor assembly.

8. Remove the two fillister head screws that hold each cylinder cover plate to the cylinder head. Then lift the plates off upward.

Installation is the reverse of removal. On 1971 and later cars, use new gaskets between the cast aluminum intake pipes and the cylinder heads. On 1970 cars, replace the gaskets in the cylinder head intake ports. Torque the nuts to 2.0 mkg (14 ft. lb.). Torque the M 6 bolts for the carburetor engine preheating pipe flanges to 1.0 mkg (7 ft. lb.).

Fig. 6-4. Nuts (arrows) that hold feed pipe on EGR valve of fuel injection engine. Carburetor engines have a pipe union nut.

14 ENGINE

6.1 Adjusting Thermostat

You should check the thermostat adjustment any time components of the cooling air system have been removed or replaced. The lower air duct below the right-hand cylinders must be removed for access to the thermostat. The engine must be completely cold when adjusting the thermostat.

To check and adjust:

1. If a new thermostat is being installed, screw it onto the thermostat rod. If the adjustment is simply being checked, unbolt the thermostat from its bracket.

2. Raise the thermostat off its bracket until the control flaps are wide open. If the top of the thermostat does not just contact the upper part of the bracket, an adjustment is required.

3. To adjust, loosen the mounting nut that holds the thermostat bracket on the crankcase. Then move the bracket as indicated in Fig. 6-5 until the upper part of the bracket just contacts the top of the thermostat with the thermostat raised to open the control flaps fully.

4. Lock the bracket in position by tightening the nut.

5. Pull the thermostat down and bolt it onto the lower part of the bracket.

Fig. 6-5. Bracket movement (double arrow) that permits thermostat to be adjusted.

6.2 Removing and Installing Crankshaft Pulley

To remove the crankshaft pulley, lock the flywheel with the flywheel holding fixture illustrated in **14.2 Removing and Installing Flywheel**. If the engine is installed, place the transmission in 4th gear and set the parking brake (manual transmission only). Then remove the M 20 bolt and washer from the center of the pulley. Use a puller, as shown in Fig. 6-6, to remove the pulley from the crankshaft.

CAUTION —
Do not attempt to pry the pulley off. You can easily damage the crankcase by doing so.

Fig. 6-6. Crankshaft pulley being removed.

Check the pulley seating surface, the Woodruff key, the belt contact surface, and the crankshaft. Remove dirt and burrs; replace bent or cracked pulleys. Clean the oil return thread on the pulley hub and coat it with molybdenum grease. Start the pulley on the crankshaft, then draw it into place with the center bolt. Torque the bolt to 4 to 5 mkg (29 to 36 ft. lb.).

7. Distributor and Drive

The distributor shaft does not engage the crankshaft gear directly. Instead, a drive dog on the distributor shaft engages a separate distributor drive shaft. The distributor drive shaft has an integral gear that meshes with the crankshaft gear.

7.1 Removing and Installing Distributor

You should remove the distributor to protect it during engine repairs. Removal will also simplify breaker point installation when performing a tune-up.

To remove:

1. Remove the distributor cap. Hand-turn the engine until the rotor tip points to the No. 1 cylinder mark on the distributor body (Fig. 7-1).

ENGINE 15

Fig. 7-1. Distributor rotor aligned with No. 1 cylinder mark on distributor body.

2. If not previously removed, disconnect all hoses and wiring to the distributor.

3. Remove the bolt that holds the distributor bracket to the crankcase. Then lift out the distributor and cover the opening in the engine to keep out dirt.

To install:

1. If necessary, hand-turn the crankshaft to align the timing mark on the crankshaft pulley with the seam on the crankcase centerline, and to position the distributor drive shaft as shown in Fig. 7-2.

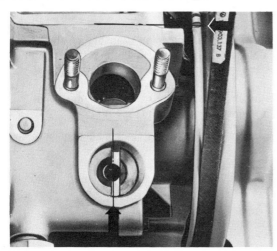

Fig. 7-2. Engine positioned for distributor installation. Arrow at upper right indicates timing mark on crankshaft pulley. Lower arrow indicates centerline of distributor mounting hole relative to distributor drive shaft groove.

2. Insert the distributor with its rotor pointing to the No. 1 cylinder mark. Then install the bolt, hoses, and wiring. Check the ignition timing.

7.2 Removing and Installing Distributor Drive Shaft

You will simplify the installation of the distributor drive shaft if you leave the engine in position to fire the No. 1 cylinder while the shaft is out.

To remove:

1. Remove the distributor. Take the spacer spring out of the recess in the top of the drive shaft.

2. Using an expansion tool or magnet, withdraw the drive shaft. Then remove the two shim washers under the drive shaft.

CAUTION —
Be careful not to drop the washers into the timing gear chamber. Getting them out could require complete engine disassembly.

To install:

1. Inspect the drive shaft for worn gear teeth and the washers for scoring or wear. The thickness of each washer should be 0.60 mm (.024 in.). Have the engine in position to fire the No. 1 cylinder.

2. Coat the washers with grease so that they will stick in place until the drive shaft is installed, then insert a stiff wire or rod in the engine. Let the washers slide down the rod or wire and into position.

3. Insert the drive shaft. It will turn slightly as it engages the crankshaft gear. Engage the teeth so that the drive shaft will be in the position shown previously in Fig. 7-2 when fully installed.

4. Install the distributor. Check the ignition timing.

NOTE —
On some rebuilt exchange engines, the drive shaft bore in the crankcase is machined out and the lower shim washer replaced by a 3-mm (.118-in.) spacer. When repairing these engines, always install the 0.60-mm (.024-in.) shim washer between the 3-mm (.118-in.) spacer and the drive shaft gear so that the spacer will be on the bottom.

8. OIL COOLER

Because of crankcase changes, the oil coolers used on the cars covered by this Manual are different from those used on previous VWs. For one thing, starting with the cars covered by this Manual, the oil cooler was made larger, and, with the introduction of the 1971 models, the oil cooler was relocated on the crankcase. Despite these modifications, the removal and installation of the oil cooler remain basically unchanged.

16 ENGINE

8.1 Removing and Installing Oil Cooler

Although some mechanics make a practice of raising the fan housing as far as possible and then removing the oil cooler with the engine in the car, there is risk of abrasive dirt falling from the fan housing and entering the oil passages. Therefore, it is best if you remove the engine and clean it before completely taking off the fan housing and removing the oil cooler.

On 1970 cars, the oil cooler is mounted directly on the crankcase. On later models, there is a light alloy intermediate flange bolted in the former oil cooler location. The oil cooler itself is mounted on the side of the intermediate flange. The oil cooler can either be removed from the flange by taking off the nuts, or it can be removed complete with the flange, following a procedure similar to that used on the 1970 engine. If the oil cooler is to be replaced, remove it from the intermediate flange. When making engine repairs, or when having the oil cooler pressure-tested by a VW dealer, remove the cooler with the intermediate flange attached.

To remove the oil cooler (1970 cars) or the oil cooler in combination with the intermediate flange (later models), remove the M 6 nut at the right-hand side of the oil cooler or flange, then remove the other two nuts from beneath the left-hand side as illustrated in Fig. 8-1.

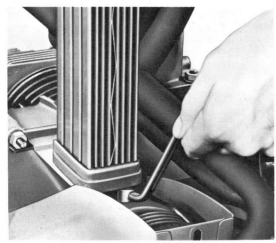

Fig. 8-1. Removing one of two nuts under left-hand side of oil cooler (or intermediate flange).

During installation, use new rubber seals between the oil cooler and the crankcase (1970 cars) or between the intermediate flange and the crankcase (later models). If the late oil cooler is removed from the intermediate flange, use new rubber seals between the two parts during assembly. Torque the mounting nuts to 0.7 mkg (5 ft. lb. or 60 in. lb.).

8.2 Testing and Replacing Oil Pressure Switch

If the oil pressure warning light does not come on when the ignition is on and the engine is not running, check the bulb and replace it if necessary. If the bulb is not at fault, remove the wire from the oil pressure switch and ground it on the crankcase with the ignition on. If the light does not come on, the wiring is faulty. If it does come on, there is trouble in the switch. Remove the switch as shown in Fig. 8-2 and inspect it for a clogged orifice. If cleaning does not restore the switch to serviceable condition, replace it.

Fig. 8-2. Oil pressure switch being removed.

You can also test the pressure switch for accuracy by installing an oil pressure gauge between it and the crankcase as shown in Fig. 8-3. The warning light should go out at 2 to 6 psi (0.14 to 0.42 kg/cm^2). Oil pressure at 2500 rpm should be about 42 psi (3 kg/cm^2) with SAE 30 oil at a temperature of 70°C (158°F).

Fig. 8-3. Oil pressure gauge installed between crankcase and oil pressure switch.

Engine 17

9. Valve Gear and Cylinder Heads

Fig. 9-1 is an exploded view of the cylinder head and valve gear. Eight nuts hold each cylinder head on a pair of cylinders. These nuts thread onto long studs set in the crankcase casting. The rocker arm shaft assembly must be removed for access to the nuts that hold the cylinder head.

1. Bracket (not on cars covered by this Manual)
2. 1971 and later cylinder head
3. Exhaust valve
4. Intake valve
5. Exhaust valve guide
6. Intake valve guide
7. Flat washer
8. Cylinder head nut
9. 1970 cylinder head
10. Stem seal ring
11. Valve spring
12. Spring retainer
13. Valve keeper
14. Pushrod
15. Rubber seal
16. Pushrod tube
17. End clip
18. Flat washer
19. Spring washer
20. Rocker arm
21. Adjusting screw
22. Adjusting screw locknut
23. Rocker arm shaft support
24. Silicone rubber seal (1970 through 1976 only)
25. Rocker arm shaft
26. Spring washer
27. Rocker arm shaft nut
28. Gasket
29. Cylinder head cover

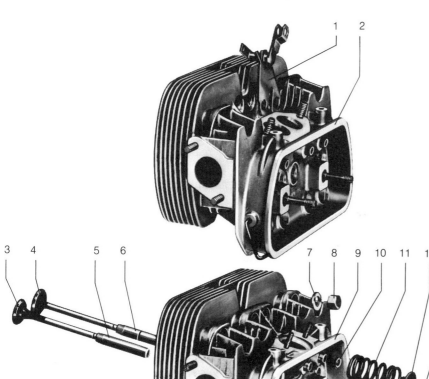

Fig. 9-1. Exploded view of cylinder head and valve gear.

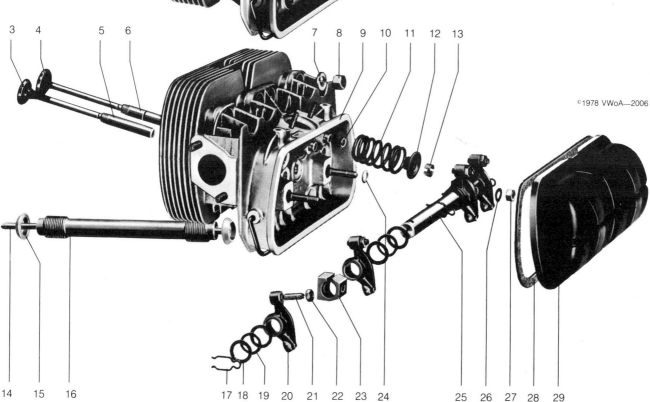

18 ENGINE

9.1 Removing and Installing Rocker Arm Shaft

Before you remove the cylinder head cover, clean all road dirt from the cover and the cylinder head. To remove the cover, pry the heavy spring wire retainer clip outward and down. If you install a new gasket, do not use gasket cement between the cylinder head cover gasket and the cylinder head. Replace leaking gaskets. The rocker arm shaft itself is held to the cylinder head by two nuts threaded onto steel studs anchored in the head casting.

To remove rocker arm shaft:

1. Remove the cylinder head cover.
2. Remove the rocker arm shaft retaining nuts.

   NOTE —
   Loosen the nuts a little at a time, working alternately to relieve spring tension evenly.

3. Lift off the rocker arm shaft assembly.

To install rocker arm shaft:

1. Where applicable, install new silicone rubber seals on the rocker arm shaft support studs.

   NOTE —
   The silicone rubber seals were eliminated beginning with the 1977 models. Consequently, the seal recesses in the cylinder heads, including replacement heads, have been discontinued.

2. If previously removed, install the pushrods.

   NOTE —
   Roll the pushrods on a flat table to see that they are not bent.

3. Carefully guiding the pushrods into the rocker arm sockets, install the rocker arm shaft assembly as indicated in Fig. 9-2.

4. Install the nuts, then torque them to 2.0 to 2.5 mkg (14 to 18 ft. lb.).

Fig. 9-2. Rocker arm shaft installation. The slotted side of the support (left arrow) should point upward. The beveled end (right arrow) should be away from the cylinder head.

9.2 Disassembling and Assembling Rocker Arm Shaft Assembly

If the rocker arms are worn or binding, you may have to take apart the rocker arm shaft assembly in order to replace individual parts or to clean away hardened dirt accumulations.

To disassemble:

1. Remove the end clips. Then slide the rocker arms, supports, and washers off the shaft.
2. Inspect the shaft, the rocker arm bores, the adjusting screws, and the rocker arm sockets for wear. Replace worn parts.
3. Using Fig. 9-3 as a guide, assemble the components.

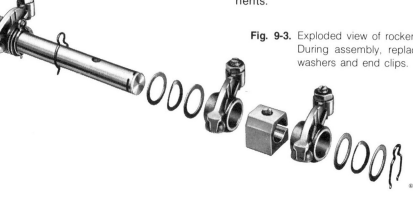

Fig. 9-3. Exploded view of rocker arm shaft assembly. During assembly, replace worn or corroded washers and end clips.

9.3 Removing and Installing Cylinder Head and Pushrod Tubes

The engine must be out of the car before you can remove the cylinder heads. Formerly, it was also necessary to remove the engine from the car and then remove the cylinder head before a faulty pushrod tube could be replaced. However, telescopic pushrod tubes have been introduced as replacement parts. These new parts make it unnecessary to remove the engine or the heads in order to replace one or more of the pushrod tubes.

To replace pushrod tubes (cylinder heads installed):

1. Clean all road dirt from the cylinder head cover and the cylinder head. Then remove the cylinder head cover by prying the heavy spring wire retainer clip outward and down.

2. Remove the rocker arm shaft retaining nuts. Loosening each nut a little at a time, work alternately to relieve spring tension evenly.

3. Lift off the rocker arm shaft assembly. Withdraw the pushrod(s) from the pushrod tube(s) that you intend to replace.

4. Using a large screwdriver, pry out the defective pushrod tube(s). Remove the old pushrod tube seals and wipe clean the recesses in the cylinder head and the crankcase.

5. Install new seals on the telescopic pushrod tube(s). Then squeeze together each telescopic tube and install it so that the small-diameter end is toward the cylinder head.

6. Install the pushrods.

    NOTE ——
    Roll the pushrods on a flat table to see that they are not bent.

7. Install new silicone rubber seals on the rocker arm shaft support studs of 1970 through 1976 cylinder heads, if there are recesses for the seals.

    NOTE ——
    The silicone rubber seals were eliminated beginning with the 1977 models. Consequently, the seal recesses in the cylinder heads, including replacement heads, have been discontinued.

8. Carefully guiding the pushrods into the rocker arm sockets, install the rocker arm shaft assembly on the cylinder head.

    NOTE ——
    Be sure that the rocker arm shaft supports have their slotted sides up and beveled ends outermost, as shown previously in Fig. 9-2.

9. Loosely install the rocker arm shaft nuts. Tighten each nut a little at a time, working alternately in order to compress the valve springs evenly. Torque the nuts to 2.0 to 2.5 mkg (14 to 18 ft. lb.).

To remove head:

1. Remove the exhaust system as described in **5. Exhaust System.**

2. Remove the fan housing, the cylinder cover plates, and the intake manifold as described in **6. Cooling Air System and Intake Manifold.**

3. Remove the rocker arm shaft assembly, as described in **9.1 Removing and Installing Rocker Arm Shaft.**

4. Withdraw the pushrods.

5. Using a 15-mm ($^{19}/_{32}$-in.) socket wrench, remove the cylinder head nuts.

6. Pull the cylinder head off the cylinders.

    NOTE ——
    If the cylinders are to be left in the engine, use some kind of retaining device to keep them from pulling free along with the heads.

To install head:

1. Check the head for cracks in the combustion chamber and valve ports. If there are cracks, replace the head.

2. Check the rocker arm shaft support studs and cylinder studs for tightness. Take any parts that have pulled studs to an Authorized VW Dealer or a qualified automotive machine shop for repair with special threaded inserts. See **16.4 Repairing Tapped Holes.**

3. Remove the pushrod tubes and measure them as indicated in Fig. 9-4. Dimension **a** must be 190 to 191 mm ($7^{31}/_{64}$ to $7^{33}/_{64}$ in.) If it is not, stretch the tube bellows slightly.

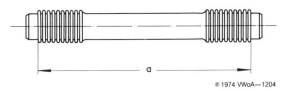

Fig. 9-4. Pushrod tube length (**a**). Check for cracks and pinholes. During cylinder head installation, you can use either the one-piece pushrod tube shown here or the new-type telescopic replacement pushrod tubes.

20 ENGINE

4. Making sure that the convex surfaces are outermost, install new rubber seals on the ends of the pushrod tubes as shown in Fig. 9-5.

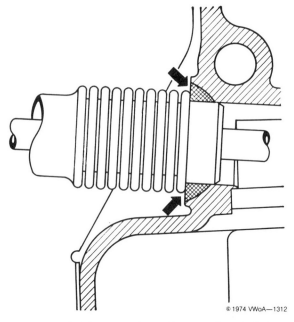

Fig. 9-5. Rubber seal (arrows) correctly installed on pushrod tube. Convex face of seal engages recess in crankcase or cylinder head.

5. Install the head on the cylinder studs. Install the pushrod tubes with their seams facing upward, then push the head onto the cylinders to hold the tubes.

6. Loosely install the cylinder stud washers and nuts—using oil resistant sealer on the four lower washers, nuts, and studs of fuel injection engines.

7. Following the sequence given in Fig. 9-6, torque the nuts to 1 mkg (7 ft. lb.).

   NOTE
   This sequence is for initial tightening only. It is not the final sequence.

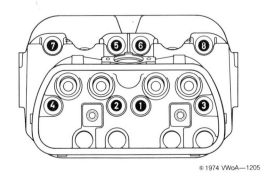

Fig. 9-6. Initial tightening sequence.

8. Following the sequence given in Fig. 9-7, torque M 10 nuts to 3.0 to 3.2 mkg (22 to 23 ft. lb.) and torque M 8 nuts to 2.5 mkg (18 ft. lb.).

   NOTE
   The M 8 nuts are used only with the thinner studs installed since early 1973. Both the early M 10 and the late M 8 cylinder head nuts have the same (15-mm) hex size.

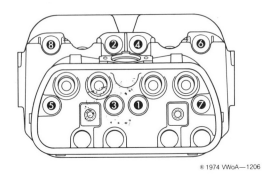

Fig. 9-7. Final torque sequence.

9. Install the pushrods and the rocker arm shaft assembly.

10. Using a new gasket, install the cylinder head cover.

10. VALVES

The components of a valve assembly for the VW engine are shown in Fig. 10-1.

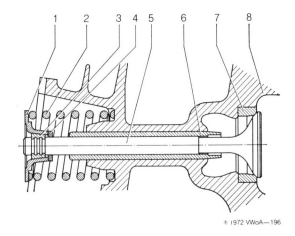

1. Spring retainer
2. Valve spring
3. Valve keeper
4. Stem seal ring
5. Valve
6. Valve guide
7. Valve seat insert
8. Cylinder head

Fig. 10-1. Cross section of valve assembly.

ENGINE 21

10.1 Removing and Installing Valves

A proper valve spring compressing tool is necessary to remove the valves from the cylinder head.

To remove valves:

1. Remove the cylinder head.
2. Compress the valve spring as shown in Fig. 10-2.

Fig. 10-2. Valve spring compressing tool being used to push spring retainer down valve stem against spring tension.

3. Remove the split keeper halves from the valve stem.
4. Release the compressing tool. Then take off the spring retainer, the valve spring, and the stem seal ring.
5. Remove the valve from the cylinder head.

> **CAUTION** ———
> If the keeper grooves in the valve stem are burred, file them smooth before you remove the valve. If forced out, the burred stem will ruin the valve guide.

To install:

1. Using a valve spring tester, check the valve spring tension. The springs should require a load of 53.2 to 61.2 kg (117 to 135 lb.) to compress them to a loaded length of 31.0 mm (1 7/32 in.).

> **NOTE** ———
> If you do not have a valve spring tester, have the springs checked by an Authorized VW Dealer or a qualified automotive machine shop.

2. To check the keepers, oil them and fit them to the valve stem. Hold the keeper halves together while you turn the valve. The valve stem should rotate freely in the assembled keeper.
3. Check the valve seat and valve facings. If necessary, recondition (grind) the seats and the facings.
4. Check the valve guides as described in **10.3 Checking and Replacing Valve Guides**.

> **NOTE** ———
> If a guide is within specifications but the valve can be moved from side to side in it, check the valve stem for wear. See **Table a**.

Table a. Valve Stem Diameters

	New valves mm (inch)	Wear limit mm (inch)
Intake valve	7.94–7.95 (.3126–.3130)	7.90 (.3110)
Exhaust valve	7.91–7.92 (.3114–.3118)	7.88 (.3102)
Exhaust valve for fuel-injected engine	8.91–8.92 (.3508–.3512)	8.88 (.3496)

5. Install the valve and the stem seal ring. Install the spring so that its closely-spaced coils are against the cylinder head.
6. Install the spring retainer. Then compress the spring with a valve spring compressing tool and install the keeper.

10.2 Removing and Installing Valve Springs (head installed)

A worn stem seal ring, broken valve spring, damaged keeper, or spring retainer can be replaced without removing either the engine or the cylinder head. To replace these parts, use the procedure that follows.

> **CAUTION** ———
> To carry out this work successfully, you will need a compressed air supply that will maintain a constant air pressure of at least 85 psi (6 kg/cm²) in the cylinder. Otherwise, the valve may not be held against its seat with sufficient force. The valves and the piston rings must be in good condition to ensure an air seal.

To remove spring:

1. Remove the spark plug. Remove the cylinder head cover and the rocker arm shaft as described in **9.1 Removing and Installing Rocker Arm Shaft**.
2. Install an air hose adapter in the spark plug hole. Apply a constant air pressure of at least 85 psi (6 kg/cm²) to the cylinder.

22 ENGINE

3. Attach the special valve spring compressor as shown in Fig. 10-3.
4. Compress the spring retainer and the spring. Then remove the keeper.

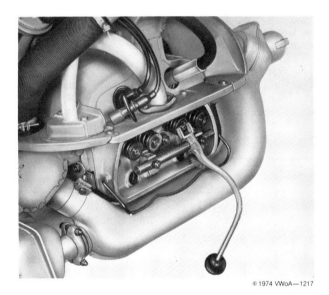

Fig. 10-3. VW tool 653/2 being used to remove valve spring with cylinder head installed.

10.3 Checking and Replacing Valve Guides

Use a dial indicator as shown in Fig. 10-4 to check the valve guides for wear. With the valve stem flush with the end of the guide, it should not be possible to rock the valve head sideways more than 0.23 to 0.27 mm (.009 to .011 in.) with new parts. The wear limit for this measurement is 0.80 mm (.031 in.).

Fig. 10-4. Valve head rock being checked with dial indicator.

If the valve guide permits excessive rocking motion even with a new valve, replace the guide. If a new valve's stem fits too tightly in the guide, the guide probably has "varnish" deposits inside it. You can correct this with either a guide hone or a brass bristle brush and carburetor cleaning solvent.

The valve guides are a press fit in the cylinder head. They should be removed only with special equipment available at an Authorized VW Dealer or a qualified automotive machine shop. The proper tool is a device that cuts threads inside the old guide so that a threaded pulling tool can be fitted into it. With the help of a powerful press, the old guide can then be withdrawn.

CAUTION —
The practice of driving valve guides in or out with a hammer and a tool similar to a bushing driver is common in many shops. However, this technique is unsuitable for the lightweight aluminum cylinder heads of the VW engine. There is a grave possibility of damage to surrounding parts of the cylinder head. Such damage might make it impossible to fit a new valve guide or might cause the head dimensions to be distorted.

The valve seats should be reconditioned as described in **10.5 Refacing Valve Seats** after new guides have been installed. This will ensure that the valve seats are concentric with the valve guide bores.

CAUTION —
If you lack the skills, tools, or a suitable workshop for reconditioning the valves, we suggest you leave such repairs to an Authorized VW Dealer or other qualified shop. We especially urge you to consult your Authorized VW Dealer before attempting repairs on a car still covered by the new-car warranty.

10.4 Replacement of Valve Seats

It might appear that the valve seat inserts in the cylinder head can be replaced. Some specialty shops that rebuild VW engines for racing do make this modification. However, the job is impossible to carry out successfully without special equipment and is done only when valves of increased head diameter are to be used.

Such work is expensive and, done properly, costs more than a new replacement cylinder head. It is therefore not economically wise when standard valve diameters are retained. Moreover, there is always the possibility of a replacement insert coming out. In salvaging an unserviceable cylinder head, an entire engine may be ruined. Therefore, we strongly recommend that you replace the entire cylinder head if the valve seats are no longer serviceable.

10.5 Refacing Valve Seats

Pitted or gas-cut valve seats can usually be reconditioned by refacing. A cylinder head may undergo this operation many times before it becomes unserviceable. Irreparable seat damage is usually the result of continued driving long after a compression test has indicated that the valves are leaking.

The valve seat contact facing should be cut to 45° for both intake and exhaust valves. The width of the contact facing, dimension **a** in Fig. 10-5, must be kept to 1.30 to 1.60 mm (.051 to .063 in.) for intake valves or 1.70 to 2.00 mm (.067 to .079 in.) for exhaust valves. This width is adjusted by cutting a 15° chamfer at the outer edge of the seat and a 75° chamfer at its inner edge. Seats cannot be refaced if too little material remains for a 15° chamfer to be cut without going beyond the boundary of the insert. If the 15° chamfer already extends to the edge of the insert, replace the cylinder head.

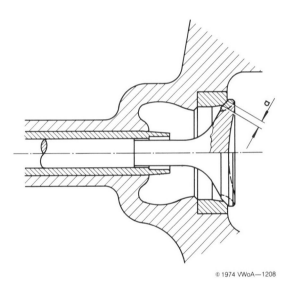

Fig. 10-5. Seat contact width (**a**).

CAUTION
Check the condition of the valve guides before refacing the valve seats. If new valve guides are fitted, they may not be concentric with a seat cut while the original guides were in place. Reface the valve seats only after installing new guides.

To obtain a true surface on valve seats, the refacing tools must be in good condition. Following the tool manufacturer's instructions will help you avoid chatter marks, out-of-round seats, or concentric scoring. Do not rely on lapping to correct careless work done earlier with a seat cutter or stone.

To reface:

1. Cut the valve seat contact facing to a 45° angle, as shown in Fig. 10-6. Remove no more metal than is necessary to remove pitted or burned areas.

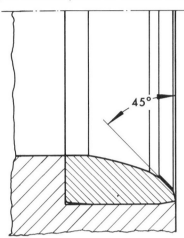

Fig. 10-6. Contact facing of valve seat. Work to obtain a glass-smooth finish.

2. Using a 75° cutter, chamfer the valve seats to true and narrow the seats from their inner edge. See Fig. 10-7.

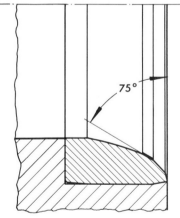

Fig. 10-7. Lower edge of valve seat. Chamfer to 75°.

CAUTION
Do not neglect narrowing the valve seats to specifications. Overly wide seats produced by using only a 45° cutter tend to trap carbon particles and other deposits.

3. Cut a 15° chamfer, as shown in Fig. 10-8, to narrow the seat from its outer edge. The chamfer must not extend to the aluminum material of the cylinder head.

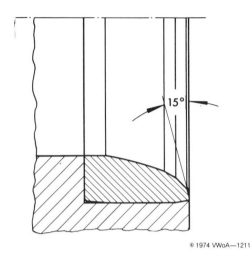

Fig. 10-8. Outer edge of valve seat. Chamfer to 15°.

10.6 Refacing Valves

The correct dimensions of new valves are given in Fig. 10-9. You can reface the valves to the indicated angles as long as dimension **b** is not reduced below the minimum listed margin width.

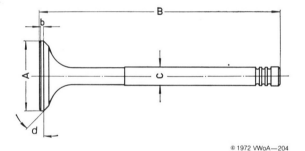

Dimension **A**, valve head diameter	Intake: 35.60 mm (1.400 in.) Exhaust: 32.10 mm (1.260 in.) Fuel injection engine Exhaust: 30.10 mm (1.185 in.)
Dimension **B**, valve length	112.00 mm (4.409 in.)
Dimension **C**, valve stem diameter	Intake: 7.94–7.95 mm (.3126–.3130 in.) Exhaust: 7.91–7.92 mm (.3114–.3118 in.) Fuel injection engine Exhaust: 8.91–8.92 mm (.3508–.3512 in.)
Dimension **b**, valve head margin	Intake: 0.80–1.50 mm (.031–.059 in.) Exhaust: 1.00–1.70 mm (.040–.067 in.)
Dimension **d**, facing angle	Intake: 44° Exhaust: 45°

Fig. 10-9. Dimensions for new valves.

Carefully check used valves before you decide to reuse them.

To check valve:

1. Using a motor-driven wire brush, remove the combustion chamber deposits from the valves.

2. Examine the seat contact area for pits, burns, and other signs of wear. If the damage is too extensive to be cured by refacing, replace the valve.

   NOTE
   Because of the extreme conditions under which exhaust valves operate, many experienced mechanics routinely replace any exhaust valves that have been in service for 25,000 mi. (40,000 km) or more.

3. Discard any valve with damaged keeper grooves or with a stem that has been warped or galled by seizure.

4. Discard any valve that exceeds dimension **B** shown in Fig. 10-9. It has lost strength and is stretching.

After you have selected new valves or refaced used valves, check the valve fit in the refaced seats you have cut into the cylinder head.

To check valve seating:

1. Coat the face of the valve lightly with Prussian blue.

2. Insert the coated valve into the valve seat. Applying light pressure, rotate the valve about a quarter turn.

3. Remove the valve and examine the contact pattern. If the seating is correct, the valve will leave an even coating of Prussian blue on the seat.

   NOTE
   If areas of bare metal show on the valve seat, a more careful refacing of the valve or seat is required. Minor irregularities can be corrected by lapping.

If the valve leaves an irregular contact pattern on the seat, inspect the seat first. A poorly fitting stone or cutter pilot may have caused you to cut the seat out-of-round. A correctly fitting pilot is especially important to the accuracy of the seat produced.

Another possible cause of an out-of-round seat is a tapered valve guide that allows the cutter pilot to enter it at an angle. Check the guides. If they are tapered or out-of-round, replace them before again attempting to reface the valve seat. If you find that the valve itself is warped, replace it. Minor warping can sometimes be corrected by refacing the valve. However, if the remaining margin is noticeably irregular, replace the valve.

10.7 Lapping Valves

Using a tool such as that shown in Fig. 10-10, you can lap valves into their seats for a perfect seal. However, modern precision refacing machines that are in good condition will usually produce an excellent fit without significant lapping.

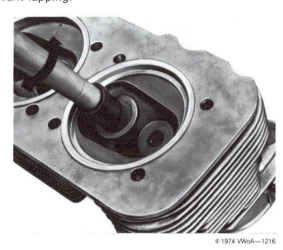

Fig. 10-10. Valve being lapped with suction cup tool. Tool holds valve while you turn it against seat.

To lap valve:

1. Coat the seat with a small amount of valve grinding compound. Then insert the valve in the guide.

2. Using the suction cup tool, turn the valve clockwise and counterclockwise against the seat.

   NOTE ——
   By lifting the valve off the seat every few turns and shifting its position slightly, you will avoid cutting concentric grooves into the seat.

3. Clean away every trace of grinding compound. Check the valve seating with Prussian blue.

10.8 Valve Clearance

Valve clearance is specified at 0.15 mm (.006 in.) for both intake and exhaust valves. Measure this clearance between the adjusting screw tip of the rocker arm and the tip of the valve stem. For accurate measurement, the engine must be completely cold (oil temperature no more than 50°C [122°F]) and the cylinder in position to fire.

If the valve clearance is insufficient, the valves will open early and close late, resulting in inadequate cooling and compression loss. Engine performance will be poor or uneven and the valves will burn prematurely. Too much clearance will cause excessive noise, accelerated wear, and reduced performance.

10.9 Adjusting Valves

Adjust the valve clearance in cylinder sequence 1-2-3-4 (the reverse of the firing order).

To adjust:

1. Remove the cylinder head covers and the distributor cap. Hand-turn the engine until the distributor rotor is adjacent to the No. 1 cylinder notch in the top edge of the distributor body.

   NOTE ——
   When the rotor is in this position, both valves for the No. 1 cylinder are closed, the piston is at approximately top dead center on its compression stroke, and the spark plug is ready to fire.

2. Using a 0.15-mm (.006-in.) feeler gauge, check the clearance of both valves for the No. 1 cylinder. You should just be able to insert and withdraw the gauge with moderate drag.

   NOTE ——
   No. 1 cylinder is at the right front of the engine, No. 2 behind it. No. 3 is at the left front of the engine, No. 4 behind it.

3. If the clearance is incorrect, loosen the locknut. Then turn the adjusting screw clockwise to reduce clearance or counterclockwise to increase it. (See Fig. 10-11.) When the clearance produces moderate drag on the feeler gauge, tighten the locknut while holding the adjusting screw stationary. Then recheck the clearance.

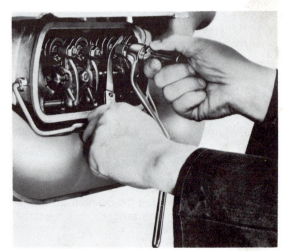

Fig. 10-11. Clearance between rocker and valve stem tip being adjusted. Note feeler gauge and combination wrench and screwdriver being used.

26 ENGINE

> **NOTE** ——
> If you lack experience in adjusting valves, a set of "go/no-go" feeler gauges may help you to determine clearances more accurately. Such gauges are stepped so that when clearance is correct, the gauge will slide in about an inch and no further.

4. Turn the engine until the rotor moves counterclockwise 90° from the No. 1 cylinder firing position. This will correctly position the No. 2 cylinder for valve adjustments.

5. Check and adjust the No. 2 cylinder's valves as you did on the No. 1 cylinder. Turn the engine so that the rotor moves counterclockwise in 90° increments. Check and adjust the valves of the No. 3 and the No. 4 cylinders in the same way.

11. CYLINDERS

You can replace cylinders individually on the VW engine. However, you must replace the piston at the same time. Select a piston that fits the new cylinder bore and matches the weights of the other pistons in the engine. Weight and fitting dimensions are indicated by marks on the piston crown.

11.1 Removing and Installing Cylinders

The engine must be out of the car before you can remove the cylinders. Prior to disassembly, clean the engine to keep dirt out of the working parts. Normally, you will remove the cylinders without further engine disassembly only if you are inspecting for wear or correcting oil leaks between the cylinders and the crankcase.

To remove cylinder:

1. Remove all parts preparatory to cylinder head removal. Then remove the cylinder head.

2. Remove the sheet metal air deflector plates below the cylinders. Then, if more than one cylinder is to be removed, mark each cylinder so that you will be able to reinstall it in its original position.

3. Pull the cylinder off the crankcase and the piston.

To install cylinder:

1. Check the cylinder for wear. You must measure the piston diameter and the cylinder bore and compute the clearance. The procedure is described in **12.3 Piston Clearance**.

    > **NOTE** ——
    > If the clearance exceeds 0.20 mm (.008 in.), install a new cylinder and matching piston. The clearance between new parts should be 0.04 to 0.06 mm (.0016 to .0024 in.).

2. Inspect the cylinder sealing surfaces shown in Fig. 11-1.

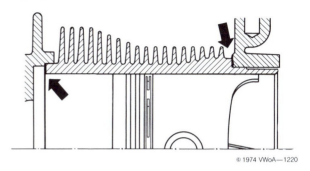

Fig. 11-1. Cylinder sealing surfaces. These areas must be perfectly clean and true prior to cylinder installation.

3. Stagger the ring gaps 90° apart so that the oil ring gap faces upward when the cylinder is installed.

4. Oil the piston, piston rings, and piston pin.

5. Install the paper sealing ring that fits between the cylinder and the crankcase.

6. Compress the piston rings with a piston ring compressor.

    > **NOTE** ——
    > Use a piston ring compressor that can be taken apart. Otherwise, you will not be able to get it off the piston after the cylinder has been installed.

7. Oil the cylinder walls lightly and install the cylinder as shown in Fig. 11-2.

8. To complete the assembly, install the cylinder head and the other engine parts. Then adjust the valves.

Fig. 11-2. Cylinder being installed. Note design of ring compressor.

12. Pistons and Rings

Piston size, weight, and installation position are marked on the crown as shown in Fig. 12-1.

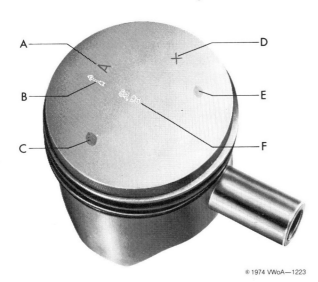

A. Index letter of piston part number
B. Arrow that must point toward flywheel
C. Paint spot indicating matching cylinder size (blue, pink, green)
D. Weight grading (+ or −)
E. Paint spot indicating weight grading (brown for − weight, gray for + weight)
F. Piston size in mm

Fig. 12-1. Piston markings used as fitting guide.

12.1 Removing and Installing Pistons

You can remove the pistons once you have taken off the cylinders. It is unnecessary either to remove the connecting rods from the crankshaft or to separate the crankcase halves. The pistons are usually removed as a step in engine disassembly, in order to replace worn piston rings, or to replace the pistons themselves.

To remove pistons:

1. Remove the cylinders.
2. Mark the pistons so that you will be able to reinstall them in their original positions.
3. Using suitable pliers, remove the piston pin circlips as shown in Fig. 12-2.
4. Push out the piston pin. Then take the piston off the connecting rod.
5. Using piston ring expanding pliers, remove the piston rings. Remove the top ring first, then work downward.

Fig. 12-2. Piston pin circlip being removed or installed.

To install pistons:

1. Clean carbon and other deposits from the crown, skirt, and ring grooves of each piston.
2. Check the clearance as described in **12.3 Piston Clearance**.

   NOTE
   If the clearance between piston and cylinder is excessive, install a new piston and cylinder combination. The new piston must be of the same weight grade as the original or within 10 grams of original piston weight.

3. Check the pin fit in each piston. It must be a light push fit at room temperature or with the piston heated to 75°C (167°F) in an oil bath. If not, replace both the piston and the pin.
4. Check the piston pin fit in each connecting rod. The wear limit is 0.04 mm (.0016 in.). New clearance should be 0.01 to 0.02 mm (.0004 to .0008 in.).

   NOTE
   Some pistons have oversize pin bores. Oversize pins, marked in green, are available for these pin bores. You should not, however, use oversize pins to reduce the clearance in worn pistons. If the clearance in the rod is excessive, fit new pins and new rod bushings and check them for proper clearance.

5. Install one circlip in each piston in the side toward the flywheel, as shown by the arrow on the piston crown.
6. Install each piston on its rod so that the arrow on the piston crown points toward the flywheel. If necessary, heat the piston to 75°C (167°F) in an oil bath, then push in the piston pin and install the remaining circlip. Install pistons for No. 1 and No. 3 cylinders first, then install pistons for No. 2 and No. 4 cylinders.
7. Install the cylinders.

28 ENGINE

12.2 Piston Rings

New piston rings must always be fitted with reference to the markings on the piston crown. The various cylinder-bore/piston diameter grades each have similarly graded piston ring sets to match them. It is highly important that the rings fit properly in the cylinder and piston ring grooves, and that they are correctly installed on the piston.

To check ring gap:

1. Carefully hand-compress the ring and insert it in the bottom of the cylinder.
2. Using the matching piston as a pusher to prevent tilt, push the ring into the cylinder about 5 mm (¼ in.).
3. Measure the ring gap as shown in Fig. 12-3.

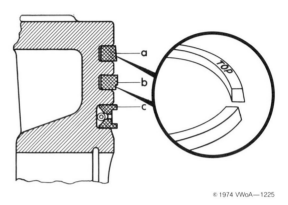

NOTE ——
The rings must be installed as shown in Fig. 12-4. If installed otherwise, they will not produce a proper seal with the cylinder wall.

a. Upper compression ring c. Oil scraper ring
b. Lower compression ring

Fig. 12-4. Proper ring installation. Word **TOP** must be toward piston crown.

2. Check the ring side clearance, as shown in Fig. 12-5.

NOTE ——
The side clearance for a new upper compression ring is 0.07 to 0.10 mm (.003 to .004 in.), and 0.05 to 0.07 mm (.002 to .003 in.) for a new lower compression ring. The side clearance for a new oil scraper ring should be 0.03 to 0.05 mm (.001 to .002 in.). The wear limit is 0.12 mm (.005 in.) for the upper compression ring and 0.10 mm (.004 in.) for the lower compression and oil scraper rings.

Fig. 12-3. Ring gap being measured with feeler gauge.

NOTE ——
The proper gap for the upper and lower compression rings—when new—is 0.30 to 0.45 mm (.012 to .018 in.). A new oil scraper ring should gap at 0.25 to 0.40 mm (.010 to .016 in.). The wear limit for the two compression rings is 0.90 mm (.035 in.) and for the oil scraper ring, 0.95 mm (.037 in.).

To check side clearance:

1. Using piston ring expanding pliers, install the piston rings in the piston grooves.

Fig. 12-5. Ring side clearance being measured. Insert feeler gauge between ring and piston land.

ENGINE

12.3 Piston Clearance

Piston clearance must not exceed the prescribed range or the pistons will rock in their bores during engine operation. This would cause the piston rings to develop an elliptical wear pattern that would impair their sealing effectiveness. Some clearance, however, is essential. If the clearance is inadequate, the aluminum pistons could expand and seize in their cylinders.

To keep the necessary clearance from causing excessive rocking motion (and piston slap noises), the piston pin bores are offset slightly from the piston centerline. This offset, shown in Fig. 12-6, is the reason that the pistons must be installed so that the arrow on their crown points toward the flywheel.

Fig. 12-7. Cylinder bore being measured with a special dial indicator.

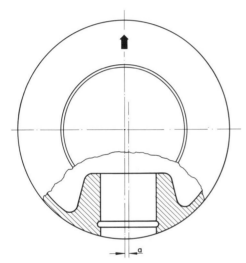

Fig. 12-6. Piston pin offset. Arrow must point toward flywheel to position offset properly. Dimension **a** is 1.5 mm (.059 in.).

2. Record the maximum cylinder diameter on paper.
3. Measure the piston diameter, as illustrated in Fig. 12-8.

Fig. 12-8. Piston diameter being measured. A micrometer is the best tool for this task.

The clearance between the piston and the cylinder wall is computed from measurements taken on the piston and in the cylinder. The clearance for new parts is specified at 0.04 to 0.06 mm (.0016 to .0024 in.) with a wear limit of 0.20 mm (.008 in.).

To check clearance:

1. Using a dial indicator device, measure the cylinder bore as shown in Fig. 12-7.

   NOTE
   When checking the cylinder bore, make your measurements at several points throughout the cylinder and at right angles to one another. If there are wide variations, the cylinder and piston may have to be replaced. To compute piston clearance, use the largest reading you obtain.

4. To obtain the piston clearance, subtract the piston diameter from the cylinder diameter.

   NOTE
   If wear proves to be almost entirely on the piston, the piston alone can be replaced with one of matching size. Only pistons of the same size and weight grade should be installed in the same engine.

30 Engine

13. Torque Converter

Cars equipped with automatic transmission have a torque converter instead of a clutch and a drive plate rather than a conventional flywheel. The torque converter is held to the drive plate by four bolts that are taken out when the engine is removed.

13.1 Removing and Installing Drive Plate

When the engine must be disassembled, it is best to remove the drive plate before separating the crankcase halves.

To remove:

1. Mark the position of the drive plate so that you can install it in the exact original position on the crankshaft.
2. Attach the drive plate holding fixture as shown in Fig. 13-1. Using a 36-mm (1 7/16-in.) socket, unscrew the gland nut in the center of the drive plate.
3. Remove the drive plate from the dowels on the crankshaft.

Fig. 13-1. Holding fixture installed on drive plate.

To install:

1. Check the converter mounting bolt holes for cracks and wear. Replace the drive plates if faulty.
2. Check the dowel pins and holes for wear. Inspect the oil seal in the crankcase for cracks and wear. Replace faulty parts.
3. Install the dowel pins and the drive plate on the crankshaft so that the marks made earlier line up.
4. Attach the drive plate holding fixture. Then torque the gland nut to 35 mkg (253 ft. lb.).
5. If necessary, check and adjust the crankshaft end play as described in **17.4 Crankshaft End Play**.

14. Clutch and Flywheel

Cars with manual transmissions have diaphragm spring clutches. However, some of the illustrations show the coil spring clutch used previously. The repair procedures are the same for both. The pressure plate assembly introduced on the 1973 models is shown in Fig. 14-1. It can be used as a replacement for the earlier diaphragm spring pressure plate assembly shown in Fig. 14-2.

Fig. 14-1. Late-type diaphragm spring clutch.

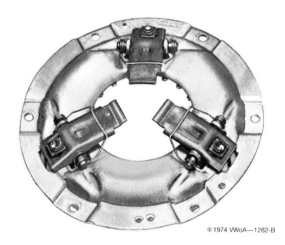

Fig. 14-2. Diaphragm spring clutch used on 1970 through 1972 models.

14.1 Removing and Installing Clutch

The pressure plate assembly and driven plate remain on the flywheel when you remove the engine. The clutch release bearing and related parts stay in the transmission.

To remove clutch:

1. Remove the engine. Mark the position of the pressure plate so that you can install it in the exact original position on the flywheel.

2. If you have a tool such as the one shown in Fig. 14-3, use it to release the clutch. Then remove the bolts that hold the pressure plate to the flywheel. If you lack the special tool, loosen the mounting bolts a quarter turn at a time. Work around the flywheel until the pressure is relieved, then remove the bolts completely.

Fig. 14-3. Pressure plate bolts being removed. The special tool is in place to release clutch pressure. Note the fixture used to prevent flywheel rotation.

3. Remove the pressure plate assembly from the flywheel.

To install:

1. Clean the friction surface on the flywheel and inspect it for wear, cracks, and grooves. Minor defects may be correctable by regrinding or machining at a specialty shop. If the flywheel is unserviceable, replace it.

2. Inspect the pressure plate for wear, cracks, and grooves. Alternate bright and dull areas indicate a warped plate. Shake the pressure plate assembly. The diaphragm spring should be under tension and should not rattle loosely. Also see that none of the release levers is out of line with the others. If the pressure plate assembly is in any way unserviceable, replace it.

3. Inspect the clutch release bearing. If it feels gritty when you turn it, or if it has been making noise, replace it. Never wash the bearing in solvent since this will remove the factory-installed lubricant. If the release bearing is unserviceable, replace it as described in **14.4 Replacing Clutch Release Bearing**.

4. Using solvent, clean the needle bearings inside the flywheel gland nut. Check to see that the needles are not flattened by wear. Pack the bearings with 1 gram (1/32 oz.) multipurpose grease (just enough to coat all needles lightly).

5. Inspect the driven plate for wear. There should be at least 2.00 mm (about 1/16 in.) of friction material remaining above the rivet heads. Check the plate for runout as shown in Fig. 14-4. Runout should not exceed 0.50 mm (.020 in.).

Fig. 14-4. Dial indicator being used to check driven plate for runout.

6. Inspect the splines in the driven plate and on the transmission's rear driveshaft. They must not be broken or distorted. Lubricate the splines with molybdenum disulfide powder. Then see that the driven plate slides freely on the rear driveshaft without undue radial play. If the driven plate is in any way unserviceable, replace it.

7. Apply motor oil to the felt ring in the gland nut. Wipe away all excess lubricants.

8. Using a centering tool or a spare transmission rear driveshaft, install the driven plate against the flywheel.

9. Install the pressure plate assembly according to the reference marks made earlier.

NOTE
On some replacement clutches, a white paint spot is used as a balance mark. This should be positioned 180° from the 5-mm (3/16-in.) countersunk hole or from the white paint balance mark on the flywheel.

10. Loosely install all the mounting bolts.

11. Working diagonally, tighten the mounting bolts a turn or two at a time as shown in Fig. 14-5.

Fig. 14-5. Mounting bolts being tightened. Notice the driven plate centering tool being used.

12. Torque the bolts to 2.5 mkg (18 ft. lb.).

   CAUTION ——
   To simplify centering and mounting, replacement clutches are pretensioned by clips between the release levers and the clutch plate. Be sure to remove these clips following installation.

14.2 Removing and Installing Flywheel

Remove the flywheel from the crankshaft before you separate the crankcase halves.

To remove:

1. Hold the flywheel with the fixture illustrated earlier in Fig. 14-3 and Fig. 14-5. Mark the position of the flywheel so that you can install it in the exact original position on the crankshaft.
2. Using a 36-mm (1 7/16-in.) socket, remove the gland nut from the center of the flywheel.
3. Pull the flywheel off the dowels on the crankshaft.

To install:

1. Using solvent, clean the gland nut. Then inspect the needle bearings indicated in Fig. 14-6. Replace the gland nut if the needles are flattened by wear. If the bearings are unworn, pack them with 1 g (1/32 oz.) multipurpose grease (just enough to coat all needles lightly). Apply motor oil to the felt ring, then wipe away all excess lubricants.

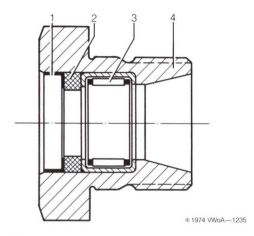

1. End ring
2. Felt ring
3. Needle bearing
4. Gland nut

Fig. 14-6. Flywheel gland nut.

2. Inspect the starter gear teeth for wear and damage.

   NOTE ——
   Up to 2 mm (1/16 in.) can be removed from the clutch side of the flywheel to deburr the starter gear. Chamfer the teeth after you have machined the flywheel.

3. Inspect the dowel holes in the crankshaft and flywheel for wear. See that the dowels fit tightly.
4. Install the dowel pins and the flywheel on the crankshaft so that the reference marks made earlier line up. Then install the gland nut, but do not torque it.

   NOTE ——
   Always use a new sealing ring (gasket) between the flywheel and the crankshaft. If the spring-type lock washer under the gland nut is deformed, replace it.

5. Check the crankshaft end play as described in **17.4 Crankshaft End Play**. If necessary, adjust the end play.

   CAUTION ——
   Crankshaft end play must be within the specified range. Incorrect end play will cause damage to the bearings or to other internal engine parts.

6. Install the flywheel locking fixture. Then torque the gland nut to 35 mkg (253 ft. lb.).
7. Lubricate the tip of the transmission's rear driveshaft with a very light coating of multipurpose grease. Install the clutch. Then install the engine in the car.

ENGINE 33

14.3 Removing and Installing Crankshaft Oil Seal

If the oil seal in the flywheel end of the crankcase leaks, the clutch will be contaminated by engine oil. A leaking seal is easy to mistake for a leak in the crankcase joint. This has led to many needless major repairs. So be sure to check the oil seal before concluding that the crankcase has to be taken apart.

To remove seal:

1. Remove the flywheel. Inspect the oil seal lip and the contact surface on the flywheel shoulder. If the lip is worn or cracked, replace the seal.

   NOTE
   If the flywheel shoulder contact surface is deeply grooved, you will have to replace the flywheel in order to obtain a good seal.

2. Pry out the old oil seal, being careful not to scratch or gouge the magnesium alloy crankcase casting.

To install:

1. If the outer edge of the crankcase is sharp, chamfer it as illustrated in Fig. 14-7. Clean away the metal shavings.

Fig. 14-7. Scraper being used to chamfer edge of seal recess in crankcase. Such chamfering will prevent damage to the seal during installation.

2. Clean the oil seal recess to remove dirt, old sealer, and traces of oil. Then apply a thin film of sealing adhesive.

3. Start the seal into the chamfered recess by hand. The seal lip must point into the crankcase.

4. Press the new seal into position (Fig. 14-8).

Fig. 14-8. Seal being installed with VW special tool. A similar tool can be made with an M 28 × 1.5 bolt, a nut, washer, and a steel disk.

NOTE
The installing tool centerbolt is screwed into the crankshaft. The guide piece is threaded and can be advanced along the bolt by using a wrench as shown in Fig. 14-8. The oil seal must be seated squarely in the crankcase recess.

5. Remove the installing tool.

6. Using engine oil, lubricate the oil seal contact surface on the flywheel. Then install the flywheel on the engine.

   CAUTION
   You should check the crankshaft end play during flywheel installation, although adjustment is seldom necessary if only the crankshaft oil seal is being replaced. By checking crankshaft end play whenever you make repairs to the flywheel or related parts, you guard against possible bearing failure or other engine damage.

7. Install the clutch. Then install the engine in the car.

   NOTE
   The transmission shaft pilot bearing in the gland nut should be cleaned, inspected, and repacked as described in **14.2 Removing and Installing Flywheel** whenever the clutch and flywheel are being serviced.

14.4 Replacing Clutch Release Bearing

The clutch release bearing is situated at the rear of the transmission. It is fastened to the clutch operating shaft by two retaining springs. On 1971 and later models, the clutch release bearing is supported by a central guide sleeve as shown in Fig. 14-9.

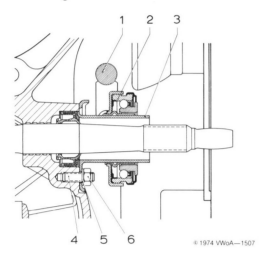

1. Operating shaft
2. Release bearing
3. Sleeve
4. Stud (M 7)
5. Lock washer
6. Nut (M 7)

Fig. 14-9. Clutch release bearing and related parts. The sleeve, stud, lock washer, and nut are not installed on 1970 cars.

The clutch release bearing is a maintenance-free item. However, if it is noisy, the plastic facing ring should be lubricated through a hole drilled in the transmission case, as shown in Fig. 14-10.

Fig. 14-10. Lubrication access. Drill 10 mm (0.4 in.) hole in transmission case as shown. Dimension **a** is 67 mm (2.6 in.); **b** is 10 mm (0.4 in.).

Use a spray can of molybdenum disulfide lubricant with a plastic hose attached to its nozzle. Insert the hose through the hole you have drilled in the transmission case and direct the spray against the plastic facing ring.

If the engine has been removed, simply roughen the surface of the plastic facing ring on the clutch release bearing with coarse emery cloth. Then rub in a small amount of molybdenum grease. Do this routinely whenever the engine is removed.

To remove the clutch release bearing, pry the retaining springs off with a screwdriver as shown in Fig. 14-11. Then remove the bearing, sliding it off the guide tube on 1971 and later cars.

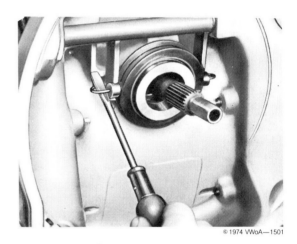

Fig. 14-11. Clutch release bearing retaining springs being removed. Hold a rag over the springs to prevent them from flying off and becoming lost when tension is relieved.

To install:

1. Check the general condition of the bearing and, if necessary, replace it. Never wash the bearing in solvent since this will remove the factory-installed lubricant.

   NOTE ——
   Hand-turn the bearing. It should not feel gritty or be difficult to turn. Make sure the plastic facing ring has not worn through.

2. Lightly lubricate the operating shaft bushings with multipurpose grease. On 1971 and later cars, lightly lubricate the guide sleeve with molybdenum grease.

3. Install the clutch release bearing and the retaining springs. Make certain that the hooked ends of the springs engage behind the levers on the operating shaft. Replace loose-fitting retaining springs.

4. After you have installed the engine, check the clutch pedal freeplay. If necessary, adjust the freeplay.

ENGINE 35

14.5 Adjusting Clutch Pedal Freeplay

As the clutch linings wear, the clearance between the release bearing and the release levers is reduced. If this condition progresses until there is no clearance at all, clutch pressure will decrease and permit slippage that can lead to burned linings. When checking for the proper clearance, you should be able to depress the clutch pedal 10 to 20 mm (3/8 to 3/4 in.) before encountering working resistance (Fig. 14-12). If not, adjust the clutch cable to obtain this amount of clearance.

Fig. 14-12. Clutch pedal freeplay (**a**). After making adjustments, pump the pedal several times before measuring the distance.

To adjust the cable, raise the car on a lift or support it on safety stands. Grip the clutch cable with a pair of pliers to keep it from twisting, then adjust the wing nut shown in Fig. 14-13 until there is 2.0 mm (about 1/16 in.) clearance between the clutch lever and the wing nut. Recheck freeplay at the pedal to make sure it is within specifications.

Fig. 14-13. Clutch cable adjustment. Clean and lubricate cable end, clutch lever, and wing nut. Engage nut wings with recesses in clutch lever to prevent loosening.

Replacing Clutch Cable

To replace the clutch cable, first remove the pedal cluster as described in **BRAKES AND WHEELS**. Then pull out the old cable toward the front of the car. Before installation, inspect the threads on the rear end of the replacement cable to make certain that they are not burred or deformed.

Grease the entire replacement cable with molybdenum grease. Then insert it through the pedal opening in the frame tunnel and into the cable guide tube. Following installation of the pedal cluster, you must adjust the brake master cylinder's pushrod length as described in **BRAKES AND WHEELS**. Also adjust clutch pedal freeplay as previously described.

When replacing a clutch cable, always check the condition of the rubber sleeve and the position of the flexible guide tube. Cracked or loose-fitting rubber sleeves must be replaced. The flexible guide tube should sag 25 to 45 mm (1 to 1 3/4 in.), as indicated at **B** in Fig. 14-14. The sag can be adjusted by adding or subtracting spacer washers at the point indicated by arrow **A** in the same illustration.

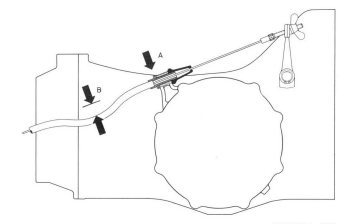

Fig. 14-14. Flexible guide tube sag (between arrows at **B**). Spacer washers are at arrow **A**.

Excessive sag in the flexible guide tube will cause the cable to bind, creating noise, difficult clutch pedal operation, or even cable breakage. If the sag is inadequate, there will be insufficient preload on the cable. This can cause poor clutch pedal feel and accelerated wear.

In early 1972, a modified flexible guide tube was introduced. It has a hole in its front ferrule that permits grease to be forced into the guide tube with a standard lever-type hand-operated grease gun equipped with a tapered nozzle. Grease should be pumped into the guide tube only if the clutch cable creaks, jams, or is stiff in operation. The rubber boot over the flexible guide tube ferrule must be pushed forward against the frame cross member to uncover the lubrication hole.

36 ENGINE

14.6 Troubleshooting Clutch

Road testing is an important part of troubleshooting the clutch because it lets you base your diagnosis on first-hand information.

For example, trouble described as "lack of power" might turn out to be a slipping clutch and not inadequate engine output. Similarly, a complaint that "it's hard to shift gears" may mean a dragging clutch, not transmission trouble. Noises and clutch chatter are often described simply as "vibrations." It is important that such inaccurate descriptions do not result in troubleshooting errors or in a misinterpretation of the problem's real origin.

Troubleshooting Table

Table b lists the most common clutch complaints. They are grouped into four classes of symptoms. The actual causes, however, may vary. The numbers in bold type in the Remedy column refer to the headings in this section where the suggested repairs are described.

CAUTION —
If you lack the skills, tools, or a suitable workshop for clutch repairs, we suggest that you leave them to an Authorized VW Dealer or other qualified shop. We especially urge you to consult your Authorized VW Dealer before attempting repairs on a car still covered by the new-car warranty.

Table b. Clutch Troubleshooting

Problem	Probable Cause	Remedy
1. Clutch noise	a. Needle bearing in gland nut worn	a. Replace gland nut; lubricate with 1 gram (1/32 oz.) multipurpose grease. See **14.2**.
	b. Driven plate fouling pressure plate	b. Replace driven plate. See **14.1**.
	c. Release lever springs weak or tension uneven	c. Replace pressure plate assembly. See **14.1**.
	d. Release bearing defective	d. Fit new bearing. See **14.4**.
2. Clutch grabbing	a. Transmission mountings loose	a. Tighten mounting bolts and nuts. See **TRANSMISSION AND REAR AXLE**.
	b. Bend in cable guide too slight or excessive	b. Correct the bend to between 25 and 45 mm (1 and 1¾ in.). See **14.5**.
	c. Pressure plate contacting unevenly	c. Replace pressure plate assembly. See **14.1**.
	d. Release plate not running true	d. Replace pressure plate assembly. See **14.1**.
	e. Spring segments deformed	e. Replace driven plate. See **14.1**.
3. Clutch dragging	a. Excessive pedal freeplay	a. Reduce pedal freeplay to lower limit of 10 mm (3/8 in.). See **14.5**.
	b. Excessive bend in cable guide	b. Reduce bend to 25 to 45 mm (1 to 1¾ in.). See **14.5**.
	c. Driven plate not running true	c. Replace driven plate. See **14.1**.
	d. Spring segments excessively or unevenly set	d. Replace driven plate. See **14.1**.
	e. Plate linings broken	e. Replace driven plate. See **14.1**.
	f. Main drive shaft not running true with gland nut	f. It is sometimes sufficient to loosen the engine mounting bolts, move the engine slightly, and retighten the bolts. Otherwise check gland nut. If the thread is damaged or if there is excessive play between inner and outer threads, the gland nut is not centered correctly. In that case, replace the gland nut. See **14.2**.
	g. Needle bearing in gland nut defective or insufficiently greased	g. Replace gland nut or lubricate needle bearing with 1 gram (1/32 oz.) of multipurpose grease. See **14.2**.
	h. Splines on main drive shaft or clutch driven plate dirty or burred	h. Clean splines. Remove burr.
	i. Sticky clutch linings (lining dust)	i. Replace driven plate. See **14.1**.
	j. Felt ring in gland nut too tight on main drive shaft	j. Replace the gland nut with one having a better fitting felt ring. See **14.2**.
	k. Stiffness in pedal cluster, clutch cable, and operating shaft	k. Grease the parts thoroughly with universal grease. See **BRAKES AND WHEELS** or **TRANSMISSION AND REAR AXLE**.
4. Clutch slipping	a. Insufficient pedal freeplay because of lining wear	a. Adjust freeplay to 10 to 20 mm (3/8 to 3/4 in.) at clutch pedal. Replace driven plate. See **14.5**, **14.1**.
	b. Oily linings	b. Replace driven plate. If necessary, replace engine oil seal or transmission oil seal. See **14.1**, **14.3**.

© 1974 VWoA

ENGINE

15. OIL PUMP

A larger oil pump and a different camshaft with camshaft gear were introduced in May 1971. The larger pump can be used in earlier engines provided the camshaft and gear are also replaced. Never use the earlier pump with a late engine.

Cars with Automatic Stick Shift are equipped with a dual oil pump serving both the engine and the torque converter. The components of the dual pump are shown in Fig. 15-1.

15.1 Removing and Installing Oil Pump

The oil pump can be removed without draining the oil or separating the crankcase halves. Be careful not to let dirt enter the engine while the pump is out.

To remove:

1. Remove the engine. Remove the four sealing nuts that hold the oil pump cover, then take off the cover.
2. On cars with Automatic Stick Shift, remove the ATF pump gears, intermediate plate with gaskets and oil seals, and oil pump gears.
3. On cars with manual transmissions, remove the oil pump gears and cover gasket.
4. Attach the puller shown in Fig. 15-2, then use it to remove the oil pump housing.

   NOTE —
   A different puller claw must be used for the larger pump introduced on 1971 cars.

5. Remove the housing gasket.

Fig. 15-2. Puller being used to remove oil pump housing from crankcase.

NOTE —
Thoroughly clean and inspect the pump before installing it.

To install:

1. While hand-turning the oil pump upper shaft to engage its drive dog in the camshaft, install the oil pump with a new housing gasket.
2. Hand-turn the crankshaft through two revolutions so that the pump housing will align itself with the camshaft.
3. Using new gaskets and (where applicable) oil seals, install the remaining pump parts on the studs.
4. Install new sealing nuts with their plastic facings toward the pump cover. Then torque them to 2.0 mkg (14 ft. lb.).

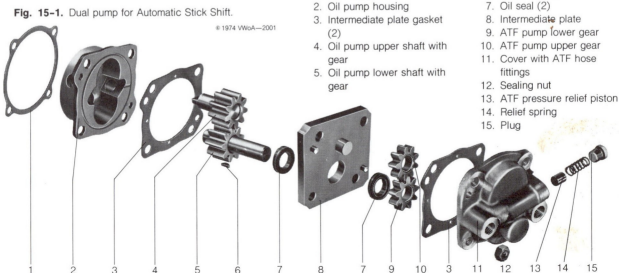

Fig. 15-1. Dual pump for Automatic Stick Shift.

1. Housing gasket
2. Oil pump housing
3. Intermediate plate gasket (2)
4. Oil pump upper shaft with gear
5. Oil pump lower shaft with gear
6. Woodruff key
7. Oil seal (2)
8. Intermediate plate
9. ATF pump lower gear
10. ATF pump upper gear
11. Cover with ATF hose fittings
12. Sealing nut
13. ATF pressure relief piston
14. Relief spring
15. Plug

38 ENGINE

15.2 Checking Oil Pump

Check the oil pump thoroughly whenever you remove it. If faulty, it will not supply adequate oil pressure and must be replaced.

To check:

1. Inspect the housing for wear, especially in the gearseating areas.

2. On cars with manual transmissions, check the driven gear shaft for a tight fit. If it is loose, peen in the area indicated by the arrow in Fig. 15-3 or replace the housing.

3. On all pumps, check the fit of the shafts that turn in the housing.

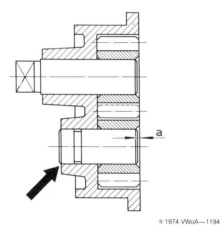

Fig. 15-3. Driven gear shaft in housing. Dimension **a** must be 0.50 to 1.00 mm (.020 to .040 in.).

4. Check the gears for wear. Backlash must not exceed 0.20 mm (.008 in.), and it should be even less.

5. Measure the end play of the gears as shown in Fig. 15-4. It must not exceed 0.10 mm (.004 in.) without the gasket.

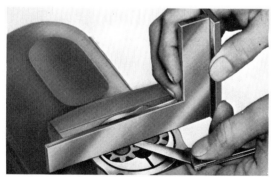

Fig. 15-4. End play being measured with a feeler gauge and straightedge.

15.3 Removing and Installing Oil Pressure Valves

Unlike earlier VW engines, there are two oil pressure valves used in the engines of cars covered by this Manual. The oil pressure valves are spring-loaded pistons that move in bores machined into the left-hand crankcase castings. Fig. 15-5 is a schematic view of the oiling system.

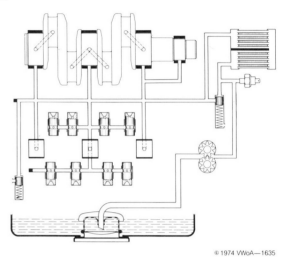

Fig. 15-5. Schematic view of oiling system. One oil pressure valve is at the right, below and to the left of the oil cooler. The other is at the lower left-hand side of the illustration.

The oil pressure valve at the right-hand side of Fig. 15-5 is known as the pressure relief valve. When the oil is cold and thick, the resulting increased oil pressure forces the valve down against spring tension. This allows oil from the pump to go directly to the engine bearings. In by-passing the oil cooler, warm-up is speeded and the oil cooler is protected against excessive pressures that might cause it to burst.

The oil pressure valve at the lower left-hand side of Fig. 15-5 is known as the oil pressure control valve. It is located at the extreme end of the main oil passage. When oil pressure rises above the point necessary to supply lubrication to the bearings, the valve's piston is forced down against the spring tension. This allows oil to be returned directly to the crankcase sump.

The oil pressure control valve ensures that oil pressure will remain constant at the bearings, regardless of engine speed. Thus, at high engine speeds, when pump output is greatest, the excess oil is allowed to escape. If the surplus oil were delivered to the engine's working parts, it might spray out onto the cylinder walls in excessive quantities and cause increased oil consumption. Avoiding excess oil pressure also reduces the power lost by the engine in driving the oil pump.

ENGINE 39

A cracked or otherwise leaking oil cooler should prompt the removal and inspection of both oil pressure valves. If either valve is stuck or sticks intermittently in its closed position, it can cause excessive oil pressure. Usually, the oil pressure valves become sticky only in neglected engines. If the oil is not changed according to the schedule given in **LUBRICATION AND MAINTENANCE**, foreign matter or corrosive agents may attack the valves and prevent them from moving freely.

Valve removal is illustrated in Fig. 15-6. Because the slotted plugs for the oil pressure valves are a tight fit in the crankcase, you should not try to remove them with an undersize screwdriver. Use a tool that will completely fill the slot. Otherwise, the alloy plug may be ruined or deformed.

Table c. Spring Specifications

Model	Loaded length	Loaded tension
Pressure relief spring	44.10 mm (1¾ in.)	5.6-7.3 kg (12⁵⁄₁₆-16 lb.)
Pressure control spring	20.20 mm (¾ in.)	3.1-3.8 kg (7-8½ lb.)

Fig. 15-6. Oil pressure relief valve (**1**) and oil pressure control valve (**2**).

To remove oil pressure valve:

1. Using a properly fitting screwdriver, remove the slotted plug.

2. Remove the spring and plunger. If the plunger is stuck, remove it by screwing a 10-mm (⅜-in.) tap into it.

To install:

1. Check the plunger and bore for signs of seizure. Carefully remove rough spots and deposits, then install a new plunger if necessary.

2. Examine the spring. It should not be deformed or damaged by abrasion or corrosion.

   CAUTION —
   If both valves are removed, be careful not to mix up the springs. The springs for the pressure relief valve (rear of engine) and the pressure control valve (front of engine) are not interchangeable.

3. Check the spring tension and length. Then compare your findings with those given in **Table c**.

4. Install the plunger and spring as indicated in Fig. 15-7.

   CAUTION —
   Do not allow the spring to scratch the wall of the bore as you install it. Burrs on the wall could either cause the plunger to jam or could limit its travel.

1. Plunger   3. Gasket
2. Spring   4. Plug

Fig. 15-7. Order in which oil pressure valve parts are installed in engine. Although the oil pressure relief valve is illustrated, the parts of the oil pressure control valve are installed in the same order.

5. Using a new gasket, install the plug.

   CAUTION —
   Do not overtighten the plug. You could strip the threads in the crankcase casting. Also do not apply sealing compound to the plug or gasket. Doing so could make future removal difficult or impossible.

40 ENGINE

16. CRANKCASE

All disassembly procedures described up to this point must already have been carried out before you can disassemble the crankcase. After you have removed the various components, the crankcase will appear as it does in Fig. 16-1. The flywheel or drive plate can be left installed on the crankshaft for some repairs. However, if the flywheel or drive plate is to be removed, it should be done prior to separating the crankcase halves.

16.1 Removing and Installing Oil Strainer

The oil strainer cover and oil strainer must be removed from the bottom of the crankcase before any other nuts or bolts are removed. To remove the oil strainer, remove the six cap nuts and seal washers. Then take off the cover, strainer, and the two oil strainer gaskets. Use new seal washers during installation. Full details on oil strainer servicing and installation are given in **LUBRICATION AND MAINTENANCE**.

Fig. 16-1. Crankcase ready for disassembly.

1. M 8 galvanized nut (4)
2. Galvanized lock washer (4)
3. Generator support
4. Oil deflector plate
5. Generator support gasket (2)
6. M 8 galvanized bolt
7. Galvanized lock washer (6)
8. M 8 galvanized nut (6)
9. M 8 bolt (2)
10. Lock washer (11)
11. M 8 nut (9)
12. Crankcase
13. Washer (6)
14. M 12 nut (6)
15. Crankshaft oil seal
16. Shim (3)
17. Steel dowel pin (4)
18. Rubber seal
19. Flywheel
20. Spring-type lock washer
21. Gland nut with bearing
22. Drive plate
23. Spring-type lock washer
24. Gland nut without bearing

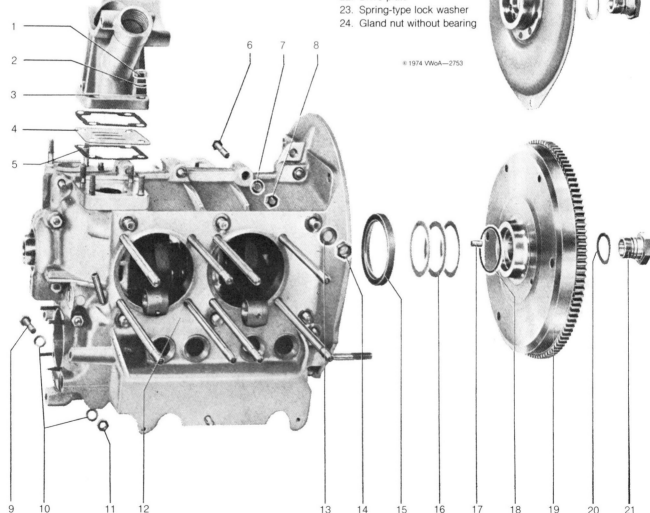

16.2 Disassembling and Assembling Crankcase

The crankcase must be taken apart to replace the connecting rods, the connecting rod bearings, and the main bearings. It is also necessary to "split" the crankcase in order to remove and install the crankshaft, camshaft, and camshaft bearings.

To disassemble:

1. Remove the oil strainer and the cover plate below the flywheel pulley. (First remove the pulley as described in **6.2 Removing and Installing Crankshaft Pulley**.) Then remove the six M 12 sealing nuts from the main bearing studs. Remove the nine M 8 nuts and the one M 8 bolt and nut from the crankcase flange.

2. Using spring clips, clamp the cam followers in the right-hand crankcase half in place. Then lift off the right-hand crankcase half as illustrated in Fig. 16-2.

Fig. 16-2. Crankcase halves being separated. Note the spring clips holding the cam followers.

CAUTION
Never insert any tools between the crankcase flanges to separate the halves. The slightest scratch can produce an oil leak. If the halves are stuck together with sealer, use a rubber hammer to loosen the right-hand half from the left.

To assemble:

1. Make sure that the mating surfaces of the crankcase halves are absolutely clean.

2. Clean any hardened sealer or dirt from the studs as such matter can cause inaccurate torque wrench readings.

3. Check the studs for tightness where they thread into the crankcase. Check the oil suction pipe for tightness at the points marked by the arrows in Fig. 16-3.

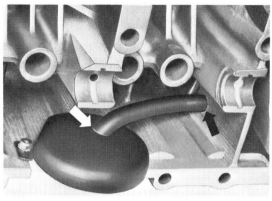

Fig. 16-3. Checkpoints (arrows) for oil pipe tightness. If necessary, secure them with a peening tool.

4. Using sealing compound all around it, install the camshaft plug as shown in Fig. 16-4. Replace the plug if it is rough or deformed.

Fig. 16-4. Camshaft end plug installed in left half of crankcase.

5. Using spring clips, clamp the cam followers in the right-hand half of the crankcase.

6. Spread an even film of sealing compound over the mating surfaces of the crankcase halves.

CAUTION
Do not let the sealing compound get into the oil passages for the crankshaft and camshaft bearings. It could block lubrication.

7. Install new O-rings around the main bearing studs as indicated in Fig. 16-5.

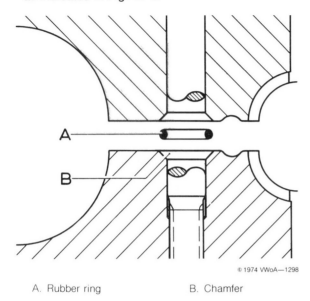

A. Rubber ring B. Chamfer

Fig. 16-5. O-ring installation. As the crankcase halves are joined, the silicone rubber O-rings will be compressed into the chamfers around the studs.

8. Join the crankcase halves. Install the nuts loosely.

9. Torque the M 8 nut near the lower main bearing stud adjacent to the flywheel to 2.0 mkg (14 ft. lb.).

10. Torque the six M 12 main bearing stud nuts.

    NOTE
    If you are using sealing nuts on the main bearing studs, as are standard on 1971 to 1973 models, torque them to 2.5 mkg (18 ft. lb.). If you are using plain nuts, as originally installed on 1970 engines, torque them to 3.5 mkg (25 ft. lb.). If you service-install sealing nuts on a 1970 engine, do not install the steel washers used under the original nonsealing nuts.

11. Turn the crankshaft to check for free movement.

    NOTE
    If the latest crankcase with thin, M 8 cylinder studs (manufactured since early 1973) is used with an older crankshaft, check the crankshaft for free rotation before joining the crankcase halves. If the crankshaft contacts the center main bearing web, file or grind the web to obtain clearance.

12. Torque the remaining M 8 nuts and the single M 8 bolt to 2.0 mkg (14 ft. lb.).

13. Remove the spring clips holding the cam followers.

14. Check the crankshaft end play, as described in **17.4 Crankshaft End Play**. Then install a new crankshaft oil seal.

15. Install the crankcase oil strainer. Then torque the cap nuts to 0.7 mkg (5 ft. lb.).

16. Install the remaining engine parts to complete the assembly.

16.3 Removing and Installing Oil Filler and Breather

On fuel injection engines, the oil filler is held to the alternator stand by four nuts. Filler removal does not require the special tool shown in Fig. 16-6. Use a new gasket when you install the filler on the alternator stand. On carburetor engines, you seldom need to remove the oil filler and breather assembly from the generator or alternator stand but, if you must, remove the assembly before you unbolt the stand from the engine.

To remove the oil filler and breather, first take off the oil filler cap. Then, using the special tool VW 170 (Fig. 16-6), remove the gland nut that holds the oil filler and breather on the generator stand.

Fig. 16-6. Special tool being inserted into oil filler to remove gland nut.

Installation of the oil filler and breather on the generator stand is the reverse of removal. Torque the gland nut to 2.0 mkg (14 ft. lb.).

If you removed the generator stand by taking off the four M 8 nuts that hold it on the crankcase, use two new gaskets during installation—one above the deflector plate and one below it. The deflector plate must be installed on the crankcase so that the louvers face downward with the slightly longer end of the central louver to the rear, as shown in Fig. 16-7.

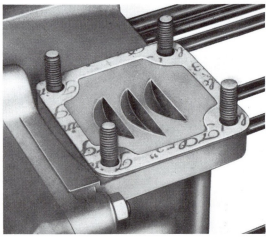

Fig. 16-7. Oil deflector plate and gasket properly installed. The upper face of the deflector plate is marked top.

16.4 Repairing Tapped Holes

If the threads for the cylinder studs have been stripped, you can return the crankcase to service by installing steel inserts in the casting. The repair kit, containing 16 inserts plus the required tool set, may be uneconomical if you are repairing only one crankcase. If so, it is probably best to turn the job over to an Authorized VW Dealer or other qualified shop.

Sequence of operations:

1. Read the installation instructions supplied with the tool set.

    NOTE ———
    The tool set contains a drill and tap jig, a drill pilot, a $^{33}/_{64}$-in. drill, a special .5570-in. tap, and an insert driving tool.

2. After removing the studs, install the jig on the crankcase. Enlarge the stud holes to 13 mm ($^{33}/_{64}$ in.) following the instructions supplied with the tool set.

3. Using the jig as a guide, cut threads into the enlarged holes with the special .5570-in. tap.

4. Place each insert, tangs first, in the insert driving tool. Then thread them into the holes tapped in the crankcase.

5. Strike the insert driving tool with a hammer to drive the tangs flush with the insert, thereby locking the insert in the crankcase casting.

17. Camshaft and Crankshaft

After disassembling the crankcase, you can remove the camshaft and crankshaft with their bearings and associated parts. Removal involves nothing more complicated than lifting the shafts from the left-hand half of the crankcase after you have removed the right-hand half. Inspection and installation procedures are therefore the major consideration when servicing internal engine parts.

17.1 Checking and Installing Camshaft and Bearings

Two camshafts and gears have been used during the model years covered by this Manual. On the early type, three rivets hold the gear onto the camshaft, whereas the later type has four rivets. The change was made to accommodate the larger oil pump introduced in May 1971.

CAUTION ———
The late-type oil pump, which has 26-mm (1.024-in.) gears, must be used with the four-rivet camshaft. The early-type pump, which has 21-mm (.827-in.) gears, must be used with the three-rivet camshaft. If the parts are mixed, the oil pump will either not fit or not supply lubrication.

To check and install camshaft:

1. Check the camshaft gear rivets for tightness.
2. Check the cam bearings and lobes for wear (Fig. 17-1).

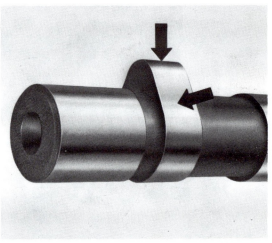

Fig. 17-1. Cam lobes. Toe of cam (top arrow) should not be scored or worn unevenly. There should be no sign of wear at second arrow.

44 Engine

3. Check the camshaft for runout as in Fig. 17-2. Runout for new camshafts should not exceed 0.02 mm (.0008 in.). Runout for used camshafts should not exceed 0.04 mm (.0016 in.).

Fig. 17-2. Camshaft runout being measured.

4. Examine the camshaft gear for wear and for a properly centered tooth contact pattern.

5. Clean the bearing saddles in the left-hand half of the crankcase. Check the oil holes for cleanliness.

6. Install the three camshaft bearing halves in their saddles, making sure that the oil holes in the bearing shells line up with those in the crankcase half.

7. Lightly coat the bearings with assembly lubricant. Then install the camshaft as illustrated in Fig. 17-3.

Fig. 17-3. Camshaft installation. The tooth marked with a small circle (arrow) goes between the two teeth marked with centerpunches on the camshaft gear.

8. Check the axial play as shown in Fig. 17-4. The play permitted by the thrust bearing should be 0.04 to 0.13 mm (.0016 to .0051 in.) with a wear limit of 0.16 mm (.0063 in.).

Fig. 17-4. Dial indicator being used to check axial play.

9. Check the backlash around the entire circumference of the gear. It should be between 0.00 and 0.05 mm (.000 and .002 in.). The cam gear size is correct if backlash is hardly noticeable and if the camshaft does not lift from its bearings when you turn the crankshaft rearward.

NOTE

Camshaft gears are marked -1, 0, $+1$, or $+2$, etc. on their inner face. These numbers indicate in $1/100$ mm how much their pitch radius departs from the standard pitch radius denoted by 0. Do not confuse a zero on the inner face with the circular timing mark on the outer face.

10. Coat the upper halves of the bearing shells with assembly lubricant. Then install them atop the camshaft journals. Also coat the cam lobes and followers with assembly lubricant.

CAUTION

If you are installing a new camshaft, install new cam followers as well. The existing wear pattern of used followers may prevent a proper wear pattern from developing on the cam lobes.

11. Clean and install the right-hand half of the crankcase as described in **16.2 Disassembling and Assembling Crankcase**.

12. Install the remaining engine parts to complete the assembly.

17.2 Removing and Installing Crankshaft

As with camshaft removal, crankshaft removal merely requires that you lift out the crankshaft after you have separated the crankcase halves.

To install crankshaft:

1. Inspect the crankshaft main bearing journals for scoring and wear. They should not be deeply grooved. The wear limit is 0.03 mm (.0012 in.). Maximum runout is also 0.03 mm (.0012 in.) measured at journals No. 2 and 4 with journals No. 1 and 3 on V-blocks.

   NOTE
   Using a micrometer, check journals No. 1, 2, and 3 for out-of-round. Their specified diameter is 54.97 to 54.99 mm (2.1641 to 2.1649 in.). Then check No. 4 journal. Its specified diameter is 39.98 to 40.00 mm (1.5740 to 1.5748 in.). Replacement crankshafts are available in three undersizes of 0.25 mm (.010 in.) each.

2. Inspect the oil passages in the crankshaft and bearings to see that there are no sharp edges. Burrs can be removed with a half-round needle file.

   NOTE
   Although foreign matter embedded in the bearing material can be removed with a sharp scraper, most experienced mechanics prefer to replace bearings that are in such condition, rather than to reuse them.

3. Check the main bearing dowels (Fig. 17-5) for tightness.

Fig. 17-5. Main bearing dowels in bearing saddles. These dowels are found in the left-hand side of the crankcase only.

4. Liberally coat the bearings with assembly lubricant. Then install bearings No. 1, 3, and 4 on the crankshaft. Install the lower half of the split No. 2 bearing in the crankcase, carefully engaging the dowel with the hole in the bearing shell.

5. Turn the bearings on the crankshaft to properly position the oil holes and dowel holes. Then install the crankshaft assembly in the left-hand crankcase half, aligning the timing marks on the crankshaft and camshaft gears.

   CAUTION
   Do not force the crankshaft into position, as doing so could damage the bearings. Instead, lift the crankshaft slightly and hand-turn the bearings until their holes align with the dowels in the bearing saddles.

6. Install the right-hand half of the crankcase as described in **16.2 Disassembling and Assembling Crankcase**.

   NOTE
   Main bearing clearance is partially determined by the preload (crush) imparted to the bearing shells as the crankcase halves are bolted together. Strict adherence to the prescribed torque figures is therefore of extreme importance. Bearing clearance is basically determined by the proper selection of bearings and by the use of a crankshaft within prescribed tolerances. However, you can measure individual clearances either with Plastigage ® or with a sensitive dial indicator. Clearances, as measured with the crankcase halves together, are given in **Table d**.

Table d. Bearing Clearance

Crankshaft bearing(s)	New	Wear limit
1 and 3	0.04-0.10 mm (.0016-.004 in.)	0.18 mm (.007 in.)
2	0.03-0.09 mm (.001-.0035 in.)	0.17 mm (.0067 in.)
4	0.05-0.10 mm (.002-.004 in.)	0.19 mm (.0075 in.)

7. Check the crankshaft end play as described in **17.4 Crankshaft End Play**.

8. Install the remaining engine parts to complete the assembly.

Plastigage®, which is used for measuring installed bearing clearances, is available at auto supply stores. However, it is not well suited for checking clearances of ring-type bearings, such as those used on journals No. 1, 3, and 4.

46 ENGINE

17.3 Disassembling and Assembling Crankshaft

When you remove the crankshaft from the crankcase, the connecting rods remain installed on the crankshaft. In addition to connecting rod removal, it is necessary to press off the distributor drive gear and crankshaft gear to get No. 3 main bearing on or off its journal. The positions of the crankshaft assembly's components are shown in Fig. 17-6.

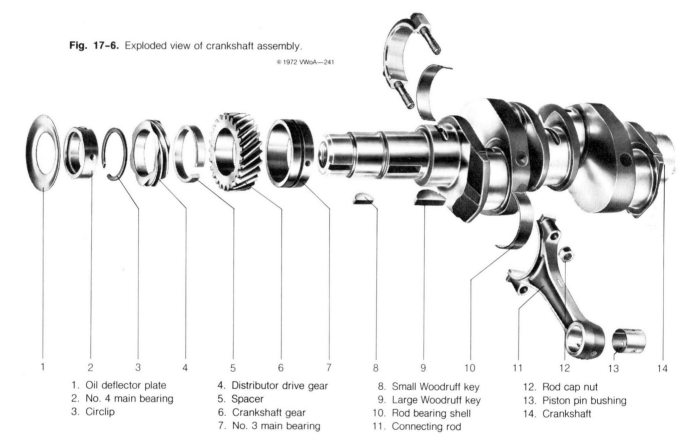

Fig. 17-6. Exploded view of crankshaft assembly.

© 1972 VWoA—241

1. Oil deflector plate
2. No. 4 main bearing
3. Circlip
4. Distributor drive gear
5. Spacer
6. Crankshaft gear
7. No. 3 main bearing
8. Small Woodruff key
9. Large Woodruff key
10. Rod bearing shell
11. Connecting rod
12. Rod cap nut
13. Piston pin bushing
14. Crankshaft

To assemble:

1. Clean the crankshaft and blow out its oil passages with compressed air.

2. Check the gear teeth. Inspect the gear seating surfaces for seizure marks.

   NOTE ———
   Light seizure marks can be removed, provided the press fit is not affected.

To disassemble:

1. Remove the nuts from the rod cap studs on each connecting rod. See **18.1 Removing Connecting Rods**.

2. Remove the connecting rods and the rod bearing shells.

3. Using circlip pliers, remove the distributor drive gear circlip.

4. Using a hydraulic press, press the crankshaft gear, spacer, and distributor drive gear off the crankshaft. Put a support plate under the crankshaft gear, then position the press ram against the end of the crankshaft.

5. Remove the No. 3 main bearing. Take the large and small Woodruff keys out of their recesses.

3. After coating No. 3 journal with assembly lubricant, install the No. 3 main bearing. Install the large Woodruff key in its recess.

4. Heat the crankshaft gear to about 80°C (176°F) in an oil bath. Then press the gear onto the crankshaft. The chamfer in the gear bore must face the No. 3 main bearing journal. The keyway must align with the Woodruff key on the crankshaft.

5. Slip on the spacer, then insert the small Woodruff key in its recess. Install the distributor drive gear as you did the crankshaft gear.

6. Install the circlip, being careful not to scratch the No. 4 bearing journal on the crankshaft.

7. Check to see that the gears fit tightly when cool.

© 1974 VWoA

Engine 47

17.4 Crankshaft End Play

Crankshaft end play is controlled by three shims between the flywheel and the No. 1 main bearing flange. The position of these three shims is shown in Fig. 17-7.

CAUTION—
Keep the crankshaft end play within the prescribed range. Excessive end play can cause early wear or failure of the main bearings and the connecting rod bearings and could thereby damage other engine parts.

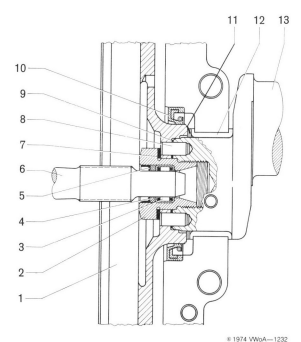

Fig. 17-8. Setup for measuring crankshaft end play.

To adjust end play:

1. Remove the flywheel and reinstall it with only two shims.

   NOTE—
   Shims come in five thicknesses: 0.24 mm (.0095 in.), 0.30 mm (.0118 in.), 0.32 mm (.0126 in.), 0.34 mm (.0134 in.), and 0.36 mm (.0142 in.). Sizes are etched on new shims. You may have to check the thickness of old shims with a micrometer.

2. Measure end play again.

3. Compute how thick the third shim must be in order to bring the end play within specifications.

4. Install the three-shim combination. Then install the oil seal and the flywheel.

5. Give the end play a final check.

17.5 Crankshaft Reconditioning

Crankshafts damaged by inadequate lubrication or crankshafts that have journals that are worn to a taper or that are out-of-round can sometimes be reconditioned by grinding. Properly ground crankshaft journals are reduced in diameter by 0.25, 0.50, or 0.75 mm (.010, .020, or .030 in.) to match the oversize bearing sets available.

Crankshaft grinding requires expensive precision machinery and is never done in repair shops. This job is left to specialty shops or engine reconditioning facilities. Factory reconditioned crankshafts are of the same quality as new crankshafts. It is always necessary, however, that new bearings of the proper dimensions be fitted in order to correct for the slight undersize of the reground crankshaft. Clean the oil passages thoroughly before you install a reground crankshaft.

1. Flywheel
2. Gland nut
3. Needle bearing
4. Felt ring
5. Retaining ring
6. Main drive shaft
7. Lock washer
8. Dowel pin
9. Rubber sealing ring
10. Crankshaft oil seal
11. Shims
12. Crankshaft bearing
13. Crankshaft

Fig. 17-7. Cross section of flywheel and crankshaft.

To check end play:

1. Attach a dial indicator as shown in Fig. 17-8.

2. Move the crankshaft in and out while watching the gauge. End play should be between 0.07 and 0.13 mm (.003 and .005 in.). The wear limit is 0.15 mm (.006 in.).

3. If necessary, make adjustments to bring end play within specifications.

48 ENGINE

18. CONNECTING RODS

Several connecting rod designs have been used in VW engines. Some rods shown in the illustrations may not look the same as those in your engine. However, the service and installation procedures are the same.

18.1 Removing Connecting Rods

The connecting rod caps have pressed-in studs that pass through holes in the connecting rod. Nuts are installed on the studs to join the connecting rod and cap.

CAUTION —
Do not drive the connecting rod studs out of the cap. Doing so could lead to connecting rod damage or failure during engine operation. If the studs are damaged, replace the entire connecting rod.

To remove connecting rod:

1. Remove the crankshaft assembly from the engine. Mount it on a crankshaft-holding fixture or in an old flywheel clamped in a vise.
2. Remove the connecting rod nuts.
3. Remove the connecting rod and cap complete with bearing shells from the crankshaft. Be careful not to scratch the crankshaft journals with the rod cap studs.

18.2 Checking and Installing Connecting Rods

Unless the connecting rods are in perfect condition, have uniform weight, and are perfectly fitted, there is always the possibility of rod failure and serious engine damage.

CAUTION —
If you lack the skills, tools, or special equipment needed to check, fit, or recondition connecting rods, we suggest you leave such repairs to an Authorized VW Dealer or other qualified automotive machine shop.

To check rods:

1. Examine the connecting rod and cap for obvious external damage or damage to the bearing bore.
2. Using steel blocks or a fixture such as that shown in **18.3 Reconditioning Connecting Rods**, check the connecting rod for bends and twists.
3. Check the weight of the connecting rod without the bearing shells but with the rod cap and nuts installed.

   NOTE —
   The difference in weight between connecting rods in an engine must not exceed 10 g. You can remove metal from heavier rods at the points shown in Fig. 18-1. Replacement rods come in two weights. Those identified by a brown paint mark weigh from 580 to 588 g. Those marked with gray paint weigh from 592 to 600 g.

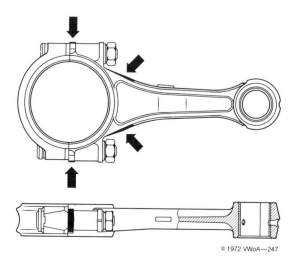

Fig. 18-1. Metal removal from overweight connecting rod. Arrows indicate where metal can be ground away.

CAUTION —
When grinding material from connecting rods to remove weight, be careful not to grind away the numbers on the side of the cap and rod. Nor should you grind off the projection forged into the side of the rod between the big and small ends. These are reference marks for installation.

4. Check the piston pin fit as shown in Fig. 18-2. Correct a loose fit by installing a new pin or a new bushing in the connecting rod. See **18.3 Reconditioning Connecting Rods** and **12. Pistons and Rings**.

   NOTE —
   When you check the pin fit, be sure the parts are at room temperature. The pin and bushing in the connecting rod must be clean and unlubricated. A new pin will sometimes correct a loose fit, but you will get best results by installing a new bushing as well.

ENGINE 49

Fig. 18-2. Piston pin fit being checked. The pin should be a push fit in a new bushing.

To install rods:

1. Install the bearing shells in the connecting rod and cap so that the tangs on the shells engage the notches in the rod bore.

2. Coat the bearing surfaces with assembly lubricant. Then install the connecting rod and cap on the crankshaft as shown in Fig. 18-3.

   NOTE
   When the rod is properly installed, the numbers stamped into the connecting rod and into the cap will be on the same side.

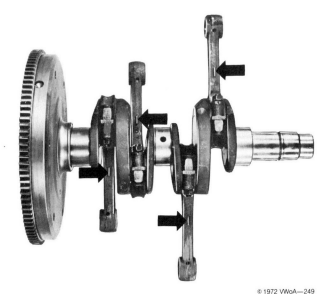

Fig. 18-3. Connecting rod installation. Forged marks at arrows must face upward when crankshaft is positioned as it is in when in the engine.

3. Check the alignment of the rod and cap, then torque the nuts to 3.0 to 3.5 mkg (22 to 25 ft. lb.).

   NOTE
   The bearing clearance should be 0.02 to 0.07 mm (.0008 to .0028 in.). This can be checked with Plastigage ®. The wear limit is 0.15 mm (.006 in.). A slight preload may develop between bearing halves as you torque the nuts. You can relieve this by lightly tapping both sides of the connecting rod with a mallet near the joint.

4. Check the axial play (side clearance) as illustrated in Fig. 18-4. It should be 0.10 to 0.40 mm (.004 to .016 in.). The wear limit is 0.70 mm (.028 in.).

Fig. 18-4. Side clearance being checked with feeler gauge.

NOTE
Prior to the autumn of 1972, connecting rod nuts were of a special type that could be locked by peening their edges into recesses on the connecting rod. With the new nuts, no locking step is necessary and tightening torque remains unchanged. This modified procedure can be applied to all earlier engines.

CAUTION
Do not reuse nuts that have been locked by peening. Always install new nuts of the later type to avoid having them come loose. The old-type nuts can be reused, however, if they have never been peened.

18.3 Reconditioning Connecting Rods

You can straighten slightly bent or twisted connecting rods if you have the proper equipment. You can also replace piston pin bushings.

To recondition rod:

1. Using a repair press and suitable driving tools, press out the bushing.

2. Place the connecting rod in a checking-and-straightening device such as that shown in Fig. 18-5.

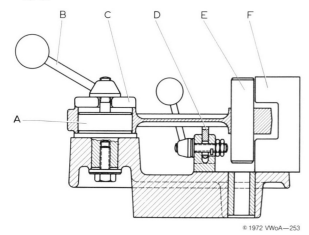

A. Eccentric mandrel D. Support
B. Locking lever E. Pin
C. Washer F. Gauge

Fig. 18-5. Connecting rod reconditioning tool VW 214f. Rod is installed without piston pin bushing.

3. Turn the 7-mm eccentric mandrel to position it as shown in Fig. 18-6.

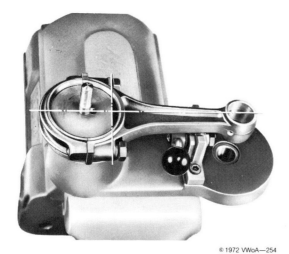

Fig. 18-6. Mandrel turned so that machined face is at a right angle to connecting rod centerline.

4. Install the washer over the mandrel. Tighten the locking lever until the connecting rod can just barely be turned on the mandrel.

5. Without tilting the rod, insert the pin in the connecting rod bushing. Move the gauge around the pin, as shown in Fig. 18-7, to check for bends and twists.

Fig. 18-7. Using gauge block to check for bends and twists. The piston pin should be perpendicular to the machined measuring surface of the device.

6. If you detect deviations, fully tighten the locking lever. Using the bar, straighten the connecting rod.

7. Using a proper tool in a repair press, press in the new bushing. Using the hole in the rod eye as a guide, drill the 3.5-mm (9/64-in.) oil hole.

8. Using a pin-fitting machine or hand reamer (Fig. 18-8), fit the bushings to the pin. Clearance must be 0.01 to 0.02 mm (.0004 to .0008 in.).

Fig. 18-8. New bushing being reamed. Afterward, using piston pin, recheck rod alignment.

ENGINE 51

19. Ignition System

The VW ignition is a conventional coil and battery system. Its components and the wires that connect them are shown in Fig. 19-1. The primary circuit, carrying 12-volt battery current, is indicated by striped lines. The secondary circuit cables, carrying high voltage from the coil to the spark plugs, are shown by solid gray lines. The solid black lines represent the chassis of the car, which carries negative-polarity (ground) current.

19.1 Coil

By the process of electrical induction, the coil steps up the primary circuit's 12-volt current to as much as 20,000 volts in the secondary circuit. This high voltage is required to ionize the air in the spark plug gaps so that a spark can jump across them. Despite the painful electrical shock that spark current can produce, its low amperage (current flow) renders it harmless. When you test the spark, you can avoid getting shocks either by wearing rubber gloves or by holding the secondary cables with a spring-type clothespin. Cables with sound insulation, however, are safe to handle.

When removing and installing the coil, please be sure you connect the wire from the ignition switch to terminal 15 and the wire from the distributor to terminal 1. Reversing the connections changes the polarity of the secondary current from positive to negative. This reversed polarity limits the ability of the spark voltage to ionize the air in the spark plug gaps and, therefore, causes misfiring at high speeds or during hard acceleration.

Keep the coil tower clean and dry and make sure that the rubber boot surrounding the high tension cable is tight-fitting and waterproof. Otherwise, spark current might begin arcing from terminal 4 to ground. Such a short could prevent the engine from running well or possibly from running at all. Such a short might also burn irreparable carbon tracks into the coil tower.

Ignition coils seldom produce trouble. However, if all the other ignition components are sound and the car still misses or is hard to start, check the coil. First test the battery as described in **ELECTRICAL SYSTEM** to make sure it is not run down. Then disconnect the high tension cable from the center terminal of the distributor cap and hold it about 10 mm (⅜ in.) from the engine's crankcase. Have someone run the starter while you observe the spark produced. If the spark is weak and yellowish, fires only when the cable is moved close to the crankcase, or fails to fire at all, the coil is probably weak and should be replaced.

Since faulty wiring between the ignition switch and coil, distributor faults, or a weak condenser can produce similar symptoms, it is best to have the coil tested before investing in a new one. If the coil proves satisfactory, use a voltmeter to check the voltage between terminals 15 and 1. It must not be below 9.6 volts.

19.2 Distributor

Servicing the distributor includes checking, replacing, and adjusting the breaker points; checking and possibly replacing the rotor and cap; checking the spark advance mechanism; and adjusting the ignition's timing. The condenser should not be replaced routinely but only when electrical testing indicates that it is defective. The breaker

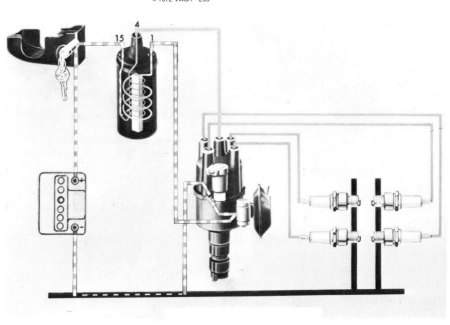

Fig. 19-1. Ignition system shown schematically.

52 ENGINE

points should be checked and replaced at the mileage intervals specified in **LUBRICATION AND MAINTENANCE**. Look for the following conditions when you check the breaker points:

1. Moderate pitting and build-up with bright point surfaces can be considered normal wear.
2. Gray point surfaces indicate that the point gap is too narrow, that the spring tension is inadequate, or both. Normal tension is 400 to 600 g (14 1/8 to 21 1/8 oz.).
3. Blued points indicate a defective coil or condenser.
4. Points fouled by grease, oil, or dirt will display yellow or black markings.

To replace breaker points:

1. Take off the distributor cap and rotor.
2. Remove the screw that holds the fixed contact to the distributor's advance plate, then pull the point wire, shown in Fig. 19-2, off the connector.
3. Lift the breaker point assembly out of the distributor.

Fig. 19-2. Breaker points installed in distributor. The point assembly's wire can be pulled off the connector at the left side of the distributor housing.

To install:

1. Wipe clean the inside of the distributor.
2. Install the new points, making sure the contacts meet squarely. If necessary, align them by bending the fixed contact.
3. Apply one drop of engine oil to the breaker point pivot and a small amount of multipurpose grease to the distributor cam and breaker point rubbing block. Wipe off any excess lubricants so that they will not be thrown onto the point contacts.

Point Adjustments

The breaker points must be adjusted after you have installed them. You can use either a feeler gauge or a dwell meter.

To adjust with feeler gauge:

1. Remove the distributor cap and rotor.
2. Gripping the generator pulley or fan belt, hand-turn the crankshaft until the breaker point rubbing block is on a high point on the distributor cam (points wide open).

   NOTE
   If you are doing a complete tune-up, adjust the points while the spark plugs are out. This adjustment will make it easier to hand-turn the crankshaft.

3. Insert a 0.40-mm (.016-in.) feeler gauge in the point gap and determine whether the points are gapped correctly—too close together, or too far apart. The tolerance range is 0.40 to 0.50 mm (.016 to .020 in.).
4. If necessary, loosen the fixed contact mounting screw. Then, using the tip of a screwdriver as shown in Fig. 19-3, move the fixed contact one way or the other until you get the specified breaker point gap.

Fig. 19-3. Breaker points being adjusted. Screwdriver tip goes between projections on advance plate, with one edge engaged in the notch on fixed contact mount. Note feeler gauge inserted in gap.

5. Tighten the fixed contact mounting screw, then recheck the gap.
6. Install the distributor rotor and cap.
7. Adjust the timing.

ENGINE 53

To adjust with dwell meter:

1. If the points have not already been gapped with a feeler gauge, do so now to make sure the engine will start. The adjustment can be approximate.

2. Connect the dwell meter. The black clip should be attached to ground on any bare metal part and the colored clip attached to terminal 1 at the coil.

3. Switch the dwell meter to the tachometer position. Start the engine and run it at 1000 rpm.

4. Move the switch to measure dwell angle. Read the dwell.

   NOTE
   Adjust new breaker points to a dwell angle of 44° to 50°. Used points can be retained without adjustment as long as they are within 42° to 58° of dwell. Despite wear, further adjustment is unnecessary if the dwell falls within this range.

5. Run the engine at 2000 rpm. The dwell should not vary more than ± 1° from the first reading. Larger deviations indicate a worn distributor shaft.

6. If necessary, remove the distributor cap and rotor and adjust the dwell. To do this, loosen the fixed contact's mounting screw. Then, using the tip of a screwdriver, move the fixed contact either to widen the gap and decrease dwell or to narrow the gap and increase dwell.

7. Fully tighten the fixed contact's mounting screw. You can run the engine with the starter to check the dwell during adjustment but always make the final check with the engine running.

8. Install the distributor rotor and cap. Then adjust the timing.

Adjusting Timing

You should adjust the ignition timing whenever the distributor has been removed and reinstalled or after installing or adjusting the breaker points.

To adjust timing:

1. Following the instrument manufacturer's instructions, install a tachometer and a stroboscopic timing light.

2. On all 1970 cars, and on 1973 and 1974 cars that have a single vacuum hose attached to the distributor (beginning with chassis No. 113 2 674 897 for cars with manual transmissions and chassis No. 113 2 690 032 for cars with Automatic Stick Shift), disconnect and plug the vacuum hose that goes to the distributor. On other models (those with two hoses) leave both hoses attached—including cars with fuel injection.

3. Let the engine warm up until the oil temperature is between 30° and 70°C (86° and 158°F).

4. By turning the bypass air screw (see **FUEL SYSTEM** or **FUEL INJECTION**), adjust the idle of carburetor engines to 800 to 900 rpm. Adjust fuel injection engines with manual transmissions to 800 to 950 rpm; adjust fuel injection engines with Automatic Stick Shift to 850 to 1000 rpm.

5. On 1971 through early 1973 engines with two hoses to the distributor, the throttle valve must close adequately for accurate timing adjustments. To check, disconnect the green (retard) hose from the distributor with the engine idling. If the timing mark does not move 15 to 18 mm (19/32 to 23/32 in.) to the left, the carburetor requires adjustment.

6. Aim the timing light at the top of the crankshaft pulley. The timing mark should fall relative to the seam in the crankcase centerline as shown in **Table e**.

Table e. Ignition Timing

Model	Ignition setting	Marking
1970 cars and fuel injection cars with Automatic Stick Shift	0° before TDC	0° (1970) 0° (Fuel Injection)
Carburetor cars from August 1970 to spring 1973 and fuel injection cars with manual transmissions	5° after TDC	5° AFTER
Carburetor cars only from spring 1973 (engines with a single vacuum hose to the distributor)	7.5° before TDC	7.5° BEFORE

7. If necessary, adjust the timing by loosening the through bolt in the distributor clamp just enough so that you can hand-turn the entire distributor clockwise or counterclockwise in the engine.

   WARNING
   Be careful of the fan belt and pulleys. Do not wear a necktie or necklace. Tie back long hair and do not wear loose-fitting clothing. Serious injury will result if you are caught in the moving machinery.

8. Hand-turn the distributor clockwise or counterclockwise in the crankcase until the timing marks, illuminated by the timing light, are correctly aligned. Then, while holding the distributor in the correct position, tighten the through bolt to lock the distributor in place.

© 1976 VWoA

Checking Spark Advance Mechanism

The distributors used on cars covered by this Manual have both vacuum and centrifugal advance mechanisms—except for early 1970 models with Automatic Stick Shift, which have a vacuum advance only. Special timing lights are available that incorporate a built-in meter you can use to measure spark advance. The instructions supplied tell you how to use the device to check the centrifugal advance mechanism. You can also check the vacuum advance mechanism by installing a vacuum gauge with a T-fitting in the vacuum hose(s) between the carburetor and the distributor.

Table f lists the rpm and vacuum levels at which critical changes in the spark advance curves take place. Comparing actual distributor operation to the data in the table will help you locate dirty, binding advance mechanisms or mechanical faults in the distributor. The distributor number will be found stamped on the distributor itself.

Table f. Ignition Timing with Distributor Installed

Distributor	Centrifugal timing range						
	Begin					End	
	rpm	rpm	degrees	rpm	degrees	rpm	degrees
113 905 205 T	—	—	—	—	—	—	—
113 905 205 AD	1050-1200	1700	13-15°	2200	13-16°	3900	25-28°
113 905 205 AJ	1000-1250	1500	6-12°	—	—	3800	22-25°
113 905 205 AH	1000-1200	1600	12-15°	2200	12-16°	3800	22-25°
113 905 205 AN	1000-1250	1500	6-12°	—	—	3800	22-25°
043 905 205	1050-1250	1600	7-12°	—	—	3800	20-25°
043 905 205 A	1050-1200	1600	12-14°	2300	13-15°	3900	22-25°
043 905 205 H	1000-1250	1500	6-12°	—	—	3500	20-23°
043 905 205 J	1025-1250	1500	6-11°	—	—	3500	20-23°

Distributor	Vacuum timing range				Timing direction
	Begin		End		
	mm Hg (in. Hg)	mm Hg (in. Hg)	degrees	mm Hg (in. Hg)	degrees	
113 905 205 T	4-8 (.2-.3)	30 (1.1-1.3)	17-19°	80 (3.2)	32-35°	advance
113 905 205 AD	70-120 (2.8-4.7)	—		240 (9.4)	8-12°	advance
	60-100 (2.4-3.9)	—		170 (6.7)	6-8°	retard
113 905 205 AJ	110-160 (4.3-6.3)	—		210-240 (8.3-9.4)	9-12°	advance
	80-130 (3.2-5.1)	—		160-230 (6.3-9.1)	11-13°	retard
113 905 205 AH	110-160 (4.3-6.3)	—		150-170 (5.9-6.7)	2-5°	advance
	80-130 (3.3-5.1)	—		160-230 (6.3-9.1)	11-13°	retard
113 905 205 AN	110-160 (4.3-6.3)	—		180-200 (7.1-7.9)	5-8°	advance
	80-150 (3.2-5.9)	—		160-230 (6.3-9.1)	11-13°	retard
043 905 205	55-100 (2.2-3.9)	—		200 (7.9)	8-12°	advance
043 905 205 A	60-100 (2.4-3.9)	—		200 (7.9)	8-12°	advance
043 905 205 H	115-155 (4.5-6.1)	—		180-200 (7.1-7.9)	5-8°	advance
	60-150 (2.4-5.9)	—		170-230 (6.7-9.1)	11-13°	retard
043 905 205 J	100-160 (3.9-6.3)	—		200-220 (7.9-8.7)	8-12°	advance
	60-150 (2.4-5.9)	—		140-200 (5.5-7.9)	6-8°	retard

© 1976 VWoA

Engine 55

If testing reveals irregularities in the spark advance curve, first clean and lubricate the moving parts of the distributor. If the vacuum advance still fails to conform to specifications, install a new vacuum chamber unit. However, on the early distributor with a vacuum advance only, the advance curve can be adjusted by turning the eccentric pin on to which the breaker plate return spring hooks. Any discrepancies revealed in retesting the centrifugal advance curve indicate worn internal parts. In such cases, the distributor body and its internal parts should be replaced as a unit.

Even if you do not have the equipment for checking the spark advance curve, you can still quick-check the spark advance mechanisms. To check the vacuum advance, hand-turn the breaker plate counterclockwise. It should move without grittiness and spring back solidly to its original position when released. To check the diaphragm for leaks, turn the breaker plate as far as it will go counterclockwise, take off the vacuum hose(s), and cover the hose connection(s) with your finger(s). Then release the breaker plate. The vacuum in the chamber should keep the breaker plate from returning fully to its original position until you uncover the hose connection(s). If the breaker plate returns fully before you uncover the hose connection(s) there is a leak in the diaphragm or in the chamber.

To quick-check the centrifugal advance, hand-turn the distributor rotor clockwise. When you release it, the rotor should return automatically to its original position. If it does not, either the mechanism is dirty or the distributor body and internal parts are faulty and should be replaced as a unit.

19.3 Disassembling and Assembling Distributor

The distributor is shown disassembled in Fig. 19-4. The individual components shown are those available as replacement parts. If the distributor body assembly must be cleaned, disassemble the distributor as shown. Then wash the distributor body assembly thoroughly in clean solvent. Hand-turn the distributor shaft and make certain that it revolves smoothly in its bearings.

After you have cleaned the distributor body assembly, blow it dry with compressed air. Be careful not to direct a strong blast of air into the housing at close range, since this could damage the calibrated springs in the centrifugal advance mechanism. Thoroughly lubricate the shaft and bearings with engine oil by applying oil to the space between the drive dog and the distributor body. Hand-turn the shaft with the distributor body inverted until the oil has worked its way all along the shaft. Apply one drop of motor oil only to each pivot point in the centrifugal advance mechanism and to the breaker plate bearing. If necessary, you can remove the breaker plate to gain access to the centrifugal advance.

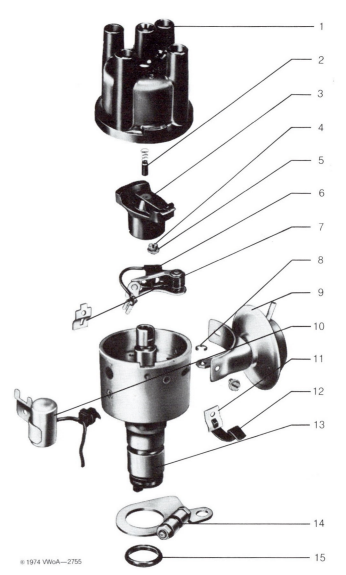

1. Distributor cap
2. Carbon brush and spring
3. Distributor rotor
4. Screw (2)
5. Washer (1)
6. Breaker point assembly
7. Retaining clip mount with cap locating lug
8. E-clip for pull rod
9. Vacuum chamber unit
10. Condenser
11. Retaining clip mount
12. Retaining clip (2)
13. Distributor body assembly
14. Distributor clamp
15. Rubber sealing ring

Fig. 19-4. Distributor disassembled.

19.4 Secondary Circuit

The secondary circuit of the ignition system consists of the distributor rotor, the distributor cap, the spark plug cables, and the high tension cable that links the coil and the distributor cap. Secondary circuit resistance, for radio suppression and for reduction of plug gap erosion, is built into the distributor rotor and spark plug connectors. This resistance allows the use of metallic conductor ignition cables.

You can check the rotor resistor with an ohmmeter as shown in Fig. 19-5. You can test individual plug connectors the same way. Rotor resistance should not exceed 10,000 ohms. Connector resistance should not exceed 5000 to 10,000 ohms.

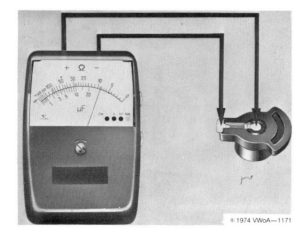

Fig. 19-5. Rotor resistance being checked.

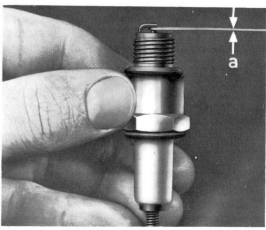

Fig. 19-6. Spark plug gap being set. Dimension **a** should be 0.60 mm (.024 in.) for carburetor engines, or 0.70 mm (.028 in.) for fuel injection engines.

Keep the distributor rotor and cap clean and dry to prevent spark flashover, which could cause irreparable carbon tracks. Also keep the ignition cables clean and replace them when the insulation shows signs of cracking or deteriorating. The rubber boots at the coil tower, distributor cap towers, and plug connectors must be sound and tight-fitting if the secondary circuit is to remain waterproof.

19.5 Spark Plugs

Spark plugs must be of the correct heat range and physical dimensions. The plug reach (length of the threaded portion of the shell) should be 12 mm (.472 in.). The threaded diameter should be 14 mm. Suitable spark plugs are listed in **20. Engine Technical Data**.

Check or replace the spark plugs at the mileage interval specified in **LUBRICATION AND MAINTENANCE**. Before you install the plugs, adjust their gaps to the dimensions given in Fig. 19-6. The deposits on the plug's firing tip may vary in color, indicating the following:

1. Gray or light tan deposits indicate good combustion, proper fuel mixtures, and consistently good spark plug performance.
2. Light gray or chalky-white deposits indicate too lean a fuel/air mixture or an overheated spark plug.
3. Soft, fluffy, black deposits indicate too rich a fuel/air mixture or a plug that is misfiring.
4. A spark plug fouled by oil indicates severe oil leakage past the piston rings or valve guides, or a spark plug that is no longer firing.

Clean spark plugs with a sharp stick or plastic rod, or in a sandblast machine. Never sandblast an oil-fouled plug until after the oily deposit is removed with solvent. After sandblasting, always file the spark plug electrodes to the same sharp, square profiles of a new plug. Such filing is also an advisable step when regapping used plugs that have not been sandblasted. Never use a wire brush to clean spark plugs. The bristles will leave electrically conductive "pencil marks" on the insulator nose.

You can safely clean the spark plug threads with a thread-cutting die. The spark plug hole in the cylinder head can be cleaned with a thread chaser made especially for the task. Do not clean the plug threads with a power-driven wire brush. This can deform the threads so that they will damage the cylinder head.

Another important precaution during spark plug installation is to coat the threads lightly with a good quality anti-seize compound. This coating will prevent corrosion damage to the soft aluminum threads in the cylinder head. If nothing else is available, lubricate the spark plug threads with multipurpose grease or even with motor oil. Never install the plugs dry.

Anti-seize compounds usually contain graphite, which must not come into contact with the plug electrodes or insulator. The graphite's metallic composition could short out the spark plug. Therefore, when coating the threads apply the compound with your fingertip. Don't dip the plug into the compound.

Start the plug in the cylinder head with your fingertips or by hand-holding the spark plug socket. This will give improved feel and help prevent accidental cross-threading. Turn the spark plug in until you feel the gasket contact the cylinder head. Then install a torque wrench on the spark plug socket and torque the plug to 3.0 to 4.0 mkg (22 to 29 ft. lb.).

ENGINE

20. Engine Technical Data

The technical data included here is based on the most up-to-date factory specifications and workshop bulletins available at the time of publication. This data should be adhered to even though it may conflict with data published by non-VW sources or with data that was current at the time the car was manufactured.

As improvements are made in manufacturing, new parts and specifications are introduced to replace older parts and specifications. Sometimes new specifications are applied retroactively to improve the performance of older models.

Although all dimensions are given in both the metric system and the U.S. equivalent, the engine is built to metric specifications. Since the metric dimensions are precise, the U.S. equivalent should be used only if metrically-calibrated tools are not available.

II. General Engine Data

Engine code letter	from August 1969—B
	from August 1970—AE
	from October 1972—AK
	from August 1971 (California only)—AH
	fuel injection engines—AJ
Dry weight	120 kg (265 lb.)
No. of cylinders	4
Cylinder layout	horizontally opposed pairs
Bore	85.5 mm (3.366 in.)
Stroke	69.0 mm (2.716 in.)
Piston displacement	1584 cc (96.7 c.i.)
Compression ratio	code letter B—7.5
	code letter AE (up to July 1971)—7.5
	code letter AE (from August 1971)—7.3
	code letter AK—7.3
	code letter AH—7.3
	code letter AJ—7.3
Fuel requirement	91 octane RON—lead-free only for cars with catalytic converters
Horsepower (DIN)	code letter B—47 @ 4000 rpm
(SAE)	code letter B—57 @ 4400 rpm
(DIN)	code letter AE
	(up to July 1971)—50 @ 4000 rpm
(SAE)	code letter AE—60 @ 4400 rpm
(DIN)	code letter AE
	(from August 1971)—48 @ 4000 rpm
(SAE net)	code letter AE—46 @ 4000 rpm
(DIN)	code letter AK—48 @ 4000 rpm
(SAE net)	code letter AK—46 @ 4000 rpm
(DIN)	code letter AH
	(California only)—48 @ 4000 rpm
(SAE net)	code letter AH—46 @ 4000 rpm
(SAE net)	code letter AJ—48 @ 4200 rpm
Torque (DIN)	code letter B—10.6 mkg @ 2200 rpm
(SAE)	code letter B—18.7 ft. lb. @ 3000 rpm
(DIN)	code letter AE (up to July 1971)
	—10.8 mkg @ 2800 rpm
(SAE)	code letter AE—18.7 ft. lb. @ 3000 rpm
(DIN)	code letter AE (from August 1971)
	—10.2 mkg @ 2000 rpm
(SAE net)	code letter AE—72.0 ft. lb. @ 2000 rpm
(DIN)	code letter AK—10.2 mkg @ 2000 rpm
(SAE net)	code letter AK—72.0 ft. lb. @ 2000 rpm
(DIN)	code letter AH (California only)
	—10.2 mkg @ 2000 rpm
(SAE net)	code letter AH—72.0 ft. lb. @ 2000 rpm
(SAE net)	code letter AJ—73.1 ft. lb. @ 2800 rpm

I. Tightening Torques

Location	Designation	Thread	Quality grade	mkg	ft. lb.
Connecting rods	connecting rod nut	M 9 x 1	8 G	3.0-3.5	22-25
Crankcase halves	nut	M 8	5 S	2.0	14
Crankcase halves	sealing nut	M 12 x 1.5	35 S 20 K	2.5	18
Cylinder head	nut	M 10	35 S 20 KV	3.2	23
		M 8	35 S 20 KV	2.5	18
Rocker shaft to cylinder head	nut	M 8	8 G	2.0-2.5	14-18
Oil pump to crankcase	nut	M 8	—	2.0	14
Oil drain plug	plug	M 14 x 1.5	9 S 20 K	3.5	25
Oil strainer to crankcase	nut	M 6	—	0.7	5
Flywheel to crankshaft	bolt	M 28 x 1.5	45 S 20 KN	35.0	253
Clutch to flywheel	bolt	M 8 x 1.5	8 G	2.5	18
Spark plugs	spark plug	M 14 x 1.25	—	3.0-4.0	22-29
Engine to transmission	nut	M 10	—	3.0	22
Converter to drive plate	bolt	M 8	8 G	2.5	18
Generator pulley	nut	M 12 x 1.5	—	5.5-6.5	40-47
Special nut for fan	nut	M 12 x 1.5	9 S 20 K	5.5-6.5	40-47
Crankshaft pulley	bolt	M 20 x 1.5	9 S 20 K	4.0-5.0	29-36

© 1976 VWoA

58 Engine

III. Tolerances, Wear Limits, and Settings

Designation	New Parts on Installation mm (in.)	Wear Limit mm (in.)
A. Crankcase bores		
1. Bores for main bearings		
a. Bearings 1, 2, and 3 .... diameter	65.00-65.02 (2.5591-2.5598)	65.03 (2.5602)
b. Bearing 4 .... diameter	50.00-50.03 (1.9685-1.9697)	50.04 (1.9701)
2. Bore for oil seal/flywheel end .... diameter	90.00-90.05 (3.5433-3.5453)	—
3. Bores for camshaft bearings .... diameter	27.50-27.52 (1.0826-1.0834)	—
4. Bore for oil pump housing .... diameter	70.00-70.03 (2.7559-2.7571)	—
5. Bores for cam followers .... diameter	19.00-19.02 (.7480-.7488)	19.05 (.7500)
B. Crankshaft		
1. Journal dimensions		
a. Main journals 1, 2, and 3 .... diameter	54.97-54.99 (2.1641-2.1649)	—
b. Main journal 4 .... diameter	39.98-40.00 (1.5740-1.5748)	—
c. Connecting rod journals .... diameter	54.98-55.00 (2.1646-2.1654)	—
d. Three undersizes of 0.25 mm (.010 in.) each		
2. Crankshaft at No. 2 and 4 main journals (No. 1 and 3 journals on V-blocks) .... runout	—	0.03 (.0012)
3. Crankshaft .... imbalance	max. 12 cmg	—
4. Main bearing journal .... out-of-round	—	0.03 (.0012)
5. Connecting rod journal .... out-of-round	—	0.03 (.0012)
6. Crankshaft/main journals (taking housing preload into account):		
a. Bearings 1 and 3 .... radial play	0.04-0.10 (.0016-.004)	0.18 (.007)
b. Steel bearing 2 .... radial play	0.03-0.09 (.001-.0035)	0.17 (.0067)
c. Bearing 4 .... radial play	0.05-0.10 (.002-.004)	0.19 (.0075)
7. Crankshaft/main journal 1 .... end play	0.07-0.13 (.0027-.0051)	0.15 (.006)
8. Connecting rod journal/connecting rod .... radial play	0.02-0.07 (.0008-.0028)	0.15 (.006)
end play	0.10-0.40 (.004-.016)	0.70 (.028)
C. Connecting rods		
1. Weight difference between connecting rods in one engine	max. 5 g	max. 10 g
Weight of replacement connecting rods		
− Weight (brown)	580-588 g	—
+ Weight (gray)	592-600 g	—
2. Piston pin .... diameter	21.996-22.00 (.8658-.8661)	—
3. Piston pin bush .... diameter	22.009-22.017 (.8665-.8668)	—
4. Piston pin/pin bush .... radial play	0.01-0.02 (.0004-.0008)	0.04 (.0016)
D. Camshaft		
1. Bearings 1, 2, and 3 .... diameter	24.99-25.00 (.9839-.9843)	—
2. Measured at center bearing (bearings 1 and 3 on V-blocks) .... runout	0.02 (.0008)	0.04 (.0016)
3. Camshaft/camshaft bearings (taking housing preload into account) .... radial play	0.02-0.05 (.0008-.002)	0.12 (.0047)
Thrust bearing .... end play	0.04-0.13 (.0016-.0051)	0.16 (.0063)
4. Camshaft gear .... backlash	0.00-0.05 (.000-.002)	—
5. Cam follower .... diameter	18.96-18.98 (.7465-.7472)	18.93 (.7453)
6. Bore/cam follower .... radial play	0.02-0.06 (.0008-.0024)	0.12 (.0047)
7. Pushrod .... runout	max. 0.30 (.012)	—
E. Lubrication		
1. Oil pressure (for SAE 30 only) at an oil temperature of 70°C (158°F) and 2500 rpm	min. 28 psi (2 kg/cm²)	—
2. Spring for pressure relief valve, loaded length 44.1 mm (1.73 in.) ..... load	5.6-7.3 kg (12.3-16.1 lb.)	—
3. Spring for oil pressure control valve, loaded length 20.2 mm (0.79 in.) load	3.1-3.8 kg (6.8-8.4 lb.)	—
4. Oil pressure switch opens at .... pressure	2.1-6.4 psi (0.15-0.45 kg/cm²)	—
F. Flywheel		
1. Flywheel (measured at center of friction surface) .... lateral runout	max. 0.30 (.012)	—
imbalance	max. 20 cmg	—
2. Shoulder for oil seal .... outside diameter	69.9-70.1 (2.752-2.760)	—
3. Drive plate .... imbalance	max. 5 cmg	—
G. Cylinders and pistons		
Two oversizes of 0.5 mm (.020 in.) each		
1. Cylinders .... out-of-round	max. 0.01 (.0004)	—

ENGINE

III. Tolerances, Wear Limits, and Settings (continued)

Designation	New Parts on Installation mm (in.)	Wear Limit mm (in.)
2. Cylinder/piston.. clearance	0.04-0.06 (.0016-.0024)	0.20 (.008)
3. Ring side clearance		
a. Upper piston ring...	0.07-0.10 (.0028-.0039)	0.12 (.0047)
b. Lower piston ring...	0.05-0.07 (.0020-.0028)	0.10 (.004)
c. Oil scraper ring...	0.03-0.05 (.0012-.0020)	0.10 (.004)
4. Ring gap		
a. Upper piston ring...	0.30-0.45 (.012-.018)	0.90 (.035)
b. Lower piston ring...	0.30-0.45 (.012-.018)	0.90 (.035)
c. Oil scraper ring...	0.25-0.40 (.010-.016)	0.95 (.037)
5. Piston weight		
− Weight (brown)..	398-410 g	—
+ Weight (gray)..	406-418 g	—
− Weight (brown)....................................... from August 1971	402-412 g	—
+ Weight (gray)... from August 1971	410-420 g	—
6. Weight difference between pistons in one engine	max. 5 g	max. 10 g
H. Cylinder head and valves		
1. Cylinder seating depth in cylinder head	13.45-13.55 (.530-.533)	—
2. Rocker arm .... inside diameter	18.00-18.02 (.7087-.7095)	18.04 (.7102)
3. Rocker shaft... diameter	17.97-17.99 (.7075-.7083)	17.95 (.7067)
4. Valve spring tension at a loaded length of 31.0 mm (1.22 inches)..... load	53.2-61.2 kg (117.3-134.9 lb.)	—
5. Valve seats		
a. Intake .... width	1.3-1.6 (.051-.063)	—
b. Exhaust.. width	1.70-2.00 (.067-.079)	—
c. Intake .... seat angle	45°	—
d. Exhaust... seat angle	45°	—
e. Outer correction angle	15°	—
f. Inner correction angle	75°	—
6. Valve guides		
a. Intake .... inside diameter	8.00-8.02 (.3150-.3158)	8.06 (.3173)
b. Exhaust (carburetor engines)................. inside diameter	8.00-8.02 (.3150-.3158)	8.06 (.3173)
c. Exhaust (fuel injection engines).............. inside diameter	9.00-9.02 (.3543-.3551)	9.06 (.3507)
7. Valve stem		
a. Intake... diameter	7.94-7.95 (.3126-.3130)	7.90 (.3110)
b. Exhaust (carburetor engines)........................ diameter	7.91-7.92 (.3114-.3118)	7.88 (.3102)
c. Exhaust (fuel injection engines)..................... diameter	8.91-8.92 (.3508-.3512)	8.88 (.3496)
out-of-round	0.01 (.0004)	—
8. Valve head		
a. Intake... diameter	35.6 (1.40)	—
b. Exhaust (carburetor engines)........................ diameter	32.1 (1.26)	—
c. Exhaust (fuel injection engines)..................... diameter	30.1 (1.19)	—
9. Valve facing		
a. Intake... angle	44°	—
b. Exhaust... angle	45°	—
10. Valve guide/valve stem (intake and exhaust)..................... rock	0.23-0.27 (.009-.011)	0.80 (.032)
11. Valve clearance (cold)		
a. Intake... setting	0.15 (.006)	—
b. Exhaust.. setting	0.15 (.006)	—
12. Compression		
a. Compression pressure........................ from Aug. 1967	114-142 psi (8.0-10.0 kg/cm²)	100 psi (7 kg/cm²)
from Aug. 1971	107-135 psi (7.5-9.5 kg/cm²)	85 psi (6 kg/cm²)
b. Pressure difference between cylinders	max. 28 psi (2 kg/cm²)	—
I. Cooling		
1. Thermostat................................... opening temperature	65-70°C (149-158°F)	—
2. Fan and pulley .... out of balance	max. 5 cmg	—
J. Clutch		
1. Total pressure	380-420 kg (838-926 lb.)	—
2. Complete clutch .... out of balance	max. 15 cmg	—
3. Pressure plate .... runout	—	0.10 (.004)
4. Clutch plate (measured at 195 mm [7.677 in.] dia.).......... lateral runout	max. 0.50 (.020)	—
5. Play at clutch pedal	10-20 (⅜-¾)	—

© 1976 VWoA

60 ENGINE

IV. Basic Tune-up Specifications

Idle speed (carburetor engines)	Manual transmission: 850 rpm ± 50 Automatic transmission: 950 rpm ± 50
Idle speed (fuel injection engines)	Manual transmission: 800 to 950 rpm Automatic transmission: 850 to 1000 rpm
Ignition	12 volt
Ignition timing (carburetor engines)	Through July 1970: 0° before TDC with vacuum hose disconnected and plugged From August 1970: 5° after TDC with vacuum hose(s) connected From spring 1973: 7.5° before TDC with vacuum hose(s) connected
Ignition timing (fuel injection engines)	Manual transmission: 5° after TDC with vacuum hose(s) connected Automatic transmission: 0° before TDC with vacuum hose(s) connected
Firing order	1-4-3-2
Cylinder location	No. 1: front cylinder on right side of car No. 2: rear cylinder on right side of car No. 3: front cylinder on left side of car No. 4: rear cylinder on left side of car
Point gap	0.40 to 0.50 mm (.016 to .020 in.)
Dwell angle	New points: 44° to 50° Used points: 42° to 58°
Spark plugs (carburetor engines)	Normal service: Champion L-88-A Bosch W 145 T1 Beru 145/14 High-speed service: Champion L-85 or L-86 Bosch W 175 T1 Beru 175/14 Electrode gap: 0.60 mm (.024 in.)
Spark plugs (fuel injection engines)	Normal service: Champion L-288 Bosch W 145 M1 Beru 145/14/L Electrode gap: 0.70 mm (.028 in.)
CO volume % at idle	34 PICT-3 or 34 PICT-4 carburetor: 3% ± 1% Fuel injection: 0.2% to 2.0%

Section 2

FUEL SYSTEM

Contents

Introduction 2	
1. General Description 3	
Fuel Storage 3	
Pump and Lines 3	
Air Cleaner 3	
Carburetor 3	
Emission Controls 3	
2. Maintenance 4	
3. Fuel Tank 4	
3.1 Removing and Installing Fuel Tank 4	
3.2 Fuel Gauge Sending Unit and Fuel Pickup 4	
3.3 Treating Corroded Fuel Tank 4	
4. Fuel Pump and Lines 5	
4.1 Fuel Pump Troubleshooting 5	
4.2 Removing and Installing Fuel Pump 6	
Checking Fuel Pump Pressure and Delivery Capacity 7	
4.3 Disassembling and Assembling Fuel Pump 7	
Fuel Cutoff Valve 8	
5. Carburetor 8	
5.1 Carburetor Troubleshooting 9	
5.2 Adjusting Carburetor 10	
Adjusting Throttle Valve Positioner 11	
Adjusting Dashpot 11	
5.3 Removing, Rebuilding, and Installing Carburetor 11	

Removing and Installing Throttle Valve Positioner 12	
Disassembling and Assembling Carburetor 12	
Adjusting Fuel Level in Float Bowl 14	
Adjusting Accelerator Pump 15	
Checking Electromagnetic Cutoff Valve . . 15	
Carburetor Installation 15	
6. Air Cleaner 16	
Checking and Adjusting 1970 Intake Air Preheating System 16	
Checking and Adjusting 1971 Intake Air Preheating System 16	
Checking and Adjusting 1972 and Later Intake Air Preheating System 16	
7. Checking, Repairing, and Adjusting Emission Controls 17	
7.1 Exhaust Emission Control Troubleshooting 17	
7.2 Evaporative Emission Control 17	
7.3 Exhaust Gas Recirculation 18	
7.4 Throttle Valve Positioner 19	
8. Fuel System Technical Data 20	
I. Carburetor Settings and Jets 20	
II. Fuel Pump 20	
III. 34 PICT-4 Carburetor Part Numbers . . . 20	

TABLES

a. Fuel Pump Troubleshooting 5	
b. Carburetor Troubleshooting 9	
c. Exhaust Emission Control Troubleshooting . . . 17	

Fuel System

The fuel system, as covered in this section of the Manual, applies mainly to the 1970 through 1974 cars which have carburetors. The electronic fuel injection system, introduced on the 1975 models, is covered separately in **FUEL INJECTION**. The data given in this section under **3. Fuel Tank** and **7.2 Evaporative Emission Control** apply to both cars with fuel injection and cars with carburetors. However, you should refer to **FUEL INJECTION** for all other information related to fuel injection engines—including emission control data and air cleaner servicing. Whether they have carburetors or fuel injection, VWs covered by this Manual are designed to operate on regular (91 octane) gasoline. Fuel injection cars that are equipped with catalytic converters require lead free gasoline.

The carburetor engine's fuel system handles five main tasks necessary for proper engine operation: (1) it provides storage space for the gasoline; (2) it includes the components necessary for delivering gasoline to the engine; (3) it is responsible for admitting the proper amount of filtered air to the engine; (4) it incorporates a carburetor and distribution system for mixing fuel and air in precisely controlled proportions and delivering this mixture to the cylinders; and (5) it modifies the density of the incoming air so the combustion process does not produce an excess of undesirable exhaust emissions. The fourth function mentioned above, that of mixing the fuel with air, is handled by one single-venturi downdraft carburetor.

The carburetor is mounted atop a tubular welded-steel intake manifold that has an exhaust-warmed preheating pipe. (The 1974 models sold in California have a manifold with two preheating pipes, through which exhaust gasses flow in opposite directions.) On 1971 and later models, the two outer ends of the manifold are joined to bifurcated cast aluminum intake pipes that conduct the fuel/air mixture into the cylinder head intake ports. On these late models, the cylinder heads have dual intake ports rather than the single siamesed intake port used in the heads of the 1970 engine.

Because many of the repairs and tune-up procedures described in this section of the Manual have a direct influence on exhaust emissions, these jobs should not be undertaken unless all prescribed equipment is available. If you lack the skills, special equipment, or tools needed for servicing and adjusting the fuel system, we suggest you leave such repairs to an Authorized VW Dealer or other qualified shop. We especially urge you to consult your Authorized VW Dealer before attempting repairs on a car still covered by the new-car warranty.

FUEL SYSTEM 3

1. GENERAL DESCRIPTION

As noted earlier, the fuel system may conveniently be divided into five subsystems, each with a separate function. For brevity, these will be called fuel storage, pump and lines, air cleaner, carburetor, and emission controls.

Fuel Storage

The fuel tank is housed under the cargo liner in the front luggage compartment. It has a capacity of 10.6 U.S. gallons (8.8 Imperial gallons or 40 liters) in Beetles and Karmann Ghia models and 11.1 U.S. gallons (9.2 Imperial gallons or 42 liters) in 1971 and later Super Beetles and Convertibles.

The tank is equipped with a sending unit for the electrical fuel gauge, a pickup tube for transfer of fuel to the engine, and a vent tube—routed, on 1970 and later cars, into the evaporative emission control system. The unvented filler cap is concealed beneath a flap in the right front side of the car.

Pump and Lines

Three different fuel pump designs have been used on cars covered by this Manual. The 1970 models have a pump with a cast metal top that houses a fuel cutoff diaphragm. Beginning with the 1971 models, a pump with a stamped steel cover has been installed. This unit underwent two significant modifications in the 1973 models. First, a built-in cutoff valve was installed to prevent the gradual flow of fuel to the carburetor when the engine is not running. On 1971 and 1972 cars, this task is handled by a separate inline cutoff valve. Then, beginning in January 1973, a mechanically identical pump with an angled body was introduced for use on engines equipped with alternators.

Fuel reaches the pump through a steel line housed in the frame tunnel. Hoses connect this line to the fuel tank at the front of the car and to the fuel pump at the rear. Another hose links the fuel pump with the carburetor.

Air Cleaner

Several different air cleaners have been installed on the models covered by this Manual. These air cleaners differ mainly in the design and function of the intake air preheating system. Originally incorporated to prevent carburetor icing, the intake air preheating system has become an important auxiliary to the emission control system. In its latest form, the air cleaner is equipped with a thermostatic valve that controls a flap operated by intake manifold vacuum. This makes the intake air preheating regulator valve sensitive to both temperature and engine load. The result is good performance and economy under all driving conditions with no adverse effect on exhaust gas content.

©1978 VWoA

Carburetor

The 1970 models are equipped with Solex 30 PICT-3 carburetors. All later models through 1973 have Solex 34 PICT-3 carburetors. The 1974 models sold in California have the 34 PICT-4 carburetor which incorporates a thermostatically regulated accelerator pump. Other 1974 cars retain the 34 PICT-3 carburetor.

On all models, the carburetor has a built-in automatic choke operated by a thermostatic spring. The thermostatic spring closes the choke valve when the engine is cold. When the ignition is turned on, an electric heating element warms the thermostatic spring, causing the choke to open at a predetermined rate. Whenever the throttle valve is closed—as at idle or during deceleration—a vacuum diaphragm overrides the thermostatic spring and causes the choke to open slightly.

Emission Controls

The evaporative emission control system prevents gasoline fumes from escaping into the atmosphere. The fuel tank is vented into a system that traps and contains fuel vapors until they can be burned by the engine.

The ventilation hose from the tank is connected to an expansion chamber located under the cowl panel. Liquid fuel, expelled from a full tank by expansion, is stored in this chamber and then returned to the tank as the fuel level falls. Evaporated fuel escapes from the expansion chamber into a ventilation line that carries it to an activated charcoal canister located at the rear of the car. The activated charcoal absorbs the vapor, preventing it from entering the atmosphere. When the engine is started, air from the engine's cooling fan is directed into the canister, blowing trapped fumes from the activated charcoal into the air cleaner. The filtered vapors are drawn through the carburetor and into the cylinders of the engine.

A throttle valve positioner is installed on all 1970 and 1971 cars equipped with manual transmissions. For 1972, it was installed only on cars sold in California equipped with manual transmissions. The positioner automatically adjusts the throttle closing rate for minimum exhaust emissions and prevents the carburetor throttle valve from closing suddenly when the accelerator pedal is released—again to reduce emissions. Beginning with the 1971 models, this latter function is supplemented on some cars by a dashpot installed on the carburetor.

Instead of a throttle valve positioner, an Exhaust Gas Recirculation system is used on 1972 and later cars sold in California with Automatic Stick Shift, on all 1973 and later cars with Automatic Stick Shift sold in the U.S., and on all 1974 cars sold in the U.S. The system diverts a portion of the exhaust gases into the intake manifold below the carburetor, modifying the density and content of the incoming air. The recirculated exhaust gases lower the flame peaks in the combustion process, thereby reducing the formation of oxides of nitrogen—a major air pollutant.

4 Fuel System

2. Maintenance

There are only a few fuel system maintenance operations that must be carried out at a specified mileage or after a certain period of service. These are listed below and covered briefly in **LUBRICATION AND MAINTENANCE**.

1. Checking the throttle positioner (where fitted)
2. Servicing the air cleaner
3. Replacing the air filter (paper type)
4. Checking the exhaust gas recirculation (EGR) valve
5. Cleaning the cyclone-type EGR filter (1972 cars sold in California with Automatic Stick Shift)
6. Replacing the element-type EGR filter (1973 and later cars)
7. Replacing the activated charcoal filter canister.

3. Fuel Tank

If the fuel tank must be removed for cleaning or repairs, it is important that the connections leading to the tank be reinstalled correctly and in their original locations. Errors can lead to fuel starvation, or to the improper venting of fumes.

3.1 Removing and Installing Fuel Tank

To avoid the hazard of spilled fuel, the tank should be no more than half full during removal. Surplus fuel should be drained off.

> **WARNING —**
> *Disconnect the battery ground strap. Do not smoke or work near heaters or other fire hazards. Have a fire extinguisher handy.*

To remove:

1. Using a pinch clamp or clamps, squeeze the fuel hose(s) shut at a point near the tank connection(s). Then detach the hose(s) from the tank.
2. Remove the fiberboard luggage compartment liner.
3. Disconnect the wire(s) from the fuel gauge sending unit.
4. Detach the ventilation hose(s) from the fuel tank.
5. Loosen the large hose clamps on the fuel filler neck hose. Then remove the hose.
6. Remove the bolts from the four fuel tank retaining plates. Lift off the retaining plates and take the fuel tank out of the car. Be careful not to damage the rubber sealing strip between the tank and the body.

Installation is the reverse of removal. However, new clips should be used on the ventilation hose(s) and the hoses must not be installed in a twisted position or bent so sharply that they collapse.

3.2 Fuel Gauge Sending Unit and Fuel Pickup

The removal, installation, and testing of the fuel gauge sending unit is covered in **ELECTRICAL SYSTEM**. Only the fuel pickup is discussed here.

On Beetles and Karmann Ghias, the fuel pickup in the bottom of the tank consists of a 6-mm (¼-in.) diameter feed pipe held onto the tank by a union nut. The feed pipe's inner end is covered by a wire gauze filter. To remove the filter, unscrew the union nut, remove the feed pipe, and withdraw the wire gauze filter from the opening in the tank.

The fuel pickup on Super Beetles and Convertibles is a tube welded into the bottom of the tank. However, there is a drain plug nearby for access to a wire gauze filter on the end of the fuel pickup tube. After removing the drain plug, you can pull the filter off the pickup tube with a wire hook. Install the filter by inserting it in a tube that will bear against the flange around the top of the filter. Using the tube as a handle, insert the filter through the drain opening and slide it over the end of the fuel pickup tube. The wire gauze filter was discontinued as of chassis No. 133 2802 561. It is replaced by an in-line filter close to the fuel tank.

3.3 Treating Corroded Fuel Tank

Water is heavier than gasoline, so condensation or water from contaminated fuel will collect in the bottom of the tank where it may cause rusting.

> **WARNING —**
> *Tanks containing holes caused by corrosion should not be welded or soldered with an open flame. Even an empty tank contains fumes that make it a potential bomb.*

There are two standard agents for cleaning rusted fuel tanks:

A. Derusting phosphate agent (commercially available, as Rodine 50® or a similar product), that is mixed either in a solution of one part agent to ten parts water or according to the manufacturer's directions.

© 1976 VWoA

B. Aqueous solution of hydrochloric acid (this is industrial muriatic acid, specific gravity 1.190) in the proportion of 20 parts hydrochloric acid solution, 80 parts water, and one part inhibitor (such as Rodine 50).

Cleaning agent **A** is the preferred treatment. It is milder and leaves a protective phosphate film. Both methods require immediate rinsing with a soluble oil mixture (one part machine coolant to 20 parts water). Rust will form again if the tank is not rinsed thoroughly.

To remove rust:

1. Seal the fuel pickup and vent pipes. Then remove the fuel gauge sending unit.

2. Fill the tank with solution **A** or **B**. Make sure the solution completely fills the tank so that acid fumes will not attack the tank wall above the fluid level.

   NOTE
   Leave the solution in the tank until the rust is removed. Forty minutes may be sufficient for a very small accumulation of rust, but severe rusting may require as long as eight hours. For best results, let the solution stand overnight.

3. Pour out the solution, and pour in 4 to 5 U.S. quarts (6.5 to 8.5 Imperial pints or 4 to 5 liters) of rinsing solution. Rock the tank vigorously to slosh the rinse over the entire interior.

4. Drain the tank and dry it with compressed air.

The cleaning solution can be used 10 or 15 times. Store it only in a glass container with a glass or rubber stopper that will not be attacked by vapors from the acid in the solution.

4. Fuel Pump And Lines

The fuel pump is mounted on the engine. It is a mechanical pump, operated by a cam on the engine's distributor drive shaft. The stroke of the pump is determined by the number of gaskets installed under the black intermediate flange that goes between the pump and the engine's crankcase. It is important that the stroke be limited to its specified length.

4.1 Fuel Pump Troubleshooting

Table a lists possible fuel pump trouble symptoms with their probable causes and remedies. When more than one probable cause or remedy is given, check them in the order in which they are listed. The numbers in bold type in the Remedy column refer to the headings in this section where the prescribed repairs are described.

NOTE
Carburetor flooding may also result from a defective in-line fuel cutoff valve on 1971 and 1972 cars. On earlier or later cars, the cutoff valve is an integral part of the fuel pump.

Table a. Fuel Pump Troubleshooting

Symptom	Probable Cause	Remedy
1. Fuel leaking at joint faces of pump	a. Slotted screws loose b. Diaphragm cracked	a. Tighten screws. b. Replace diaphragm. See **4.3**.
2. Fuel leaking at diaphragm rivets	Diaphragm damaged during assembly	Replace diaphragm. See **4.3**.
3. Fuel leaking through diaphragm itself	Diaphragm material damaged by solvent in fuel	Replace diaphragm. See **4.3**.
4. Diaphragm damaged, apparently from excessive pump stroke	Pump incorrectly installed, gasket too thin	a. Install pump correctly with additional gasket. Replace diaphragm. See **4.2**, **4.3**. b. Check pushrod stroke. See **4.2**.
5. Pump pressure low	a. Pump incorrectly installed, gasket too thick b. Spring pressure low	a. Install pump correctly, removing one gasket if necessary. See **4.2**. b. Stretch spring to lengthen or, if necessary, replace. See **4.3**.
6. Carburetor flooding	a. Pump pressure excessive, forcing needle valve down. Pump gasket too thin b. Spring pressure excessive	a. Install pump correctly. Check pushrod stroke. Add gasket if needed. See **4.2**. b. Press spring together to shorten or, if necessary, replace spring. See **4.3**.
7. Insufficient fuel delivery	Valves leaky or sticking	Free valves or replace pump. See **4.2**, **4.3**.

6 Fuel System

4.2 Removing and Installing Fuel Pump

The installed position of the fuel pump is shown in Fig. 4-1. The pump is located between the ignition distributor and the generator (or alternator) support, on the center of the crankcase just below the carburetor.

Fig. 4-1. A view of the engine compartment that shows the position of the fuel pump in relation to other engine parts.

To remove pump:

1. Pull the fuel hoses off the pump, quickly plugging them to prevent the escape of gasoline.

    WARNING —
    Disconnect the battery ground strap. Do not smoke or work near heaters or other fire hazards. Have a fire extinguisher handy.

2. Remove the nuts on the pump flange and take off the pump.

3. Remove the pushrod from the center of the intermediate flange. Then remove the intermediate flange and gaskets from the crankcase.

To adjust pump stroke:

1. Place two new gaskets on the pump mounting studs. Then install the intermediate flange on top of them.

    CAUTION —
    Always install the intermediate flange on the crankcase before you insert the pushrod. Otherwise, the pushrod may slip through the flange and into the crankcase.

2. Install the pushrod in the intermediate flange with the tapered end down.

3. Turn the engine by hand until the pushrod rises to its highest point of travel.

4. Measure the distance that the pushrod projects above the intermediate flange as shown in Fig. 4-2. The pushrod should project 13 mm (½ in.). If it does not, adjust it to this dimension by removing or installing gaskets under the intermediate flange.

Fig. 4-2. Depth gauge being used to check the length of the pushrod where it projects from the intermediate flange.

To install pump:

1. Fill the lower part of the pump with universal grease.

2. Install a new gasket over the studs where they project from the intermediate flange. Then insert the pushrod with the tapered end down.

3. Place the pump over the studs and install the two spring washers and nuts. Torque the nuts to 2.5 mkg (18 ft. lb.).

4. Using new hose clamps, install the fuel hoses on the fuel pump.

    NOTE —
    If the original hoses are to be reinstalled, carefully inspect their ends to see that they have not been weakened or deformed by the previous installation of hose clamps. A new clamp may not seal hoses that have lost resiliency. In such cases, a new hose should always be installed.

5. Check the seating of the fuel line rubber grommet in the engine's front cover plate.

6. Reconnect the battery ground strap.

Fuel System

Checking Fuel Pump Pressure and Delivery Capacity

You can check the fuel pump's pressure by installing a "T" fitting between the pump and the carburetor so that a pressure gauge can be installed. To check the fuel pump's delivery capacity, run the engine on an auxiliary fuel supply while you collect the fuel pump output in a container so it can be measured.

Maximum delivery pressure should be between 3.0 and 5.0 psi (0.20 and 0.35 kg/cm^2). Minimum delivery capacity should be 400 cc per minute at 4000 rpm (3800 rpm for 1970 cars).

4.3 Disassembling and Assembling Fuel Pump

Fig. 4-3 is a cutaway view of the fuel pump installed on 1970 cars. The fuel filter can be removed for cleaning or replacement by taking off the plug that covers it. The main diaphragm can be replaced separately and a repair kit is available to rebuild the pump.

The late-type pumps can be only partially disassembled, as illustrated in Fig. 4-4. This permits cleaning and replacement of the filter and its gasket. These pumps should be replaced if they are faulty or fail to deliver the specified fuel pressure and capacity.

Fig. 4-4. Late-type fuel pump disassembled. Note the sealing washer under the head of the screw that holds the cap and filter in place.

NOTE ——

The fuel pump used on engines equipped with alternators is the same as the pump shown in Fig. 4-4, but with a body that is angled at 15°. The pushrod for the angled pump is 100 mm (3$^{15}/_{16}$ in.) long rather than 108 mm (4$^{1}/_{4}$ in.) long as for other pumps.

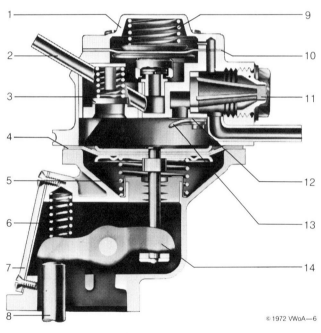

1. Fuel pump cover
2. Delivery pipe
3. Delivery valve
4. Main diaphragm
5. Main diaphragm spring
6. Spring
7. Inspection cover
8. Pushrod
9. Cutoff diaphragm spring
10. Cutoff diaphragm
11. Fuel filter
12. Suction valve
13. Suction valve retainer
14. Pump operating lever

Fig. 4-3. Positions of main parts in 1970 fuel pump.

8 Fuel System

Fuel Cutoff Valve

The 1971 and 1972 cars have a fuel cutoff valve that is separate from the fuel pump. (This device is built into the fuel pumps of other models.) The separate fuel cutoff valve is shown properly installed in Fig. 4-5. Note especially the arrow that is cast into the angled fuel passage to indicate flow direction.

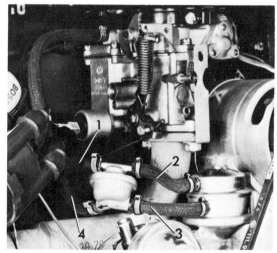

1. Hose from tank
2. Suction line to pump
3. Pressure line from pump
4. Hose to carburetor

Fig. 4-5. Cutoff valve, fuel pump, and carburetor. All fuel hoses are shown properly installed.

5. Carburetor

The Solex 30 PICT-3 carburetor installed on 1970 cars and the Solex 34 PICT-3 carburetor installed on later models are fundamentally the same. The later unit has a somewhat larger venturi and some of the parts and subsystems have been relocated. The 34 PICT-4 carburetor installed on 1974 cars sold in California is basically identical to the 34 PICT-3. However, the 34 PICT-4 has a thermostatic device which varies the accelerator pump's injection quantity in relation to the temperature of the carburetor body.

The electromagnetic cutoff valve consists of a cutoff plunger and an electrically operated solenoid. The cutoff valve has no influence on carburetion while the engine is running. It simply prevents the engine from "running on" or "dieseling" after the ignition is turned off.

On 34 PICT-3 and 34 PICT-4 carburetors, a spring inside the electromagnetic cutoff valve causes the cutoff plunger to snap out of the valve and block the idle air bypass drilling when the ignition is turned off. This prevents air and fuel from reaching the intake manifold while the carburetor's throttle valve is closed—as the valve would be when the driver's foot is off the accelerator.

When the ignition is turned on, a wire to the electromagnetic cutoff valve energizes the solenoid. The solenoid withdraws the cutoff plunger to the position shown in Fig. 5-1. Notice that a mixture of fuel and air is permitted to flow from the drilling while the throttle valve is closed, permitting the engine to idle normally.

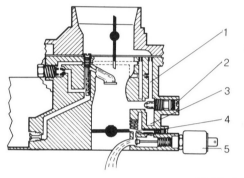

1. Bypass air drilling
2. Bypass screw
3. Bypass drilling
4. Volume control screw
5. Electromagnetic cutoff valve

Fig. 5-1. Electromagnetic cutoff valve, 34 PICT-3.

On 30 PICT-3 carburetors, the electromagnetic cutoff valve controls fuel delivery to the idle system as shown in Fig. 5-2.

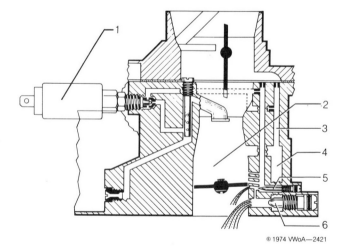

1. Electromagnetic cutoff valve
2. Carburetor throat
3. Bypass air drilling
4. Bypass drilling
5. Volume control screw
6. Bypass screw

Fig. 5-2. Electromagnetic cutoff valve, 30 PICT-3.

FUEL SYSTEM 9

5.1 Carburetor Troubleshooting

Table b lists carburetor troubles with their probable causes and remedies. The numbers in bold type in the Remedy column refer to the headings in this section where the prescribed repairs are described.

Table b. Carburetor Troubleshooting

Symptom	Probable Cause	Remedy
1. Engine does not start (tank has fuel, and ignition is working)	a. Choke valve sticking b. Automatic choke not working properly c. Bimetal spring unhooked or broken d. Ceramic plate broken e. Float needle valve sticking and carburetor flooding	a. Apply penetrating fluid to free choke valve shaft. b. Check vacuum diaphragm for freedom of movement. c. Reconnect spring or, if broken, replace complete ceramic plate (when installing, match index marks). See **5.3**. d. Replace ceramic plate (when installing, match index marks). See **5.3**. e. Clean or replace float needle valve. See **5.3**.
2. Engine runs continuously at fast idle	a. Automatic choke not switching off b. Choke valve or fast idle cam sticking	a. Check heater element and connections. b. Apply penetrating fluid to free choke valve or cam.
3. Engine idles unevenly or stalls (ignition dwell correct)	a. Idle adjustment incorrect b. Pilot jet blocked c. Loose or disconnected vacuum line on Automatic Stick Shift d. Air leak in manifold	a. Adjust idle to 900 rpm. See **5.2**. b. Clean jet. See **5.3**. c. Tighten connection or replace line. d. Check carburetor flange gasket, manifold sleeves, and intake pipe gaskets. See **5.3**.
4. Engine "runs on" when ignition is switched off	a. Faulty electromagnetic cutoff valve or wire b. Idle speed too fast c. Faulty electromagnetic cutoff valve	a. Test wire for current. Test solenoid for continuity. Replace faulty cutoff valves. See **5.3**. b. Adjust idle speed. See **5.2**. c. Check electromagnetic cutoff valve; replace if faulty. See **5.3**.
5. Exhaust backfire when car is overrunning the engine	Idle mixture slightly weak	Enrich mixture by turning volume control screw counterclockwise. See **5.2**.
6. Poor transition from idle to operating speed	a. Accelerator pump passages blocked or ball sticking b. Torn accelerator pump diaphragm c. Idle adjustment incorrect d. Amount of fuel injected is incorrect	a. Clean accelerator pump and check operation. See **5.3**. b. Replace diaphragm. See **5.3**. c. Adjust idle. See **5.2**. d. Adjust accelerator pump volume. See **5.3**.
7. Engine stalls when accelerator pedal is released suddenly	Idle mixture too rich	Adjust idle. See **5.2**.
8. Engine runs unevenly (surges) with black smoke at low idle and smokes badly as idle speed increases. Spark plugs soot up quickly and misfire	a. Excessive pressure on float needle valve b. Float leaking c. Float needle valve not closing	a. Check fuel pump pressure and reduce if necessary. See **4.2**. b. Replace float. See **5.3**. c. Clean or replace float needle valve. See **5.3**.
9. Engine runs unevenly at full throttle, misfires, and either cuts out or lacks power (ignition dwell, spark advance, and spark plugs all right)	a. Fuel starvation at driving speeds b. Low fuel level in float bowl c. Low fuel pressure d. Dirt in fuel system	a. Clean main jet. b. Clean float needle valve. c. Check fuel pump pressure and increase if necessary. d. Clean fuel tank. Flush lines. Clean carburetor and fuel pump.
10. Fuel consumption excessive	a. Jet sizes not properly matched b. Excessive pressure at float needle valve c. Float leaking d. Float needle valve not closing e. Automatic choke not working properly	a. Install correct set of jets. Check spark plugs for soot fouling. b. Check fuel pump pressure and reduce if necessary. c. Replace float. d. Clean needle valve or replace float. e. Apply penetrating fluid to free choke valve shaft.

© 1974 VWoA

10 Fuel System

5.2 Adjusting Carburetor

Because even minor carburetor adjustment changes will affect exhaust emissions, please follow the instructions carefully and use the specified equipment. Professional mechanics should find out whether state authorization is required before a repair shop can make adjustments that influence exhaust emissions.

During routine maintenance, you should adjust the idle speed only. Do this by turning the bypass screw, as described in the last step in each of the following procedures. Do not adjust the mixture by turning the volume control screw unless (1) you have installed a different carburetor, (2) you have removed, repaired, or rebuilt the carburetor, or (3) the engine is producing excessive emissions.

In troubleshooting the engine, eliminate all other possible trouble sources before you touch the carburetor adjustments (Fig. 5-3). Also, adjusting the idle rpm should be the last step in a tune-up. Otherwise, valve and ignition adjustments will upset the previously-made idle adjustment.

CAUTION
If you lack the skills, tools, or test equipment for adjusting the carburetor, we suggest you leave such repairs to an Authorized VW Dealer or other qualified shop. We especially urge you to consult your Authorized VW Dealer before attempting repairs on a car still covered by the new-car warranty.

1. Throttle valve adjustment
2. Volume control screw
3. Bypass screw

Fig. 5-3. Adjustments that influence idle speed.

To adjust idle (30 PICT-3):

1. Clean and regap or replace the spark plugs as necessary. Make certain that valve clearances, ignition dwell angle, and ignition timing are correct.

2. Connect a dwell meter/tachometer to the ignition system. Replace the oil dipstick with a thermometer to measure the oil temperature.

3. Start the engine and run it until the oil temperature reaches 50° to 70°C (122° to 158°F). Check to make sure that the automatic choke is fully open.

4. Adjust the idle speed by turning the bypass screw. The idle speed for cars with manual transmissions should be adjusted to 800 to 900 rpm. The idle speed for cars with Automatic Stick Shift should be adjusted to 900 to 1000 rpm.

CAUTION
Do not adjust the idle speed on the 30 PICT-3 carburetor by turning the throttle valve adjustment. Doing so would adversely affect exhaust emissions.

To adjust idle (34 PICT-3):

1. Clean and regap or replace the spark plugs as necessary. Make certain that valve clearances, ignition dwell angle, and ignition timing are correct.

2. Replace the oil dipstick with a thermometer to measure the oil temperature.

3. Start the engine and run it until the oil temperature reaches 50° to 70°C (122° to 158°F). Check to make sure that the automatic choke if fully open, then stop the engine.

4. Turn the throttle valve adjustment out until there is clearance between its tip and the fast idle cam. Then turn the throttle valve adjusting screw in until it just touches the fast idle cam.

5. From this position, turn the throttle valve adjusting screw in an additional one-quarter turn.

6. Slowly turn the volume control screw in until it comes to a stop, and then turn it back 2½ to 3 complete turns.

7. Connect a dwell meter/tachometer to the ignition system and start the engine.

8. By turning the bypass screw, adjust the idle speed to 800 to 900 rpm.

9. By turning the volume control screw, adjust to the fastest obtainable idle. Then turn the volume control screw slowly clockwise until the engine speed drops by 20 to 30 rpm.

10. By turning the bypass screw, reset the idle to 800 to 900 rpm.

NOTE —
If you are using a CO tester to check the concentration of CO in the exhaust, follow the instrument manufacturer's instructions. With the idle speed adjusted to specifications, CO should be 3% ± 1%. If it is not, turn the volume control screw to obtain a reading in that range.

Adjusting Throttle Valve Positioner

A tachometer with a range of at least 0 to 3000 rpm must be installed to adjust the throttle valve positioner. Start the engine. Then check the fast idle speed by pulling the fast idle lever back against the adjusting screw in the fast idle lever stop as shown in Fig. 5-4.

Fig. 5-4. Fast idle lever being pulled back with fingertip until it contacts the adjusting screw (arrow).

The fast idle should be 1550 rpm ± 100 rpm. If it is not, turn the adjusting screw in the fast idle lever stop to bring it within this range. After a warmup drive, the fast idle should not exceed 1700 rpm.

Next, check the throttle valve closing time. To do this, pull the throttle valve lever away from the fast idle lever until the engine is running at 3000 rpm. Release the lever. It should take 3.5 seconds ± 1 second for the engine to return to its normal idle. If the throttle valve closing time is not within specifications, adjust it by turning the screw on the altitude corrector (Fig. 5-5). Turning the screw clockwise will increase throttle valve closing time. Turning it counterclockwise will decrease throttle valve closing time.

After a warmup drive, the throttle valve closing time should not exceed 6 seconds. If the adjustment produces erratic results, check the condition of the hoses between the altitude corrector and the throttle valve positioner's diaphragm unit.

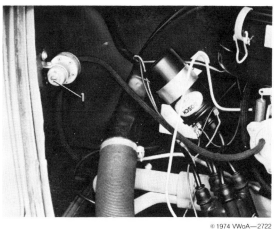

Fig. 5-5. Adjustment screw **1** on the altitude corrector.

Adjusting Dashpot

Late-model cars with manual transmissions are equipped with a dashpot. It is adjusted by loosening the two locknuts and repositioning the dashpot in its mounting bracket. Dimension **a** in Fig. 5-6 should be adjusted to 1.00 mm (.040 in.) with the dashpot's plunger fully in and the throttle fully closed on the warm running position of the fast idle cam.

Fig. 5-6. Clearance (at dimension **a**) between the plunger and throttle valve arm with the dashpot in its closed position.

5.3 Removing, Rebuilding, and Installing Carburetor

If a carburetor must be replaced, it is important that the new carburetor have the same part number as the original, or that the new carburetor be the correct replacement for the car model being serviced. Always obtain replacement carburetors and carburetor parts with reference to the carburetor part number and engine number.

12 Fuel System

To remove carburetor:

1. Remove the air cleaner as described in **LUBRICATION AND MAINTENANCE**.

2. Disconnect the vacuum hose(s) and fuel hose from the carburetor.

   WARNING—
   Disconnect the battery ground strap. Do not smoke or work near heaters or other fire hazards. Have a fire extinguisher handy.

3. Disconnect the wires from the automatic choke heating element and the electromagnetic cutoff valve.

4. Loosen the clamp screw in the accelerator cable pivot pin, then pull the cable out of the pin.

5. Remove the two nuts from the studs on the carburetor flange. Remove the carburetor and gasket.

Removing and Installing Throttle Valve Positioner

On cars equipped with a throttle valve positioner, the positioner's diaphragm unit will come off along with the carburetor. Disconnect the pull rod from the carburetor's fast idle lever by prying off the retaining pin for the pull rod clevis and not by unscrewing the pull rod from the clevis or the diaphragm. This will avoid the need for adjusting the pull rod when the unit is reinstalled. If any part of the throttle valve positioner is replaced, installation should be followed by the adjustments described in **7.4 Throttle Valve Positioner** and in **5.2 Adjusting Carburetor**.

Disassembling and Assembling Carburetor

The 30 PICT-3 carburetor is shown disassembled in Fig. 5-7. Disassembly of the 34 PICT-3 carburetor is illustrated in Fig. 5-8.

Fig. 5-7. Disassembled 30 PICT-3 carburetor.

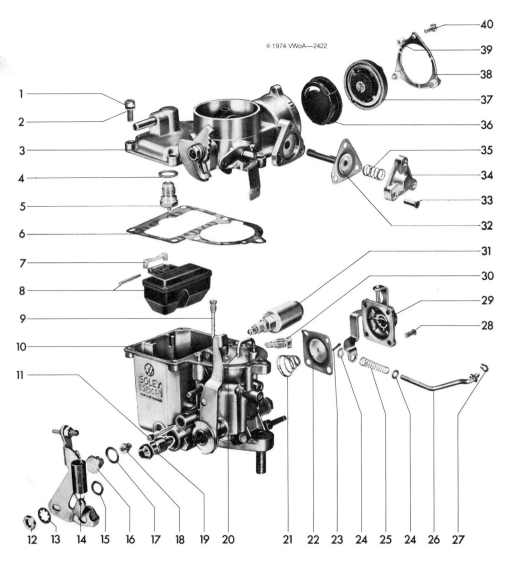

1. Fillister head screw
2. Spring washer
3. Carburetor upper part
4. Float valve washer
5. Float valve
6. Gasket
7. Float pin retainer
8. Float with pivot pin
9. Air correction jet with emulsion tube
10. Carburetor body
11. Volume control screw
12. Nut
13. Lockwasher
14. Throttle return spring
15. Spring washer
16. Main jet cover plug
17. Main jet cover plug seal
18. Main jet
19. Bypass screw
20. Accelerator pump injector
21. Pump diaphragm spring
22. Pump diaphragm
23. Cotter pin
24. 1-mm (.040 in.) thick washer
25. Connecting rod spring
26. Connecting link
27. Clip
28. Screw
29. Pump cover
30. Pilot jet
31. Electromagnetic cutoff valve
32. Vacuum diaphragm
33. Oval head screw
34. Vacuum diaphragm cover
35. Vacuum diaphragm spring
36. Plastic cap
37. Choke heating element
38. Cover retaining ring
39. Retaining ring spacer
40. Small fillister head screw

Fuel System

As can be seen by comparing Fig. 5-7 and Fig. 5-8, the parts of the 34 PICT-3 carburetor are quite similar to those of the earlier unit. Jet sizes and other specifications for the two units can be determined from the table appearing in **8. Fuel System Technical Data**. The following basic disassembly sequence applies to both carburetors:

1. Remove the carburetor upper part from the carburetor body.
2. Remove the various jets and adjustment screws from the carburetor body. Remove the float valve from the carburetor upper part.
3. Disassemble the accelerator pump and linkage.
4. Disassemble the throttle valve shaft assembly.
5. Disassemble the automatic choke assembly.

Assembly is the reverse of disassembly. Obtain a rebuilding kit and install the components it contains. With the exception of the choke heating element, the pump diaphragm, the float, and the vacuum diaphragm, wash all old parts in lacquer thinner, acetone, or a commercial carburetor cleaner.

WARNING —
Do not smoke or work near heaters or other fire hazards. The cleaning agents are highly flammable.

Blow out all jets, valves, and drillings with compressed air. Do not clean them with pins or pieces of wire, which could upset precise calibration of these parts.

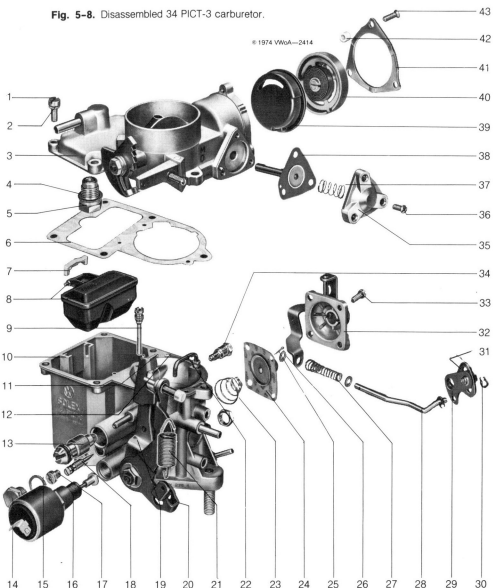

Fig. 5-8. Disassembled 34 PICT-3 carburetor.

1. Fillister head screw
2. Spring washer
3. Carburetor upper part
4. Float valve washer
5. Float valve
6. Gasket
7. Float pin retainer
8. Float with pivot pin
9. Air correction jet with emulsion tube
10. Carburetor body
11. Pilot air drilling
12. Auxiliary air drilling
13. Bypass screw
14. Main jet cover plug
15. Main jet cover plug seal
16. Electromagnetic cutoff valve
17. Main jet
18. Volume control screw
19. Fast idle lever
20. Throttle valve lever
21. Throttle return spring
22. Accelerator pump injector
23. Pump diaphragm spring
24. Pump diaphragm
25. Cotter pin
26. 1-mm (.040 in.) thick washer
27. Connecting rod spring
28. Connecting link
29. Adjustable bellcrank
30. Circlip
31. Adjusting segment
32. Pump cover
33. Screw
34. Pilot jet
35. Vacuum diaphragm cover
36. Oval head screw
37. Vacuum diaphragm spring
38. Vacuum diaphragm
39. Plastic cap
40. Choke heating element
41. Cover retaining ring
42. Retaining ring spacer
43. Small fillister head screw

14 Fuel System

Install the choke heating element so that its mark is in line with the middle mark on the housing (Fig. 5-9).

Fig. 5-9. Distance **a** from upper to middle mark. The element mark on some 1970 carburetors is at the upper housing mark, as shown, rather than at the middle mark. Check before disassembly.

The body joint gasket must be the correct one. Those for the 30 PICT-3 are marked with yellow stripes. Those for the 34 PICT-3 are marked with black stripes as indicated by the arrows in Fig. 5-10.

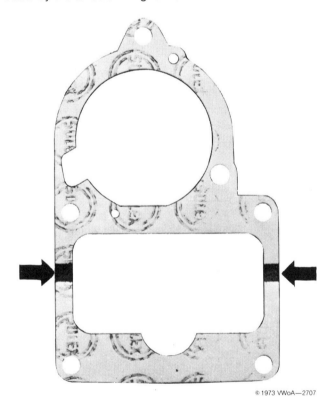

Fig. 5-10. Carburetor gasket identification stripes.

When assembling the accelerator pump linkage on a 30 PICT-3 carburetor, install the cotter pin in its original hole in the connecting link. On carburetors of either type, the opposite end of the connecting link should have at least 0.30 to 0.50 mm (.012 to .020 in.) of axial play in the bellcrank.

Check the float valve for binding and leakage. It should not be possible to blow air through the valve while the needle is pressed lightly onto its seat. To check the float for leaks, immerse it in hot water. If bubbles appear, replace the float. Also make certain that an 8.5 g float is installed and that the washer used under the float valve is the correct thickness. Select the washer as described in the procedure that follows.

Adjusting Fuel Level in Float Bowl

Either the carburetor must be level, or the car positioned on a level surface. If the carburetor is installed, idle the engine briefly to ensure that the float bowl is full. If the carburetor is not installed, fill the float bowl using a piece of hose attached to the fuel inlet pipe. Then remove the carburetor upper part and the gasket so that the fuel level can be measured as shown in Fig. 5-11.

WARNING —
Disconnect the battery ground strap. Do not smoke or work near heaters or other fire hazards. Have a fire extinguisher handy.

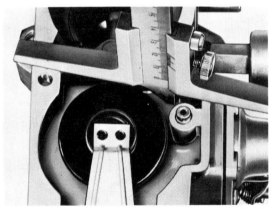

Fig. 5-11. Fuel level being measured using a depth gauge. Gauge bridge contacts the carburetor body. Slide tip just touches fuel. Carburetor is a type installed before 1970.

The distance from the top of the carburetor body to the surface of the fuel should be 19.5 ± 1.0 mm (.767 ± .040 in.). If the fuel level is too high, use a thicker washer under the float valve. If the fuel level is too low, use a thinner washer. Washers are available in thicknesses of 0.50 mm (.020 in.), 0.80 mm (.031 in.), 1.00 mm (.040 in.), and 1.50 mm (.059 in.).

Adjusting Accelerator Pump

The accelerator pump's injection quantity is adjustable. On 30 PICT-3 carburetors, the adjustment is made by installing the cotter pin through a different hole in the connecting link. The 34 PICT-3 carburetors have a bellcrank with an adjusting segment (Fig. 5-12). A spring-loaded screw is used to adjust the quantity on 34 PICT-4 carburetors.

Fig. 5-12. Injection quantity adjustment on 34 PICT-3 carburetor. Adjustment is as shown on engines with alternators, or in the position indicated by dotted lines on engines with generators.

To measure the injection quantity, first make certain that the float bowl is filled with fuel. Then attach a length of hose or tubing to the discharge end of the accelerator pump injector so that the expelled gasoline can be caught and measured in a 25-cc glass graduate. Hold the glass graduate under the end of the tubing and operate the throttle valve rapidly exactly ten times. Divide the amount caught by ten to get the average quantity of a single injection pulse.

The average quantity should be 1.20 to 1.35 cc for 30 PICT-3 carburetors; the average should be 1.45 to 1.75 cc for 34 PICT-3 carburetors installed on 1971 and 1972 cars with manual transmissions, except those sold in California; and the average should be 1.3 to 1.6 cc on all other models with 34 PICT-3 carburetors. At temperatures above 23° to 24°C (73° to 75°F), the average injection quantity should be approximately 1.1 cc per stroke for the 34 PICT-4 carburetor installed on 1974 cars sold in California. At lower temperatures, the average should be 1.7 cc. The specified injection quantity for carburetors of a particular part number can be found in **8. Fuel System Technical Data**.

On 30 PICT-3 carburetors, move the cotter pin to the outer hole to decrease injection quantity. Move the cotter pin to the inner hole to increase injection quantity. On 34 PICT-3 carburetors, loosen the retaining screw and turn the adjusting segment clockwise to decrease injection quantity or counterclockwise to increase the injection quantity. Then tighten the retaining screw and recheck the injection quantity. On 34 PICT-4 carburetors, turn the spring-loaded screw to adjust the injection quantity.

Checking Electromagnetic Cutoff Valve

The electromagnetic cutoff valve can be checked while it is installed. Turn on the ignition without starting the engine, then remove the wire from the terminal on the electromagnetic cutoff valve. Touch the wire to the terminal several times. The valve should make a clicking sound each time contact is made.

The same test can be carried out with the electromagnetic cutoff valve removed from the car. Connect negative battery current to the valve's outer casing, and apply positive battery current to the terminal. On the 34 PICT-3 carburetor, you must apply slight finger pressure to the electromagnetic cutoff valve before the plunger will retract into the solenoid.

Carburetor Installation

Lightly lubricate the choke valve shaft and throttle valve shaft with engine oil and the external linkage with molybdenum grease. Using a new gasket, install the carburetor on the intake manifold, then torque the nuts to 2.0 mkg (14 ft. lb.). Secure the fuel hose with a new hose clip. If new parts have been installed on the throttle positioner's operating diaphragm unit (where fitted), adjust the pull rod as described in **7.4 Throttle Valve Positioner**.

Have someone hold the accelerator pedal to the floor while you connect the accelerator cable. Install the cable end so that gap **a** indicated in Fig. 5-13 is 1.00 mm (.040 in.). Then adjust the carburetor as described in **5.2 Adjusting Carburetor**.

Fig. 5-13. Full-throttle clearance between throttle valve lever and the stop on the carburetor body.

16 Fuel System

6. Air Cleaner

Several different air cleaner designs have been used on the cars covered by this Manual. The intake air preheating flap on 1970 models is regulated by a cable linked to the engine's cooling air system thermostat. On 1971 models, the flap is controlled by a separate thermostat built into the air cleaner. On 1972 and later cars, the flap is regulated by a thermostatically controlled vacuum unit, making the intake air preheating system responsive to both temperature and engine load.

Cars with oil bath air cleaners also have a weighted control flap. It is important that you check the weighted arm for free movement after you have installed the air cleaner. If the arm is blocked by a carelessly installed hose or by contact with the engine's cooling air fan housing, oil may be drawn from the crankcase into the air cleaner. Removing, installing, and servicing the air cleaner are described in **LUBRICATION AND MAINTENANCE**.

Checking and Adjusting 1970 Intake Air Preheating System

Adjust the intake air preheating control cable after the air cleaner has been installed and also before the carburetor is adjusted when performing a tune-up.

To install and adjust:

1. Push the cable housing as far as it will go into the retainer on the air cleaner intake. Then tighten the clamp screw in the retainer.

2. Place the cable eye over the cranked end on the flap shaft, then secure it with the steel clip.

3. With the engine cold, check to see that the coil spring portion of the cable is slightly compressed and that the flap is completely closed. If they are not, replace the cable.

Checking and Adjusting 1971 Intake Air Preheating System

The thermostat locations for the sedan and the Karmann Ghia are shown in Fig. 6-1 and Fig. 6-2.

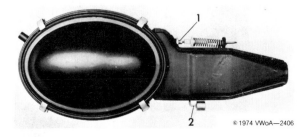

Fig. 6-1. Sedan air cleaner (1971). Thermostat is at **1**. Weighted control flap at **2**.

Fig. 6-2. Karmann Ghia air cleaner (1971). Thermostat is at **1**.

To check the thermostat, temporarily remove the activated charcoal filter hose from the air cleaner and, install a thermometer and rubber stopper in its place. The intake air preheating flap should open between 27.5° and 32.5° C (81.5° and 90.5°F). Minor adjustments can be made by bending the looped portion of the preheating flap arm.

Checking and Adjusting 1972 and Later Intake Air Preheating System

The temperature-sensitive and load-sensitive intake air preheating flap is shown schematically in Fig. 6-3.

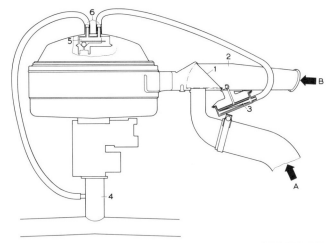

1. Control flap
2. Air intake
3. Vacuum unit
4. Intake manifold
5. Thermostatic valve
6. Vacuum hose connections

Fig. 6-3. Air cleaner introduced with the 1972 models. Cool air enters at arrow **B**. Preheated air enters through hose at arrow **A**.

The thermostatic valve installed after December 1972 differs from the thermostatic valve used on early 1972 models. It can be identified by a brass hose connection

FUEL SYSTEM

for the vacuum unit. On the early valve, both connections are black plastic. The later valve keeps the flap closed to cool air regardless of engine load during the warm-up period. The early valve allows the flap to admit cool air under heavy engine loads. If the thermostatic valve or vacuum unit is seen to be defective, it should be replaced.

7. Checking, Repairing, And Adjusting Emission Controls

The VW engine will not run properly unless the emission controls are kept in proper working order. Be especially careful not to mix up the emission control hoses when you install them. Doing so will adversely affect both performance and fuel economy.

Although all models covered by this Manual are equipped with an efficient closed positive crankcase ventilation (PCV) system, it should be noted by those who are not familiar with the VW engine that the system does not include a PCV valve—a fixture that is common on other makes. This is because the design of the VW engine limits crankcase pressure fluctuations and introduces almost no oil vapor into the PCV hose that leads to the air cleaner.

7.1 Exhaust Emission Control Troubleshooting

Table c lists emission control trouble symptoms along with their probable causes and remedies. The numbers in bold type in the Remedy column refer to headings in this section where repairs are described.

7.2 Evaporative Emission Control

The activated charcoal filter canister must be replaced after 48,000 mi. (80,000 km) of service. It is located in the engine compartment on Karmann Ghias (Fig 7-1) and under the right rear fender on sedans (Fig. 7-2).

F. Release clip
H. Weighted control flap on air cleaner
C. Air hose
B. Warm air hose to air cleaner

Fig. 7-1. Activated charcoal filter canister (foreground) and nearby parts on the Karmann Ghia air cleaner.

To replace the canister, remove the hoses that are connected to it, then take out the Phillips head screw in the canister mounting bracket. Note the positions of the hose installations, so the hoses will be installed on the new canister in their correct positions.

Table c. Exhaust Emission Control Troubleshooting

Symptom	Probable Cause	Remedy
1. Idle poor, resists adjustment	Dirt in idle system	Clean carburetor, then adjust idle as prescribed.
2. Idle too fast	a. Throttle valve sticking b. Throttle valve positioner out of adjustment c. Throttle valve positioner can not be adjusted	a. Free throttle valve lever and pull rod. A bent pull rod must be replaced. See **7.4**. b. Adjust throttle valve positioner. See **5.2**. c. Replace throttle valve positioner. See **7.4**.
3. Backfiring when car is coasting	Throttle valve positioner out of adjustment	Adjust throttle valve positioner or replace it. See **5.2**, **7.4**. In extreme cases, set cut-in speed to 1900 rpm max.
4. Engine stalls at idle (1972 cars sold with Automatic Stick Shift in California, all 1973 cars with Automatic Stick Shift, and 1973 cars sold in California with manual transmission)	a. Vacuum hose between exhaust gas recirculation (EGR) valve and base of intake manifold cracked, loose, or off b. Faulty EGR valve	a. Replace hose. b. Check EGR valve and replace if necessary. See **7.3**.

18 Fuel System

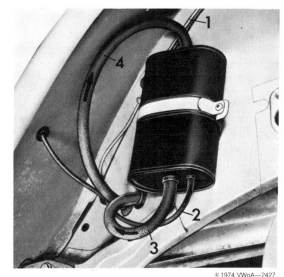

1. Steel ventilation line from tank
2. Plastic tube from steel line to canister
3. Exit hose to air cleaner
4. Air hose from engine fan

Fig. 7-2. Activated charcoal filter canister (sedan).

If the engine has been removed and installed, it is important that the hoses from the activated charcoal filter canister be reinstalled as shown in Fig. 7-3. Replace cut, cracked, or loose-fitting hoses. Otherwise, gasoline vapors may escape into the engine compartment.

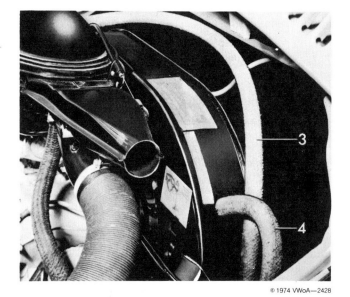

Fig. 7-3. Hoses to engine. Hose **3** brings vapors from the exit end of the canister (the end that also receives the line from the tank). Hose **4** carries air to the top of the canister on sedans, and to the left end of the canister on Karmann Ghias.

7.3 Exhaust Gas Recirculation

Fig. 7-4 is a drawing of the exhaust gas recirculation (EGR) system as installed on 1973 cars with Automatic Stick Shift transmissions and on all 1974 cars. Exhaust gas is drawn from the exhaust flange of Number 4 cylinder into a filter that cools the gas and removes particulates. The introduction of exhaust gas into the intake manifold is controlled by a vacuum operated exhaust gas recirculation valve.

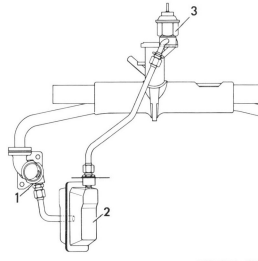

1. Exhaust flange of Number 4 cylinder
2. Element-type filter
3. Exhaust gas recirculation valve

Fig. 7-4. Exhaust gas recirculation system.

The 1972 cars sold in California with Automatic Stick Shift also have EGR, but with the filter shown in Fig. 7-5.

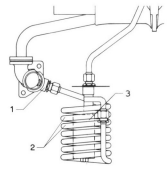

1. Exhaust flange of Number 4 cylinder
2. Cooling coil
3. Cyclone-type filter

Fig. 7-5. The 1972 California EGR filtering system.

Fuel System 19

The operation of the exhaust gas recirculation valve should be checked every 24,000 mi. (40,000 km) at which time the cyclone filter is removed and cleaned or the element-type filter is replaced. It should also be checked whenever trouble is suspected in the system.

NOTE —
For easier servicing, replace the cyclone filter and cooling coil on 1972 California cars with the later element-type filter.

To check the EGR valve, run the engine at an idle. Pull the vacuum hose off the EGR valve. Then, in its place, push on the black hose from the intake air preheating thermostat. Stalling, or a noticeable drop in engine speed indicates that the valve is working properly. If the idle speed does not change, the valve is faulty or the hose is blocked. The 2-stage valve used on 1974 cars sold in California has a visible pin. Instead of using the preceding test on 1974 cars, simply note whether the pin moves in and out normally according to engine speed.

To check the throttle valve switch used on 1974 California cars with Automatic Stick Shift, run the engine at an idle and operate the switch by hand. The engine should stall or noticeably slow down. If not, there is a fault in the switch, 2-stage EGR valve, or in the wiring. The switch should be adjusted to produce an audible click at 12° and at 72° of throttle valve shaft rotation.

7.4 Throttle Valve Positioner

Individual parts for the throttle valve positioner can be replaced on cars equipped with this emission control device. Fig. 7-6 is an exploded view that shows the components of the diaphragm unit and the altitude corrector. The gasket should be replaced whenever the altitude corrector is reassembled.

Fig. 7-6. Exploded view of the two main parts of the throttle valve positioner. Parts 1 and 2 are mounted on a bracket on the carburetor.

When a new diaphragm unit is installed, you must adjust the pull rod length after the diaphragm unit has been installed on the carburetor. To adjust, loosen the locknuts at each end of the pull rod. One end of the pull rod has right-hand threads. The other end has left-hand threads. The effective length of the pull rod can be adjusted by turning the rod in one direction or the other. Adjust the length until the fast idle lever does not touch either the carburetor body or the throttle valve lever when the throttle valve is closed (Fig. 7-7).

Fig. 7-7. Turning pull rod with fingertips to adjust pull rod's effective length. Arrows indicate space between fast idle lever and carburetor body and between fast idle lever and throttle valve lever.

After adjusting the pull rod length, tighten the two locknuts to hold it in place. The throttle valve positioner should then be adjusted with the engine running as described in **5.2 Adjusting Carburetor**. However, it is usually best to perform a tune-up before carrying out fine throttle valve positioner adjustments.

1. Diaphragm unit
2. Diaphragm unit retaining ring
3. Control part cover
4. Gasket
5. Control part
6. Control diaphragm
7. Plastic foam filter
8. Altitude corrector bellows
9. Altitude corrector housing
10. Phillips screw (6)
11. Setscrew
12. Mounting clamp
13. Rubber mounting (2)

20 Fuel System

8. Fuel System Technical Data

The specifications for the various carburetors and fuel pumps installed on cars covered by this Manual are given here in table form. It is important that the correct carburetor jet sizes be used when making replacements.

When obtaining replacement parts, always have the engine number and carburetor part number with you so that the parts department can locate the correct jet or rebuild kit in the VW parts list. The carburetor type designation alone is not sufficient for identifying the correct replacement part.

III. 34 PICT-4 Carburetor Part Numbers

Manual transmission (California only)	113 129 033 E
Automatic Stick Shift (California only)	113 129 033 F

I. Carburetor Settings and Jets

Car model	1970	1971-1972 except Calif.	1972 Calif.	1972-1973 Calif.	1973-1974 except Calif. 1972 Calif. Auto. only	1974 Calif.
Carburetor type	30 PICT-3	34 PICT-3	34 PICT-3	34 PICT-3	34 PICT-3	34 PICT-4
Part no.	113 129 029 D Manual*	113 129 029 S Manual	113 129 031 K Manual	113 129 031 N Manual/Auto.	113 129 029 T Manual/Auto.	113 129 033 E Manual
	113 129 029 E Auto.*	141 129 029 T Manual	141 129 031 G Manual	141 129 031 K Manual/Auto.	141 129 029 T Manual/Auto.	113 129 033 F Auto.
Venturi (mm diameter)	24	26	26	26	26	26
Main jet	x 112.5	x 130	x 127.5/ x 130	x 112.5	x 127.5/ x 127.5	x 127.5
Air correction jet	125 Z (Manual) 140 Z (Auto.)	75 Z/80 Z	75 Z/80 Z	75 Z/70 Z	75 Z/80 Z	75 Z (Manual) 70 Z (Auto.)
Pilot jet	65	g60	g55	g55	g55	g55
Pilot jet air bleed	135	147.5	147.5	147.5	147.5	147.5
Auxiliary fuel jet	45	47.5	42.5	42.5	42.5	42.5
Auxiliary air jet	130	90	90	90	90	90
Relief drilling (mm)	1.8	—	1.2	1.2	1.2	1.2
Power fuel jet	100	100	100	100	100	100
Float needle valve (mm diameter)	1.5	1.5	1.5	1.5	1.5	1.5
Washer under float needle valve (mm)	1.5	0.5	0.5	0.5	0.5	0.5
Float weight	8.5	8.5	8.5	8.5	8.5	8.5
Pump injection quantity (cc/stroke)	1.05-1.35	1.45-1.75	1.3-1.6	1.3-1.6	1.3-1.6	1.1 or 1.7

* Manual = Manual transmission; Auto. = Automatic Stick Shift

II. Fuel Pump

From engine No.	Part No.	Minimum delivery capacity	rpm	Maximum delivery pressure
B 6 000 001	113 127 025 A	400 cm^3/min.	3400	3.5 psi (0.25 kg/cm^2)
AE 0 000 001 AH 0 000 001 AK 0 000 001	113 127 025 C (D)	400 cm^3/min.	4000	3.0–5.0 psi (0.21-0.35 kg/cm^2)

© 1974 VWoA

Section 3

TRANSMISSION AND REAR AXLE

Contents

Introduction 3	6. **Disassembling Rear Axle** 14
	6.1 Removing and Installing Differential in Case 16
1. **General Description** 4	
Transmission Case 4	7. **Reverse Gears** 19
Clutch Release Mechanism 4	7.1 Removing and Installing Reverse Gears and Rear Driveshaft 19
Transmission Gears and Synchronizers 4	
Final Drive Gears 4	8. **Differential Servicing** 19
Rear Axle and Suspension 4	8.1 Disassembling and Assembling Differential 20
2. **Maintenance** 4	8.2 Removing and Installing Tapered-Roller Bearings 22
3. **Rear Axle Repairs (engine and transmission in car)** 5	9. **Transmission** 24
3.1 Removing and Installing Driveshafts 5	9.1 Disassembling and Assembling Shift Housing 24
3.2 Removing and Installing Constant Velocity Joints 6	9.2 Removing and Installing Transmission Gears in Case 25
3.3 Servicing Constant Velocity Joints 8	9.3 Disassembling and Assembling Gear Carrier 28
3.4 Removing and Installing Gearshift Lever and Shift Rod 10	1973 Changes 28
	1976 Changes 28
4. **Transmission Repairs (transmission in car, engine removed)** 11	Repair Procedures 29
4.1 Replacing Rear Driveshaft Oil Seal 11	9.4 Adjusting Shift Forks 32
4.2 Removing and Installing Starter Bushing 12	9.5 Removing and Installing Reverse Shaft and Needle Bearings 34
4.3 Removing and Installing Clutch Operating Shaft 12	9.6 Disassembling and Assembling Main Drive Shaft 35
	1972 Changes 35
5. **Removing and Installing Transmission and Rear Axle (engine removed)** 13	1976 Changes 36
	Repair Procedures 36

©1978 VWoA

2 Transmission and Rear Axle

9.7 Disassembling and Assembling Drive
    Pinion 39
    1970 Changes 39
    1976 Changes 39
    Repair Procedures 39

10. **Adjusting Final Drive** 43
    10.1 Adjusting Drive Pinion 46
         Determining S_3 Thickness 48
    10.2 Adjusting Ring Gear 50

11. **Rear Suspension** 54
    11.1 Removing and Installing Shock
         Absorbers 54
    11.2 Removing and Installing Torsion Bars
         and Spring Plates 54
    11.3 Spring Plate Adjustment 57
    11.4 Disassembling and Assembling
         Diagonal Arm 58

12. **Rear Wheel Alignment** 60
    12.1 Adjusting Rear Wheel Alignment ... 61

13. **Technical Data** 62
    I. Constant Velocity Joints 62
    II. Double-Jointed Axle Driveshaft 62
    III. General Data 62
    IV. Tightening Torques 62
    V. Transmission Rebuilding Data 63
    VI. Ratios 63
    VII. Torsion Bar Adjustment 64

TABLES

a. Differential Spacer Sleeve Lengths 22
b. 3rd-gear Circlip Sizes 42
c. Necessary Adjustments for Replaced Parts 45
d. Symbols Used in Ring Gear/Pinion
   Adjustment 45
e. S_3 Shim Measurements 48
f. Available S_3 Shim Sizes 49
g. S_1 Shim Thicknesses 53
h. S_2 Shim Sizes 53
i. S_1 and S_2 Shim Combinations 53

Transmission And Rear Axle

The VW four-speed manual transmission, standard equipment on all Type 1 vehicles, is fully synchronized in all forward gears. A lightweight alloy case houses both the transmission and the final drive gearsets. The transmission receives power from the engine by means of a single-plate clutch and then transmits it to the rear wheels through double-jointed driveshafts. Many parts for the manual transmission are interchangeable with those of the Automatic Stick Shift Transmission.

The information in this section should be of interest to nearly every VW Type 1 owner, even though much of the repair data may be of practical value only to the professional mechanic. A general knowledge of the hows and whys of transmission operation and repair can often be helpful when troubleshooting. It can also lead to better understanding between you and your VW Service Advisor or mechanic.

Many of the repair procedures outlined in this section require tools and skills that only a trained VW mechanic or transmission specialist is likely to have. So if you do not feel completely confident in these areas, we recommend that such repairs be left to an Authorized VW Dealer or other qualified repair facility. We especially urge you to consult an Authorized VW Dealer before attempting any repairs on a car still covered by the new-car warranty.

Though the home mechanic often lacks the tools and experience necessary to actually repair his transmission, he may be perfectly capable of removing and installing the unit. If possible, deliver the transmission to an Authorized VW Dealer or other qualified repair shop after giving it a thorough exterior cleaning but without doing any disassembly work. A partially disassembled transmission in a "basket" is a mechanic's nightmare, so the car owner is not likely to be greeted with sympathy after trying to tackle a job over his head and then giving up. Also, on some transmissions several critical measurements must be made before disassembly.

Cleanliness and a careful approach are imperative when repairing the transmission. Familiarizing yourself with procedures, notes, cautions, and warnings before beginning work is good insurance. Clean and lay out the parts. When necessary, mark them to show their proper assembly order. We also suggest you make sure your tools are of the specified sizes and that you give preference to metric tools. Improper tools often damage fasteners. With appropriate care, the job should go smoothly.

4 Transmission and Rear Axle

1. General Description

Forward gears and reverse are selected with the gearshift lever which is located between the front seats. The conventional H-shift pattern shown in Fig. 1-1 is used. To engage reverse, press down on the lever to overcome a lockout detent, then move it all the way to the left and back toward you.

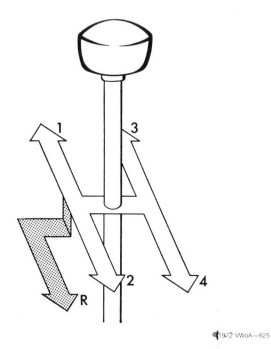

Fig. 1-1. Manual transmission gearshift pattern. Depress the clutch pedal all the way when shifting. Do not move the gearshift diagonally as it can damage the shift linkage.

Transmission Case

The transmission case, which contains the transmission and final drive gearsets, is a one-piece die casting of light magnesium alloy. It is secured to the car by three rubber mountings. An integral bellhousing at the rear of the case houses the clutch and clutch release mechanism. The forward end of the case is closed off by the gear carrier, which holds the selector shafts and front bearings for the transmission shafts. A partition near the midpoint of the case separates the transmission from the final drive. The partition also houses the rear bearings for both the drive and pinion shafts. There are openings for the axle flanges on either side of the final drive portion of the case.

Clutch Release Mechanism

The clutch release mechanism consists of the clutch release bearing, the operating shaft, and—on 1971 and later models—the guide sleeve. The guide sleeve is bolted to the rear part of the transmission case at the point where the clutch operating shaft is supported by two bushings pressed into the case. In operation, the clutch cable pulls the lever on the end of the operating shaft, which in turn forces the release bearing against the appropriate parts of the clutch pressure-plate assembly. The guide sleeve installed on 1971 and later models aids in aligning the release bearing and pressure plate.

Transmission Gears and Synchronizers

The gears and synchronizer units are located on the main drive shaft and the pinion shaft near the front of the transmission case. The gears are in constant mesh to provide quiet operation and long service. Balk-ring type synchronizers on the main drive shaft (3rd and 4th gears) and on the pinion shaft (1st and 2nd gears) match gear wheel speed and permit smooth and clashless shifting.

Final Drive Gears

A helically cut drive pinion transmits power from the engine to the rear wheels via a ring gear and a differential gearset. The pinion and ring gear are in precise alignment for quiet operation and durability. The differential gearset allows the rear wheels to turn at different speeds, which is necessary for making turns (the outside wheel must travel farther than the inside wheel in the same amount of time).

Rear Axle and Suspension

Each rear axle is independently sprung by means of torsion bars, trailing links (spring plates), and hydraulic shock absorbers. The wheels are driven by double-jointed axles that have two constant velocity joints on each shaft.

> **WARNING —**
> Do not re-use fasteners such as nuts, bolts, washers, self-locking nuts, circlips, cotter pins that are worn or deformed in normal use. Always replace.

2. Maintenance

Regular transmission maintenance at specified intervals will extend the life of moving parts and reduce the possibility of component malfunction. Complete details on lubricants and necessary checks are given in **LUBRICATION AND MAINTENANCE**. The following procedures are covered there:

1. Checking transmission and final drive leaks
2. Checking constant velocity joint screws
3. Checking constant velocity joint seals
4. Changing transmission and final drive hypoid oil
5. Lubricating rear wheel bearings

TRANSMISSION AND REAR AXLE

3. REAR AXLE REPAIRS

(engine and transmission in car)

The following repair jobs can be performed without removing the engine or the transmission from the car. Cleanliness and order are essential because seals and bearings can be ruined by even the most minute particles of dirt and road grime.

Remember that you are working with driveline components that are subject to great stress during operation. Therefore, caution and attentiveness to details are vital to proper servicing.

3.1 Removing and Installing Driveshafts

Driveshafts of different lengths are installed according to the transmission type. A code number is stamped in the end of the shaft to denote dimension **b**. A second color-coded number stamped on the side of the shaft denotes dimension **a**, (Fig. 3.1). Dimension **a** can vary on shafts with the same dimension **b**, depending upon the location of two dished washers.

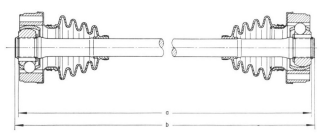

Fig. 3-2. Removing socket head screws (transmission end). When removing the driveshaft, also check the condition of the rubber boots.

Fig. 3-1. Driveshaft dimensions. On Type 1 cars with manual transmission, dimension **a** is 405.3 mm (15.957 in.) and dimension **b** is 415.5 mm (16.358 in.).

Should it be necessary to replace a driveshaft, it is imperative that both dimension **a** and dimension **b** of the replacement shaft match those of the original. Shafts used on all Type 1 cars (Automatic Stick Shift or manual transmission) are identical.

To remove shaft:

1. Engage the parking brake and remove the socket head screws at the transmission end of the driveshaft, as shown in Fig. 3-2.

    CAUTION ——
    Do not remove twelve-point driveshaft screws, if installed, with a hex-key as it may damage the screws. Use a twelve-point driver.

2. Remove the socket head screws at the wheel end of the driveshaft as shown in Fig. 3-3. Now remove the driveshaft.

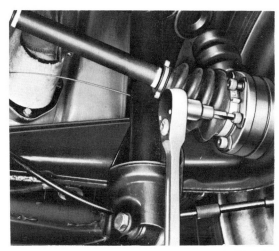

Fig. 3-3. Removing socket head screws (wheel end).

To install, simply reverse the removal procedure. Insert the socket head screws without any lockwashers. Then torque them to 3.5 mkg (25 ft. lb.). The factory-installed lockwashers have been discontinued and are unnecessary. Turn the driveshaft to make certain that the protruding ends of the socket head screws do not contact the transmission case.

If there is contact, replace the screws with shorter ones (available as VW part No. 113 501 229D). The newer socket head screws, shorter by 1 mm, are standard equipment on cars produced since July 1972.

6 Transmission and Rear Axle

3.2 Removing and Installing Constant Velocity Joints

The double-jointed rear axle with constant velocity joints was introduced on Type 1 cars with manual transmissions in 1969. Fig. 3-4 is an exploded view of the driveshaft and constant velocity joint components.

WARNING —
Do not re-use fasteners that are worn or deformed in normal use. They may fail when used a second time. This includes nuts, bolts, washers, self-locking nuts or bolts, circlips, cotter pins. Always replace.

CAUTION —
After removing the protective cap, do not tilt the ball hub more than 20° in the joint outer ring. If you do, the balls may fall out.

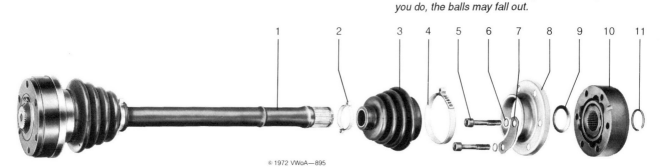

1. Driveshaft
2. Pinch clamp (2)
3. Driveshaft boot (2)
4. Hose clamp (2)
5. Socket head screw (12)
6. Lockwasher (12)
7. Plate (6)
8. Cap (2)
9. Concave washer (2)
10. Constant velocity joint (2)
11. Circlip (2)

Fig. 3-4. Driveshaft and constant velocity joint components. The hose clamps, pinch clamps, cap, and lockwashers are no longer used (see text).

To remove:

1. Remove the driveshaft.
2. Loosen the boot clamps and peel back the boot.
3. Remove the circlip from the ball hub.
4. Tap the cap off with a drift, as shown in Fig. 3-5.
5. Slide the outer ring, together with the balls, onto the ball hub.
6. While supporting the hub, press out the driveshaft and remove the concave washer (see Fig. 3-6).

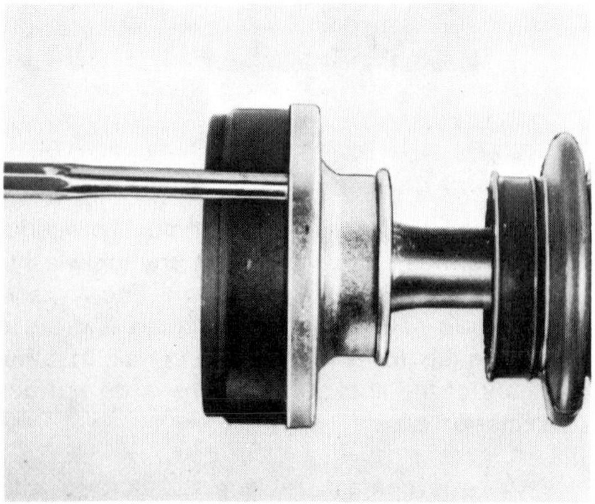

Fig. 3-6. Pressing out driveshaft. The hub must be supported by a plate during this operation.

Fig. 3-5. Removing constant velocity joint cap.

To install:

1. Examine all parts for wear or damage, and replace as necessary.

2. Install new hose and pinch clamps.

   NOTE ―
   As of July 1972, the hose clamps, pinch clamps, and separate cap and boot were discontinued. Instead, the driveshaft was given an additional ridge to secure the small end of the boot. The cap has been combined with the boot for a more effective seal. The new boot and cap combination, shown in Fig. 3-7, can be installed on earlier models. The pinch clamp must be retained unless the new-type driveshafts are installed.

Fig. 3-7. New combined boot and protective cap. Arrow indicates clampless rolled seal.

3. To protect the sealing surface of the boot, either install a guide sleeve or cover the driveshaft splines with tape. Then slide the boot onto the shaft.
4. Assemble the driveshaft, cap, and joint, as shown in Fig. 3-8.

Fig. 3-8. Driveshaft, cap, and joint assembly. The wider portion of the joint (arrow) must face the boot.

5. Position the concave washer. Then press the constant velocity joint onto the shaft as shown in Fig. 3-9.

Fig. 3-9. Constant velocity joint being pressed onto driveshaft. Note beam supporting lower end of driveshaft.

6. Install a new circlip.
7. Using pliers, squeeze the circlip until it is fully seated all around its groove.
8. Repack the joint with 90 g (3.2 oz.) of molybdenum grease.

   NOTE ―
   Pack two-thirds of the grease between the joint, cap, and boot. Pack the remaining third into the open front of the joint.

9. Position both clamps and tighten them. If the new-type driveshaft is installed, make certain the small end of the boot is firmly seated between the ridges.

   NOTE ―
   When seating the clamps on the boot, take care to position the clamp projections so that they do not interfere with the removal and installation of the driveshaft socket head screws.

10. Hand-squeeze the boot tightly. This ensures that grease is also pressed into the joint from the rear.
11. Install the driveshaft, then tighten the socket head screws to 3.5 mkg (25 ft. lb.).

8 Transmission and Rear Axle

3.3 Servicing Constant Velocity Joints

The following procedure is included solely to provide a method of cleaning and examining the constant velocity joints. The components of each joint are factory-matched—as a group—for mated tolerances. Therefore, it is not possible to repair the joint or replace individual parts.

To disassemble joint:

1. Remove the constant velocity joint from the driveshaft as previously described in **3.2 Removing and Installing Constant Velocity Joints**.

2. Using hand pressure, push the ball hub and ball cage out of the outer ring, as shown in Fig. 3-10.

Fig. 3-11. Removing ball hub from or installing in ball cage. Arrows indicate ball hub groove alignment.

Fig. 3-10. Separation of ball hub and ball cage from outer ring. Position ball hub and ball cage perpendicular to outer ring, then push in direction of arrow.

3. Push the balls out of the cage, taking care not to drop them.

   NOTE —
   The ball cage, the outer ring, and the balls themselves are selected for matching tolerances. When disassembling more than one joint, do not mix parts.

4. Inspect the outer ring, ball cage, and balls for burrs, pitting, or excessive wear. If any are in poor condition, the entire unit must be replaced.

5. Rotate the ball hub into the position illustrated in Fig. 3-11. The groove in the ball hub must be in line with the outer edge of the ball cage. The ball hub now has sufficient clearance to be tipped out of the ball cage.

6. Inspect the ball hub for excessive wear and general condition. If it is worn or damaged in any way, replace the entire joint.

To assemble the joint:

1. Clean all parts thoroughly and coat them with molybdenum grease.

2. To install the ball hub in the ball cage, insert it with the ball hub groove aligned with the outside edge of the ball cage (see Fig. 3-11).

3. Install the balls in the ball cage. If necessary, add more grease to hold the balls in place (see Fig. 3-12).

Fig. 3-12. Installing balls in the ball cage.

4. Insert the ball hub and ball cage in the outer ring. Be sure that the chamfer on the inner splines of the ball hub and the larger diameter portion of the outer ring both face the driveshaft.

Transmission and Rear Axle

NOTE —

Insert the ball hub and ball cage in the outer ring at a 90° angle as shown in Fig. 3-13. When you slip the ball hub and ball cage into the outer ring, make certain that the narrow ball hub grooves **b** and the wide outer ring grooves **a** are positioned as illustrated.

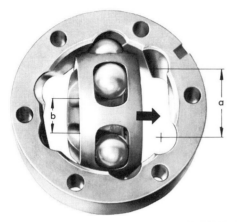

Fig. 3-13. Positioning ball hub and ball cage in outer ring. Arrow indicates direction in which the ball cage must be turned to complete the installation.

5. Hold the ball cage steady and push the ball hub in the direction shown in Fig. 3-14 (left arrow). This will align the balls with their respective grooves in the outer ring (right arrows).

Fig. 3-14. Aligning balls with grooves in outer ring. Note that the ball spacing is determined by the position of the ball hub in the ball cage.

6. When the balls are aligned with their grooves in the outer ring, firmly press the ball cage, as shown in Fig. 3-15, until it swings fully into position. Excessive pressure should not be required.

Fig. 3-15. Installing ball hub and ball cage in outer ring. Arrow indicates the point where hand pressure should be applied. Again, make certain that the chamfered splines inside the ball hub and the larger diameter portion of the outer ring are both on the same side of the assembled joint.

CAUTION —

Excessive force will result in improper joint assembly. Therefore, double-check all previous steps to make sure the assembly goes smoothly and without undue pressure. An improperly assembled joint will lock solid in operation and render the unit unserviceable.

7. Check the operation of the joint. You should be able to hand-turn the ball hub smoothly throughout its entire range of travel.

8. Pack the joint with enough additional molybdenum grease to make a total of 90 g (3.2 oz.). Two-thirds of the grease should be packed between the joint, cap, and boot. Pack the remaining third into the open joint from the front.

9. Press the constant velocity joint onto the driveshaft.

10. Install the driveshaft and tighten the socket head screws to 3.5 mkg (25 ft. lb.).

10 Transmission and Rear Axle

3.4 Removing and Installing Gearshift Lever and Shift Rod

The construction of the gearshift lever is shown in Fig. 3-16. To remove the shift lever, simply move the floor mat out of the way and remove the two M 8 × 20 bolts. When installing, be sure to place the stop plate with the tabs up and the narrow tab towards the driver seat. The bracket and stop plate can be variously positioned on their elliptical holes for best shift lever angle relative to the driver and shift rod socket. The pin on the shift lever ball must engage the slot in the shift lever socket.

To install shift rod:

1. Examine all parts and replace as necessary.

   NOTE
   If the guide sleeve must be replaced, first install the retaining ring on the guide sleeve, then press the sleeve into the bracket. Grease the sleeve following installation. The sleeve is correctly installed if the slot is toward the left side of the car. (Fig. 3-17.)

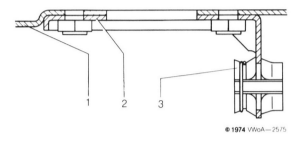

1. Frame tunnel
2. Reinforcement bracket
3. Guide sleeve

Fig. 3-17. Cross section of guide sleeve and mounting. Make certain sleeve is correctly installed in bracket.

2. Grease the entire length of the shift rod and install it through the frame tunnel.

   NOTE
   If the gearshift lever is angled, a shift rod with slotted ball socket must be used.

3. Push the shift rod to the rear through the front guide until the ball socket is in the center of the hole in the frame tunnel.

4. Install the body apron cover plate and the frame head cover plate and gasket. On Super Beetles, you must also install the deformation element.

5. Install the shift rod coupling with the spring pin and the sheet metal screw.

6. If it has been previously removed, install the coupling on the inner shift lever and secure it with the square head setscrew. Use lockwire to prevent the setscrew from loosening. Install the inspection plate.

7. Assemble and install the gearshift lever.

   NOTE
   Lubricate all parts with multipurpose grease before you install the gear shift lever assembly. If the lever rattles in the ball socket of the shift rod, you can quiet it by filling the ball socket with a very heavy grease.

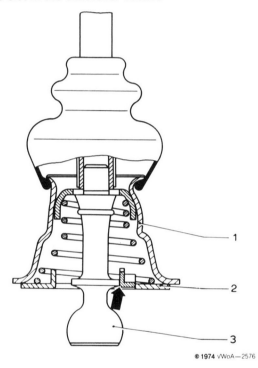

1. Bracket
2. Stop plate
3. Gearshift lever

Fig. 3-16. Gearshift lever construction.

To remove shift rod:

1. Remove gearshift lever and associated parts.

2. Working in the area under the rear seat, remove the inspection cover from the top of the frame fork and remove the shift rod coupling's hexagon screw.

3. Remove the cover plates from both the frame head and the front body apron. Working with a pair of pliers through the gearshift lever opening, push the shift rod out through the front of the car.

   NOTE
   On Super Beetles, you must also remove the deformation element in front of the frame head plate.

Transmission and Rear Axle 11

4. Transmission Repairs

(transmission in car, engine removed)

The following repairs can be performed with the engine removed and the transmission remaining in the car. Maintain order and cleanliness, as seals and bearings can be ruined by minute particles of dirt or road grime.

4.1 Replacing Rear Driveshaft Oil Seal

The rear driveshaft oil seal is located behind the clutch release bearing at the rear of the transmission case. The seal can be replaced without removing the driveshaft.

The removal and installation of the clutch release bearing is described in **ENGINE**. It is necessary to remove the clutch release bearing before prying the rear driveshaft oil seal out of its recess in the transmission case. On 1970 cars, the clutch release bearing is secured solely by the two retaining spring clips that hold it to the clutch operating shaft. To remove the clutch release bearing simply remove the spring clips and lift the release bearing off the operating shaft. On 1971 and later models, a new type clutch release bearing will be encountered. These release bearings are mounted on a central guide sleeve as shown in Fig. 4-1. This guide sleeve must also be taken off before the rear driveshaft oil seal can be pried out.

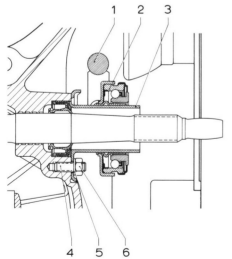

1. Operating shaft
2. Release bearing
3. Sleeve
4. Stud (M 7)
5. Lockwasher
6. Nut (M 7)

Fig. 4-1. Cutaway of clutch release bearing and guide sleeve. The sleeve improves the engagement of the release bearing with the clutch levers.

CAUTION —
Never wash a clutch release bearing in solvent. Doing so will destroy the lubricant built into the bearing during assembly at the factory. It is not possible to repack release bearings with grease.

To remove oil seal:

1. Remove the clutch release bearing.

2. If the car is equipped with a clutch release bearing guide sleeve, remove the securing nuts and the sleeve.

3. Carefully pry out the damaged oil seal as shown in Fig. 4-2.

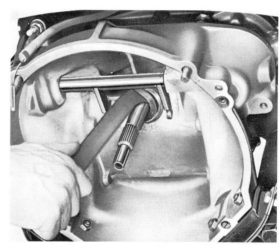

Fig. 4-2. Removing rear driveshaft oil seal. Be careful not to damage the seal bore in the transmission case.

To install oil seal:

1. Lightly coat the outer surface of the new seal with sealing compound. Lubricate the rear driveshaft and seal lip with transmission oil.

   #### NOTE —
   Before you drive the seal into the transmission case, make certain the spring is correctly positioned around the seal lip.

2. Slide the oil seal onto the rear driveshaft. Then drive it into its bore with a mallet and properly fitting sleeve.

3. Install the clutch release bearing guide sleeve, if your model is so equipped. Torque the securing nuts to 1.5 mkg (11 ft. lb.).

4. Install the clutch release bearing.

12 Transmission and Rear Axle

4.2 Removing and Installing Starter Bushing

The starter bushing is located in the rear of the transmission bellhousing. It is accessible only after the engine and starter motor have been removed. The starter bushing can be pulled out of the transmission case with an extractor, as shown in Fig. 4-3.

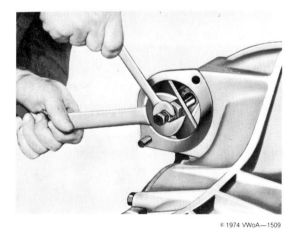

Fig. 4-3. Removing starter bushing from transmission case. Note the extractor being used. One wrench keeps the extractor holder from rotating while the other turns the extractor bolt counterclockwise to withdraw the bushing from the case.

When installing the new bushing, make certain that you start it properly into the hole in the transmission case casting. If it enters at an angle, both the bushing and transmission case may be damaged. Drive the new starter bushing into place using a suitable bushing driver, as shown in Fig. 4-4.

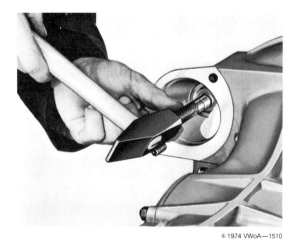

Fig. 4-4. Installing starter bushing. Make certain the new bushing is flush with the edges of the bore.

4.3 Removing and Installing Clutch Operating Shaft

The clutch operating shaft is located in the bellhousing portion of the transmission case. It supports the clutch release lever and bearing.

To remove shaft:

1. Remove the clutch release bearing.
2. Remove the clutch operating lever circlip.
3. Remove the lever, return spring, and return spring seat.
4. Remove the screw that secures the left-side bushing (see Fig. 4-5).

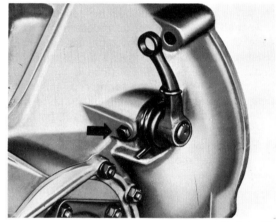

Fig. 4-5. Left-side bushing securing screw (arrow).

5. Slide the shaft to the left and remove the bushing.
6. Remove the shaft as shown in Fig. 4-6.

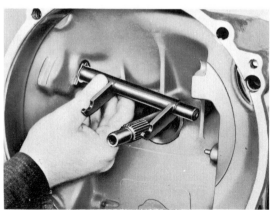

Fig. 4-6. Removing clutch operating shaft.

Transmission and Rear Axle 13

NOTE —
Since January 1972, all Type 1 cars with manual transmission have been equipped with new clutch mechanism parts that improve the leverage ratio but also increase pedal travel. These parts are not interchangeable on earlier models unless the new transmission case (part No. 111 301 051D) and other additional parts are installed. For details, consult your Authorized VW Dealer or other qualified shop.

To install shaft:

1. Examine the bushings and washers for nicks, burrs, or excessive wear. Replace as necessary.

   CAUTION —
   The original fit of gaskets and seals cannot be duplicated once they have been compressed or seated. Therefore, they should be replaced routinely to guard against oil leakage.

2. Lubricate the operating shaft with multipurpose grease. Then install the shaft and the right-side bushing.

3. Install the left-side bushing and secure it with the screw, as shown in Fig. 4-7.

Fig. 4-7. Installing left-side bushing. Make certain that the securing screw and the hole in the bushing are properly aligned.

4. Check the condition of the return spring. If it is weak or badly rusted, replace it.

5. Install the return spring seat, return spring, and clutch operating lever.

6. Install a new clutch operating lever circlip.

7. Install the clutch release bearing.

8. Check clutch pedal freeplay. Adjust if necessary.

5. Removing and Installing Transmission and Rear Axle

(engine removed)

Use common sense when handling the transmission/rear axle unit. Carelessness or the use of an unsuitable floor jack could lead to expensive damage or to personal injury.

To remove:

1. Remove the frame tunnel cover located beneath the rear seat.

2. Remove the square head setscrew that secures the shift rod coupling (Fig. 5-1). Move the gearshift in the passenger compartment in order to disengage the coupling.

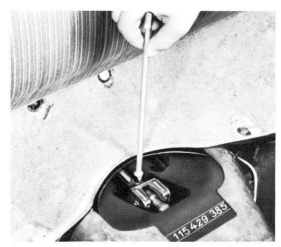

Fig. 5-1. Removing shift rod coupling setscrew.

3. Remove the socket head screws from the transmission end of the driveshafts.

4. Remove the socket head screws from the wheel end of the driveshafts. Remove the shafts.

   NOTE —
   If the car does not have to be moved after the transmission has been taken out, the driveshafts can be disconnected only at the transmission end and then wired to the frame of the car.

5. Disconnect the cable at the clutch lever, then remove the cable and guide from the bracket on the left-side final drive cover.

6. Pull back the rubber caps and disconnect the back-up light wires.

14 Transmission and Rear Axle

7. Disconnect the wires and battery cable from the starter solenoid (Fig. 5-2).

Fig. 5-2. Starter terminals 30 and 50.

8. Remove the nuts from the front transmission mounting as shown in Fig. 5-3.

Fig. 5-3. Removing front transmission mounting nuts.

9. Position a suitable floor jack under the transmission and raise it until it supports the transmission.

10. Remove the transmission carrier bolts. Then, using the floor jack, pull the transmission to the rear. Then lower it from the car.

Installation is basically the reverse of removal. First torque the front mounting nuts to 2 mkg (14 ft. lb.). Grease the carrier bolts. Then fully tighten them to 23 mkg (166 ft. lb.). Torque the driveshaft socket head screws to 3.5 mkg (25 ft. lb.).

6. Disassembling Rear Axle

Fig. 6-1 is a cutaway view of the transmission and final drive components. Familiarize yourself with the part names and their locations, since they will be referred to in the text.

Disassembly of the transmission and final drive must be performed in a neat and orderly manner. In some cases, precision measurements must be made before parts are removed, so carelessness always carries a high penalty. Approach each step thoughtfully and make certain you have clear in your mind what is to be done before actually beginning the work.

It is also important to use either the specified tools or suitable substitutes for them. Incorrect or makeshift tools can damage fasteners and other parts.

Unless specifically instructed not to do so, thoroughly clean all metal parts in a suitable solvent. Lubricate the parts with hypoid oil before you assemble them. Otherwise, the transmission may seize when it is put in service.

The use of a torque wrench in assembling the transmission and rear axle is of particular importance. The transmission case itself is made of a very light magnesium alloy, which is very susceptible to crossthreading. Also, all tolerances within the transmission and final drive are of a critical nature. Bearing preloads and pinion bearing retaining ring torque must be exact. You can use a torque wrench calibrated in either meter-kilograms (mkg) or foot pounds (ft. lb.). Specifications are given for both systems of measurement.

CAUTION —
If you lack the skills, tools, or suitable workshop for servicing the transmission and rear axle, we suggest repairs be left to an Authorized VW Dealer or other qualified shop. We also urge you to consult your Authorized VW Dealer before attempting any repairs on a car still covered by the new-car warranty.

Referring again to Fig. 6-1, you will note that swing-type axle shafts are illustrated, along with final drive ball bearings. These parts are not found on cars covered by this Manual. 1970 through 1974 models have double-jointed axles and tapered-roller bearings. Also, a four-bolt type pinion bearing retainer is shown, whereas the transmission installed in the cars covered by this Manual utilizes a retainer ring. Other than these modifications, Fig. 6-1 illustrates the transmission used in 1970 and later Type 1 cars.

Transmission and Rear Axle 15

Fig. 6-1. Cutaway view of transmission.

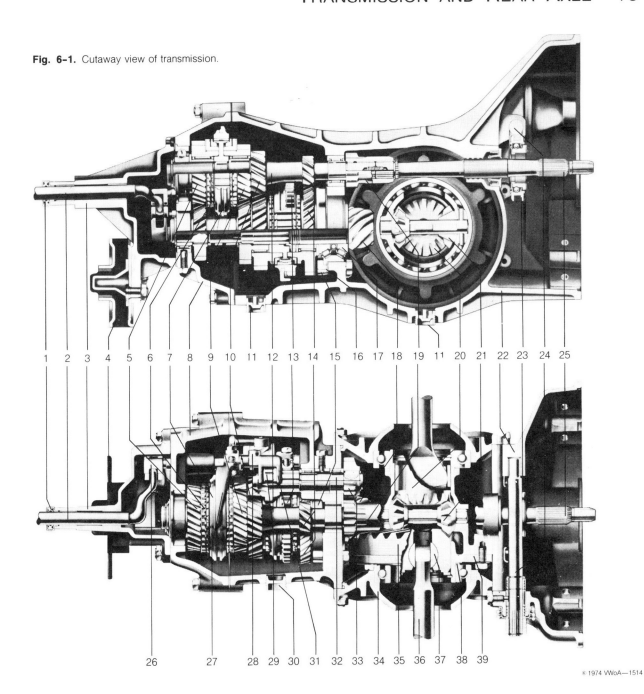

1. Bushing with seal
2. Inner shift lever
3. Gearshift housing
4. Front bonded rubber mounting
5. 4th gear
6. Synchronizer ring for 4th gear
7. Synchronizer hub for 3rd/4th gear
8. Gear carrier
9. 3rd gear
10. 2nd gear
11. Drain plug (magnetic)
12. Main drive Shaft
13. Synchronizer hub for 1st/2nd gear with reverse gear
14. 1st gear
15. Bearing retainer
16. Pinion adjusting shim
17. Double tapered-roller bearing
18. Pinion
19. Reverse gear
20. Differential gear
21. Differential pinion
22. Housing
23. Clutch release bearing
24. Clutch operating shaft
25. Rear driveshaft
26. Selector shaft for 3rd/4th gear
27. Shift fork for 3rd/4th gear
28. Reverse gear lever
29. Reverse sliding gear
30. Oil filler plug
31. Reverse shift fork
32. Reverse shaft
33. Drive gear for reverse
34. Side cover
35. Ring gear
36. Axle shaft
37. Fulcrum plate
38. Spacer
39. Differential housing

16 Transmission and Rear Axle

6.1 Removing and Installing Differential in Case

When removing the differential, be sure you keep separate the parts for the left and the right final drive covers. Mark them for reference if necessary. Clean all metal parts in solvent, then dry them with compressed air.

> **CAUTION** ―
> *If you lack the skills, tools, or workshop to service the differential, we suggest you leave such repairs to an Authorized VW Dealer or other qualified shop. We especially urge you to consult your Authorized VW Dealer before attempting repairs on a car still covered by the new-car warranty.*

To remove differential:

1. Remove the starter motor.
2. Mount the transmission in a suitable repair stand and fasten it securely.
3. Puncture the right axle flange cap with a screwdriver, then pry it out carefully.
4. Remove the retaining circlip and pry off the flange, as shown in Fig. 6-2.

Fig. 6-2. Prying off axle flange. Use extreme care to avoid damaging the final drive cover or flange. For leverage, use wooden dowels as shown.

5. Turn the transmission over and remove the spacer ring.
6. Remove the left axle flange cap, circlip, and flange. Remove the spacer ring.

7. Remove the securing nuts gradually in a cross pattern. Then lift off the left final drive cover with a slide hammer and puller, as shown in Fig. 6-3.

Fig. 6-3. Removing left final drive cover. Note that a bridge-type puller is used.

8. Press the oil seal out of the left final drive cover, as shown in Fig. 6-4.

Fig. 6-4. Pressing out oil seal. Firmly support the final drive cover.

Transmission and Rear Axle

9. Press the tapered bearing outer race out of the left final drive cover, as shown in Fig. 6-5.

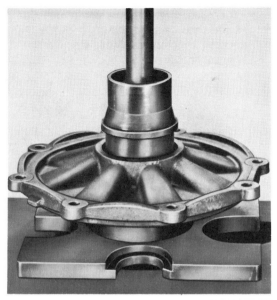

Fig. 6-5. Pressing out bearing outer race. If the race is to be reused, do not interchange it with that for the right final drive cover.

10. Remove the shims and the O-ring. Keep track of the thickness and location of shims to facilitate proper installation.
11. Lift out the differential assembly.
12. Remove the right final drive cover.

NOTE
Some 1972 and all later cars have the new-type transmission case. This case differs from the earlier one in that the right final drive cover is an integral part of the case. See Fig. 6-6.

Fig. 6-6. New-type transmission case.

13. To replace the roller bearing outer race on transmissions with integral right final drive cover, use a rubber mallet and driver as shown in Fig. 6-7.

Fig. 6-7. Driver used to remove and install roller bearing outer race on new-type transmission case.

To install differential:

1. Inspect the bearings and seals for burrs, nicks, and wear. Replace as necessary.
2. Install the original shims in the final drive covers. Drive or press the bearing outer races into place (see Fig. 6-7 or Fig. 6-8).

Fig. 6-8. Installing roller bearing outer race. Firmly support the final drive cover.

18 Transmission and Rear Axle

NOTE —
If the repair job you are doing involves adjusting the tapered-roller bearing preload and gear backlash, first install the final drive covers without the shims (for measurement purposes). See **10.3 Adjusting Ring Gear**.

3. Lightly lubricate the final drive cover oil seals, then press them into place, as indicated in Fig. 6-9.

Fig. 6-9. Installing final drive cover oil seal.

4. Lightly lubricate and install new final drive cover O-rings.

CAUTION —
The original fit of gaskets and seals cannot be duplicated once they have been compressed or seated. Therefore, they should be replaced routinely to guard against oil leakage.

5. On earlier-type transmissions, install the right-final drive cover. Then tighten the retaining nuts to 2 mkg (14 ft. lb.).

CAUTION —
The right final drive cover must not be installed on the ring gear side. Doing so may damage the cover.

6. Liberally lubricate the tapered-roller bearings with hypoid oil.

7. Install the differential assembly, taking special care to avoid damaging the ring gear teeth.

8. Install the final drive cover on the ring gear side. After the differential is installed, it may be necessary to reseat the oil seal.

9. Insert the spacer rings with their flat sides facing toward the flanges. Install the axle flanges.

10. Secure the axle flanges with new circlips. It may be necessary to lift the differential side gear and press the flange down (as shown in Fig. 6-10) in order to compress the spacer washer enough to install the circlip.

Fig. 6-10. Compressing spacer washer to install flange circlip. Install a crossbar similar to that shown. Then tighten the M 10 bolt to pull up the gear.

11. Drive new plastic caps into the flanges with the driver shown in Fig. 6-11.

Fig. 6-11. Installing axle flange caps. Make certain seating surfaces are clean.

Transmission and Rear Axle 19

7. Reverse Gears

The two reverse gear wheels and the rear driveshaft are contained in the final drive portion of the transmission case. They can be removed only after the differential has been taken out. Inspect the gears for burrs, cracking, or excessive wear. Replace as necessary.

7.1 Removing and Installing Reverse Gears and Rear Driveshaft

The following procedure applies to all Type 1 cars up to the late 1972 models. Later models are equipped with a splined reverse gear shaft (formerly slotted for a Woodruff key), a modified needle bearing, and two lockrings that prevent axial bearing movement. See Fig. 7-1; see also **9.5 Removing and Installing Reverse Shaft and Needle Bearings.**

Fig. 7-1. Exploded view of new reverse drive gear and reverse gear shaft components.

To remove reverse gears:

1. Remove the circlip for the reverse drive gear on the rear driveshaft. Slide the gear rearward and unscrew the shaft (see Fig. 7-2).

Fig. 7-2. Unscrewed rear driveshaft and reverse drive gear. Slip gear off and pull shaft out from the rear, being careful not to damage the oil seal.

2. Remove the retaining circlip. Take the reverse drive gear off the reverse shaft as shown in Fig. 7-3.

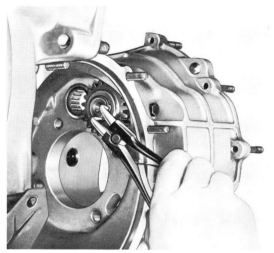

Fig. 7-3. Removing reverse drive gear circlip.

Installation is the reverse of removal. Lubricate all parts with hypoid oil and check operation after installation.

8. Differential Servicing

Once the differential assembly has been removed, it can be disassembled on the bench for repair. Clean all parts thoroughly in solvent and blow them dry with compressed air. Liberally lubricate all parts with hypoid oil before installation.

The differential is the heart of power transmission to the rear wheels. It consists of the ring gear, differential housing, differential gears, tapered-roller bearings, and associated small parts. It should be handled with great care to avoid damage that may result in noisy operation or premature wear.

The differential gearset (ring gear and pinion) is factory-adjusted for quiet operation and durability. Abuse or neglect, however, can shorten service life. For example, mounting different sized tires on the rear wheels causes the differential to work continuously. Another common error is continuing to drive the car after axle damage has been sustained. If the car is operated properly, according to the owner's handbook, the differential should last many thousands of miles. If the ring gear is replaced, it will be necessary to readjust the final drive. See **10. Adjusting Final Drive.**

20 Transmission and Rear Axle

8.1 Disassembling and Assembling Differential

Secure the differential in a soft-jawed vise, making certain that the ring gear teeth are protected from damage. An exploded view of the differential is shown in Fig. 8-1.

1. Ring gear
2. Tapered-roller bearing
3. Differential housing
4. Retaining pin
5. Thrust washer
6. Differential gear (long shaft)
7. Differential pinion shaft
8. Differential pinion
9. Spacer sleeve
10. Differential gear (short shaft)
11. Differential housing cover
12. Spring washer
13. Bolt

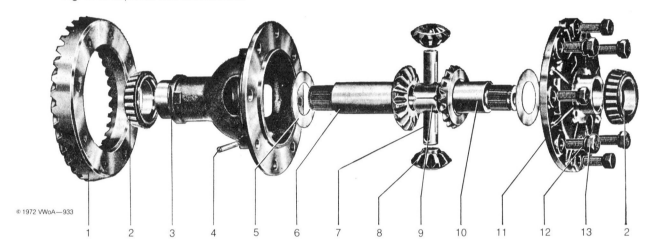

Fig. 8-1. Exploded view of differential.

To disassemble differential:

1. Remove ring gear.

2. Using a slide hammer as shown in Fig. 8-2, lift the cover from the differential housing.

Fig. 8-2. Removing differential housing cover.

3. Remove the differential gear (short shaft) and thrust washer.

   NOTE —
   If necessary, the tapered-roller bearing and the inner race can now be removed. See **8.2 Removing and Installing Tapered-Roller Bearings**.

4. Drive the retaining pin and differential pinion shaft out of the housing (see Fig. 8-3).

Fig. 8-3. Differential pinion shaft removal. Tap shaft out in direction of arrow.

5. Remove the spacer sleeve, differential pinions, differential gear (long shaft), and thrust washer.

6. If necessary, remove the differential housing tapered-roller bearing. If it is damaged or worn, replace it.

Transmission and Rear Axle

To assemble differential:

1. Inspect the gear teeth for burrs or excessive wear. Replace as necessary. Examine all thrust surfaces on differential housing, cover, ring gear, spacer sleeve, and thrust washers for wear and scoring. Replace as necessary.

2. Using a hydraulic press, install the inner race and tapered-roller bearing on the differential housing. Make sure the housing is flat and firmly supported.

3. Install the differential gear (long shaft) and thrust washer in the housing.

    NOTE
    If you notice gear-tooth damage (usually the result of insufficient backlash) on models not equipped with a spacer sleeve, the sleeve should be service-installed. If you make this modification, you must use differential gears with ground end faces.

4. Install the differential pinions, spacer sleeve, and shaft in the differential housing.

5. Install a new retaining pin. Secure the pin by carefully peening both ends with a drift as illustrated in Fig. 8-4.

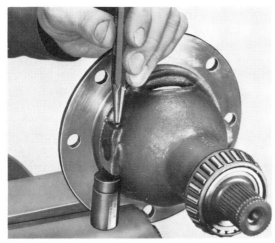

Fig. 8-4. Peening retaining pin. Note the support used under the pin. To avoid damaging the differential housing, use a soft-jawed vise or aluminum jaw covers.

NOTE
If you replace the differential housing, cover, side gear, or thrust washer, you must adjust the axial play between the side gears before installing the differential parts. The procedure for adjusting axial play follows the remainder of the differential assembly procedure.

6. Install the ring gear and assemble the differential housing and cover. Torque the securing bolts diagonally to 6.0 mkg (43 ft. lb.).

To adjust side gear axial play:

1. Position the differential gear (short shaft) and both thrust washers in the differential cover. Then clamp the gear tightly against the cover as shown in Fig. 8-5.

Fig. 8-5. Clamping differential gear, cover, and thrust washers. Note clamping device used.

2. Using a micrometer, measure the shortest spacer sleeve available, as shown in Fig. 8-6. Mark the length on the sleeve and set it aside.

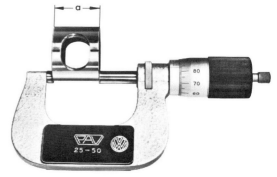

Fig. 8-6. Measuring spacer sleeve length. Mark the length on the sleeve permanently with a scribe.

3. Install the differential gear (long shaft).
4. Insert the marked spacer sleeve. Secure the differential housing to the cover with four bolts, washers, and nuts. Position the washers on the ring gear side.
5. Install a dial indicator on the differential housing as shown in Fig. 8-7. The pin must rest on the housing neck. Measure axial play by moving the differential gear up and down.

Fig. 8-7. Measuring differential gear axial play.

6. Add the measured play to the known spacer sleeve length to obtain dimension **X**. Select the correct spacer sleeve according to **Table a**.

Table a. Differential Spacer Sleeve Lengths

Sleeve Part No.	X-Range mm (in.)	Sleeve Length mm (in.)
004 517 241	28.82-28.90 (1.135-1.137)	28.77-28.82 (1.133-1.135)
004 517 242	28.91-28.99 (1.138-1.141)	28.86-28.91 (1.136-1.138)
004 517 243	29.00-29.08 (1.142-1.145)	28.95-29.00 (1.139-1.142)
004 517 244	29.09-29.18 (1.146-1.149)	29.04-29.09 (1.144-1.146)

7. Disassemble the differential and install the correct spacer sleeve. Reassemble and recheck axial play. It should be from 0.00 to 0.14 mm (.000 to .005 in.).

8.2 Removing and Installing Tapered-Roller Bearings

The inner races for the differential's tapered-roller bearings are a press-fit on the differential housing cover and on the differential housing itself. Each inner race consists of a hard steel cone surrounded by a set of caged, tapered rollers. Removal of the inner races requires disassembling the differential.

CAUTION —
If you lack the skill, work space, or proper press for removing and installing the differential tapered-roller bearings, we suggest that you leave such repairs to an Authorized VW Dealer or other qualified shop. We especially urge you to consult your Authorized VW Dealer before attempting repairs on a car still covered by the new-car warranty.

To remove bearings:

1. Using the tools and support plate illustrated in Fig. 8-8, press the roller bearing and inner race off the differential housing cover.

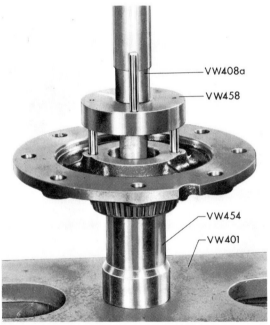

Fig. 8-8. Removing tapered-roller bearing from differential housing cover. Do not drop the bearing.

2. Install the differential housing in a support stand. Then press the housing neck out of the tapered-roller bearing inner race as shown in Fig. 8-9.

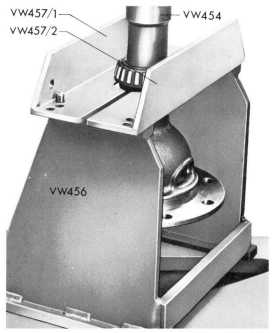

Fig. 8-9. Removing tapered-roller bearing and inner race from differential housing. Make certain the inner race is firmly supported. Also, you may want to put a piece of wood underneath the differential housing to prevent possible damage.

To install bearings:

1. Examine the bearings for pitting, galling, or excessive wear. Replace as necessary. If the bearings are to be reused, clean them thoroughly in solvent and dry them with compressed air.

   CAUTION —
   Before installing the tapered-roller bearings, heat them to attain the prescribed interference fit. First immerse the bearings in a pan of oil. Then, as a safety precaution, put the pan of oil into a pan of boiling water. Using a torch to heat the bearing inner races will distort and damage them.

2. Heat the tapered-roller bearing to approximately 100°C (212°F) and position it on the differential housing. Press the bearing fully to its seat, using three tons pressure (see Fig. 8-10).

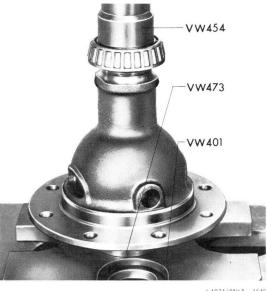

Fig. 8-10. Installing tapered-roller bearing on differential housing.

3. Clean the seating surface on the differential housing cover. Remove minor burrs with a fine oilstone.

4. Heat the tapered-roller bearing to approximately 100°C (212°F) and position it on the differential housing cover.

5. Using three tons pressure, press the bearing onto the cover (see Fig. 8-11).

Fig. 8-11. Tapered-roller bearing being installed on differential housing cover.

24 Transmission and Rear Axle

9. Transmission

Fig. 9-1 is an exploded view of the transmission case and gear train. Knowing the names and locations of the parts will make repairs easier.

CAUTION ━
If you lack the skills, tools, or workshop necessary to service the transmission, we suggest that you leave such repairs to an Authorized VW Dealer or other qualified shop. We especially urge you to consult your Authorized VW Dealer before attempting repairs on a car still covered by the new-car warranty.

9.1 Disassembling and Assembling Shift Housing

The shift housing is a separate casting at the extreme front end of the transmission. It contains the inner shift lever that engages the three selector shafts in the gear carrier. The shift housing is removed from the transmission gear carrier by taking the nuts off the seven studs.

Fig. 9-1. Exploded view of transmission and case.

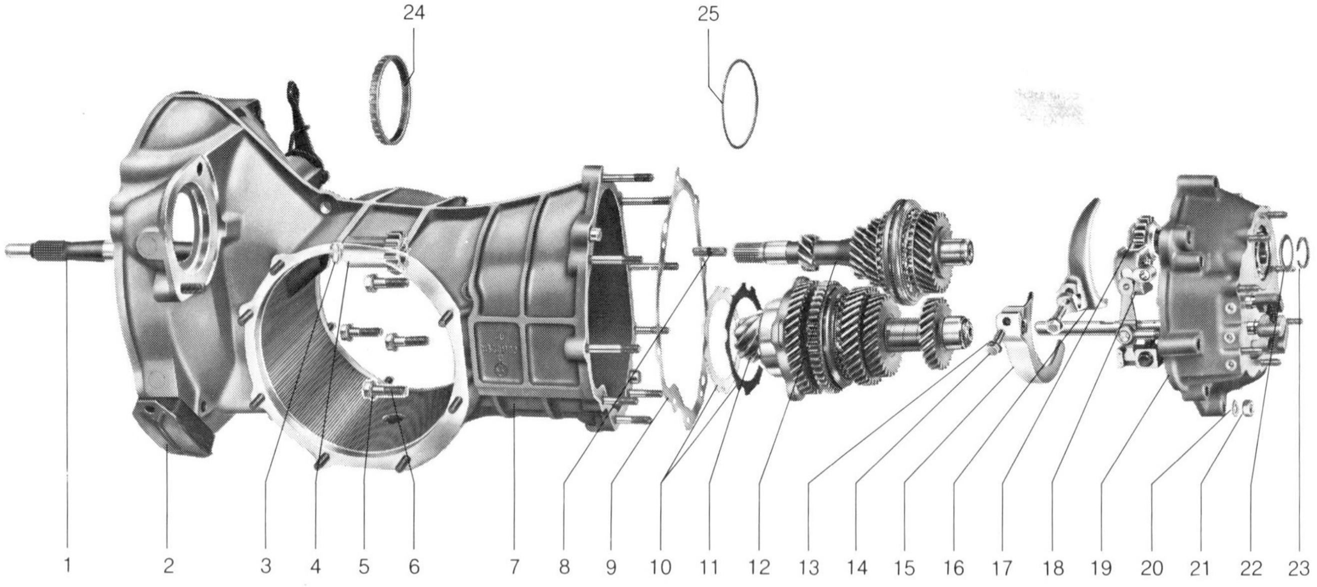

To disassemble shift housing:

1. Using water pump pliers, twist out the bushing with seal. See exploded view in Fig. 9-2.
2. Press the guide bushing together at the slot, then remove it with a screwdriver.

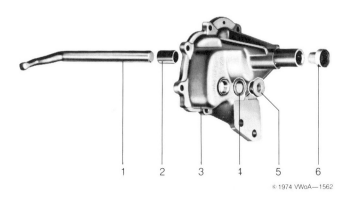

1. Inner shift lever
2. Guide bushing
3. Shift housing
4. Washer
5. Back-up light switch
6. Bushing with seal

Fig. 9-2. Exploded view of shift housing.

1. Rear driveshaft
2. Bonded rubber mounting
3. Circlip
4. Reverse drive gear
5. Washer B 10 (used prior to August 1969)
6. Bolt (used prior to August 1969)
7. Transmission case
8. Stud M 6
9. Gear carrier gasket
10. Shim S_3 (used prior to August 1969)
11. Drive pinion
12. Main drive shaft
13. Spring washer B 6
14. Bolt
15. 1st/2nd gear shift fork
16. 3rd/4th gear shift fork
17. Reverse sliding gear
18. Reverse shift fork
19. Gear carrier
20. Spring washer B 8
21. Nut
22. Dished washer
23. Circlip
24. Pinion retaining nut
25. Shim S_3

Transmission and Rear Axle 25

3. Unscrew the back-up light switch or the plug and washer.

To assemble shift housing:

1. Press in a new guide bushing as shown in Fig. 9-3.

Fig. 9-3. Installing new guide bushing. If inner shift lever operates stiffly, ream bushing to 15.03 to 15.05 mm (.591 to .592 in.) with a 15-mm (.590-in.) reamer. Do not ream seal.

2. Install the back-up light switch or the plug and washer.
3. Replace the bonded rubber mounting if necessary.
4. Press a new bushing with seal into the shift housing as shown in Fig. 9-4.

Fig. 9-4. Installing new bushing with seal.

9.2 Removing and Installing Transmission Gears in Case

The transmission gears can be taken out of the case only after the differential assembly has been removed.

To remove transmission gears:

1. Remove the seven mounting nuts, then remove the gearshift housing.
2. Remove the nine nuts from the studs that hold the gear carrier on the transmission case. (See Fig. 9-1 for part names and locations.)
3. Remove the pinion retaining nut.

   NOTE
   The ring-shaped pinion retaining nut is best removed with a special C-wrench, such as the one illustrated in **AUTOMATIC STICK SHIFT**, or with the special tool VW 381/14. On 1972 and later cars, the pinion retaining nut is locked by peening, as indicated in Fig. 9-5. Only peened nuts are supplied as replacement parts. If a replacement nut must be fitted to an earlier transmission, it is necessary to grind a recess in the edge of the threaded portion of the bearing so that the retaining nut can be peened into it.

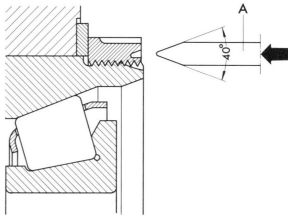

Fig. 9-5. Peening pinion retaining nut. A blunt punch (arrow), ground to a 40° angle, is driven into a groove in the nut to expand it into a recess in the bearing. No unlocking step is required with these nuts.

   CAUTION
   If you lack the special tools for removing and installing the retaining ring, we suggest you leave such repairs to an Authorized VW Dealer or other qualified shop.

4. Using a lever (VW 296), press the transmission gears out of the case as shown in Fig. 9-6.

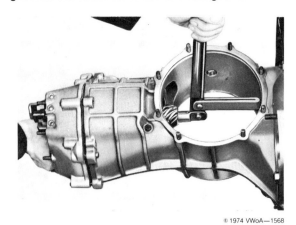

Fig. 9-6. Transmission gears being pressed out of case.

5. Remove the reverse sliding gear and the shift fork from the relay lever.

6. Remove the drive pinion adjusting shims S_3 from the double tapered-roller bearing and note their number and thickness. See **10. Adjusting Final Drive.**

7. Position the gear carrier in a soft-jawed vise or other holding device. Remove the 1st/2nd and the 3rd/4th gear shift fork clamping bolts (see Fig. 9-7).

Fig. 9-7. Removing shift fork clamping bolts. Do not damage any gear teeth.

8. Remove the 1st/2nd gear shift fork.

9. Pull the selector shaft for the 3rd/4th gear out of the shift fork.

10. Using circlip pliers, remove the drive shaft circlip, as shown in Fig. 9-8.

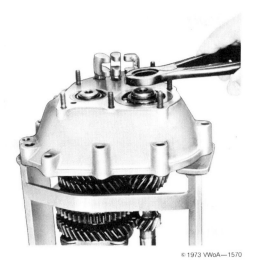

Fig. 9-8. Removing drive shaft circlip. Dished washer beneath circlip is under tension.

11. Place the gear carrier in a support as shown in Fig. 9-9. Carefully press out the drive shaft and pinion by applying pressure to the drive shaft.

CAUTION —
When performing this procedure, guide the pinion and drive shaft carefully so that you do not damage the gear teeth. Also, make certain the 3rd/4th gearshift fork does not jam.

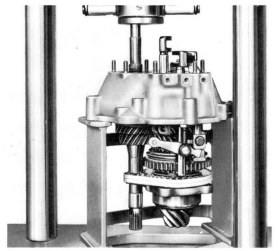

Fig. 9-9. Removing drive shaft and pinion. Note rubber bands used to help hold shafts together.

12. Remove the drive shaft and pinion along with the gear clusters.

To install transmission gears:

1. Examine all parts and make necessary repairs. See **9.6 Disassembling and Assembling Main Drive Shaft** and **9.7 Disassembling and Assembling Drive Pinion**.

2. Lubricate all parts liberally with hypoid oil.

3. Press the assembled main drive shaft and pinion into the gear carrier.

4. Install the dished washer and a new circlip. Press the circlip into its groove with a repair press and a sleeve-type driver such as the one shown in Fig. 9-10.

Fig. 9-10. Driver used to install circlip.

5. Raise the press tool. Using pliers, squeeze the circlip all the way around until it bottoms fully in its groove (Fig. 9-11).

Fig. 9-11. Seating circlip in groove.

6. Install and adjust the shift forks. See **9.4 Adjusting Shift Forks**.

7. Attach the reverse shift fork and sliding gear to the relay lever. Engage reverse gear.

8. Clean the gear carrier and transmission case mating surfaces. Install a new paper gasket over the gear carrier mounting studs on the transmission case.

9. Install the original number and thickness of S_3 shims over the double tapered-roller bearing.

10. On 1972 and later cars, position the recess in the threaded part of the double tapered-roller bearing so that it is toward the side of the transmission case.

    NOTE
    If a peened-type pinion retaining nut is being installed on a 1970 or 1971 car, a recess must be ground into the threaded part of the double tapered-roller bearing so that the pinion retaining nut can be peened into it. Position the recess toward the side of the transmission case.

11. Applying a rubber mallet alternately to the main drive shaft and the pinion shaft, tap the transmission gears fully into place.

12. Install the pinion retaining nut. Torque it to 22.0 mkg (160 ft. lb.), loosen it, then again torque it to the same specification.

13. Peen the late-type pinion retaining nut into the recess in the double tapered-roller bearing as indicated in Fig. 9-12.

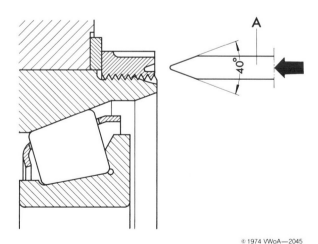

Fig. 9-12. Peening pinion retaining nut. A blunt punch (arrow), ground to a 40° angle, is driven into the groove in the pinion retaining nut. This expands the nut into the recess in the bearing.

28 Transmission and Rear Axle

NOTE —

Because of a modification to the 1st gear on the pinion shaft, the width of the inner races of the double tapered-roller bearing was increased during February 1972. If one of these later-type bearings must be installed in an earlier transmission, it is also necessary to discard the 1st gear thrust washer and install the late-type 1st gear. Transmissions having the late-type 1st gear can be identified by the 2-mm (.080-in.) washer between the pinion retaining nut and the transmission case (Fig. 9-13).

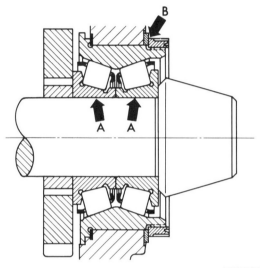

Fig. 9-13. Double tapered-roller bearing with wider inner races (arrows **A**). A 2-mm (.080-in.) washer is factory-installed at arrow **B** in transmissions fitted with this bearing.

14. Install the nine spring washers and nuts on the gear carrier mounting studs. Working diagonally, gradually torque the nuts to 2.0 mkg (14 ft. lb.).

15. Install the gearshift housing with a new paper gasket. Working diagonally, gradually torque the nuts to 1.5 mkg (11 ft. lb.).

16. If it was removed earlier, install the transmission's bonded rubber mounting on the gearshift housing. Torque the nuts to 2.0 mkg (14 ft. lb.).

9.3 Disassembling and Assembling Gear Carrier

The gear carrier has undergone some significant redesigns during the model years covered by this Manual. It is therefore important to examine and identify correctly the kind of gear carrier used on the transmission before you select replacement parts or begin disassembly and reassembly work on the transmission's gear train.

1973 Changes

Beginning with the 1973 models, the ball bearing for the main drive shaft and the gear carrier itself were altered to accommodate a larger main drive shaft. At the same time the shoulder, previously on the bearing, was moved to the bore in the gear carrier.

The locking screw for the drive pinion's needle bearing was eliminated also beginning with the 1973 cars. The needle bearing is now held axially by a shoulder on the gear carrier and by two projections in the gearshift housing.

Another important 1973 change was the introduction of a new reverse gear relay lever. The three new lever components are shown in Fig. 9-14. Because of this change, it is important that you know the transmission number when you obtain replacement parts for the relay lever and other gear carrier components.

1. Relay lever   2. Relay lever guide   3. Support plate

Fig. 9-14. New reverse gear relay lever parts.

1976 Changes

Another important change was made to the gear carrier and the main drive shaft ball bearing at about the beginning of the 1976 model year. On the transmission's main drive shaft, the 4th gear thrust washer is eliminated and a new main drive shaft ball bearing is installed in the gear carrier.

You can install the new main drive shaft ball bearing (Part No. 091 311 123) and eliminate the 4th gear thrust washer from an earlier transmission only if you install the new-type gear carrier. For this reason, the old-type ball bearing and the old-type 4th gear thrust washer remain available as spare parts for the repair of 1970 through 1975 transmissions.

Transmission and Rear Axle 29

The new main drive shaft ball bearing, introduced in 1976, is shown in Fig. 9-15. This is the only kind of main drive shaft ball bearing that should be installed in 1976 and later cars.

Fig. 9-15. New main drive shaft ball bearing with wide shoulder (arrow) on inner race.

Fig. 9-16. New gear carrier with added oil groove (arrow). Bearing shown in Fig. 9-15 must not be installed in a gear carrier that does not have this groove.

The redesigned gear carrier introduced in 1976 can be identified by an oil groove in the casting. This groove is adjacent to the bore for the main drive shaft ball bearing and is indicated by the arrow in Fig. 9-16.

Repair Procedures

Fig 9-17 shows the gear carrier used through 1972. Though this gear carrier is shown in the following disassembly and assembly procedures, specifications and illustrations for the redesigned gear carriers are given whenever they are necessary for correct repair of the transmission.

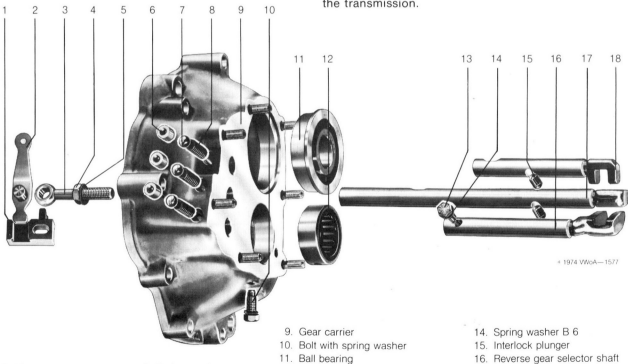

1. Guide
2. Relay lever
3. Relay lever support
4. Nut M 10
5. Spring washer
6. Plug
7. Detent ball
8. Detent spring
9. Gear carrier
10. Bolt with spring washer
11. Ball bearing
12. Needle bearing outer race
13. Bolt
14. Spring washer B 6
15. Interlock plunger
16. Reverse gear selector shaft
17. 1st/2nd gear selector shaft
18. 3rd/4th gear selector shaft

Fig. 9-17. Exploded view of early-type gear carrier.

Transmission and Rear Axle

To disassemble gear carrier:

1. Fasten the gear carrier in a soft-jawed vise.
2. Pull out the forward gear selector shafts. Remove the reverse gear selector shaft guide (see Fig. 9-18) and shaft.

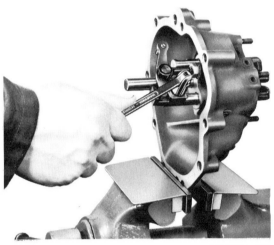

Fig. 9-18. Removing reverse gear selector shaft guide.

3. Remove the detent spring plugs (see Fig. 9-19).

NOTE —
The plugs are best removed by tapping an M 6 (or a 1/4-in. × 24) thread into them and then pulling them out with a screw.

Fig. 9-19. Removing detent spring plugs. Note screw used as an extractor. The aluminum covers on the vise protect the carrier.

4. Remove the detent springs with a small screwdriver.
5. Remove the relay lever and support.
6. On transmissions built before November 1972, remove the securing bolt with its spring washer from the needle bearing retainer.
7. Press out the needle bearing as shown in Fig. 9-20.

Fig. 9-20. Pressing out needle bearing.

8. Press out the ball bearing as shown in Fig. 9-21.

Fig. 9-21. Pressing out ball bearing.

To assemble gear carrier:

1. Clean all parts thoroughly and blow them dry with compressed air.
2. Examine the bearings, selector shafts, interlock plungers, and detent springs and balls for wear and damage. Replace as necessary. Standard detent spring free length is 23 to 25 mm ($^{29}/_{32}$ to $^{31}/_{32}$ in.).

Transmission and Rear Axle

3. If the transmission had a shifting problem before its disassembly, use a spring scale to check the force required to overcome the detent springs, as illustrated in Fig. 9-22.

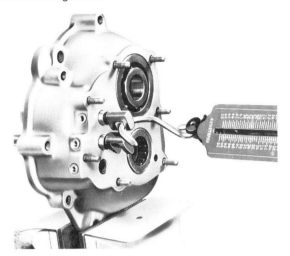

Fig. 9-22. Measuring force required to overcome detents. Spring scale should read between 15 and 20 kg (33 and 44 lb.).

NOTE ─
You can measure the force needed to overcome the detents only after you have installed the interlock plungers, the selector shafts, and the detent balls, springs, and plugs. Therefore, make the measurement before you disassemble the carrier.

4. Press the needle bearing into the gear carrier as shown in Fig. 9-23. Then, on transmissions built before November 1972, install the locking screw.

Fig. 9-23. Installing needle bearing. Make certain the carrier has firm support.

5. Inspect the gear carrier casting. If it has the oil groove indicated in Fig. 9-24, you must install the latest-type main drive shaft ball bearing, Part No. 091 311 123.

Fig. 9-24. Latest-type gear carrier with oil groove (arrow). This gear carrier can be installed on earlier transmissions as a replacement part if the correct main shaft ball bearing is used.

6. Using a press tool that will apply pressure to only the outer race of the main drive shaft ball bearing, press in the ball bearing as shown in Fig. 9-25. Press the bearing in only until the shoulder makes contact.

Fig. 9-25. Installing ball bearing.

7. Install detent springs, balls, and plugs. Using a screwdriver to hold the balls against the spring pressure, insert the selector shafts and interlock plungers.

32 Transmission and Rear Axle

8. Install the reverse selector shaft along with the guide for the relay lever. Adjust the lever as shown in Fig. 9-26 or Fig. 9-27.

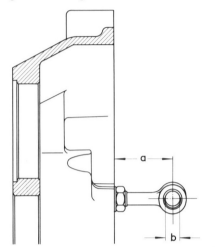

Fig. 9-26. Relay lever support adjustment for transmissions through October 1972. Dimension **a** is 38.60 ± 0.40 mm (1.520 ± .016 in.); dimension **b** is 10.00 mm (.394 in.).

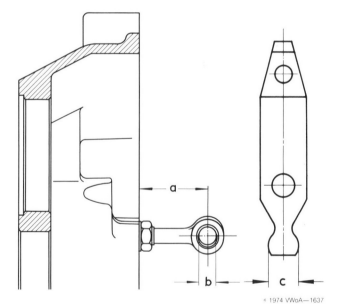

Fig. 9-27. Relay lever support adjustment for transmissions from November 1972. Dimension **a** is 43.40 ± 0.40 mm (1.709 ± .016 in.); dimension **b** is 10.00 mm (.394 in.); dimension **c** is 14 mm (.551 in.).

9.4 Adjusting Shift Forks

To adjust the shift forks properly, a special VW jig is almost a necessity. The gear train and carrier must be positioned exactly as they will be in the transmission case. To do this, select the adjusting shim S_3 beforehand and install it between the pinion bearing retainer and the jig. See **10. Adjusting Final Drive.** If you are using the older VW fork adjusting jig VW 294a, also install the paper gasket between the gear carrier and the jig spacer sleeves. If the spacer sleeves are marked, install them with the marking "Type 1.2. 3/69" toward the jig plate.

To adjust shift forks:

1. Secure the gear train and carrier to the jig, as shown in Fig. 9-28.

Fig. 9-28. Transmission gears and gear carrier installed in jig VW 294b. Arrows indicate securing nuts.

2. With a special wrench, lightly install the pinion retaining nut. Do not install the 2-mm (.080-in.) washer indicated at **B** in Fig. 9-29.

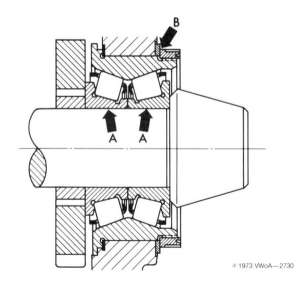

Fig. 9-29. Double tapered roller pinion bearing. Bearing races are at arrows **A**, washer at arrow **B**.

NOTE —

The 2-mm (.080-in.) washer is factory-installed only on late transmissions but is sometimes service-installed on earlier transmissions. Do not use this washer when you mount the pinion bearing in the jig.

3. Insert the 3rd/4th gear selector shaft through the shift fork and tighten the clamp bolt.

4. Position the 1st/2nd gear selector shaft in the 2nd gear notch. Slide the operating sleeve along with the fork over the coupling teeth until the sleeve rests against 2nd gear.

5. Center the shift fork in the operating sleeve groove and tighten the clamp bolt (see Fig. 9-30).

Fig. 9-30. Centering 1st/2nd gear shift fork. The fork must not rub against either side of the operating sleeve.

CAUTION —

When a gear is engaged or when the transmission is in neutral, there must always be clearance between the shift fork and the groove in the operating sleeve. The fork must not exert pressure on or rub against the operating sleeve.

6. While turning the transmission, engage both 1st and 2nd gears and check the clearance between the fork and the operating sleeve. If necessary, adjust the fork position so that the clearance is the same in both positions.

7. Tighten the fork clamp bolt to 2.5 mkg (18 ft. lb.).

8. Position the 3rd/4th gear selector shaft in the 3rd gear notch. Adjust the fork in the same way as described for the 1st/2nd gear shift fork. Tighten the clamp bolt to 2.5 mkg (18 ft. lb.).

NOTE —

To adjust the 3rd/4th gear shift fork correctly, the gear carrier ball bearing must be pressed fully into position.

9. Set the reverse gear shift fork with 2nd gear engaged so that the sliding gear is centered on the drive shaft between the operating sleeve and 2nd gear. When reverse is engaged, make sure the gear fully engages reverse gear on the drive pinion. Tighten the clamp bolt to 2 mkg (14 ft. lb.) after satisfying these two conditions.

CAUTION —

When positioning the reverse sliding gear make certain there is no axial play at the shift fork. If necessary, loosen the relay lever support and turn the relay lever in the direction of the sliding gear until only running clearance remains (see Fig. 9-31).

Fig. 9-31. Correcting axial play in reverse shift fork. Arrow indicates direction in which to press the relay lever.

10. Engage reverse gear and check the tooth alignment between the sliding gear and the operating sleeve. If necessary, readjust the relay lever until the teeth mesh fully.

11. Check the selector shaft interlock. When one gear is engaged, you should not be able to engage another. If you can engage another gear, check the selector shafts and interlock plungers. See **9.3 Disassembling and Assembling Gear Carrier.**

34 Transmission and Rear Axle

9.5 Removing and Installing Reverse Shaft and Needle Bearings

Since July 1972, all VW models have been equipped with a new reverse gear shaft and needle bearing. This gear shaft has splines for locating reverse gear (the earlier shaft used a Woodruff key and slot). The spacer sleeve and bearing locking bolt used earlier have also been discontinued, in favor of two lockrings that secure the bearing. Fig. 9-32 is an exploded view of the new assembly.

Fig. 9-32. Exploded view of new reverse gear shaft and needle bearing. These new parts are used in conjunction with a new transmission case and several other internal changes detailed elsewhere in this Manual.

To remove:

1. Remove the differential and transmission.

2. Using hot water or a steam cleaner, heat the transmission case surrounding the reverse gear shaft to approximately 80°C (176°F).

3. Remove the reverse gear shaft assembly as a unit.

4. Remove the locking bolt for the driveshaft needle bearing. Then drive out the bearing with a drift as shown in Fig. 9-33.

Fig. 9-33. Removing driveshaft needle bearing. Be careful not to damage the bearing bore in the case.

5. To disassemble the reverse gear shaft, press the drive gear off the shaft. Then remove the lockrings and take out the needle bearing.

To install:

1. Inspect all parts for damage or excessive wear. Replace as necessary.

2. Install the needle bearing and the front lockring on the reverse gear shaft.

3. Heat the reverse gear to 100°C (212°F) and press it onto the shaft.

4. Press the front lockring into place as shown in Fig. 9-34.

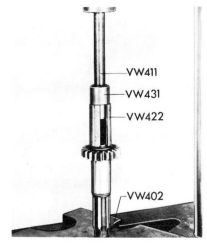

Fig. 9-34. Installing reverse gear shaft lockring. Make certain the shaft is firmly supported.

5. Using water pump pliers, squeeze the lockring all around and seat it firmly in the groove. Be careful not to burr the lockring.

6. Heat the transmission case surrounding the reverse gear shaft bore to approximately 80°C (176°F).

CAUTION ——

Do not use a torch to heat the transmission case. This could cause uneven expansion and distortion or heat damage to the casting. Use hot water or a steam cleaner locally to bring the shaft bore to the proper temperature. It is not necessary to heat the entire transmission case to install the reverse gear shaft and the needle bearing.

7. Lightly lubricate the reverse gear shaft with transmission oil.

8. Using a plastic hammer or rubber mallet, drive the reverse gear shaft assembly into the case as far as it will go (see Fig. 9-35).

Fig. 9-35. Installing reverse gear shaft assembly. Arrow indicates where to apply mallet.

9. Using a drift, drive the driveshaft needle bearing into its bore, as shown in Fig. 9-36.

CAUTION —
Do not attempt to drive in the needle bearing with an improper tool. Doing so can bend the bearing out-of-round or cause it to enter the transmission case bore at an angle.

Fig. 9-36. Drift and driveshaft needle bearing. The drift must be applied to the lettered end of the bearing to prevent damage.

10. Install the driveshaft needle bearing locking bolt.
11. Install the transmission and differential.

9.6 Disassembling and Assembling Main Drive Shaft

Late in the 1970 model year an important change was made to the transmission gear train. New 3rd and 4th gears with finer teeth became standard from Chassis No. 110 2912 968. The angle of the teeth was also changed.

1972 Changes

Beginning in 1972, grooves were machined into one part of the 3rd and 4th gear synchronizer rings. This gives the ring an imbalance that, in conjunction with centrifugal force, keeps the ring in continual contact with the gear wheel. This helps prevent 4th gear howling, which sometimes resulted from a cold transmission. See Fig. 9-37.

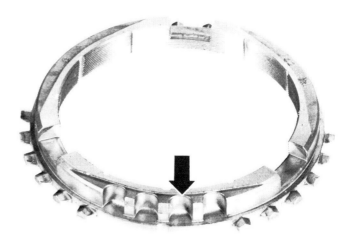

Fig. 9-37. Redesigned 3rd and 4th gear synchronizer ring. Arrow indicates machined grooves that produce ring imbalance.

Midway through the 1972 model year, a second kind of redesigned synchronizer ring was introduced in production. This ring resembles the one that was previously shown in Fig. 9-37, but is for use with 4th gear only. Subsequently, a redesigned 3rd gear synchronizer ring made its appearance. The new 3rd gear ring was introduced gradually and, therefore, may not be found except in transmissions manufactured later in the 1972 model year. However, both of these redesigned synchronizer rings can be used as replacement parts in earlier transmissions.

NOTE —
The latest-type synchronizer rings should always be used for 4th gear repairs. The earlier-type 3rd gear rings, however, are adequate as replacement parts if the latest-type rings are not available.

36 Transmission and Rear Axle

1976 Changes

Beginning with the 1976 models, the 4th gear thrust washer is eliminated from the transmission's main drive shaft. The end thrust is now taken by a wide shoulder that is integral with the inner race of the main drive shaft ball bearing assembly. This new-type bearing, Part No. 091 311 123, is shown in Fig. 9-38.

Fig. 9-38. New main drive shaft ball bearing with wide shoulder (arrow) on inner race.

You can install the new main drive shaft ball bearing and eliminate the 4th gear thrust washer from an earlier transmission only if you install the new-type gear carrier. For this reason, the old-type ball bearing and the 4th gear thrust washer remain available as spare parts for the repair of 1970 through 1975 transmissions.

Repair Procedures

Fig. 9-39 is an exploded view of the main drive shaft. We suggest that you familiarize yourself with the names of the various components since these names will be used frequently in the instructions. The fine-tooth gears at the bottom of the illustrations are the redesigned gears that became standard late in the 1970 model year. In replacing any gear on the main drive shaft, it is also necessary to replace the meshing gear on the drive pinion. Replacement gears are always supplied as a factory-matched pair.

CAUTION ——

Please read **10. Adjusting Final Drive** *before you attempt repairs to the drive pinion. Precision adjustment is required following certain repairs. If you lack the skills, special tools, or a clean workshop for making transmission repairs, we suggest that you leave this work to an Authorized VW Dealer or other qualified shop. We especially urge you to consult your Authorized VW Dealer before attempting repairs on a car still covered by the new-car warranty.*

To disassemble:

1. Remove the circlip from the end of the main drive shaft.

2. Remove the dished washer (where fitted), the 4th gear thrust washer (where fitted), and 4th gear. Then remove the needle bearing and the 4th gear synchronizer ring.

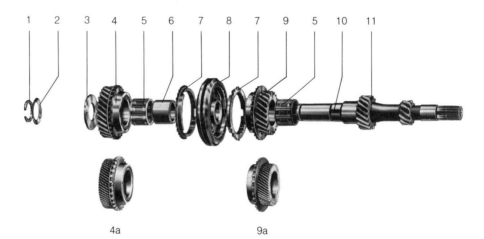

1. Circlip
2. Dished washer (used only until October 1972)
3. 4th gear thrust washer (used only until October 1975)
4. 4th gear (earlier-type)
4a. 4th gear (new-type)
5. 3rd/4th gear needle bearing (2)
6. 4th gear needle bearing inner race (used only until October 1972)
7. 3rd/4th gear synchronizer ring (2)
8. Synchronizer hub
9. 3rd gear (earlier-type)
9a. 3rd gear (new-type)
10. Woodruff key
11. Main drive shaft

Fig. 9-39. Exploded view of main drive shaft.

3. On transmissions built through October 1972, press off the inner race, synchronizer hub, and 3rd gear, as shown in Fig. 9-40. On later transmissions, remove these parts by hand.

Fig. 9-40. Driver (VW 439) being used to press off inner race, synchronizer hub, and 3rd gear (early models only). Make certain hub is firmly supported.

4. Remove the 3rd gear needle bearing.

5. Remove the operating sleeve, the synchronizer shifting plates, and the synchronizer plate springs from the synchronizer hub.

To assemble:

1. Examine the shaft, needle bearings, races, and gears for excessive wear or damage. Replace as necessary.

2. Hand-press the synchronizer rings over the gear cones and measure gap **a** as shown in Fig. 9-41.

Fig. 9-41. Gap between synchronizer ring and gear wheel. The nominal dimension for **a** is 1.00 to 1.90 mm (.040 to .075 in.). Wear limit is 0.60 mm (.024 in.).

3. Examine the synchronizer rings for wear and damage; examine the synchronizer teeth on the gears. Replace worn or damaged parts.

4. Examine the bearings and the washers. Obtain replacement parts for worn or damaged components. Check the condition of the synchronizer hub and the operating sleeve. If either component is faulty, both should be replaced.

5. Assemble the 3rd/4th gear synchronizer hub with the 1-mm groove up and the hub chamfer down, aligning the mark as indicated in Fig. 9-42. Replacement parts are not marked, so mesh the teeth in various positions until you find a free-sliding fit.

Fig. 9-42. Synchronizer hub and operating sleeve alignment marks (arrow).

6. Check the synchronizer springs (Fig. 9-43). If necessary, carefully bend them to the correct dimensions.

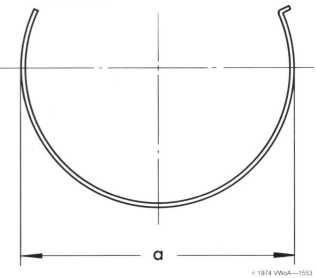

Fig. 9-43. Synchronizer spring measurement. Dimension **a** is 78 mm ($3\,5/64$ in.) for 1st/2nd gears, 74 mm ($2\,29/32$ in.) for 3rd/4th gears.

38 Transmission and Rear Axle

7. Install the synchronizer springs with the ends offset 120°. The spring ends must fit tightly over the synchronizer keys. See Fig. 9-44.

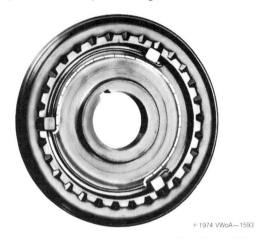

Fig. 9-44. Synchronizer spring installation. Dotted line shows second spring position.

NOTE —
When you assemble the main drive shaft components, we suggest you liberally lubricate all moving parts with hypoid oil.

8. Press the assembled synchronizer hub onto the main drive shaft as far as it will go (see Fig. 9-45).

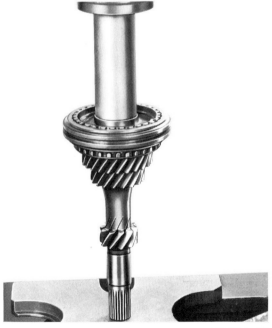

Fig. 9-45. Installing assembled synchronizer hub. Chamfer on operating sleeve must face 3rd gear.

9. On transmissions built through October 1972, heat the inner race of the 4th gear needle bearing to approximately 100°C (212°F) and start it onto the shaft. Then position the shaft on an arbor press. After the inner bearing race has cooled to room temperature, press it fully into place, as illustrated in Fig. 9-46.

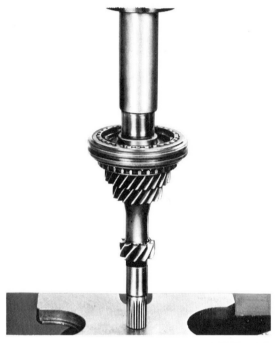

Fig. 9-46. Installing 4th gear needle bearing inner race. Make certain the inner race and main drive shaft are absolutely clean and free of burrs and nicks. Also make sure the main drive shaft is firmly supported in the press.

10. Install the needle bearing, synchronizer ring, and thrust washer for 4th gear (where fitted).

11. Install a new dished washer (where fitted) and a new circlip. Once used, the circlip and the dished washer should not be reinstalled, so replace them routinely to avoid metal fatigue.

NOTE —
On main drive shafts that are secured by a circlip and dished washer, you may find that the thrust washer causes a squeaking noise because the washer turns on the shaft. Washers with a smaller-than-standard inside diameter—part No. 113 311 151 C—24.98 to 25.03 mm (.983 to .985 in.), are factory-installed. These washers must be pressed onto the shaft when the shaft is at its upper tolerance and the washer hole is at its lowest tolerance.

9.7 Disassembling and Assembling Drive Pinion

Several changes have been made to the transmission gear train during the model years covered by this Manual. These must be taken into account when repairs are being made. Be sure to have the transmission number written down for reference in obtaining the correct replacement parts.

1970 Changes

Late in the 1970 model year an important change was made to the gears. New 3rd and 4th gears with finer teeth are standard from Chassis No. 110 2912 968. This redesign resulted in a better mesh and reduced transmission noise.

The seating of 3rd gear on the drive pinion was also modified. The splined shoulder formerly used to seat the gear was discontinued, necessitating a modification to the 3rd gear splines. The new gears must be used with the new gear wheels discussed in **9.6 Disassembling and Assembling Main Drive Shaft.**

If the transmission you are repairing is equipped with a spacer spring between the 3rd and 4th gears, the fine-tooth 3rd gear gearset must be installed. We also advise that you install the new-type 4th gear gearset in these transmissions.

The redesigned gears can also be installed in earlier transmissions that have a spacer sleeve rather than a spring. In earlier transmissions, the spacer spring, needle bearing, and circlips must also be installed along with the redesigned gears.

1976 Changes

A thicker synchronizer hub is used in the transmissions of 1976 and later cars (Fig. 9-47). The thicker hub has made it possible to eliminate the spacer washer that formerly had been installed between the round nut and the 1st/2nd gear synchronizer hub.

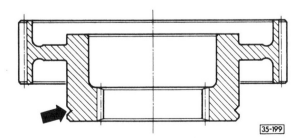

Fig. 9-47. Thicker 1st/2nd gear synchronizer hub with identification groove (arrow).

In disassembling and assembling the drive pinion, you can install the new, thicker 1st/2nd gear synchronizer hub (Part No. 113 311 243 C) if you discard the spacer washer on 1970 through 1975 transmissions. (Always install the thicker hub on 1976 and later transmissions.) Because of new precision manufacturing techniques, it is no longer necessary to adjust the 1st gear axial play if you install the thicker synchronizer hub.

Repair Procedures

The drive pinion can be disassembled after it has been removed from the transmission case. It is important that the parts be kept dirt-free. We therefore suggest that you clean all the parts with solvent and dry them with compressed air. When you reassemble the drive pinion, lubricate all moving parts liberally with hypoid oil.

CAUTION —

Please read **10. Adjusting Final Drive** *before you attempt repairs to the drive pinion. Precision adjustment is required following certain repairs. If you lack the skills, special tools, or a clean workshop for making transmission repairs, we suggest that you leave this work to an Authorized VW Dealer or other qualified shop. We especially urge you to consult your Authorized VW Dealer before attempting repairs on a car still covered by the new-car warranty.*

To disassemble:

1. Remove the drive pinion retaining circlip or nut. If necessary, press 4th gear down, as shown in Fig. 9-48, to compress the spacer spring.

Fig. 9-48. Compressing spacer spring. After the spring has been sufficiently compressed, remove the circlip. Be careful not to damage the gear.

40 Transmission and Rear Axle

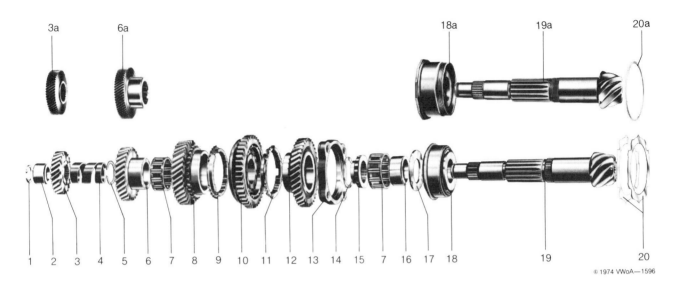

1. Circlip
2. Inner bearing race
3. 4th gear (earlier-type)
3a. 4th gear (1970 on)
4. Spacer spring
5. 3rd gear circlip
6. 3rd gear (earlier-type)
6a. 3rd gear (1970 on)
7. 1st/2nd gear needle cage (2)
8. 2nd gear
9. 2nd gear synchronizer ring
10. 1st/2nd gear synchronizer hub
11. 1st gear synchronizer ring
12. 1st gear
13. Retaining ring (before 1970)
14. Spacer washer (before 1976)
15. Round nut
16. Inner bearing race
17. 1st gear thrust washer
18. Double tapered-roller bearing (earlier-type)
18a. Double tapered-roller bearing (1970 on)
19. Drive pinion (earlier-type)
19a. Drive pinion (1970 on)
20. Shim S_3 (earlier-type)
20a. Shim S_3 (1970 on)

Fig. 9-49. Exploded view of drive pinion. Parts denoted by letter **a** after key numbers are installed on all cars covered by this Manual except some early 1970 models.

NOTE —

Fig. 9-49 shows the position of the spacer spring relative to the other parts of the drive pinion assembly. The illustration includes the names of the parts in the drive pinion. Familiarize yourself with them, since they appear frequently in repair procedures.

2. Press off the inner bearing race along with 4th gear, as shown in Fig. 9-50.

3. Take off the spacer spring and remove the 3rd gear circlip. Note the color and thickness of the circlip to ensure proper assembly.

4. Remove 3rd gear, 2nd gear, synchronizer hub with operating sleeve, synchronizer rings, and 1st gear.

NOTE —

Damaged or worn 3rd and 4th gears must be replaced in matched sets only. However, 1st and 2nd gears need only be replaced in sets if there is evidence of tooth damage.

5. Fasten the drive pinion in a suitable holder and remove the pinion round nut. Make certain the drive pinion is securely fastened because it will take considerable leverage to break the nut free.

6. Remove the 1st gear needle bearing.

Fig. 9-50. Pressing off inner bearing race and 4th gear. Support the drive pinion underneath the press.

Transmission and Rear Axle 41

7. Press the double tapered-roller bearing and 1st gear needle bearing inner race off the drive pinion together, as illustrated in Fig. 9-51.

Fig. 9-51. Removing double tapered-roller bearing and 1st gear needle bearing inner race.

8. Remove the synchronizer springs, the operating sleeve, and the keys.

To assemble:

1. Examine the drive pinion, roller bearing, needle bearings, races, and gears for damage or excessive wear. Replace as necessary.
2. Press the synchronizer rings over the gear cones and measure gap **a** shown in Fig. 9-52.

Fig. 9-52. Synchronizer ring gear wheel gap. The nominal dimension **a** is 1.00 to 1.90 mm (.040 to .075 in.). Maximum wear limit is 0.60 mm (.024 in.).

3. Check the condition of the synchronizer teeth on the gears, operating sleeve, and synchronizer rings. Replace as necessary.
4. Heat the inner races of the double tapered-roller bearing and 1st gear needle bearing to 100°C (212°F) and start them onto the drive pinion.

5. When the double tapered-roller bearing and 1st gear needle bearing races have cooled to room temperature, press them fully into position, as shown in Fig. 9-53.

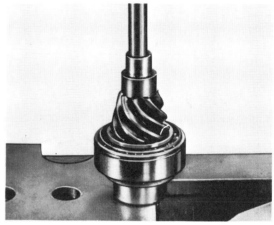

Fig. 9-53. Installing double tapered-roller bearing and 1st gear needle bearing inner race. Use a force of about 3 tons.

6. Install the 1st gear needle bearing.
7. Install a new pinion round nut and tighten it to 20.0 mkg (145 ft. lb.), as shown in Fig. 9-54.

   NOTE ——
   If the double tapered-roller bearing, transmission case, or drive pinion is replaced, the drive pinion must be readjusted. See **10.1 Adjusting Drive Pinion**. If this is the case, install and tighten the round nut, but do not lock it in position.

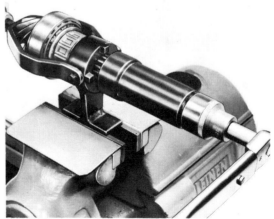

Fig. 9-54. Torquing pinion bearing round nut. Note the holding fixture (VW 293) used. Drive pinion must be firmly secured because considerable force is needed for tightening.

42 Transmission and Rear Axle

8. Check the double tapered-roller bearing preload. See **10.1 Adjusting Drive Pinion.** It is very important that you do this if you have replaced any parts that affect the drive pinion.

9. After checking the bearing preload, lock the shoulder of the round nut by tapping it into the drive pinion splines at three points 120° apart. See Fig. 9-55. Exercise caution because the round nut is easily damaged.

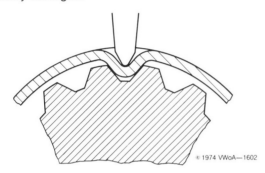

Fig. 9-55. Locking pinion bearing round nut. Use a chisel that has been rounded on the end. If this operation cracks or damages the round nut in any way, replace the nut.

10. Install the shims that control 1st gear axial play. Slide on 1st gear along with the synchronizer ring and the preassembled hub. Then measure the axial play with a feeler gauge as shown in Fig. 9-56. The play must be between 0.10 and 0.25 mm (.004 and .010 in.).

Fig. 9-56. Measuring 1st gear axial play. After determining actual clearance, remove synchronizer hub, ring, and gear. Then install appropriate shims if needed. Shims are available in thicknesses of 0.55, 0.60, 0.65, 0.70, 0.75, 0.80, and 0.85 mm.

11. Install the 2nd gear synchronizer ring, 2nd gear with needle bearing, and 3rd gear.

12. Measure the 3rd gear axial play with a feeler gauge as shown in Fig. 9-57. Clearance should be between 0.10 and 0.25 mm (.004 and .010 in.), preferably closer to the lower limit. To adjust, install the correct circlip. See **Table b** for circlip selection.

Fig. 9-57. Measuring 3rd gear axial play. After determining actual clearance, select the correct circlip from the table below.

Table b. 3rd-gear Circlip Sizes

Shim S mm (in.)	Part No.	Color
1.45 (.0571)	113 311 381	plain steel
1.60 (.0630)	113 311 382	black
1.75 (.0689)	113 311 383	blue
1.90 (.0748)	113 311 384	brown
2.05 (.0807)	113 311 385	gray
2.20 (.0866)	113 311 386	copper

13. Install the new circlip and make certain it is fully seated in the groove.

   CAUTION —
   Once used, an old 3rd gear circlip should not be reinstalled. Replace the circlip routinely to guard against metal fatigue.

14. Install the spacer spring. Make certain there are no nicks or burrs to impair operation.

15. Heat the needle bearing inner race to 100°C (212°F). Start it, along with 4th gear, onto the drive pinion.

16. Press the needle bearing and 4th gear fully into position while they are hot, as shown in Fig. 9-58.

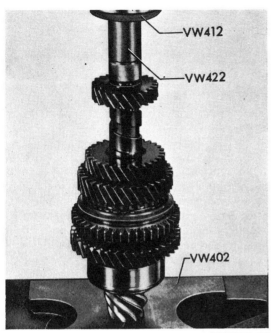

Fig. 9-58. Installing 4th gear and needle bearing inner race.

17. Install the new retaining circlip. Using water pump pliers, squeeze the circlip all around until it is fully seated.

18. If the inner race is not tight, press 3rd gear down as far as it will go. Heat the race and force it down against spacer spring tension, then install the circlip or retaining nut. See Fig. 9-59.

Fig. 9-59. Pressing down 4th gear to compress spacer spring and allow retaining circlip to be installed.

10. Adjusting Final Drive

Careful adjustment of the ring gear and pinion is essential to ensure rear axle durability and quiet operation. For this reason, the ring gear and pinion are factory-matched. Special testing equipment is used to find the position of quietest operation. This is done by moving the pinion axially, with the ring gear lifted far enough out of the no-play mesh position so that backlash is within the specified limits of 0.15 and 0.25 mm (.006 and .010 in.). The deviation **r** from the master gauge R_o of the test machines is measured and stamped on the outer face of the ring gear. Each hypoid gearset is marked with a matching number and may be replaced only as a pair.

The gearset requires adjustment only when parts directly affecting the adjustment have been replaced. When the differential housing, a final drive cover, or a differential tapered-roller bearing is replaced, you need reset only the ring gear. If the transmission case or the complete gearset is replaced, however, both the pinion and the ring gear must be adjusted. If the pinion double tapered-roller bearing is replaced, only the pinion will require adjustment. The object of the adjustment is to restore the pinion and ring gear to the original quiet running position achieved at the factory.

The first step in the adjustment procedure is to locate the pinion by fitting shims between the double tapered-roller bearing and the transmission case. The distance measured from the ring gear centerline to the end face of the pinion should correspond exactly to the dimension R_o + **r** obtained at the factory.

Install the ring gear and adjust it to give the specified backlash and preload between it and the tapered-roller bearings. The preload is determined by the friction in the tapered-roller bearings when the ring gear is rotated. Then select the appropriate shims and install them on either side of the ring gear so that the backlash and bearing preload are permanently set.

CAUTION —
If you lack the skills, tools, or workshop for adjusting the final drive, we suggest you leave such repairs to an Authorized VW Dealer or other qualified shop. We especially urge you to consult your Authorized VW Dealer before attempting repairs on a car still covered by the new-car warranty.

A side view of the ring gear and drive pinion is given in Fig. 10-1. The designation G358 means that the gearset is manufactured by Gleason and has a tooth ratio of 8:35. The designation 369 is the gearset matching number. The designation 26 is the deviation **r** given in hundredths of

44 Transmission and Rear Axle

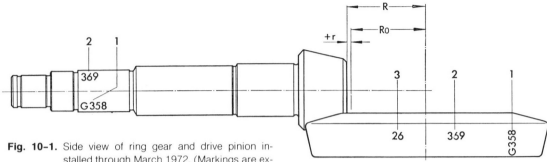

Fig. 10-1. Side view of ring gear and drive pinion installed through March 1972. (Markings are explained in text.)

millimeters. R_o is the length of the master gauge used during production, 58.70 mm (2.3110 in.). R is the distance between the ring gear centerline and the end face of the pinion that produces the least operating noise.

The above method of determining relative positions between the ring gear and the pinion applies to all cars produced before March 1972. The position of the pinion is now determined by dimension P_o, as shown in Fig. 10-2. It is the distance from the centerline of the ring gear to the back of the pinion head.

The actual adjustment of the final drive is performed by (1) finding the total shim thickness **S** needed to give the correct amount of differential bearing preload, (2) finding the thickness of shim S_3 to obtain dimension R, and (3) finding the thickness of shim S_1 and shim S_2 and adjusting the ring gear backlash. See Fig. 10-1 for shim locations and general reference.

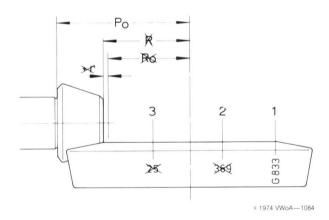

Fig. 10-2. Side view of ring gear and drive pinion after March 1972. The only stamping on these new gearsets is the manufacturer's initial and the tooth ratio. Other designations no longer appear.

Before you disassemble one of these transmissions, it is essential that you measure the position of the pinion. After taking off the left-hand final drive cover, insert a small mirror into the differential housing and examine the edge of the ring gear for markings. Replacement gearsets still have the former markings so measurement is unnecessary if the gearset is to be replaced.

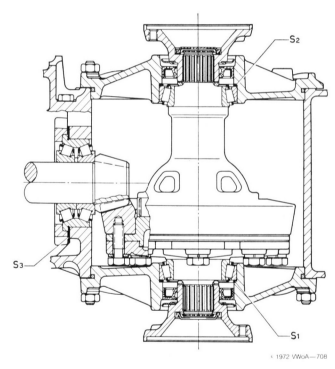

Fig. 10-3. Cutaway view of double-jointed axle differential. Shim S_3 is installed between the pinion double tapered-roller bearing and the transmission case. Shim S_1 is installed between the differential tapered-roller bearing (on the ring gear side) and the transmission case. Shim S_2 is installed between the differential tapered-roller bearing (opposite ring gear) and the transmission case.

Transmission and Rear Axle   45

CAUTION —
Cleanliness and correct procedure must be maintained when adjusting the final drive. The measurements are critical. Dirt particles or grime on the mating surfaces may make these measurements inaccurate.

As stated earlier, it is necessary to adjust the ring gear or pinion or both only if you are replacing parts that directly affect the adjustment. Parts replacement and the corresponding necessary adjustments are given in **Table c**. The ring gear setting includes the adjustment of the differential bearings.

Table c. Necessary Adjustments for Replaced Parts

Part Replaced	Parts to Be Adjusted		
	Pinion	Shift Forks*	Ring Gear
Transmission case	X	X	X
Final drive covers			X
Differential tapered-roller bearings			X
Pinion double tapered-roller bearing	X	X	
Ring gear and pinion	X	X	X
Differential housing			X
Differential housing cover			X

* See 9.4 Adjusting Shift Forks.

CAUTION —
The final drive adjustment requires considerable skill and precision. So if you lack the special tools discussed in the following procedures, we recommend that you leave this adjustment to an Authorized VW Dealer or other qualified shop.

Several formulas containing symbols are used to determine the shim thicknesses needed to make final drive adjustments. **Table d** lists the meanings of the symbols. Once familiar with them, you will find the formulas easy to use. You will encounter them in the text.

NOTE —
The terms listed below apply to all manual transmissions except those manufactured after March 1972. In the case of transmissions manufactured after March 1972, P_o (defined in Fig. 10-2) is used rather than R, R_o, and r.

Table d. Symbols Used in Ring Gear/Pinion Adjustment

Symbol	Designation	Dimension
S	Total shim thickness ($S_1 + S_2$) is found by subtracting length of differential housing either from internal housing dimension or from tapered-roller bearing preload	mm (can be from 0.95 to 2.10 mm)
S_1	Shim at ring gear end of differential	See **Table g** for shim thickness
S_2	Shim at opposite end of differential	See **Table h** for shim thickness
S_3	Shim between double tapered-roller bearing and housing	See shim table for thickness
r	Deviation (from R_o) marked on ring gear	1/100 mm (can be from 0.05 to 0.65 mm)
R_o	Length of master gauge on special test machine	58.70 mm
R	Position of pinion in relation to ring gear centerline at quietest running point (nominal dimension)	$R = R_o + r$
Sv_o	Average backlash; average of Sv_o readings	Given in 1/100 mm
e	Difference between zero setting on mandrel and actual pinion position without shims	0.10 to 0.50 mm
ΔS_1	Axial movement of ring gear to specified mean backlash (correction value)	Given in 1/100 mm
w	Correction factor for gearset	
h	Ring gear lift from no-play mesh position to mean backlash position	Given in 1/100 mm
G 833	Gear set G = Gleason 833 = ratio of 8:33	
K 833	Gear set K = Klingelnberg 833 = ratio of 8:33	

46 Transmission and Rear Axle

10.1 Adjusting Drive Pinion

When performing the following measurements, it is advisable to record all your findings. Doing so will avoid confusion and the possibility of having to repeat the measurements.

To check bearing preload:

1. Assemble the drive pinion up to the needle bearing for 1st gear. Install and tighten the round nut to 20.0 mkg (145 ft. lb.) but do not lock it.

2. Install the assembled drive pinion in the transmission case without shim S_3. Tighten the pinion retaining nut to 22.0 mkg (160 ft. lb.).

> **CAUTION** ——
> All parts must be absolutely clean. Any dirt or grime may cause errors in the measurements.

3. Check the double tapered-roller bearing preload with a torque gauge, adapter, and 32-mm (1¼-in.) socket, as illustrated in Fig. 10-4.

> **CAUTION** ——
> Before measuring bearing preload, lubricate the double tapered-roller bearing with hypoid gear oil. If checked when dry or when lubricated with any other type of oil, the readings may be incorrect.

Fig. 10-4. Checking double tapered-roller bearing preload. The turning torque for a new bearing should be 6.0 to 21.0 cmkg (5.2 to 18.2 in. lb.) and for a used bearing (used for 30 mi. [48 km] or more) 3.0 to 7.0 cmkg (2.6 to 6.1 in. lb.).

To determine pinion position (case with integral right-hand final drive cover):

1. Remove the seals and shims from the final drive covers (both integral and separate) and make certain the roller bearing outer races are pressed in fully.

2. Place the measuring plate on the drive pinion head as shown in Fig. 10-5.

Fig. 10-5. Measuring plate installed on drive pinion head. Swing axle case is shown, but installation is same for double-jointed axle transmissions.

3. Adjust the setting ring at the dial-indicator end of the universal measuring bar to dimension **a**, given in Fig. 10-6. Position the other setting ring near the end of the bar, then tighten the setscrew until you can barely slide the ring along the bar.

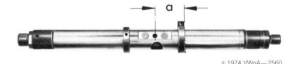

Fig. 10-6. Setting ring adjustment. Dimension **a** is approximately 70 mm (2.76 in.).

4. Slide the centering discs onto the bar until they contact the setting rings. Attach measuring pin VW 385/14 and extension VW 385/15 to the center of the bar. Then install the dial indicator.

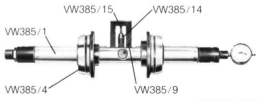

Fig. 10-7. Measuring bar with centering discs, measuring pin, and dial indicator installed. Discs shown are for swing axle transmissions. VW 385/2 discs must be used on double-jointed axle transmissions.

Transmission and Rear Axle 47

5. Install the setting gauge shown in Fig. 10-7 (VW 385/9) on the measuring bar and zero the dial indicator with a 1-mm preload.

6. Remove the setting gauge and install the measuring bar in the differential housing.

7. Carefully hand-press the left final drive cover down onto the transmission case, moving the setting ring and centering disc down with the cover. Remove the cover, then tighten the setscrew in the setting ring. Install the final drive cover and torque the nuts to 3.0 mkg (22 ft. lb.). You should just barely be able to hand-turn the measuring bar.

8. Turn the bar until the measuring pin contacts the measuring plate on the pinion and the dial indicator needle begins to reverse.

9. Note the reading, which is dimension **e**. This will be used in determining the thickness(es) of the S_3 shim(s).

To determine pinion position (transmission case with separate right-hand final drive cover):

1. Remove the oil seals and shims from both final drive covers. Install the bearing outer races.

2. Install the right final drive cover on the transmission case, then place the spindle tool (illustrated in Fig. 10-8) on top of it. Install and tighten the cover retaining nuts to 3.0 mkg (22 ft. lb.).

   NOTE —
   The spindle tool cannot be fitted on transmissions that have an integral right final drive cover. In its place, use VW universal measuring bar (VW 385). See Step 2 above.

Fig. 10-8. Spindle tool installed on right final drive cover. Tool fits over final drive cover studs.

3. Insert a dial indicator with an 18-mm extension pin into the special mandrel as shown in Fig. 10-9. Zero the dial indicator on the setting block with a preload of 1 mm.

Fig. 10-9. Special mandrel with dial indicator, setting block, and 18-mm extension.

4. Rotate the transmission on the repair stand 180°, then install the mandrel with dial indicator.

5. Install the left final drive cover and tighten the retaining nuts evenly.

6. Press the right bearing outer race against the mandrel by turning the spindle tool handle until you can just barely hand-turn the mandrel.

7. Bring the extension pin into contact with the end face of the drive pinion, then turn the mandrel until the maximum reading is obtained. Record the reading, taking into account the 1-mm preload. This reading is dimension **e**. See Fig. 10-10.

Fig. 10-10. Turning mandrel to find maximum dial indicator reading. Needle will peak and reverse when maximum reading is found.

48 Transmission and Rear Axle

8. Note the deviation **r** stamped on the ring gear. Then calculate the required thickness of shim S_3 as described below.

Determining S_3 Thickness

The following example shows how the required thickness of shim S_3 is obtained. For this particular example, R_o = 58.70 mm and the measured dimension **e** = 0.48 mm. Subtracting **e** from R_o yields the actual distance from the mandrel to the pinion face without S_3 shims.

R_o = 58.70 mm
e = 0.48 mm
R_o − **e** = 58.22 mm (actual distance).

Adding the deviation **r** to R_o gives the quietest running position for the gearset. For this particular example, **r** = 0.18 mm.

R_o = 58.70 mm
r = 0.18 mm
R_o + **r** = 58.88 mm (nominal measurement).

To find the thickness required for shim S_3, subtract the actual distance (R_o − **e**) from the nominal measurement (R_o + **r**).

R_o + **r** = 58.88 mm
R_o − **e** = 58.22 mm
S_3 = 0.66 mm (thickness required).

Combining these calculations gives the simple formula S_3 nominal = **e** + **r**. In this particular example it would be

e = 0.48 mm
r = 0.18 mm
S_3 nominal = 0.66 mm.

Locate the required S_3 shim thickness in **Table e**, then select the appropriate shims by number from **Table f**. Carefully examine the shims for nicks and burrs and discard any that are damaged. Using a micrometer, check the thickness at several points.

> **NOTE**
> We recommend that you set the preload for the differential tapered-roller bearings before you install the pinion with shims. The drive pinion can then be installed along with the transmission. In most cases the drive pinion will not have to be taken out again after the S_3 shim selection has been checked.

Table e gives the required S_3 thickness and the appropriate combination of shims available to achieve this dimension. Select shims by number from **Table f**.

Table e. S_3 Shim Measurements

Calculated Shim Thickness Required mm (in.)	Shim No.	Shim Thickness mm (in.)	S_3 Actual mm (in.)
0.33-0.37 (.0130-.0146)	2 + 1	0.20 + 0.15 (.0079 + .0059)	0.35 (.0138)
0.38-0.42 (.0150-.0165)	4	0.40 (.0157)	0.40 (.0157)
0.43-0.47 (.0169-.0185)	3 + 1	0.30 + 0.15 (.0118 + .0059)	0.45 (.0177)
0.48-0.52 (.0189-.0205)	5	0.50 (.0197)	0.50 (.0197)
0.54-0.57 (.0213-.0224)	4 + 1	0.40 + 0.15 (.0157 + .0059)	0.55 (.0216)
0.58-0.62 (.0228-.0244)	6	0.60 (.0236)	0.60 (.0236)
0.63-0.67 (.0248-.0264)	5 + 1	0.50 + 0.15 (.0197 + .0059)	0.65 (.0256)
0.68-0.72 (.0268-.0283)	7	0.70 (.0276)	0.70 (.0276)
0.73-0.77 (.0287-.0303)	6 + 1	0.60 + 0.15 (.0236 + .0059)	0.75 (.0295)
0.78-0.82 (.0307-.0323)	8	0.80 (.0315)	0.80 (.0315)
0.83-0.87 (.0327-.0342)	7 + 1	0.70 + 0.15 (.0276 + .0059)	0.85 (.0335)
0.88-0.92 (.0346-.0362)	9	0.90 (.0354)	0.90 (.0354)
0.93-0.97 (.0366-.0382)	8 + 1	0.80 + 0.15 (.0315 + .0059)	0.95 (.0374)
0.98-1.02 (.0386-.0402)	10	1.00 (.0394)	1.00 (.0394)
1.03-1.07 (.0405-.0421)	9 + 1	0.90 + 0.15 (.0354 + .0059)	1.05 (.0413)
1.08-1.12 (.0425-.0441)	9 + 2	0.90 + 0.20 (.0354 + .0079)	1.10 (.0433)
1.13-1.17 (.0445-.0461)	10 + 1	1.00 + 0.15 (.0394 + .0059)	1.15 (.0453)
1.18-1.22 (.0465-.0480)	10 + 2	1.00 + 0.20 (.0394 + .0079)	1.20 (.0472)
1.23-1.27 (.0484-.0500)	6 + 5 + 1	0.60 + 0.50 + 0.15 (.0236 + .0197 + .0059)	1.25 (.0492)
1.28-1.32 (.0504-.0520)	10 + 3	1.00 + 0.30 (.0394 + .0118)	1.30 (.0512)

Transmission and Rear Axle 49

Table f. Available S_3 Shim Sizes

Shim No.	Part No.	Thickness mm (in.)
1	113 311 391	0.15 (.0059)
2	113 311 392	0.20 (.0079)
3	113 311 393	0.30 (.0118)
4	113 311 394	0.40 (.0157)
5	113 311 395	0.50 (.0197)
6	113 311 396	0.60 (.0236)
7	113 311 397	0.70 (.0276)
8	113 311 398	0.80 (.0315)
9	113 311 399	0.90 (.0354)
10	113 311 400	1.00 (.0394)
11	113 311 401	1.20 (.0472)

To check S_3 shim selection (case with integral right final drive cover):

1. Install the drive pinion complete with S_3 shim(s). Install the measuring plate on the pinion head.

2. Tighten the pinion bearing retaining nut to 22.0 mkg (160 ft. lb.).

3. Install the measuring bar as shown in Fig. 10-11. Check the dial indicator reading when the needle peaks and begins to reverse.

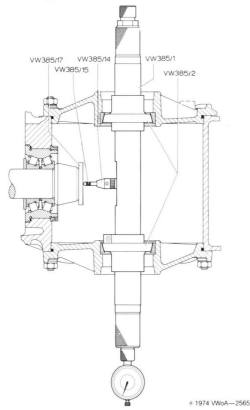

Fig. 10-11. Checking shim S_3 selection (transmission case with integral right final drive cover). The correct shim thickness selection should cause the dial indicator to read the factory-set **r** ± 0.04 mm (.0016 in.).

4. Remove the measuring bar and, if necessary, proceed with the adjustment of the ring gear.

To check S_3 shim selection (case with separate right final drive cover):

1. Install the drive pinion complete with S_3 shim(s).

2. Install and tighten the pinion bearing retaining nut to 22.0 mkg (160 ft. lb.).

3. Install the measuring mandrel with dial indicator and extension in the differential housing, as shown in Fig. 10-12. Zero the dial indicator on the special setting block (382/2) with a 1-mm preload or .050 in. on dial indicator calibrated in inches.

4. Turn the mandrel until the dial indicator pin contacts the pinion head. Note the reading when the indicator needle peaks and begins to reverse.

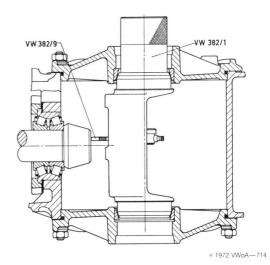

Fig. 10-12. Checking shim S_3 selection (transmission case with separate right final drive cover). Make certain all surfaces are absolutely clean. With the correct shim(s) installed, the dial indicator reading should be the factory-set **r** ± 0.04 mm (.0016 in.).

NOTE —

If the installed gearset is of the new type without **r** stamped on the ring gear, it is advisable to record the size and number of shims you find when you disassemble the final drive. This will save considerable time during future repair work.

5. Remove the measuring mandrel and, if necessary, proceed with the adjustment of the ring gear.

50 Transmission and Rear Axle

10.2 Adjusting Ring Gear

The ring gear requires adjustment only when one or more of the following parts are replaced: gearset, transmission case, differential housing, differential housing cover, final drive covers, or tapered-roller bearings. Fig. 10-13 is a cutaway view of the differential and the special tool set-up used for adjusting the ring gear. Study the illustration to familiarize yourself with the tools.

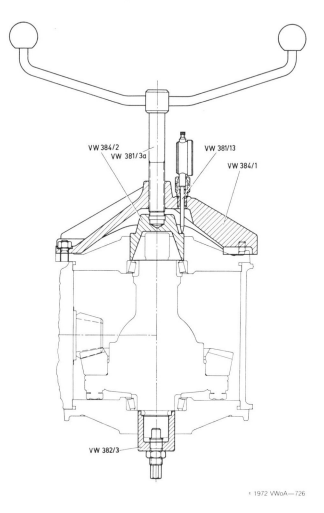

Fig. 10-13. Differential cutaway and special tools used to adjust ring gear. Tools are indicated by shaded areas.

To adjust tapered-roller bearings (pinion removed):

1. Press the oil seals and the bearing outer races out of the final drive covers. Remove the shims and carefully press the bearing outer races back into position. They must be fully seated.

2. Install the right-hand final drive cover on the transmission case if the transmission has a removable cover.

3. Install the spindle tool shown in Fig. 10-14 on the right final drive cover. Install and tighten the cover retaining nuts to 3.0 mkg (22 ft. lb).

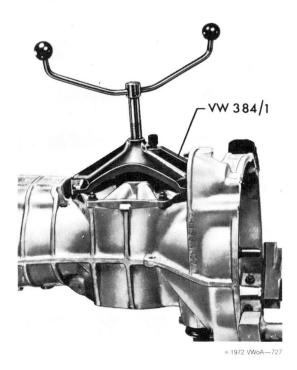

Fig. 10-14. Spindle tool installed on right final drive cover. We suggest that you remove the cover O-rings during adjustment.

4. Turn the transmission case 180° and install the differential. The ring gear should be on the left side.

   CAUTION —
   Before installing the differential, liberally lubricate the tapered-roller bearings with hypoid oil. Bearings that are dry or have been lubricated with other oils will cause incorrect readings.

5. Install the left final drive cover, then tighten the retaining nuts to 3.0 mkg (22 ft. lb.). Install the clamping sleeve (VW 382/3) without the measuring lever on the ring gear end.

6. Turn the transmission over so that the ring gear is on top. Slowly turn the spindle until the thrust piece just touches the bearing outer race without play. The outer race must not be moved at this point. If it is, the final drive cover and special tools will have to be disassembled and the bearing outer race once again pressed fully into its seat.

Transmission and Rear Axle 51

7. Install a dial indicator with a 52-mm extension on the spindle tool. Zero the indicator with a 3-mm preload (see Fig. 10-15).

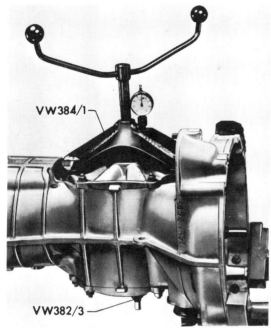

Fig. 10-15. Dial indicator installed on spindle tool.

8. Press in the bearing outer race by turning the spindle tool until there is no detectable play in the differential bearings.

9. Turn the transmission 90° and put a 10-mm socket torque gauge on the clamping sleeve. Turn the differential in both directions. See Fig. 10-16.

Fig. 10-16. Torque gauge installed on the special clamping sleeve.

10. Screw the spindle in very slowly, increasing pressure on the bearing race until the preload reaches 30 to 35 cmkg (26 to 30 in. lb.) for a new bearing or 3 to 7 cmkg (3 to 6 in. lb.) for a used bearing (in service 30 mi. [48 km] or more).

NOTE

Apply pressure to the bearing outer race very carefully until the specified preload is reached. If the preload is exceeded, the special tool will have to be removed and the outer race once again pressed fully into the final drive cover.

11. Note the dial indicator reading. This reading gives the total required shim thickness **S**. It should fall somewhere between 0.95 and 2.10 mm (.0374 and .0827 in.).

NOTE

On the new-type transmission case with integral right final drive cover, the spindle, clamping sleeve, and torque gauge must be installed on an opposite side of the transmission case. The set-up is shown in Fig. 10-17.

Fig. 10-17. Measuring tools installed on new-type transmission case with integral right final drive cover.

12. Remove the final drive cover and the differential. Press in the bearing outer race until it is again fully seated in the cover.

52 Transmission and Rear Axle

13. Install the transmission with the previously determined pinion bearing shim S_3.

To adjust ring gear backlash (transmission installed):

1. Install the right-hand final drive cover on the transmission, if the transmission has a removable cover. Then install the spindle tool. Tighten the retaining nuts to 3.0 mkg (22 ft. lb.).

2. Install the rear drive shaft. Install the differential with clamping sleeve fitted and position the final drive cover on the ring gear side with the dial indicator. Tighten the retaining nuts to 3.0 mkg (22 ft. lb.).

3. While rotating the differential, tighten the spindle tool until the dial indicator shows the same figure as it did when the bearing preload was being set.

4. Install the pinion retaining bracket on the gear carrier and tighten the nuts by hand.

5. Screw the special lever into the clamping sleeve and install a dial indicator with a 6-mm extension in the indicator holder. The end of the indicator stem should be flush with the edge of the holder.

6. Turn the differential until the lever contacts the dial indicator pin. Continue to turn the differential slowly until the indicator shows a preload of 1.50 mm (or .050 in.). See Fig. 10-18.

CAUTION
Take care not to turn the differential too far. Doing so could damage the dial indicator.

Fig. 10-18. Installed dial indicator and special lever.

7. Secure the drive pinion in position with a 1.5-mm (or .050-in.) preload by tightening the retaining bracket nuts securely. Recheck the dial indicator setting to make certain the reading is unchanged. See Fig. 10-19.

Fig. 10-19. Drive pinion secured by retaining bracket with a 1.5-mm (or .050-in.) preload on the dial indicator.

NOTE
On new-type transmissions with the integral right final drive cover, install the measuring tools in opposite positions from those indicated above. Fig. 10-20 shows the setup.

Fig. 10-20. Backlash measuring tools installed on new-type transmission case with integral right final drive cover.

8. Hand-turn the ring gear away from the dial indicator until the ring gear is stopped by the locked drive pinion. Then zero the dial indicator.

9. From the zeroed position, hand-turn the ring gear toward the dial indicator until it is stopped by the locked drive pinion. Write down the reading. This is the backlash, sometimes abbreviated Sv_o.

10. Loosen the clamping sleeve and pinion retaining bracket. Turn the ring gear, then retighten the tools and take three other readings at intervals of

Transmission and Rear Axle

90°. Add the readings together and divide by 4 to find the average.

NOTE —
The difference between backlash readings must always be smaller than 0.06 mm (.0024 in.). If there is a greater variation, something is wrong with the gearset or its installation.

11. Determine the required shim S_1 thickness by using **Table g**. Measure the shims at several points and discard any that are burred or damaged.

Table g. S_1 Shim Thicknesses

Backlash Sv_o Mean mm (in.)	Shim Thickness mm (in.)	Backlash Sv_o Mean mm (in.)	Shim Thickness mm (in.)
0.20 (.0079)	0.00 (.0000)		
0.21 (.0083)	0.01 (.0004)	0.71 (.0280)	0.51 (.0201)
0.22 (.0087)	0.02 (.0008)	0.72 (.0283)	0.52 (.0205)
0.23 (.0091)	0.03 (.0012)	0.73 (.0287)	0.53 (.0209)
0.24 (.0094)	0.04 (.0016)	0.74 (.0291)	0.54 (.0213)
0.25 (.0098)	0.05 (.0020)	0.75 (.0295)	0.55 (.0217)
0.26 (.0102)	0.06 (.0024)	0.76 (.0299)	0.56 (.0220)
0.27 (.0106)	0.07 (.0028)	0.77 (.0303)	0.57 (.0224)
0.28 (.0110)	0.08 (.0031)	0.78 (.0307)	0.58 (.0228)
0.29 (.0114)	0.09 (.0035)	0.79 (.0311)	0.59 (.0232)
0.30 (.0118)	0.10 (.0039)	0.80 (.0315)	0.60 (.0236)
0.31 (.0122)	0.11 (.0043)	0.81 (.0319)	0.61 (.0240)
0.32 (.0126)	0.12 (.0047)	0.82 (.0323)	0.62 (.0244)
0.33 (.0130)	0.13 (.0051)	0.83 (.0327)	0.63 (.0248)
0.34 (.0134)	0.14 (.0055)	0.84 (.0331)	0.64 (.0252)
0.35 (.0138)	0.15 (.0059)	0.85 (.0335)	0.65 (.0256)
0.36 (.0142)	0.16 (.0063)	0.86 (.0339)	0.66 (.0260)
0.37 (.0146)	0.17 (.0067)	0.87 (.0343)	0.67 (.0264)
0.38 (.0150)	0.18 (.0071)	0.88 (.0346)	0.68 (.0268)
0.39 (.0154)	0.19 (.0075)	0.89 (.0350)	0.69 (.0272)
0.40 (.0157)	0.20 (.0079)	0.90 (.0354)	0.70 (.0276)
0.41 (.0161)	0.21 (.0083)	0.91 (.0358)	0.71 (.0280)
0.42 (.0165)	0.22 (.0087)	0.92 (.0362)	0.72 (.0283)
0.43 (.0169)	0.23 (.0091)	0.93 (.0366)	0.73 (.0287)
0.44 (.0173)	0.24 (.0094)	0.94 (.0370)	0.74 (.0291)
0.45 (.0177)	0.25 (.0098)	0.95 (.0374)	0.75 (.0295)
0.46 (.0181)	0.26 (.0102)	0.96 (.0378)	0.76 (.0299)
0.47 (.0185)	0.27 (.0106)	0.97 (.0382)	0.77 (.0303)
0.48 (.0189)	0.28 (.0110)	0.98 (.0386)	0.78 (.0307)
0.49 (.0193)	0.29 (.0114)	0.99 (.0390)	0.79 (.0311)
0.50 (.0197)	0.30 (.0118)	1.00 (.0394)	0.80 (.0315)
0.51 (.0201)	0.31 (.0122)	1.01 (.0398)	0.81 (.0319)
0.52 (.0205)	0.32 (.0126)	1.02 (.0402)	0.82 (.0323)
0.53 (.0209)	0.33 (.0130)	1.03 (.0406)	0.83 (.0327)
0.54 (.0213)	0.34 (.0134)	1.04 (.0409)	0.84 (.0331)
0.55 (.0217)	0.35 (.0138)	1.05 (.0413)	0.85 (.0335)
0.56 (.0220)	0.36 (.0142)	1.06 (.0417)	0.86 (.0339)
0.57 (.0224)	0.37 (.0146)	1.07 (.0421)	0.87 (.0343)
0.58 (.0228)	0.38 (.0150)	1.08 (.0425)	0.88 (.0346)
0.59 (.0232)	0.39 (.0154)	1.09 (.0429)	0.89 (.0350)
0.60 (.0236)	0.40 (.0157)	1.10 (.0433)	0.90 (.0354)
0.61 (.0240)	0.41 (.0161)	1.11 (.0437)	0.91 (.0358)
0.62 (.0244)	0.42 (.0165)	1.12 (.0441)	0.92 (.0362)
0.63 (.0248)	0.43 (.0169)	1.13 (.0445)	0.93 (.0366)
0.64 (.0252)	0.44 (.0173)	1.14 (.0449)	0.94 (.0370)
0.65 (.0256)	0.45 (.0177)	1.15 (.0453)	0.95 (.0374)
0.66 (.0260)	0.46 (.0181)	1.16 (.0457)	0.96 (.0378)
0.67 (.0264)	0.47 (.0185)	1.17 (.0461)	0.97 (.0382)
0.68 (.0268)	0.48 (.0189)	1.18 (.0465)	0.98 (.0386)
0.69 (.0272)	0.49 (.0193)	1.19 (.0469)	0.99 (.0390)
0.70 (.0276)	0.50 (.0197)	1.20 (.0472)	1.00 (.0394)

12. Determine the required thickness for shim S_2 by subtracting S_1 from total shim thickness **S** found earlier. Measure the shims at several points and discard any that are burred or damaged. **Table h** lists S_1 and S_2 shim sizes. **Table i** lists combinations of shims to achieve the required thickness.

Table h. S_2 Shim Sizes

Shim No.	Part No.	Thickness mm (in.)
1	113 517 201 A	0.15 (.0059)
2	113 517 202 A	0.20 (.0079)
3	113 517 203 A	0.30 (.0118)
4	113 517 204 A	0.40 (.0157)
5	113 517 205 A	0.50 (.0197)
6	113 517 206 A	0.60 (.0236)
7	113 517 207 A	0.70 (.0276)
8	113 517 208 A	0.80 (.0315)
9	113 517 209 A	0.90 (.0354)
10	113 517 210 A	1.00 (.0394)
11	113 517 211 A	1.20 (.0472)

Table i. S_1 and S_2 Shim Combinations

Nominal Shim Thickness as Found for S_1 or S_2 mm (in.)	Shim Thickness mm (in.)	Shim No.
0.28–0.32 (.0110–.0126)	0.30 (.0118)	3
0.33–0.37 (.0130–.0146)	0.35 (.0138)	1 + 2
0.38–0.42 (.0150–.0165)	0.40 (.0157)	4
0.43–0.47 (.0169–.0185)	0.45 (.0177)	1 + 3
0.48–0.52 (.0189–.0205)	0.50 (.0197)	5
0.53–0.57 (.0209–.0224)	0.55 (.0217)	1 + 4
0.58–0.62 (.0228–.0244)	0.60 (.0236)	6
0.63–0.67 (.0248–.0264)	0.65 (.0256)	1 + 5
0.68–0.72 (.0268–.0283)	0.70 (.0276)	7
0.73–0.77 (.0287–.0303)	0.75 (.0295)	1 + 6
0.78–0.82 (.0307–.0323)	0.80 (.0315)	8
0.83–0.87 (.0327–.0343)	0.85 (.0335)	1 + 7
0.88–0.92 (.0346–.0362)	0.90 (.0354)	9
0.93–0.97 (.0366–.0382)	0.95 (.0374)	1 + 8
0.98–1.02 (.0386–.0402)	1.00 (.0394)	10
1.03–1.07 (.0406–.0421)	1.05 (.0413)	1 + 9
1.08–1.12 (.0425–.0441)	1.10 (.0433)	2 + 9
1.13–1.17 (.0445–.0461)	1.15 (.0453)	1 + 10
1.18–1.22 (.0465–.0480)	1.20 (.0472)	11
1.23–1.27 (.0484–.0500)	1.25 (.0492)	1 + 5 + 6
1.28–1.32 (.0504–.0520)	1.30 (.0512)	3 + 10

13. Install shim S_1 on the ring gear side and S_2 on the opposite side. Install the right final drive cover along with the oil seal and O-ring. Install the left final drive cover without the oil seal and O-ring.

14. Recheck the backlash at four points 90° apart. Sv_o should be 0.15 to 0.25 mm (.006 to .010 in.). Readings must not vary more than 0.05 mm (.002 in.).

15. Remove the measuring tools and install the left final drive cover along with the oil seal and O-ring. Tighten the retaining nuts to 3.0 mkg (22 ft. lb.).

54 Transmission and Rear Axle

11. Rear Suspension

The rear wheels are independently sprung by means of trailing spring plates, torsion bars, and shock absorbers. The system is fully adjustable to allow accurate rear axle alignment, wheel toe-in, and camber.

11.1 Removing and Installing Shock Absorbers

The shock absorbers are hydraulic double-acting units that dampen suspension rebound. They are mounted on the diagonal arm and body by two bolts and two nuts.

To remove:

1. Lift the car and remove the rear wheel.

2. Remove the bottom shock absorber mounting bolt, washer, and nut, as shown in Fig. 11-1. Remove the top mounting bolt, washer, and nut.

Fig. 11-1. Removing bottom shock absorber mounting bolt and nut. Make sure the car is firmly supported since considerable force may be required to free the nut.

To install:

1. Examine the shock absorber. Minor fluid leakage is acceptable. Extend and compress the shock absorber by hand. Damping should be smooth throughout the entire range. Replace the shock absorber if necessary.

2. Install the shock absorber along with mounting nuts, bolts, and washers. Tighten the nuts to 6.0 mkg (43 ft. lb.).

   NOTE —
   Install only shock absorbers intended for the rear axle. The wrong shock absorbers can impair the handling and riding comfort of the car. Damaged shock absorbers can be replaced individually regardless of manufacturer, if damping characteristics are the same for both rear units.

11.2 Removing and Installing Torsion Bars and Spring Plates

WARNING —
Do not re-use fasteners that are worn or deformed in normal use. Many are designed to be used only once and may fail when used a second time. This includes nuts, bolts, washers, self-locking nuts or bolts, circlips, cotter pins. Always use new parts.

To remove:

1. Remove the road wheel, cotter pin, and axle shaft nut.

   WARNING —
   Loosen the axle shaft nuts while the car is on the ground. The leverage needed for this job could topple the car from a lift.

2. Remove the driveshafts and protect the constant velocity joints with plastic caps.

3. Remove the lower shock absorber mounting bolt or the whole shock absorber unit for ease of access.

4. Remove the brake drum and backing plate. A puller may be required to remove the brake drum, especially if the car has a lot of mileage.

   NOTE —
   If only the wheel bearings are to be serviced, it is unnecessary to disconnect the hydraulic brake lines or the parking brake cables. Leave them connected and wire the backing plate to the frame of the car.

5. With a chisel, mark the spring plate and the diagonal arm bearing flange as shown in Fig. 11-2. The marks will be used for proper realignment during assembly.

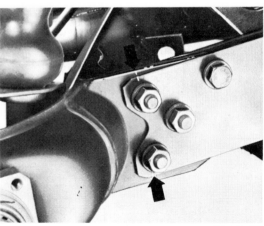

Fig. 11-2. Spring plate alignment marks (arrows).

Transmission and Rear Axle

6. Remove the nuts and bolts that secure the spring plate to the diagonal arm.
7. Remove the fitted bolt that secures the diagonal arm to the bracket, as shown in Fig. 11-3. Remove the diagonal arm.

Fig. 11-3. Diagonal arm bracket mounting. Note number and location of washers for reference when installing. Arrow indicates socket head fitted bolt.

8. Remove the securing bolts and the spring plate hub cover.
9. Using a tire iron, lift the spring plate off its lower stop as shown in Fig. 11-4.

Fig. 11-4. Lifting spring plate off lower stop. Apply pressure in direction of arrow.

10. Remove the five bolts at the front of the fender panel. Remove the spring plate and pull out the torsion bar.

CAUTION —
Unless the torsion bar is to be replaced, handle it with great care to avoid scratching the protective paint. Exposed metal will cause eventual corrosion and possibly fatigue fractures. Touch up any scratched areas immediately.

To install:

1. Examine the torsion bar for spline damage and signs of rusting. Replace as necessary. Also inspect rubber bushings for cracks and signs of excessive wear or deterioration.

   NOTE —
   Torsion bars are prestressed in only one direction, so they are not interchangeable. They are marked on their outer ends with the letter **L** or **R**, indicating left- or right-hand side.

2. Liberally grease the torsion bar splines.
3. Install the rubber bushings.

   NOTE —
   Before installing the rubber bushings, coat them with talcum powder as shown in Fig. 11-5. Do not use graphite because graphite will cause premature bushing wear. Also, make certain the bushings are installed with the word **oben** (top) facing upward.

Fig. 11-5. Rubber bushing coated with talcum powder.

4. Install the torsion bar, spring plate, and outer bushing. The spring plate and the diagonal arm should form the angle shown in Fig. 11-6.

56 Transmission and Rear Axle

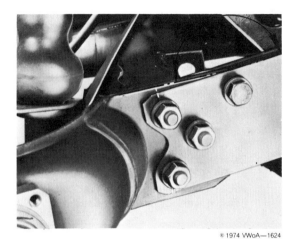

Fig. 11-6. Spring plate secured to diagonal arm. Angle should be approximately as shown to prevent binding when car is lowered and suspension has settled.

5. Adjust the spring plate angle as in **11.3 Spring Plate Adjustment**.

6. Install the spring plate hub cover and secure it with two bolts. If necessary, longer-than-standard bolts can be used temporarily. See Fig. 11-7.

Fig. 11-7. Spring plate hub cover installed. Two longer-than-standard bolts are used until original bolts can be installed.

NOTE ——
When lifting the double spring plate to attach it to the diagonal arm, fit a nut between the leaves to keep them apart. This will make it considerably easier to install the diagonal arm flange.

7. Using a tensioner, lift the spring plate onto the lower stop, as shown in Fig. 11-8. Install the original spring plate hub cover bolts and tighten them to 11.0 mkg (80 ft. lb.).

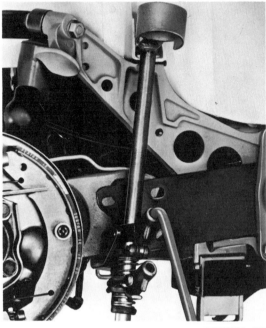

Fig. 11-8. Lifting spring plate onto lower stop.

8. Secure the diagonal arm to its bracket with the fitted bolt and washers. Tighten the bolt to 12.0 mkg (87 ft. lb.). Lock the bolt by peening it to the shoulder of the bracket, as shown in Fig. 11-9.

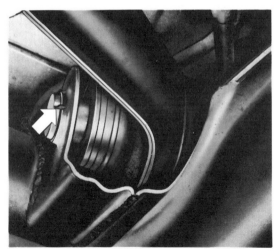

Fig. 11-9. Peened diagonal arm bracket fitted bolt (arrow).

Transmission and Rear Axle 57

NOTE —
To prevent tension in the bonded rubber mounting, tighten the fitted bolt with the diagonal arm extended.

9. Secure the spring plate to the diagonal arm. Be sure the alignment marks match. Torque the bolts to 11.0 mkg (80 ft. lb.).

NOTE —
If a torsion bar, spring plate, or diagonal arm is replaced, the rear wheel alignment must be checked. See **12. Rear Wheel Alignment**.

10. The remaining assembly procedure is simply the reverse of removal. Torque the driveshaft screws to 3.5 mkg (25 ft. lb.). Torque the axle shaft nuts to 30.0 mkg (217 ft. lb.), then install a new cotter pin. Bleed the brakes if necessary.

WARNING —
- *Tighten the axle shaft nuts while the car is on the ground. The leverage needed for this job could topple the car from a lift.*
- *Do not re-use cotter pins or any fasteners that are worn or deformed in normal use. Always replace.*

11.3 Spring Plate Adjustment

There are 40 splines at the inner end of each torsion bar and 44 splines on the outer end. This provides a vernier adjustment for setting the spring plate angle. By rotating the bar—one spline (9° 00′) at its inner end and one spline in the opposite direction at its outer end (8° 10′)—the spring plate angle is changed in increments of 0° 50′ in whichever direction the inner spline is rotated.

To adjust:

1. Using a fluid level protractor, determine the car's axis deviation from the horizontal, as in Fig. 11-10.

Fig. 11-10. Finding axis deviation from horizontal.

2. Place the spring plate on the torsion bar and position the protractor as indicated in Fig. 11-11.

3. The angle should be 21° 20′ + 50′—that is, it can be up to 50′ greater than 21° 20′ but no less. If it is outside this range, the spring plate angle should be adjusted.

NOTE —
When measuring spring plate angle, gently lift the plate lightly to eliminate any freeplay between splines.

Fig. 11-11. Measuring spring plate angle. Make certain contact surfaces are clean.

4. Depending upon the amount of error, turn the torsion bar one spline forward or backward. Turn the spring plate one spline in the opposite direction to effect a 0° 50′ change.

5. After the spring plate angle has been set, install the spring plate hub cover and connect the spring plate to the diagonal arm. Tighten the spring plate bolts to 11.0 mkg (80 ft. lb.).

6. Install the remaining parts in the reverse order of removal. With the car on the ground, tighten the driveshaft screws to 3.5 mkg (25 ft. lb.) and the axle shaft nuts to 30 mkg (217 ft. lb.).

NOTE —
Whenever possible, especially on high-mileage cars, adjust both torsion bars in order to ensure an equal and correct setting.

58 Transmission and Rear Axle

11.4 Disassembling and Assembling Diagonal Arm

Cleanliness and order are important in making the following repairs. Clean all metal parts thoroughly in solvent. Do not spin the bearings until after they have been cleaned, dried, and lubricated.

CAUTION —
If you lack the skills, tools, or workshop for servicing the diagonal arm, we suggest you leave such repairs to an Authorized VW Dealer or other qualified shop. We especially urge you to consult your Authorized VW Dealer before attempting repairs on a car still covered by the new-car warranty.

To disassemble:

1. Clamp the diagonal arm flange in a vise.
2. If you have not already done so, remove the axle shaft nut and the brake drum.
3. Remove the securing bolts and bearing cover, O-ring, outer spacer, and brake backing plate.
4. Drive out the shaft with a soft drift, or press it out with a 250-mm (10-in.) puller as shown in Fig. 11-12. Remove the inner spacer.

Fig. 11-12. Pressing out rear wheel shaft. Make certain the diagonal arm is firmly secured in the vise.

5. Pry out the inner oil seal.
6. Remove the bearing retaining circlip.
7. Using a drift, knock out the ball bearing as shown in Fig. 11-13.

Fig. 11-13. Driving out ball bearing.

8. Remove the spacer sleeve and the roller bearing inner race. Drive out the outer race with a suitable drift.
9. Using a suitable drift and arbor press, press out the inner sleeve of the bonded rubber bushing.
10. Insert an expanding internal extractor tool (Fig. 11-14) into the flanged sleeve. Open the extractor hooks in order to engage the recess between sleeves.
11. Using the arbor press, force down the extractor until the lower sleeve drops out. Turn over the diagonal arm and remove the remaining sleeve in the same way.

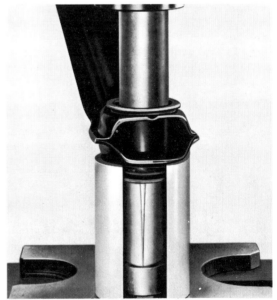

Fig. 11-14. Removing outer sleeve of bonded rubber bushing.

To assemble:

1. Examine all parts for damage or excessive wear. Replace as necessary.

Transmission and Rear Axle

2. Install one bonded rubber bushing with a suitable drift and support. Press the second bushing into place as shown in Fig. 11-15.

Fig. 11-15. Installing second bonded rubber bushing. Press it in as far as it will go.

3. Drive the ball bearing into place with a drift, as shown in Fig. 11-16, or press it into the bore.

Fig. 11-16. Installing ball bearing

4. Install the bearing retaining circlip.
5. Drive or press the oil seal into place with the same tools you used to install the ball bearing.
6. Pack the hub of the diagonal arm with about 60 g (2 oz.) of multipurpose grease. Also grease the ball bearing and the oil seal lip.
7. Drive or press the rear wheel shaft into the hub of the diagonal arm until it is seated lightly against the bearing inner race.
8. Install the spacer sleeve.
9. Grease the outer race of the roller bearing, and then drive it into place with the same tool you used to install the ball bearing and the oil seal (see Fig. 11-17).

Fig. 11-17. Roller bearing outer race installed. Lubricate the rollers liberally.

CAUTION —
It is imperative that all parts in the diagonal arm be perfectly clean and thoroughly lubricated. Before installing the parts, examine the hub itself and make certain there are no particles of grit or metal that could damage the bearings, rear wheel shaft, or spacer sleeves.

10. Using the castellated nut, the outer spacer, and the thrust piece (VW 454), as shown in Fig. 11-18, press the roller bearing inner race into place.

60 Transmission and Rear Axle

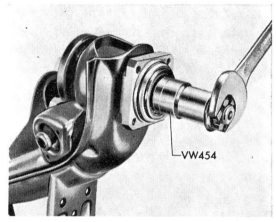

Fig. 11-18. Installing roller bearing inner race. Lubricate race before installation.

11. Fill the double lip of the oil seal with grease. Press the seal into the bearing cover as shown in Fig. 11-19.

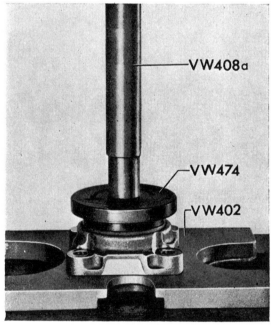

Fig. 11-19. Installing oil seal in bearing cover. Take care not to damage the seal.

12. Install the backing plate, bearing cover, spacer, and new O-ring. Tighten the securing bolts to 6.0 mkg (43 ft. lb.).
13. Install the brake drum and castellated nut. With the car on the ground, tighten the nut to 30.0 mkg (217 ft. lb.). Then install a new cotter pin.

12. Rear Wheel Alignment

Rear wheel alignment must be performed whenever you replace parts that directly affect the adjustment. These parts include the diagonal arm, diagonal arm bushings, spring plates, and torsion bar. Rear wheel alignment should also be carefully checked following accident damage to the rear or side of the car.

In order to properly align the rear wheels, special equipment is almost a necessity. The manufacturer's instructions should be followed closely.

There are three adjustments to be checked when aligning the rear wheels: wheel toe, wheel camber, and rear axle alignment relative to the longitudinal axis of the car. It is also necessary that the spring plate angle be correctly set as this has a direct effect on wheel camber.

Toe, more commonly referred to as toe-in and toe-out, is the angular relationship between the wheels and longitudinal axis of the car. If the front edge of a wheel is closer to this axis than the rear edge, the wheel is said to be toed-in or have positive toe; if the rear edge of a wheel is closer to this axis than its front, the wheel is said to be toed-out or have negative toe. This relationship is sometimes expressed in degrees, sometimes in linear measurement.

Camber is the angular relationship between the wheels and a vertical line drawn through the tire contact patch. If the top of a wheel is inclined outward from this line, the wheel is said to have positive camber; if the top of a wheel is inclined inward from the line, it is said to have negative camber. This relationship is expressed in degrees of deviation from the vertical line.

Even if camber and toe are correctly set, the rear axles may not be perpendicular to the longitudinal axis of the car. In this case, one rear wheel will be closer to the front axle than the other. This can be corrected by adjusting the axle position in the spring plate.

CAUTION —
If you lack the skills, tools, or machine for aligning the rear wheels we suggest you leave such adjustments to an Authorized VW Dealer or other qualified shop. We especially urge you to consult your Authorized VW Dealer before attempting repairs on a car still covered by the new-car warranty.

Correct rear wheel alignment is essential to proper tire wear characteristics, long drive train life, and good handling. It is also necessary for proper front wheel alignment. Suspension and alignment adjustments are interrelated.

Transmission and Rear Axle

12.1 Adjusting Rear Wheel Alignment

The following procedures apply to all Type 1 cars, whether they have manual or Automatic Stick Shift transmissions. The special equipment necessary for alignment is available through a number of manufacturers. Follow the instructions given below to measure toe, camber, and rear axle alignment.

To check and adjust rear wheel toe and axle alignment:

1. Check and, if necessary, correct tire pressures.
2. Fill the fuel tank. Except for fuel, the car must be unladen.
3. Position the car on the alignment stand and lock the parking brake. Mount the testing equipment.
4. Remove the nuts securing the spring plate but do not remove the bolts.
5. Check and adjust rear wheel toe with suitable optical alignment equipment. Toe is adjusted by moving the diagonal arm forward or backward in the slotted spring plate mounting holes. When set to specifications, install and tighten the spring plate nuts. Make certain the nuts are tightened gradually and evenly.
6. Install the centering bar shown in Fig. 12-1.
7. Check the centering of the engine/transmission with the centering bar. The pointer must be aligned with the center of the cast rib on the transmission case and the sleeve between the two tolerance marks on the bar. If not, recheck and inspect for collision damage to the axle shafts.

Fig. 12-1. Rear axle alignment centering bar. Make certain the ends of the centering bar do not rest against the lockplates on the axle shaft flanges.

To check and adjust rear wheel alignment:

1. Check and, if necessary, correct tire pressures.
2. Fill the fuel tank. Except for fuel, the car must be unladen.
3. Position the car on the alignment stand and lock the parking brake. Mount the testing equipment.
4. Check camber, individual wheel toe and track following the equipment manufacturer's instructions. The specifications given below apply to all Type 1 models:

   Rear wheel camber (with spring plates set): 50' ± 40'
   Maxirnum difference between sides: 45'
   Rear wheel (with correct camber): 0° ± 15'
   Maximum deviation in wheel alignment: 10'

5. If the track is not within a tolerance of 10', correct by moving wheels in or out.
6. If the total toe angle exceeds 15' and the camber exceeds −2°, loosen the diagonal arm/spring plate mounting bolts and shift the arm's position to obtain the largest possible positive camber.
7. Recheck the toe angle and, if necessary, adjust it by moving the diagonal arm forward or rearward in the slotted spring plate holes. If the camber varies by more than 45', the torsion bars must also be adjusted.

The relationship between all the preceding adjustments is very important. Often you may establish the correct setting for one, only to find that it has upset another. This holds for front end alignment as well as for rear wheel and axle alignment. You will learn by experience how each adjustment affects the others.

Therefore, the key to accurate wheel alignment is the consideration of all factors as a whole. Keep a record of all settings and the changes that result in them from changes made to other settings. Doing so will enable you to judge the consequences of each adjustment.

62 Transmission and Rear Axle

13. Technical Data

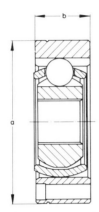

Fig. 13-1. Cross section of constant velocity joint.

I. Constant Velocity Joints (Fig. 13-1)

Part No.	113 501 331
Diameter a	90.90-91.00 mm (3.579-3.583 in.)
Width b	31.70-32.30 mm (1.248-1.272 in.)
Ball diameter	15.88 mm (.625 in.)
Grease per joint	90 g (3.2 oz.) of molybdenum grease
Shaft runout	0.50 mm (.020 in.)

II. Double-jointed Axle Driveshaft (Fig. 13-2)

Transmission type	Manual/Automatic Stick Shift
Code number on end of shaft	1
Code number on shaft assembly	1
Length of complete shaft dimension a	405.30 mm (15.957 in.)
Length of shaft dimension b	415.50 mm (16.358 in.)
Part number of shaft (without joints)	113 501 211

III. General Data

Transmission Type	Code Letter	Final Drive Ratio	Engine Displacement	Remarks	Manufacturing Dates
manual	AH	8:33	1600	double-jointed axle	through July 1972
manual	AT	8:31	1600	double-jointed axle	from August 1972

IV. Tightening Torques

Designation	Thread	mkg	ft. lb.
Engine/transmission nut	M 10 x 1.5	3.0	22
Pinion retaining nut	M 80 x 1	22.0*	160*
Gear carrier/housing nut	M 8 x 1.25	2.0	14
Final drive cover nut	M 8 x 1.25	3.0	22
Transmission/bonded rubber mounting nut	M 8 x 1.25	2.0	14
Shift housing/gear carrier nut	M 7 x 1	1.5	11
Tapered-roller bearing/drive pinion round nut	M 35 x 1.5	20.0	145
Ring gear/differential housing bolt	M 10 x 1.5	6.0	43
Selector shaft/fork bolt	M 8 x 1.25	2.5	18
Support/reverse lever nut	M 10 x 1.5	3.5	25
Bushing/clutch operating shaft lockbolt	M 6 x 1	1.0	7
Oil filler plug	M 24 x 1.5	2.0	14
Oil drain plug	M 24 x 1.5	2.0	14
Rear wheel shaft nut	M 24 x 1.5	30.0	217
Spring plate bolt	M 12 x 1.5	11.0	80
Driveshaft flange/socket head screw	M 8 x 1.25	3.5	25
Control arm fitted bolt	M 14 x 1.5	12.0	87
Spring plate bushing/cover bolt	M 10 x 1.5	3.5	25
Transmission carrier/frame bolt	M 18 x 1.5	23.0	166
Front mounting/frame/sub-frame nut	M 10 x 1.5	3.5	25
Engine carrier nut	M 8 x 1.25	2.5	18
Shock absorber/control arm nut	M 12 x 1.5	7.0	50
Shock absorber/frame nut	M 12 x 1.5	7.0	50
Bearing cover/wheel bearing bolt	M 10 x 1.5	6.0	43

* Tighten to 22.0 mkg (160 ft. lb.), loosen, and tighten finally to 22.0 mkg (160 ft. lb.).

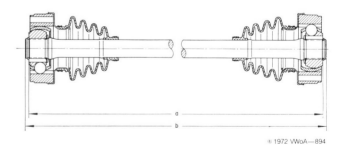

Fig. 13-2. Double-jointed axle driveshaft.

V. Transmission Rebuilding Data

Part	New Installation	Wear Limit
A. Manual transmission		
1. 1st gear . . . . end play	0.10-0.25 mm (.004-.010 in.)	—
2. 3rd gear . . . . end play	0.10-0.25 mm (.004-.010 in.)	—
3. 4th gear . . . . end play	0.10-0.25 mm (.004-.010 in.)	—
4. Synchromesh units (clearance between coupling teeth and synchronizer ring)		
1st/2nd gears . . . . clearance	1.10-1.80 mm (.043-.070 in.)	0.60 mm (.024 in.)
3rd/4th gears . . . . clearance	1.00-1.80 mm (.039-.070 in.)	0.60 mm (.024 in.)
5. Shift fork/operating sleeves for 1st/2nd and 3rd/4th gears . . . . standard clearance	0.10-0.30 mm (.004-.012 in.)	—
clearance with wide slot	0.50-0.85 mm (.019-.033 in.)	—
6. Preload of pinion double tapered-roller bearing turning torque . . . . new	6-21 cmkg (5.2-18.2 in. lb.)	—
used (in service more than 30 mi. [48 km])	3-7 cmkg (2.6-6.1 in. lb.)	—
B. Main drive shaft (surface for 3rd gear needle bearing) . . . . runout	max. 0.02 mm (.0008 in.)	—
C. Gearbox and gearshift housing		
1. Preload of final drive covers on tapered roller bearings turning torque . . . new	30-35 cmkg (26-30 in. lb.)	—
used (in service more than 30 mi. [48 km])	3-7 cmkg (2.6-6.1 in. lb.)	—
2. Shift rod shifting pressure	15-20 kg (33-44 lb.)	—
3. Gearshift housing bushings . . . . inside diameter	15.03-15.05 mm (.591-.592 in.)	15.25 mm (.600 in.)
4. Inner shift lever . . . . diameter	14.96-15.00 mm (.588-.590 in.)	14.75 mm (.580 in.)
5. Starter bushing . . . . inside diameter	12.55-12.57 mm (.494-.495 in.)	12.65 mm (.498 in.)
6. Starter shaft/bushing . . . . radial clearance	0.09-0.14 mm (.003-.005 in.)	0.25 mm (.010 in.)
D. Final drive		
1. Axial play at differential gears with differential housing bolted together . . .		
with no spacer sleeve	0.25-0.45 mm (.010-.017 in.)	0.50 mm (.020 in.)
with spacer sleeve	0.00-0.14 mm (.000-.005 in.)	0.20 mm (.008 in.)
2. Radial play between differential housing and cover/gear shaft . . . . . old	0.03-0.08 mm (.001-.003 in.)	0.12 mm (.005 in.)
new	0.025-0.06 mm (.001-.002 in.)	0.12 mm (.005 in.)
3. Double tapered-roller pinion bearing preload (turning torque)		
new bearings	6-21 cmkg (5.2-18.3 in. lb.)	—
used bearings (in service more than 30 mi. [48 km])	3-7 cmkg (2.6-6.1 in. lb.)	—
4. Tapered roller bearings for differential preload . . . . new bearings	30-35 cmkg (26-30 in. lb.)	—
used bearings (in service more than 30 mi. [48 km])	3-7 cmkg (2.6-6.1 in. lb.)	—
5. Backlash (measured at pitch circle diameter)	0.15-0.25 mm (.006-010 in.)	—
6. Rear axle shafts		
a. Flange/differential gears (measured across the convex faces) clearance	0.03-0.10 mm (.001-.004 in.)	0.20 mm (.008 in.)
b. Measured at bearing seat shaft between centers . . . . runout	max. 0.05 mm (.002 in.)	—

VI. Ratios

Gears	Model Years	No. of Teeth	Ratio
1st gear	all models through October 1972	38/10	3.80
1st gear	1973 and later models from November 1972	34/9	3.78
2nd gear	all models	35/17	2.06
3rd gear	1970 models, through April 1970	29/23	1.26
3rd gear	late 1970 and all later models	63/50	1.26
4th gear	1970 models, through April 1970	24/27	0.89
4th gear	late 1970 through 1972 models	53/60	0.88
4th gear	1973 and later models	54/58	0.93
Reverse	1970 through 1972 models	20/14 x 17/43	3.61
Reverse	1973 models through October 1972	20/14 x 15/40	3.80
Reverse	all models from November 1972	17/12 x 15/40	3.79
Final drive	all 1970 models	8/35	4.375
Final drive	Karmann Ghia, 1971 through 1974	8/31	3.875
Final drive	Beetle/Super Beetle, 1971 through 1972	8/33	4.125
Final drive	Beetle/Super Beetle, 1973 and later	8/31	3.875

©1978 VWoA

VII. Torsion Bar Adjustment (spring plates unloaded)

Model	Transmission	Installed		Torsion Bar		Setting
		from chassis No.	to chassis No.	Length mm (in.)	Diameter mm (in.)	
all	all	118 000 001	—	676 (26.614)	22 (.866)	21° 20′ (+ 50′)

Section 4

BODY AND FRAME

Contents

Introduction 2	
1. General Description 3	
Doors, Hoods, and Windows 3	
Seats 3	
Heating and Ventilation 3	
Trim, Sliding Roof, and Convertible Tops 3	
2. Maintenance 3	
3. Bumpers 4	
3.1 Removing and Installing Front Bumper 4	
3.2 Removing and Installing Rear Bumper . . 5	
Replacing Front and Rear Rubber Bumper Molding 6	
4. Fenders and Running Boards 7	
4.1 Removing and Installing Front Fender . . 7	
4.2 Removing and Installing Rear Fender . . 7	
4.3 Removing and Installing Running Board 8	
5. Hoods and Fuel Filler Flap 8	
5.1 Removing and Installing Front Hood and Hinge 9	
5.2 Removing, Installing, and Adjusting Front Hood Lock 9	
5.3 Removing and Installing Rear Hood and Hinge 11	
5.4 Removing and Installing Rear Hood Lock 11	
5.5 Removing and Installing Fuel Filler Flap and Cable 12	
6. Doors and Door Locks 12	
6.1 Removing and Installing Door 12	
6.2 Adjusting Door Striker Plate 13	
6.3 Removing and Installing Door Trim Panel 15	
6.4 Removing and Installing Door Check . . 16	
6.5 Removing and Installing Door Handle . . 16	
6.6 Removing and Installing Outside Mirror 16	
7. Seats 17	
7.1 Removing and Installing Front Seats . . 17	
7.2 Removing and Installing Rear Seat . . . 18	
8. Heating and Ventilating Systems 19	
8.1 Adjusting and Replacing Heating Cables and Flaps 19	
Front Footwell Flaps 20	
8.2 Removing and Installing Fresh Air Ventilation 20	
Ventilation System Without Fan 21	
Ventilation System With Fan (through 1972) 22	
Ventilation System With Fan (1973 on) . . 23	
9. Sliding Roof (sun roof) 25	
9.1 Removing and Installing Sliding Roof Panel 25	
9.2 Removing and Installing Runners and Cables 26	
9.3 Adjusting Cables 27	
9.4 Adjusting Top Height 28	
10. Frame 28	
10.1 Removing and Installing Body on Frame 29	

© 1974 VWoA

2 Body and Frame

Body
And Frame

The Super Beetle, Convertible, Beetle, and Karmann Ghia all have different frames. Frames for the Super Beetle and Convertible have the frame head used with the strut front suspension system. However, the Convertible's frame has additional strengthening members to compensate for the open body. Both the Beetle and the Karmann Ghia have the frame head used with the torsion bar front axle. But the shape of the Karmann Ghia's floor plate is entirely different from that of the other models.

Despite variations, the four frames share a similar basic design. All are built up from sheet metal pressings joined by electric welding. Their beam strength is derived largely from the heavy-gauge frame tunnel—a central tube-like "backbone." A channel pressed into the perimeter of the ribbed sheet-steel floor plate lends additional rigidity. This is further augmented on the Convertible's frame by deep rectangular side members.

The fuel line and guide tubes for the parking brake cables, clutch cable, accelerator cable, and heater control cables are housed in the frame tunnel. The shift rod is also inside the tunnel. The pedal cluster, gearshift lever, parking brake handle, heater controls, and seat belts are mounted on the tunnel, as is the inner seat track on cars built since the beginning of 1972. The sockets for the car jack are welded to each side of the floor plate, just inside the perimeter channel.

The body is a separate welded unit bolted to the frame at the frame head (or front axle on Beetles and Karmann Ghias), the front and rear crossmembers, the rear shock absorber mountings, and the channel along each side of the floor plate. On the Beetle, Super Beetle, and Convertible, the fenders are held by bolts, which makes fender replacement easy even for inexperienced mechanics. However, the body sheet metal of both the Karmann Ghia coupe and Karmann Ghia convertible must be cut and butt welded to replace a fender. Extensive body work of this nature is beyond the scope of this Manual and demands the skill and experience of an expert body repairman.

Although many repairs described in this section can be carried out by car owners, a number of the procedures may be of practical value only to professional mechanics. So if you lack the skills, tools, or the workshop necessary for body adjustments and repairs, we suggest that you leave such work to an Authorized VW Dealer or other qualified shop. We especially urge you to consult your Authorized VW Dealer before attempting repairs on a car still covered by the new-car warranty.

BODY AND FRAME 3

1. GENERAL DESCRIPTION

The VW body is a welded steel assembly that includes the front and rear ends, the inner and outer side panels, and the roof panel (except on the Convertibles). The sill panels are welded to the side panels, forming two box-section side members.

Doors, Hoods, and Windows

Each door is attached to the front body pillar by two hinges. The hinges and the door lock plates can be shifted on the body to adjust the doors in their openings.

The door windows and rear quarter windows are flat glass. The windshield and rear window are curved glass. The door windows can be fully raised or lowered and the vent wings adjusted throughout a wide range. The rear quarter windows open outward at their rear edges on sedans and Karmann Ghia coupes, and retract downward on Convertibles. Karmann Ghia convertibles have no rear quarter windows. The rear window can be heated electrically to prevent fogging.

The front hood covers the front luggage compartment, which houses the spare wheel, windshield washer reservoir, and brake fluid reservoir. The rear hood covers the engine compartment. Both the front and rear hoods are attached to the body by spring-loaded, concealed hinges that hold the hoods in their open position after they are raised.

Before the front hood can be opened, it must be unlocked from inside the car. This is done by pulling down on a lever inside the glove compartment. The Karmann Ghia has a pull knob in the left door's lock pillar that unlocks the rear hood.

Seats

The front seats are individually mounted and can be moved forward or backward. The angle of the seat backs is also adjustable. Locks built into the seat backs keep them upright even during hard braking.

The rear seat cushion folds upward to give access to the battery, voltage regulator, heater control cables, and relay for the heated rear window. The rear seat backrest can be folded forward for extra load space (except on Karmann Ghia models).

Heating and Ventilation

Fresh air is drawn in by the engine cooling fan and is heated as it passes through the fins of the heat exchangers. The flow of heated air into the passenger compartment is regulated by flaps at the front of each heat exchanger and at the heat outlets. The flaps are controlled by cables that link them to the two levers between the front seats.

© 1974 VWoA

A fresh-air box under the cowl panel collects outside air for cool-air ventilation. The air enters the interior through two vents below the windshield and two outlets in the dashboard. On 1973 and later Super Beetles and Convertibles, however, there are two vents near the center of the dashboard and two in the outer ends of the dashboard. Beginning with the 1971 models, Super Beetles and Convertibles were equipped with air exit slots in the rear quarter panels and also a two-speed fan to increase air flow.

Trim, Sliding Roof, and Convertible Tops

The bumpers and bumper guards, headlight rims, door handles, vent wing frames, parking light/side-marker light housings, and taillight rims are chrome plated. The rest of the bright trim is either stainless steel or anodized aluminum.

Those portions of the frame tunnel and floor plate inside the passenger compartment are soundproofed with thermoplastic damping material, which also insulates against road heat. The floor and front side panels are carpeted, except on the Beetle (Sedan 111), which has synthetic rubber floormats. The upholstery is easily cleaned vinyl that is perforated to improve air circulation.

A sliding steel roof (sun roof) is optional equipment on 1970 Beetles and on Super Beetles from 1971 on. Opening the sliding roof creates a clear space above the driver and front passenger seats. A hand crank controls the roof, which can be adjusted to any position from fully open to fully closed. For safety, the crank should be folded into its recess when not in use.

The interior trim panel for the sliding roof is made of the same perforated vinyl used for the headliner. The trim panel cannot be removed until the sliding roof and one of its side runners have been removed. The sliding roof can be adjusted, however, without completely removing the trim panel from the car.

The convertible top requires no special maintenance. If it requires replacement, adjustment, or repair, we suggest you leave such work to an Authorized VW Dealer or other qualified shop.

2. MAINTENANCE

Only one maintenance operation, lubrication of the door and hood hinges and locks, is required at regular mileage intervals. This procedure is covered in **LUBRICATION AND MAINTENANCE**. Care of the body, trim, upholstery, and windows is also described briefly in **LUBRICATION AND MAINTENANCE**. The latter maintenance is important, of course, but is not performed at prescribed intervals of mileage or time.

4 Body and Frame

3. Bumpers

On 1970 through 1973 models, you can remove only the bumper itself by taking off the nuts on the back of the bumper. The brackets and reinforcement strap can stay on the car as shown in Fig. 3-1. On 1974 and later cars, the reinforcement is welded to the bumper (Fig. 3-2). Remove these reinforced bumpers as a unit together with the spring-loaded damping elements that are used in place of brackets.

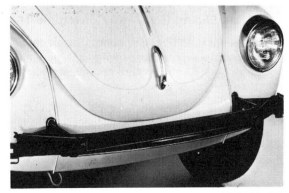

Fig. 3-1. 1973 reinforcement strap and brackets.

Fig. 3-2. 1974 and later reinforced bumper with damper element.

In addition to the welded-in reinforcement and spring-loaded damping elements, the bumpers on 1974 and later cars have plastic protectors on their ends. These protectors are intended to prevent the bodywork from being marred in the event that the damping elements are fully compressed.

3.1 Removing and Installing Front Bumper

The bumper mountings are basically the same on all Beetle, Super Beetle, and Convertible models—although minor changes have been necessitated by the damping elements used on 1974 and later cars. Two completely different bumper mountings have been used on Karmann Ghias.

To remove:

1. Using a 13-mm (½-in.) wrench, remove the bolts shown in Fig. 3-3, Fig. 3-4, Fig. 3-5, or Fig. 3-6.

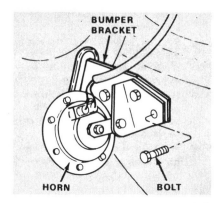

Fig. 3-3. Bumper mounting (Beetle and Super Beetle through 1973). First remove the two bolts that hold horn, then remove horn and bracket.

Fig. 3-4. Bolts that hold damping element on 1974 and later cars.

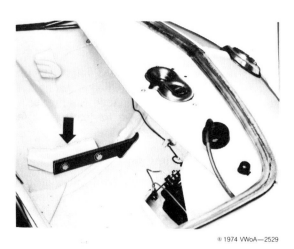

Fig. 3-5. Karmann Ghia bumper mounting (arrow) used prior to 1972 models.

BODY AND FRAME 5

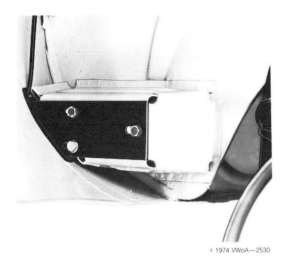

Fig. 3-6. Karmann Ghia bumper mounting (under front fender) used on 1972 and later cars.

2. Pull the bumper together with the brackets and reinforcement strap, or with the damping elements, out of the slots in the front fenders. See Fig. 3-7.

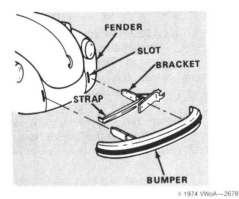

Fig. 3-7. Typical bumper removal and disassembly.

3. On 1970 through 1973 models, remove the three nuts from each bracket. Then separate the reinforcement strap from the bumper and the brackets.

4. Remove the three nuts from each bumper bracket or, on 1974 and later cars, from each damping element. Then remove the brackets or damping elements from the bumper.

Installation is the reverse of removal. Replace the rubber grommets in the front fenders if they are weathered or damaged. Lubricate the threads on the nuts and bolts before installing them.

When installing the bumper and the brackets on the car, install the bolts but do not tighten them. Then adjust the bumper to a uniform gap with the body (Fig. 3-8) before torquing the bolts to 2.0 mkg (14 ft. lb.).

Fig. 3-8. Bumper installed with an even gap between it and the body. (Karmann Ghia shown.)

3.2 Removing and Installing Rear Bumper

Karmann Ghias, Beetles, and Super Beetles have similar rear bumper mountings. But on Karmann Ghias, the rear bumper mounting bolts also mount the towing eyes.

The design of the rear bumper used on 1974 and later models is similar to that of the 1974 and later front bumper. The reinforcement is welded to the bumper itself and there are spring-loaded damping elements instead of mounting brackets. Due to the redesigned mounting, the rubber grommets and the openings in the fenders have also been changed. Nevertheless, removal and installation procedures are similar to those of previous bumpers.

To remove:

1. Using a 13-mm (½-in.) wrench, remove the bolts shown in Fig. 3-9 or Fig. 3-10.

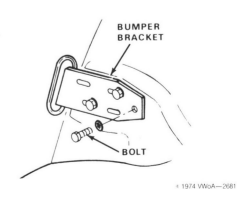

Fig. 3-9. Bumper mounting under rear fender (Beetle).

6 Body and Frame

Fig. 3-10. Karmann Ghia rear bumper mounting with towing eye. (1970 and 1971 models have only two mounting bolts.)

2. Pull the bumper together with brackets, or with the damping elements, out of the openings in the rear fenders. See Fig. 3-11.

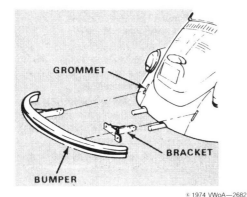

Fig. 3-11. Rear bumper removal and disassembly.

3. Remove the three nuts from each bumper bracket, or damping element. Then remove the brackets or damping elements from the bumper.

Installation is the reverse of removal. Replace the rubber grommets in the fenders if they are weathered or damaged. Lubricate the threads on the nuts and bolts before installing them.

When installing the bumper and brackets, or bumper and damping elements, install the bolts but do not tighten them. Then adjust the bumper to a uniform gap with the body (Fig. 3-12) and torque the bolts to 2.0 mkg (14 ft. lb.). On 1974 and later models, position the bumpers as far away from the body as possible.

Fig. 3-12. Rear bumper aligned to produce a uniform gap with the body (Karmann Ghia shown).

Replacing Front and Rear Rubber Bumper Molding

The procedure for installing the rubber bumper molding on 1971 and later Super Beetles and 1973 and later Karmann Ghias is shown in Fig. 3-13. Fasten the new molding first at one end of the bumper, then in the middle, and finally at the opposite end.

On 1974 and later models, the plastic protectors at the ends of the bumpers are held in place by the same nuts that hold the rubber molding, so remove the plastic protectors before the rubber molding. During installation, make sure that the rubber molding is correctly positioned, then loosely install the nuts. Slide the plastic protectors onto the bumper so that the slot flanks the stud on the rubber molding. Tighten the nuts after the protectors are in place.

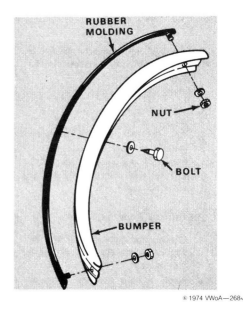

Fig. 3-13. Rubber molding installation on late-type front or rear bumpers.

BODY AND FRAME 7

4. FENDERS AND RUNNING BOARDS

Karmann Ghia fenders are welded to the body and should be replaced only by an experienced body repairman. Fenders on other models can be replaced by any car owner.

4.1 Removing and Installing Front Fender

If it is necessary to paint the fender, do so before installing it on the car.

To remove:

1. Use an 8-mm (5/16-in.) wrench to remove the two nuts under the fender. Remove the turn signal/side-marker light from the fender, then disconnect the light from its wires. Mark the wires to ensure correct reinstallation.

2. Pull the gasket for the turn signal/side-marker light down through the fender.

3. Remove the sealed beam unit from the headlight as described in **ELECTRICAL SYSTEM**. Then remove the three fillister head screws that hold the headlight assembly in the fender and remove the assembly.

4. Insert a screwdriver into the small slot below each terminal in the sealed beam connector. Then pull the wire and terminal out of the rear of the connector.

5. Pull the wires and their protective hose back and out of the fender's headlight housing.

6. Remove the front bumper as described in **3.1 Removing and Installing Front Bumper**. On left fenders only, disconnect the wires from the horn, then remove the horn.

7. Remove the nut, bolt, and washers that hold the fender to the running board. Then take out the eight fender bolts and washers indicated in Fig. 4-1 and remove the fender and beading.

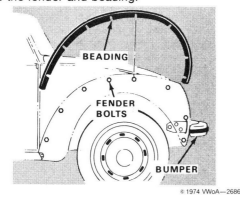

Fig. 4-1. Front fender bolts and beading. To remove the beading alone, merely loosen the fender bolts.

To install:

1. Position the fender against the car body without the beading. Loosely install a bolt at the top of the fender in order to support it.

2. Start the remaining seven bolts in their threads so that the fender hangs loosely on the car.

3. Install a new rubber spacer washer between the fender and the running board. Install but do not tighten the nut, bolt, and washers as indicated in Fig. 4-2.

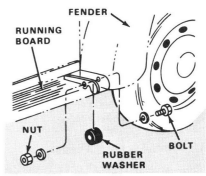

Fig. 4-2. Installation of rubber spacer washer.

4. Using your hand, press the beading into the space between the fender and the body. Make sure the beading is even and that the notches have all slipped over the bolts.

5. Gradually torque the bolts to 1.5 mkg (11 ft. lb.), checking constantly to see that the beading remains in place.

6. Torque the bolt that holds the fender to the running board to 1.5 mkg (11 ft. lb.).

The rest of installation is the reverse of removal. When installing the sealed beam terminals in their connector, the yellow wire goes in the top center slot, the white wire in the left slot, and the brown wire in the right slot (as seen from the front of the car, looking into the headlights). Further information on the installation of the headlights and their wires is given in **ELECTRICAL SYSTEM**.

4.2 Removing and Installing Rear Fender

If it is necessary to paint the fender, do so before installing it on the car.

To remove:

1. Working under the fender, use an 8-mm (5/16-in.) wrench to remove the tail light nuts, then remove the taillight assembly. Four nuts secure the taillight assembly on 1973 and later cars, two on earlier models.

8 Body and Frame

2. Disconnect the wires from the taillight assembly and mark them for correct reinstallation. Then pull the wires and the rubber grommet down and out of the fender.

3. Remove the rear bumper as described in **3.2 Removing and Installing Rear Bumper**.

4. Remove the nut, bolt, and washers that hold the fender to the running board. Then take out the nine bolts and washers indicated in Fig. 4-3 and remove the fender and beading.

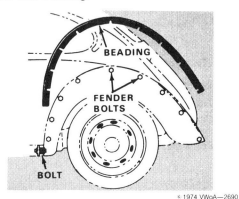

Fig. 4-3. Rear fender bolts and beading. To remove the beading alone, merely loosen the fender bolts.

To install:

1. Position the fender against the car body without the beading. Loosely install a bolt at the top of the fender to support it.

2. Start the remaining seven bolts in their threads so that the fender hangs loosely on the car.

3. Install a new rubber spacer washer between the fender and the running board. Install, but do not tighten, the nut, bolt, and washers as indicated previously in Fig. 4-2.

4. Using your hand, press the beading into the space between the fender and the body. Make sure the beading is even and that the notches have all slipped over the bolts.

5. Gradually torque the bolts to 1.5 mkg (11 ft. lb.), checking constantly to see that the beading remains in place.

6. Torque the bolt that holds the fender to the running board to 1.5 mkg (11 ft. lb.).

The remainder of installation is the reverse of removal. Check to see that the taillights, stoplights, and turn signals all work properly.

NOTE —
Because the fenders for 1973 and later cars have larger taillight mountings, they cannot be interchanged with the fenders of earlier models.

4.3 Removing and Installing Running Board

To remove the running board, remove the nuts, bolts, and washers that hold the running board to the front and rear fenders. Remove the two rubber spacer washers from between the fenders and the running board. Then loosen, but do not remove, the four bolts that hold the running board to the body. The various fasteners and their positions are shown in Fig. 4-4.

Installation is the reverse of removal. Use new rubber spacer washers at the front and rear of the running board. Torque the bolts that hold the running board to the fender to 1.5 mkg (11 ft. lb.). Torque those that hold the running board to the body to 1 mkg (7 ft. lb.).

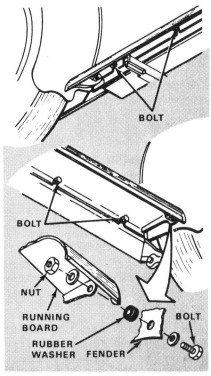

Fig. 4-4. Bolts that hold running board to fenders and to body.

5. Hoods and Fuel Filler Flap

The 1971 and later Super Beetles and Convertibles are designed around the strut front suspension system. This system, together with a longer front body section, increases luggage capacity. The front hood on these models differs from the one used on the 1970 models and on the later Beetle Sedan 111. The rear hood locks were slightly revised on all cars for model year 1972.

Body and Frame

5.1 Removing and Installing Front Hood and Hinge

To avoid damaging the finish, cover the cowl panel with a protective cloth. To save installation time, mark the hinge positions before removing the hood. You will need someone to help you with this job.

To remove:

1. Remove the two hood bolts from one of the hinges as shown in Fig. 5-1.
2. While someone holds the unbolted side, remove the two hood bolts from the other hinge.
3. Together, lift the hood up and off toward the front of the car.

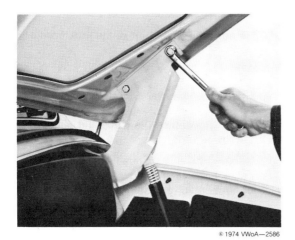

Fig. 5-1. Hood being unbolted from the hinge.

To install:

1. Before installing the hood, check the condition of the rubber weatherstrip. If necessary, reglue or replace it.

   NOTE
   Remove all old trim cement with solvent so that the new weatherstrip will adhere properly.

2. Attach the hood loosely. Move the hood in the elongated bolt holes until it contacts the weatherstrip evenly all around. Then tighten the bolts.

3. Check the lock operation by opening and closing the hood several times. If necessary, adjust the lock.

   NOTE
   If it is necessary to lower the hinges after the hood has been removed, use a lever as shown in Fig. 5-2.

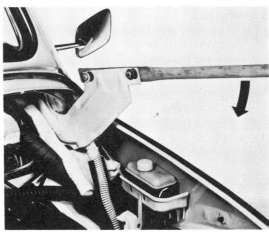

Fig. 5-2. Lever for lowering hinges. This tool is made by welding two bolts to a piece of pipe. Be careful not to let the lever snap down suddenly, as it could damage the wiring harness.

To remove the hinge from the body, remove the dashboard access panel from the rear of the luggage compartment by removing the two knurled nuts. Also remove the front luggage compartment load liner. Pry the E-clip off the bottom pivot pin for the spring mounting. Then press out the pin and swing the spring toward the front of the car until it is no longer compressed. Unbolt the hinge from its bracket under the dashboard.

5.2 Removing, Installing, and Adjusting Front Hood Lock

The replaceable components of the hood lock's upper part are shown in Fig. 5-3.

1. Hood handle 2. Packing 3. Hood lock upper part

Fig. 5-3. Replaceable components of the hood lock's upper part.

10 Body and Frame

To remove lock:

1. Remove the upper part of the lock by taking out the two bolts indicated in Fig. 5-4.

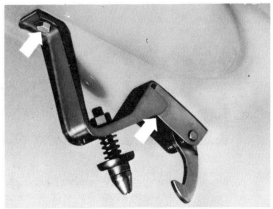

Fig. 5-4. Bolts (arrows) that hold lock's upper part.

NOTE —
If the lock cable has broken while the hood is closed, make three hacksaw cuts in the hood handle as indicated in Fig. 5-5. Unscrew the pieces of the handle from the bolts. This will free the lock's upper part so that the hood can be opened.

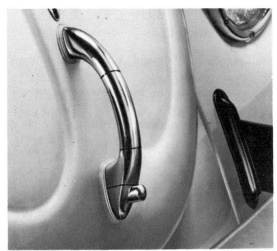

Fig. 5-5. Emergency lock removal. Cut at the three dotted lines. Remove the center and lower sections, then unscrew the remaining two.

2. Working through the opening indicated in Fig. 5-6, loosen the Phillips head clamp screw to free the lock cable from the lock's lower part.

Fig. 5-6. Clamp screw for lock cable.

3. Drill out the hollow rivets that hold the lower part of the lock to the body.
4. Remove the lower part of the lock from below.

To install:

1. Check the upper and lower parts of the lock. If either part is faulty, replace it. Prior to installation, clean and grease the pivot and rubbing points of the lower part.
2. Remove the Phillips head screws that hold the lock cable lever in the glove box. Then withdraw the control cable, grease it thoroughly, and install it in the guide tube. Reinstall the lever.
3. Temporarily install the cable end in the lock's lower part and lightly tighten the clamp screw.
4. Using pop rivets or sheet metal screws, mount the lock's lower part on the car body.
5. Loosen the clamp screw, pull the lock cable taut, and firmly tighten the clamp screw. Then bend the cable end over beyond the clamp screw.
6. Bolt the hood handle and the hood lock's upper part to the hood, making sure the packing pieces are installed properly.
7. Open and close the hood several times to check the lock's operation. If you can lift the hood slightly after it has locked, the lock pin in the upper part is too long. If the hood must be pressed down very hard to lock it, the lock pin is too short.

NOTE —
To adjust the lock pin, loosen the locknut on the top of the pin and turn the lock pin with a screwdriver. Tighten the locknut when the correct length is obtained.

8. Apply a small amount of grease to the lock pin.

Body and Frame

5.3 Removing and Installing Rear Hood and Hinge

Before removing the rear hood, mark its position on the hinges. This will make alignment easier during installation.

To remove:

1. Open the rear hood, then remove the air cleaner as described in **LUBRICATION AND MAINTENANCE**.

2. Cover the carburetor intake to prevent the entry of dirt. Then disconnect the license plate light wire and unclip it from the hood so that it will not be damaged as the hood is removed.

3. Remove the four bolts that hold the hood hinges to the curved hinge brackets on the body.

4. Pull out the hood upward against spring pressure and unhook the load spring from the bracket on the car's roof panel. The spring will stay with the hood when the hood is removed (Fig. 5-7).

Fig. 5-7. Hood being lifted off its hinge brackets so that the spring will unhook from the bracket on the roof panel.

5. If necessary, remove the curved hinge brackets from the car body by removing the three bolts at each bracket.

To install:

1. Before installing the hood, check the condition of the rubber weatherstrip. If necessary, reglue or replace it.

    NOTE
    Remove all old trim cement with solvent so that the new weatherstrip will adhere properly.

2. Loosely install one hood hinge on the curved bracket on the car body.

3. Holding the hood at an angle so that you can reach behind it, engage the spring in the bracket on the car roof panel.

4. Push the hood upward and toward the car. Loosely install the remaining hinge.

5. Move the hood in the elongated bolt holes until it contacts the weatherstrip evenly all around. Then tighten the bolts.

5.4 Removing and Installing Rear Hood Lock

To remove the lock, open the rear hood and take out the mounting screw(s). The 1972 and later cars have one screw, whereas earlier models have three screws (Fig. 5-8 and Fig. 5-9). When installing, check the handle, packing, and lock for wear. Make replacements as necessary.

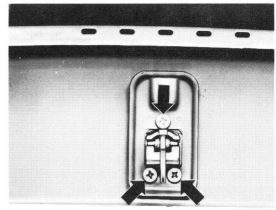

Fig. 5-8. Three Phillips head screws (arrows) that hold early-type hood lock.

Fig. 5-9. Single Phillips head screw that holds hood lock on 1972 and later models.

12 Body and Frame

5.5 Removing and Installing Fuel Filler Flap and Cable

A cable-controlled lock keeps the fuel filler flap shut. The flap can be removed only as a unit, complete with hinge and spring. To remove the flap, take out the bolts shown in Fig. 5-10.

Fig. 5-10. Filler flap bolts (arrows).

To remove cable:

1. Unscrew the handle from the cable end. On 1970 models, remove the two screws from the escutcheon plate.

2. On 1970 cars, push the cable out into the luggage compartment or (on Karmann Ghias) into the fender. On later models, pry the cable end-guide out of the fuel filler housing and withdraw the cable through the fuel filler opening.

3. On 1970 cars, remove the two screws under the fuel filler housing. Then pull the cable downward and out into the luggage compartment. Early and late cables are shown in Fig. 5-11.

Fig. 5-11. Cable for 1970 cars (**A**) and later cars (**B**).

6. Doors and Door Locks

The door hinges and locks are removed in the same way on Karmann Ghias as on the other models covered in this Manual. However, the parts are not identical.

6.1 Removing and Installing Door

If the original door is to be reinstalled, it is advisable to press out the hinge pins rather than remove the hinges from the body. This will avoid the need to align the door later. If you do decide to remove the hinges, carefully marking their original positions will make the alignment job easier.

> **CAUTION**
> *Never leave a door unsupported while one of the hinges is unscrewed or while its hinge pin is out. Doing so could bend or break the other hinge.*

To remove:

1. Remove the circlip from the lower end of the pin that holds the door check strap to the body. Then remove the door check strap pin.

2. If the same door is to be reinstalled, remove the plastic caps from the tops of the hinge pins. Then press out the hinge pins with a special tool such as the one shown in Fig. 6-1.

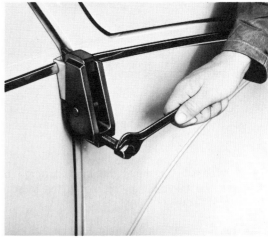

Fig. 6-1. Special tool being used to press out hinge pin. Do not attempt to hammer the pins out.

3. If the hinge pins are to be left in place, pry the plastic caps (that are adjacent to the hinges) out of the body's hinge pillar (except on Karmann Ghias).

4. Using an impact driver, loosen the Phillips head screws that hold the hinges to the body.

CAUTION—
Support the door so that the hinges will not be bent or broken.

5. Taking care not to scratch the door, body, or fender, remove the door—together with its hinges—from the body sideways.

If you are installing the original door, the procedure is simply the reverse of removal. Check the weatherseal. If it is cracked, torn, or otherwise damaged, replace it.

NOTE—
Before replacing a weatherseal, clean away all the old adhesive with solvent. Install the new seal with trim cement.

To install a replacement door:

1. Remove the lock striker plate from the lock pillar on the body.

2. Install the new door together with its hinges, but do not fully tighten the Phillips head screws.

NOTE—
The hinges are screwed to movable threaded plates. This makes it possible to shift the position of the door in its opening for alignment purposes.

3. Align the door in the door opening so that it contacts the weatherstrip evenly all around and the door's trim molding is in line with the trim molding on the side of the body.

4. After the door is aligned, tighten the Phillips head screws, then set them firmly with the impact driver.

5. Install the door check strap with the pin and circlip. Then install the lock striker plate and adjust it as described in **6.2 Adjusting Door Striker Plate**.

6.2 Adjusting Door Striker Plate

After installing a replacement door, adjust the striker plate so that the rear edge of the door aligns with the body. The striker plate should also be adjusted if the door rattles or requires excessive force to close and lock.

Because of body flexing, the door lock striker plate cannot be accurately adjusted while the car is on a lift.

A door rattle that persists even after you have made all possible adjustments indicates a worn rubber wedge. On 1970 and 1971 cars, you can correct the condition by placing a 0.50-mm to 1.50-mm (.020-in. to .060-in.) shim between the wedge and the striker plate. On 1972 and later cars, replace the wedge.

To adjust:

1. Take out the screws shown in Fig. 6-2 and remove the striker plate. Check the door alignment and, if necessary, correct it.

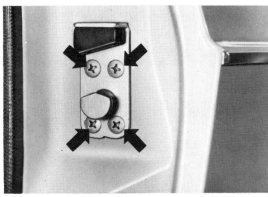

Fig. 6-2. Phillips head screws (arrows) that hold the striker plate to the car body's lock pillar.

2. Insert the striker plate, bottom first, in the latch. Press the latch down into its fully locked position. Then turn the striker plate as indicated in Fig. 6-3.

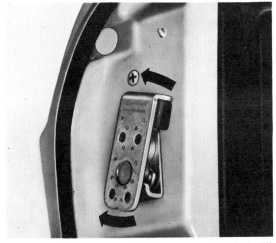

Fig. 6-3. Striker plate being positioned on door latch.

NOTE—
Although the door shown is not for a Beetle model, the procedure and the striker plate are identical.

14 Body and Frame

3. If you can move the striker plate vertically as indicated in Fig. 6-4, the rubber wedge is worn. Either add shims as shown in Fig. 6-5 (1970 and 1971 models) or replace the wedge as shown in Fig. 6-6 (1972 and later models).

Fig. 6-4. Undesirable vertical movement (arrows).

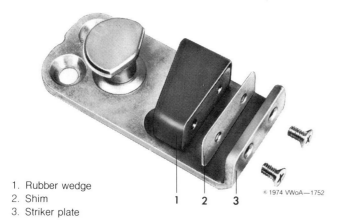

1. Rubber wedge
2. Shim
3. Striker plate

Fig. 6-5. Early striker plate with shim.

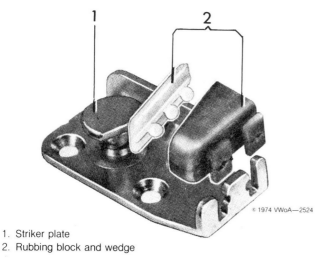

1. Striker plate
2. Rubbing block and wedge

Fig. 6-6. Late striker plate with replaceable wedge.

NOTE —

To correct either a misalignment between the door and front body or a lack of uniformity in the gap between the door and body, adjust the hinges and not the striker plate. Only misalignment (vertical or horizontal) between the door and rear body should be corrected by adjusting the striker plate.

4. After correcting any excessive wedge play, correct any misalignment between the door and rear body. Using the marks shown in Fig. 6-7 and Fig. 6-8, center the striker plate on the lock pillar.

Fig. 6-7. Striker plate center marks (arrows). Align the marks with the dots on the lock pillar.

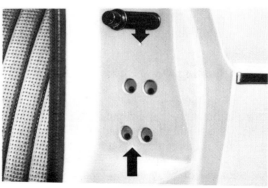

Fig. 6-8. Striker plate centering dots. Note the movable threaded plate behind the pillar panel.

Body and Frame

5. Close the door. Then see if the door aligns with the rear body. If necessary, adjust the position of the striker plate on the lock pillar.

    NOTE —
    The striker plate screws onto the movable threaded plate. By moving the plate up or down, you can correct any misalignment of the ridge in the body where the door and the rear body meet. Moving the plate in or out corrects the position of the door so that it will be flush with the side of the car when the door is closed.

6. After aligning the door, feel for play between the lock and the striker plate. If there is play or if the door will not latch, rotate the striker plate as indicated in Fig. 6-9. If the door is hard to close or if the handle works stiffly, rotate the striker plate in the opposite direction.

NOTE —
Only the door with the new inner panel is available as a spare part. If you install a new door on an earlier car, you must weld an armrest retainer bracket onto the inner panel.

1. Window crank   2. Armrest   4. Release lever

Fig. 6-10. Door trim panel.

Fig. 6-11 shows the late-type door with the weld-in armrest retainer shown as an inset. Four small protrusions on the door's inner panel mark the position for the armrest retainer. The three raised bosses with tapped holes (flanking the protrusions) are the mounting points for the late-type armrest.

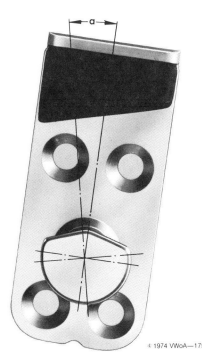

Fig. 6-9. Striker plate turned from vertical as indicated by arc **a**. This moves the wedge closer to the door latch. Notice the point at the center of the striker plate lug about which the striker plate is rotated.

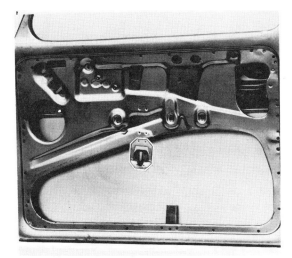

Fig. 6-11. Door introduced on 1973 models. Notice the armrest retainer that must be welded on to adapt the door to earlier models.

6.3 Removing and Installing Door Trim Panel

The door trim panel and related parts are shown in Fig. 6-10. On 1970 through 1972 models, it is unnecessary to remove the armrest from the trim panel before taking the panel off. On later models the armrest—open at the bottom—is bolted through the trim panel to the door's inner panel.

To remove the door trim panel, pry the finger plate out of the door release lever escutcheon, then pry the plastic trim off the window crank. On 1973 and later cars, remove the screws from the hollow underside of the armrest and remove the armrest from the door. The armrest stays on the trim panel on earlier models.

16 Body and Frame

Remove the screws shown in Fig. 6-12. The window crank screw is held by Loctite® and requires a properly fitting Phillips screwdriver to snap it free. Remove the crank and the escutcheon. Then, using a wooden wedge, pry off the trim panel all around the door. On 1973 and later cars, simply pull off the trim panel. On earlier cars, grasp the armrest, pull the panel away from the door slightly, and then lift the panel until it disengages from the armrest retainer bracket.

Fig. 6-12. Window crank and escutcheon screws (arrows).

Installation is the reverse of removal. If either the plastic sheet over the door inner panel or the rubber seal for the release lever have been removed, reglue them. Install the window crank screw with Loctite®.

6.4 Removing and Installing Door Check

If any part of the door check is faulty, the whole unit must be replaced. Individual components are not supplied. Replacement pins, circlips, and Phillips head screws, however, are available.

To replace:

1. Remove the circlip, then remove the check strap pin.
2. Remove the door trim panel as described in **6.3 Removing and Installing Door Trim Panel.**
3. Remove the two Phillips head screws that are above and below the opening where the door check strap projects from the front edge of the door.
4. Remove the check strap by pulling it into the door. Then take it out through one of the openings in the door's inner panel.

Installation is the reverse of removal. If the rubber check strap seal on the hinge pillar of Beetle models is damaged, replace it during installation.

6.5 Removing and Installing Door Handle

The parts involved in mounting the door handle are shown in Fig. 6-13. For removal, simply take out the Phillips head screw that passes through the edge of the door and threads into the handle. The screw is concealed by the weatherstrip, which must be moved slightly aside for access. After removing the screw, pull the handle free of the door and unhook the handle at its forward edge.

When installing, clean the lock and the door release trigger, then apply door and lock lubricant. Make sure the gaskets are in good condition and that they seat properly when you mount the door handle on the door.

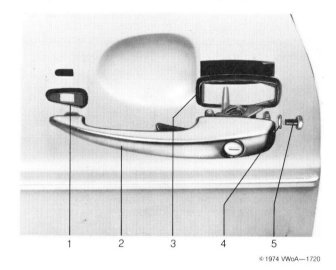

1. Small gasket
2. Handle
3. Large gasket
4. Washer
5. Phillips screw

Fig. 6-13. Parts related to door handle mounting.

6.6 Removing and Installing Outside Mirror

The outside mirror is shown in Fig. 6-14.

Fig. 6-14. Outside mirror assembled.

Body and Frame

The mounting point for the outside mirror is a tapped hole in the door. The mirror mount screws into this hole. The mirror can be removed from the door for replacement or for cleaning, lubricating, or adjusting the mount.

To remove:

1. Using an open end wrench, loosen the mounting socket. Then screw the mirror and mount out of the tapped hole in the door.

2. Make sure the joints move freely. If necessary, disassemble the mount as shown in Fig. 6–15 and lubricate it. During assembly, lock the mirror arm nut with centerpunch marks.

1. Washer
2. Mirror arm
3. Washer with hexagonal hole
4. Coil spring
5. Lock washer
6. Cap nut
7. Mirror socket
8. Sealing washer
9. Spiral spring
10. Mirror arm nut

Fig. 6–15. Components of mirror mount.

3. Installation is the reverse of removal. Check, though, that the sealing washer fits properly. Use a new sealing washer if the original is worn or damaged.

7. Seats

A new-type front seat was introduced on the 1973 models (except Karmann Ghia). The new seat has a three-point mounting system and a lever for adjusting the angle of the backrest rather than the rotary knob used formerly. The new-type seat does not fit earlier cars.

A seat belt warning system is installed on cars built since January 1972. If the ignition is turned on and a gear selected before the driver's seat belt is fastened, a warning buzzer sounds and a **Fasten Seat Belts** sign lights up until the belt is fastened. The system also functions if the ignition is turned on before a front seat passenger has fastened the passenger seat belt. The 1974 models cannot be started while the buzzer is sounding.

7.1 Removing and Installing Front Seats

Early- and late-type front seats require different procedures for removal and installation.

To remove early-type front seat:

1. Place paper in the front footwell to protect the carpet from grease stains. On cars built since January 1972, unplug the extensible wire for the seat belt warning system.

2. Lift the adjusting lever (Fig. 7–1). Slide the seat forward until its runner contacts the leaf-spring stop.

3. While standing outside the car, use a screwdriver to depress the leaf-spring stop. With the adjusting lever raised, slide the seat forward approximately 40 mm (about 1½ in.).

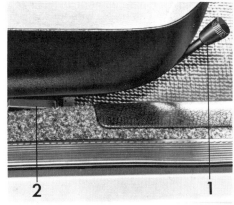

1. Adjusting lever
2. Leaf-spring stop

Fig. 7–1. Controls that hold seat in its runners.

4. Reach under the seat and unhook the coil spring as indicated in Fig. 7–2.

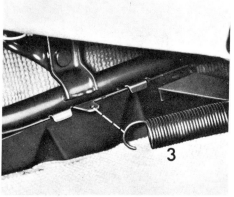

Fig. 7–2. Spring **(3)** unhooked from hole in bracket.

18 Body and Frame

5. With the adjusting lever raised, slide the seat fully off the tracks toward the front of the car. Then lift it out through the door.

To install:

1. Stand outside the car. For better balance, hold the seat with the backrest folded forward.
2. Hook the inboard seat runner on its track first. Then insert the outer runner by pulling the seat slightly toward you.
3. With the adjusting lever raised, slide the seat back on the tracks and reconnect the coil spring.
4. Reconnect the extensible wire for the seat belt warning system (where fitted).

To remove late-type front seats:

1. Unplug the extensible wire for the seat belt warning system.
2. Pull back the adjusting lever (Fig. 7-3) and push the seat back to the next to last position.

Fig. 7-3. Adjusting levers. Release as indicated by the arrows to permit the seat to be moved.

3. Pull the covers (Fig. 7-4) off the runners.

Fig. 7-4. Cover (arrow) on end of runner.

4. Using a screwdriver inserted into the bracket from the front (Fig. 7-5), press down the leaf-spring stop.

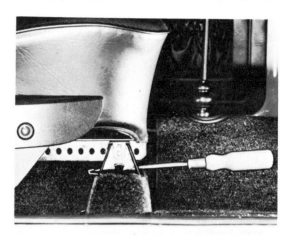

Fig. 7-5. Leaf-spring stop on front (central) seat mounting bracket being pressed down with screwdriver.

5. Pull back the seat adjusting lever on the frame tunnel and slide the seat out of its runners to the rear.

To install:

1. Check the four friction pads on the seat runners. Also check the two spring clips pressed into the upper friction pads.

   NOTE —
   If the clips are faulty, missing, or not pressed in, the seat will rattle.

2. Insert the seat into the side runners from the rear.
3. Pull back the seat adjusting lever on the frame tunnel, and slide the seat forward and into the front (central) bracket.
4. Press each cover into the rear of its side runner. Then reconnect the extensible wire for the seat belt warning system.

7.2 Removing and Installing Rear Seat

On cars built prior to January 1972, remove the rear seat by lifting it at its front edge and then pulling it forward from under the rear seat backrest. Installation is the reverse of removal. Hold the seat in a diagonal position when removing it or installing it between the side trim panels.

On cars built since January 1972, the rear seat is secured in position by a hook on the seat frame that engages a cleat on the luggage floor. These parts are indicated by the two white arrows in Fig. 7-6.

Fig. 7-6. Hook on rear seat frame (right) and cleat on floor.

To remove late-type seat:

1. Lift the seat at its front edge until it is above the front locating runner.

2. Pull the seat slightly forward, raising the front still farther to disengage the hook from the cleat.

3. Lift the seat on the passenger side of the car and remove it in this tilted position.

To install:

1. With the seat tilted downward on the driver's side, lift it into the car and place it between the rear side trim panels.

2. Lower the seat on the passenger side. Then lift the seat slightly at the front and push it back under the backrest until the hook engages the cleat.

3. Apply firm pressure at the front to press the seat down behind the front locating runner.

To remove rear seat backrest:

1. Remove the rear seat.

2. Remove the backrest pivot bolts from both sides of the car.

3. Pull the strap on the backrest lock to free the backrest from the car body.

4. Raise the bottom edge of the backrest. Lift the backrest at the passenger side until it is in a diagonal position, then remove it from the car.

Installation is the reverse of removal.

8. HEATING AND VENTILATING SYSTEMS

Heated air can be brought into the passenger compartment through vents in the front and rear footwells as well as through the windshield vents and the air vents in the dashboard. Two levers between the front seats control the amount and the temperature of heated air delivered to the car's interior.

The ventilation system admits fresh air through the windshield vents and the dashboard vents. Beginning with the 1971 models, a two-speed fan assists fresh air ventilation on Super Beetles and Convertibles. Rotary knobs on the dashboard control the amount of fresh air admitted.

8.1 Adjusting and Replacing Heating Cables and Flaps

The flaps that control air temperature are built into the heat exchangers. Their removal and repair is described in **ENGINE**. The amount of heat admitted through the rear footwell vents is controlled by flaps under the rear seat. Cables connect the heat exchanger flaps and the flaps under the rear seat to the control levers between the front seats. The amount of heated air admitted through the front footwell vents is controlled by levers built into the front footwell outlets.

To replace and adjust temperature control cables:

1. With a 9-mm wrench on the clamping sleeve bolt and a 10-mm wrench on the nut, loosen the nuts on the clamping sleeve bolts for each heat exchanger lever. (Fig. 8-1).

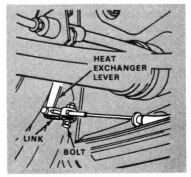

Fig. 8-1. Clamping sleeve bolt in the link on the heat exchanger lever. These parts are under the car, in front of the engine on each side.

WARNING —
Support the car properly on a lift or on support stands. It is hazardous to work under a car supported only by a jack.

20 Body and Frame

2. Pull the cables out of the clamping sleeves. Then remove the rubber sealing plugs from the guide tubes and slide them off the disconnected cables.

3. Remove the lever cover between the front seats.

4. Remove the nut from the right-hand heating control lever (marked **TEMP**).

5. Take the lever and friction disks off the mounting. Then unhook the lever from the cranked cable ends and pull the cables forward out of the guide tubes.

Installation is the reverse of removal. Install new sealing plugs in the rear of the guide tubes if the original seals no longer fit tightly. Push the temperature control lever down and make sure the heat exchanger lever is all the way to the rear. Then, using a 9-mm wrench on the bolt and a 10-mm wrench on the nut (to avoid bending or breaking the cable), tighten the nut on the clamping sleeve bolt.

To replace and adjust rear footwell flap cables:

1. Remove the rear seat as described in **7.2 Removing and Installing Rear Seat.** Loosen the cable clamp screws at both sides of the car (Fig. 8-2 or Fig. 8-3).

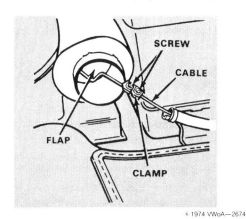

Fig. 8-2. Cable clamp (Beetle or Super Beetle). Loosen only the screw that holds the cable.

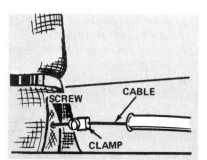

Fig. 8-3. Cable clamp (Karmann Ghia).

2. Pull the cables out of the clamps.

3. Remove the lever cover between the front seats.

4. Remove the nut from the left-hand heating control lever (marked **OFF-HEAT**).

5. Take the lever and the friction disks off the mounting. Then unhook the lever from the cranked cable ends and pull the cables forward out of the guide tubes.

Installation is the reverse of removal. Push the heat control lever down and make sure that both rear footwell flaps are closed. Then tighten the clamp screws.

NOTE —
The rear footwell flaps are held on the pivot pins by spring clips. To replace the flaps, disconnect the cables as previously described, then pull the flaps off their pivot pins with the short piece of cable that remains connected to the flap. Then snap on the new flap.

Front Footwell Flaps

The front footwell heat outlet flaps (Fig. 8-4) can be removed simply by taking out the Phillips head sheet metal screw at each end of the grille piece. The flap may then be detached from the grille.

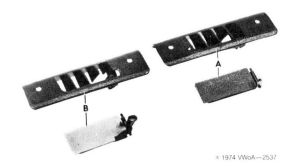

Fig. 8-4. Front footwell heat outlet flaps. Type **B** is now used. Type **A** (with remote control cable) was installed only on some 1970 Karmann Ghias.

8.2 Removing and Installing Fresh Air Ventilation

Three ventilation systems are used on the cars covered by this Manual and will be discussed separately.

The original VW fresh air ventilation system, used on all 1970 models, has been continued on subsequent Beetles and Karmann Ghias. A fresh air ventilation box with a two-speed fan was introduced on the 1971 Super Beetle and Convertible. This system was extensively revised beginning with the 1973 models.

BODY AND FRAME 4

Ventilation System Without Fan

The fresh air control box and the components attached to it are shown in Fig. 8-5.

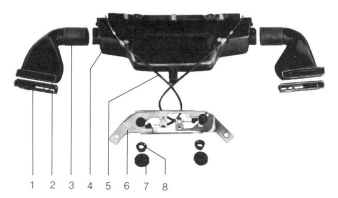

1. Fresh air vents
2. Fresh air duct
3. Hose
4. Fresh air control box
5. Cable
6. Control bracket
7. Control knob
8. Escutcheon

Fig. 8-5. Components of fresh air ventilation system (all 1970 models and all later Beetles and Karmann Ghias).

To remove fresh air box:

1. Pull off the control knobs. If they are difficult to remove, use a loop of strong string as illustrated in Fig. 8-6.

Fig. 8-6. Using string to remove control knob. Usually, this is necessary only when the knob is removed for the first time.

2. Remove the dashboard access panel from the rear of the front luggage compartment by removing the two knurled nuts.

3. Pull the two pieces of hose leading to the fresh air ducts off the fresh air control box.

4. Remove the hexagon nut indicated in Fig. 8-7. Then pull the elbow on the water drain hose off the fresh air control box.

5. Remove the three Phillips head screws (Fig. 8-7). Then take off the fresh air control box.

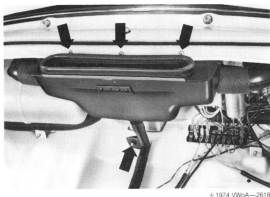

Fig. 8-7. Hexagon nut (bottom arrow) and three Phillips head screws (upper arrows).

6. Remove the two Phillips head screws that hold the control bracket to the back of the dashboard. Then remove the fresh air control box from the car together with the control bracket and cables.

   NOTE
   It is unnecessary to dismount the control bracket from the back of the dashboard if you only need to adjust a cable. Cable adjustment is described in conjunction with the installation procedure.

7. Pull the fresh air ducts and hoses downward and out of the dashboard.

8. Press the fresh air vents upward out of the dashboard.

Installation is the reverse of removal. Make sure that the flaps in the fresh air control box outlets close completely when the control knob is turned to the **OFF** position. If they do not, adjust the cable by loosening the clamp on the control box. Move the cable toward the control bracket until the flap is closed with the control knob in its **OFF** position. Then tighten the clamp.

If a cable must be replaced, remove the fresh air control box together with the control bracket and cables. Remove the cable clamp from the fresh air control box and slip the cable eye off the cranked end on the flap shaft. Then remove the spring clip that holds the cable to the control bracket and unhook the cable eye from the pin on the control knob's gear rack. Install the new cable and adjust it as described earlier.

22 BODY AND FRAME

Ventilation System With Fan (through 1972)

On 1971 and 1972 Super Beetles and Convertibles, the fresh air box with fan is installed as shown in Fig. 8-8.

Fig. 8-8. Fresh air box at rear of luggage compartment.

To remove fresh air box with fan:

1. Pull off the control knobs. If they are difficult to remove, use a loop of strong string as illustrated earlier in Fig. 8-6.

2. Remove the dashboard access panel from the rear of the front luggage compartment by removing the two knurled nuts.

3. Disconnect the fresh air ducts from the fresh air control box.

4. Free the support foot on the bottom of the fresh air control box from the car body and disconnect the water drain hose.

5. Remove the three Phillips head screws that hold the fresh air control box to the cowl panel. They are in the same approximate locations as those indicated earlier in Fig. 8-7.

6. Take the fresh air control box off the cowl panel and place it in the luggage compartment. Then remove the electrical wires from the terminals on the rear of the fan switch.

7. Remove the two Phillips head screws that hold the control bracket to the back of the dashboard. Then remove the fresh air control box from the car together with the control bracket and cables.

8. Pull the fresh air ducts and hoses downward and out of the dashboard.

9. Press the fresh air vents upward out of the dashboard.

Installation is the reverse of removal. If necessary, adjust the cables as described earlier for the fresh air control box without fan.

The fresh air box with fan is shown disassembled in Fig. 8-9.

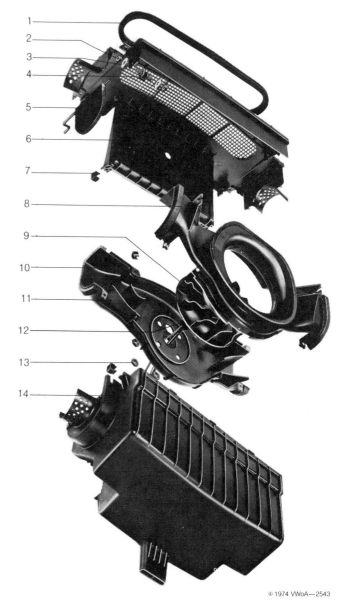

1. Weatherseal
2. Nut (8)
3. Washer (8)
4. Bonded rubber mounting (4)
5. Fresh air flap (2)
6. Upper part of fresh air box
7. Clip for fresh air box (10)
8. Front part of fan duct
9. Fresh air fan with motor
10. Rear part of fan duct
11. Sealing flap
12. Wiring harness grommet
13. Nut with washer (4)
14. Lower part of fresh air box

Fig. 8-9. Exploded view of fresh air box with fan.

The fresh air box can be disassembled to service or replace the fresh air flaps or to gain access to the motor. To disassemble the box, remove the control cables. Then pry off the clips and lift the upper part of the fresh air box off the lower part. The fan duct, fan, and motor will come out along with the upper part of the fresh air box. Take off the four nuts indicated in Fig. 8-10 in order to separate the fan duct from the upper part of the box.

Fig. 8-10. Nuts on bonded rubber mountings (arrows) that hold fan duct and upper part of fresh air box together.

The fan duct can be taken apart by prying off the clips. Four nuts hold the fan and motor assembly to the rear part of the fan duct (Fig. 8-11).

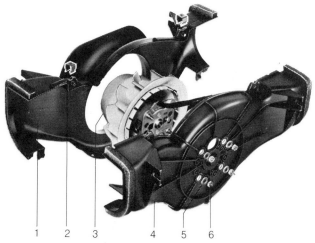

1. Front part of fan duct
2. Clip
3. Fan with motor
4. Rear part of fan duct
5. Washer
6. Nut

Fig. 8-11. Fan duct disassembly.

Assembly is the reverse of disassembly. If the fan motor is defective it can be replaced only as a complete unit. Removal and installation of dashboard switches is covered in **ELECTRICAL SYSTEM**. Consult the wiring diagrams given in **ELECTRICAL SYSTEM** if you need help in reconnecting the wiring harness from the fresh air fan motor to the switch. When adjusting the fresh air flap cables, use the procedure given earlier for the fresh air box without fan.

Ventilation System with Fan (1973 on)

Beginning with the 1973 models, the fresh air box is an integral part of the body. The two-speed fan is mounted in a redesigned duct and is installed behind the instrument panel. A panel can be removed at the rear of the luggage compartment to gain access to the inside of the fresh air box. This permits you to service the wiper motor or to inspect the air distributor box that leads into the fan and air control ducts (Fig. 8-12).

Fig. 8-12. Fresh air box **(A)** and air distributor box **(B)**.

If turning the control knobs does not change air output, remove the radio or the radio aperture cover. Then see if the control cable eyes are properly in place on the cranked ends of the fresh air flap shafts. If not, repair the cable ends working through the aperture.

If the control knobs are hard to turn or do not turn at all, remove the fuse box cover panel. Then, using a nut driver as shown in Fig. 8-13, loosen the bolts that hold the control bracket. Shift the bracket on the elongated holes until the knobs turn freely, then tighten the nuts.

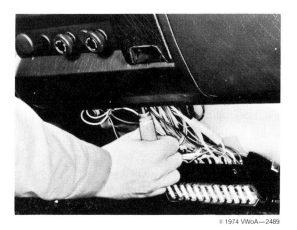

Fig. 8-13. Bolts for control bracket being loosened.

24 Body and Frame

If the knobs cannot be made to turn freely or if the flaps are stuck in the air control duct, the dashboard must be removed to gain access to the air controls.

To repair control bracket or flaps:

1. Disconnect the battery ground strap.

2. Remove the steering wheel and the steering column switch as described in **STRUT FRONT SUSPENSION**.

3. Remove the instrument panel switch cluster as described in **ELECTRICAL SYSTEM**. Then disconnect the speedometer cable from the speedometer (Fig. 8-14).

Fig. 8-14. Knurled ferrule nut that holds speedometer cable to speedometer. Access is gained by removing the instrument panel switch cluster.

4. Pry off the plastic caps. Then remove the screws on each side of the dashboard (Fig. 8-15).

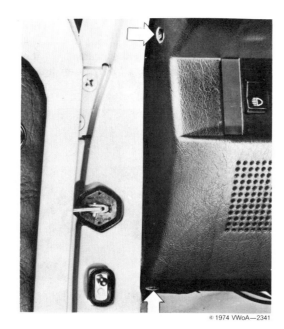

Fig. 8-15. Screws that hold dashboard.

5. Remove the dashboard support nut above the upper right corner of the fuse box cover panel. Then lift the dashboard off the support bracket and place it in the position shown in Fig. 8-16.

Fig. 8-16. Dashboard removed. Wires can remain attached while you service the control bracket or flaps.

6. If the flaps are stuck, remove the air control duct from the dashboard support bracket and repair it as described in steps 7 through 9. If the regulating control is faulty, move on to step 10.

7. Remove the cables by taking off the clips and unhooking the cable eyes from the cranked ends on the flap shafts.

8. Remove the clips on the air control duct and disassemble the duct as shown in Fig. 8-17.

9. Free-up or repair the sticking flaps. Then reassemble and install the air control duct.

Fig. 8-17. Air control duct disassembled. Right-hand part fits over the left-hand part, as indicated by the curved arrow.

10. If necessary, replace the cables. The cables can be disconnected and connected to the air control duct by working through the radio aperture. The control bracket must then be removed with the cables attached.

BODY AND FRAME 25

11. To remove the control bracket, take out the two bolts that hold it to the dashboard.

12. Remove the clips for the cable sheaths, then unhook the cable eyes from the pins indicated in Fig. 8-18.

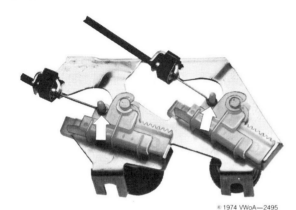

Fig. 8-18. Pins (arrows) on control racks.

13. Turn both control pinions past the locking tooth on their racks to the stops. Then install the new cables.

14. Close the flaps in the air control duct. Then attach both cables to the air control duct.

15. Check to see that the flaps open and close properly. The cables must be installed as shown in Fig. 8-19.

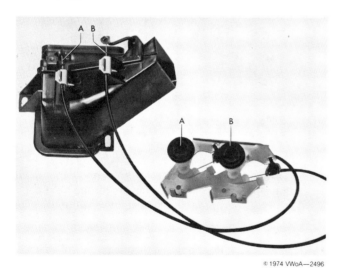

Fig. 8-19. Cable **A** for defroster control and cable **B** for central outlet control.

Installation is the reverse of removal. Make sure that the dashboard is properly hooked into the support bracket and that the air ducts all fit properly.

9. SLIDING ROOF (SUN ROOF)

The cables that operate the sliding roof (optional equipment) are connected to a crank above the windshield. Most cable malfunctions can be corrected by adjustment. However, cables that are badly stretched or broken must be replaced.

Some of the parts shown in the illustrations in the following procedures have been altered in size and shape during the model years covered by this Manual. Nevertheless, the procedures remain unchanged. For safety, always fold the handle back into its recess after adjusting the position of the sliding roof.

9.1 Removing and Installing Sliding Roof Panel

The sliding roof panel must be removed in order to gain access to the cables and runners. In carrying out the following procedure, be careful not to scratch the paint on the roof panel or on the surrounding parts of the car.

To remove:

1. Open the sliding roof halfway. Using a wooden wedge, carefully pry loose the trim panel. There are five clips to be disengaged.

2. Move the trim panel out of the way by pushing it to the rear as far as it will go (Fig. 9-1).

Fig. 9-1. Trim panel being slid fully to the rear.

CAUTION —
It is not possible to remove the trim panel until the sliding roof is out and one of the side runners has been removed. See 9.2 Removing and Installing Runners and Cables. Trying to remove the trim panel before this stage could damage the trim panel.

26 Body and Frame

3. Close the roof until it is open only about 50 mm (about 2 in.). Then unscrew both front guides as shown in Fig. 9-2.

Fig. 9-2. Removing screws that hold guides.

4. Close the roof. Unhook the leaf springs near the rear guides and swing them toward the front of the car. Then pull the rear guide lifter pins out of the brackets as indicated in Fig. 9-3.

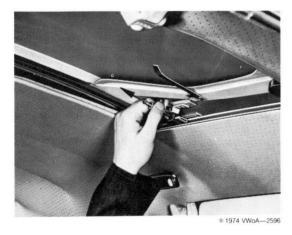

Fig. 9-3. Guide lifter pin being pulled out of bracket. Note how leaf spring is swung to front.

5. Lift out the sliding roof panel through the top of the roof.

To install:

1. Hold the trim panel down for clearance and insert the top panel, rear edge first, in the roof opening. Angle the panel upward at the front.

2. Slowly lower the front edge of the roof panel and push it to the rear.

3. When the top panel is fully lowered, pull it to the front. Insert the rear guide lifter pins in their brackets, then swing the leaf springs back into place.

4. Install the front guides. If necessary, adjust the panel flush with the top of the car as described in **9.4 Adjusting Top Height.**

5. Install the trim panel, reversing the removal procedure.

9.2 Removing and Installing Runners and Cables

If either cable is faulty, replace them both. A stretched cable could keep one side of the roof from closing fully.

To remove:

1. Remove the sliding roof panel. Then remove the five screws in the central cable guide channel cover and take off the cover (Fig. 9-4).

Fig. 9-4. Central cable guide channel cover removed to reveal gear and tensioning plate.

2. Take the tensioning plate off the gear. Then remove the eight screws from the covers for the front cable channels and from the side runners on both sides.

3. Take off the covers for the front cable channels.

4. Take the side runners with cables out of the roof opening toward the front. The sliding roof trim panel may be removed at this time also.

CAUTION ——
Be careful not to damage the paint on the roof panel as you remove the runners. Repaint any scratches around the sliding roof opening to prevent the formation of rust.

BODY AND FRAME 27

To install:

1. Check the cables for wear, and, if necessary, replace them.

2. Put the runners in place. Push the plastic locating pins (Fig. 9-5) completely into the holes in the brackets. The pins must fit tightly in both the brackets and the runners or the roof will bind.

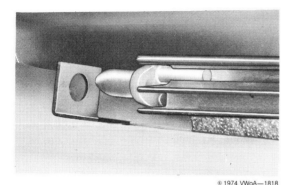

Fig. 9-5. Plastic locating pin and bracket.

3. Before installing the side runners and tightening the screws, insert the trim panel in the lower groove. Then push it fully to the rear.

4. Coat the cables with molybdenum grease. Then install the cables.

   NOTE —
   Before installing the upper left cable channel covers, cross the cables as illustrated in Fig. 9-6 and press them into the gear.

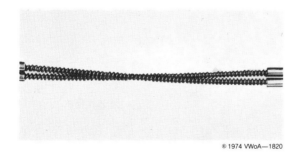

Fig. 9-6. Cables crossed to mesh with crank gear.

5. Install the front channel covers and the tensioning plate.

6. Press on the cover for the central guide channel and secure it. The three sheet metal screws go toward the front of the car.

7. Install the sliding roof panel and adjust the cables.

9.3 Adjusting Cables

The cables should be adjusted following the installation of new cables, or if the sliding roof does not open or close evenly on both sides.

To adjust:

1. Open the sliding roof halfway. Using a wooden wedge, carefully pry the trim panel loose. There are five clips to be disengaged.

2. Push the trim panel fully to the rear and close the sliding roof.

3. From inside the car, remove the screw in the center of the crank. Take off the crank and the escutcheon.

4. Loosen by about six turns the screws that hold the drive gear mounting in the roof.

5. Pull the drive gear down so that it disengages from the cables. Then pull back the two rear guides until they are in line with the brackets in the top panel. (See Fig. 9-7.)

Fig. 9-7. Arrows indicating disengagement of drive gear and rearward movement of rear guides.

NOTE —
The rear guides will already be on the brackets if you are not fitting new cables.

6. Insert the rear guide lifter pins in the holes in the brackets and swing the leaf springs to the rear. Then turn the drive gear shaft clockwise as far as it will go (maximum 12 turns).

7. Press the drive gear upward so that it engages with the cables. Tighten the mounting screws.

8. Install the escutcheon and the crank.

9. Check the cable adjustment by opening and closing the roof a few times. If the adjustment is correct, reinstall the trim panel.

28 Body and Frame

9.4 Adjusting Top Height

You can adjust the sliding roof to ensure that it will be flush with the car roof when the sliding roof is closed.

To adjust front edge:

1. Open the sliding roof halfway. Using a wooden wedge, carefully pry the trim panel loose. There are five clips to be disengaged.
2. Push the trim panel fully to the rear and close the sliding roof.
3. Loosen the screws that hold the front guides. Adjust the roof height by turning the M 10 × 1 countersunk Phillips head screws located in the front guide mountings.
4. Tighten the front guide screws. Install the trim panel.

To adjust rear edge:

1. Open the sliding roof halfway. Using a wooden wedge, carefully pry the trim panel loose. There are five clips to be disengaged.
2. Push the trim panel fully to the rear, but do not close the sliding roof.
3. Close the roof gradually until the lifters in the rear guides are just about to lift on their ramps.
4. Loosen the nuts on the upper pins as illustrated in Fig. 9-8.

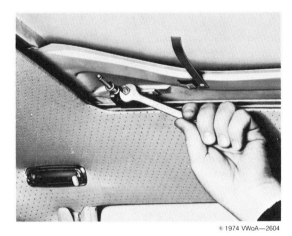

Fig. 9-8. Upper pin nuts on rear lifter being loosened.

5. Adjust the rear edge height by moving the lifter as shown in Fig. 9-9.
6. Fully close the roof and check the height. Reopen the roof if further adjustment is necessary.

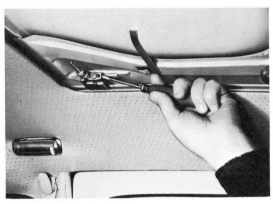

Fig. 9-9. Rear edge height adjustment.

7. When the adjustment is correct, tighten the nuts. If the roof does not open and close evenly, adjust the cables. Then refit the trim panel.

10. Frame

A number of significant frame modifications occurred at the beginning of the 1973 model year. The parking brake lever was moved 60 mm (2⅜ in.) to the rear and the gearshift lever 40 mm (1 9/16 in.) to the rear. Because of these changes, the shift rod and parking brake cables of earlier models cannot be interchanged with those of 1973 and later cars.

Because the bonded rubber mounting for the transmission was modified, the studs for the front transmission mount are now as illustrated in Fig. 10-1, not horizontal as they were on 1970 through 1972 models.

Fig. 10-1. Front transmission mount studs (arrows) as installed on 1973 and later frames.

Body and Frame

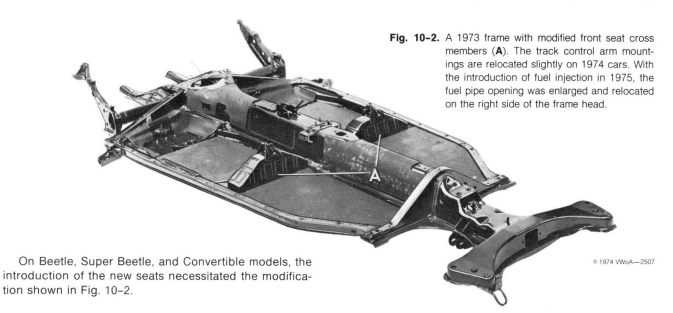

Fig. 10-2. A 1973 frame with modified front seat cross members (**A**). The track control arm mountings are relocated slightly on 1974 cars. With the introduction of fuel injection in 1975, the fuel pipe opening was enlarged and relocated on the right side of the frame head.

On Beetle, Super Beetle, and Convertible models, the introduction of the new seats necessitated the modification shown in Fig. 10-2.

10.1 Removing and Installing Body on Frame

When extensive frame work must be done, it is often advisable to completely remove the body from the frame.

To remove:

1. Remove the battery, the front seats, the rear seat, and the rear seat backrest. Remove the fuel tank as described in **FUEL SYSTEM.**

2. Remove the speedometer cable from the left front wheel steering knuckle.

3. Remove the steering column as described in **STRUT FRONT SUSPENSION** or in **FRONT AXLE.**

4. Detach the hoses between the brake fluid reservoir and the master cylinder (see **BRAKE AND WHEELS**). Disconnect the wires from the master cylinder.

5. Working through the fuel tank aperture on Beetle and Karmann Ghia models, remove the two bolts that hold the body to the front axle.

6. Place the car on a lift. Disconnect the battery cable and the wires from the starter solenoid. On 1971 and later Super Beetles and Convertibles, remove the suspension struts and the steering linkage as described in **STRUT FRONT SUSPENSION.** Then unbolt the body from the frame head.

7. Disconnect the clutch cable from the clutch operating lever and disconnect the accelerator cable from the carburetor. Pull the cables free of the engine and the transmission.

8. Disconnect the heater cables from the levers on the heat exchangers, and disconnect the heat hoses from the front of the heat exchangers.

9. From under the car, remove the 18 bolts from the floor pan side channels. Remove the two large bolts from the bottom of the front cross member.

10. Remove the rear wheels. Working under the fenders, remove the bolt on each side of the car that holds the body to the rear shock absorber mounting.

11. Install the rear wheels. Then lower the car to floor level.

12. Working in the area under the rear seat, remove the four bolts that hold the body to the rear cross member and then the two pairs of bolts that hold the body to the rear cross tube.

13. Lift the body off the frame, carefully separating the rubber weatherseal between the frame and the body from the body's lower edge.

    NOTE
    Use an overhead hoist or a gantry crane to lift the body. Attach slings to the four fenders. Alternately, have four or more helpers lift the body off while you check the separation of the weatherseal and make sure that no wires between the body and the frame have been left connected.

Installation is the reverse of removal. However, the weatherseal should be inspected and replaced if it is broken or has lost resiliency. Replace the rubber packing strip that goes between the body and the center of the rear cross member. Do not forget the rubber spacers that go between the body and the rear cross tube, the shock absorber mountings, and the front axle or the frame head.

If the perimeter weatherseal is being replaced, punch holes for the body mounting bolts in the new weatherseal before lowering the body onto the frame. Torque the body bolts to 1.5 to 2.0 mkg (11 to 14 ft. lb.).

Section 5

BRAKES AND WHEELS

Contents

Introduction 3	6.4 Checking Brake Disc 18
1. General Description 4	6.5 Removing and Installing Brake Disc . . . 19
Foot Brakes 4	6.6 Removing and Installing Brake Disc Splash Shield 19
Parking Brakes 4	6.7 Reconditioning Brake Disc 20
2. Maintenance 4	7. Drum Brakes 20
3. Brake Fluid Reservoir 4	7.1 Adjusting Drum Brakes 21
3.1 Filling Reservoir 4	7.2 Removing and Installing Brake Drums . . . 21
3.2 Removing and Installing Reservoir 4	7.3 Removing and Installing Brake Shoes . . . 22
4. Master Cylinder 5	7.4 Removing and Installing Wheel Cylinder 24
4.1 Pushrod Adjustment 5	7.5 Wheel Cylinder Repair 24
Detecting and Correcting Altered Pushrod Length 6	7.6 Removing and Installing Backing Plate . . 24
4.2 Removing and Installing Master Cylinder 6	7.7 Reconditioning Brake Drums 25
4.3 Repairing Master Cylinder 6	8. Bleeding Brakes 26
4.4 Testing and Replacing Brake Light/Warning Light Switches 8	Brake Fluids 26
	8.1 Changing Brake Fluid 26
5. Brake Lines and Hoses 9	8.2 Bleeding with Pressure Bleeder 26
5.1 Removing and Installing Brake Lines . . 9	8.3 Bleeding by Pumping 26
5.2 Removing and Installing Brake Hoses . . 10	8.4 Flushing Brake System 27
6. Disc Brakes (Karmann Ghia only) 10	8.5 Brake Cylinder Paste 27
6.1 Removing and Installing Brake Pads . . 11	9. Parking Brake 27
6.2 Removing and Installing Brake Caliper . . 15	9.1 Adjusting Parking Brake 27
6.3 Brake Caliper Repair 15	9.2 Removing and Installing Parking Brake Lever 28
	9.3 Removing and Installing Brake Cable . . 28
	10. Brake Troubleshooting 28

© 1974 VWoA

2 Brakes and Wheels

11. Removing, Repairing, and Installing Pedal Cluster . . . 30

12. Wheels and Tires . . . 31
    12.1 Wheels . . . 31
    12.2 Tire Types and Pressures . . . 31
           Conventional Tires . . . 31
           Radial Tires . . . 31
           Winter Tires . . . 31
           Tire Pressures . . . 31
    12.3 Normal Tire Wear . . . 32
    12.4 Removing and Installing Wheels . . . 32
    12.5 Wheel Rotation . . . 33
    12.6 Changing Tires . . . 33
    12.7 Abnormal Wear . . . 34
           Improper Inflation . . . 34
           Hard Driving . . . 35
           Road Surfaces . . . 35
           Faulty Wheel Alignment . . . 36
           Wheels Out of Balance . . . 36
           Overloading . . . 36

13. Technical Data . . . 36
    I. Tightening Torques . . . 37
    II. Tire Data . . . 37
    III. Tolerances, Wear Limits, and Settings . . . 38

TABLES

a. Brake Drum Specifications . . . 25
b. Brake Troubleshooting . . . 29

Brakes And Wheels

Beetles, Super Beetles, and Convertibles have four-wheel drum brakes. Karmann Ghias have disc brakes at the front and drum brakes at the rear. On all models, each axle has its own hydraulic brake circuit.

A dual-chamber master cylinder provides operating pressure to both brake circuits. The system is designed so that leaks in one circuit cannot affect the other circuit. An electrical warning system in the master cylinder causes a red warning indicator in the instrument panel to light up if hydraulic pressure is too low in either brake circuit. If you see this light while you are driving, it is imperative that the brake system be given a thorough check, even though braking action may still seem completely satisfactory. Complete loss of pressure in one of the brake circuits will cause the pedal to fall closer to the floor during braking and will result in abnormally long stopping distances.

Because safe vehicle operation depends very heavily on the brakes, all brake system service and repair work must be carried out with extreme cleanliness, careful attention to specifications, and proper working procedures. All necessary information is given here, although some of the operations that are described may be of practical value only to professional mechanics.

If you lack the skills, special tools, or a clean workshop for servicing the brake system, we suggest you leave such repairs to an Authorized VW Dealer or other qualified shop. We especially urge you to consult your Authorized VW Dealer before attempting repairs on a car still covered by the new-car warranty.

The pages devoted to wheels and tires are of practical value to all drivers, whether they service their own cars or not. A great many cases of abnormal tire wear or poor vehicle handling are the direct result of improperly fitted tires, tires incorrectly inflated, or driving practices that damage the tires. It is our belief that following the advice offered here will not only save you money but also make your driving safer

4 Brakes and Wheels

1. General Description

The following major parts make up the brake system:

Master Cylinder Actuated by the foot pedal and connected to the wheel cylinders by brake lines and hoses, the master cylinder generates the hydraulic pressure needed to operate the brakes. It has two pistons, one operating the front brakes and the other the rear brakes. Several different master cylinders are used on cars covered by this Manual. It is important that replacement cylinders be the correct type for the particular model.

Fluid Reservoir Located in the front luggage compartment, the fluid reservoir supplies brake fluid to the master cylinder. Two hoses connect the fluid reservoir with the master cylinder.

Wheel Cylinders These are hydraulic cylinders with two opposed pistons that press the brake shoes against the brake drums. There is one wheel cylinder in each drum brake assembly.

Brake Shoes Moved by the wheel cylinder pistons, one leading and one trailing shoe work against the inside of each brake drum.

Brake Calipers Karmann Ghias have front disc brakes. Each disc brake assembly has a brake caliper which houses the two opposed pistons that press the friction pads against the brake disc.

Hydraulic Lines These are steel tubes and hoses that connect the master cylinder to the front brake calipers or wheel cylinders and to the rear wheel cylinders.

Foot Brakes

The dual-circuit hydraulic foot brakes operate on all four wheels. Disc brakes are used on the front wheels of Karmann Ghia models only. The other cars covered by this Manual have drum brakes at all four wheels.

Parking Brakes

The cable-operated parking brake works on the rear wheels only. The hand lever is held or released by a ratchet and is centrally mounted on the frame tunnel.

2. Maintenance

The following routine maintenance operations are covered briefly in **LUBRICATION AND MAINTENANCE**. Additional information can be found in this section under the headings listed after each maintenance check.

1. Checking and changing brake fluid. **8.1**
2. Checking brake linings. **7.3**
3. Checking brake adjustment. **7.1**
4. Checking brake lines, hoses, and brake lights. **4.4**
5. Checking wheels, tires, and tire pressures. **12.**

3. Brake Fluid Reservoir

Fig. 3-1 shows the brake fluid reservoir. It is divided into two chambers so that it maintains normal fluid level for one of the brake circuits should the other fail.

Fig. 3-1. Brake fluid reservoir. The outlet pipes are connected to the master cylinder by hoses.

3.1 Filling Reservoir

Clean the cap and reservoir top before removing the cap. This will keep dirt out of the brake fluid. Once the cap is off, check to see that the vent is open. Fill the reservoir to just above the seam edge near the top, or to within 15 to 20 mm (about ¾ in.) of the top of the reservoir neck.

CAUTION ——
Use only new, unused brake fluid that meets SAE recommendation J 1703 and conforms to Motor Vehicle Safety Standard 116.

3.2 Removing and Installing Reservoir

The brake fluid reservoir is located behind the access panel at the rear of the front luggage compartment on Karmann Ghias and on the left rear side of the front luggage compartment on all other models. Empty the reservoir with a syringe, then detach the hoses. Use a cloth to catch spilled fluid. Take the screw out of the mounting bracket to detach the reservoir from the car.

WARNING ——
Do not start a siphon with your mouth or spill fluid onto the car. Brake fluid is both poisonous and damaging to paint.

Installation is basically the reverse of removal. Refill the reservoir with new fluid as described in **3.1 Filling Reservoir**. Then bleed the brakes, if necessary, as described in **8. Bleeding Brakes**.

BRAKES AND WHEELS 5

4. MASTER CYLINDER

The master cylinder has two pistons, one behind the other, as shown in Fig. 4-1. Notice that there is a small compensating port just ahead of each piston, as well as a larger port behind it. The compensating port can admit brake fluid to, or receive fluid from, the working sides of the pistons only when the pistons are in their rest position.

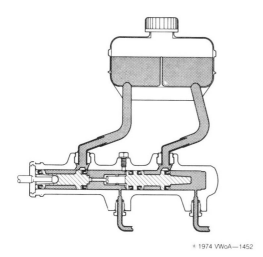

Fig. 4-1. Schematic drawing of master cylinder and fluid reservoir. Tubes at bottom of master cylinder lead to front and rear brake circuits.

4.1 Pushrod Adjustment

The two compensating ports pointed out in Fig. 4-1 are, in many ways, the most important parts of the master cylinder. Their job is to permit surplus brake fluid to return to the reservoir, and to allow the reservoir to refill the master cylinder. If the pistons block the compensating ports when the brake pedal is fully released, neither of these functions can be fulfilled. It is therefore necessary to maintain the clearance indicated in Fig. 4-2.

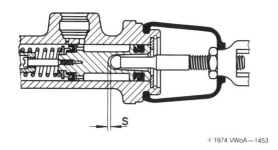

Fig. 4-2. The 1 mm (.040 in.) clearance, dimension **S**, between the end of the pushrod (connected to the brake pedal) and the master cylinder piston.

Dragging brakes, an abnormally high pedal, and brakes that lock up while driving and fail to release are symptoms of blocked compensating ports. Pushrod clearance should be checked and, if necessary, adjusted. If it is already correct, the master cylinder probably needs to be rebuilt or replaced.

The clearance must also be checked after servicing the master cylinder or after removing and installing the pedal cluster on the car. Pushrod clearance cannot be measured directly. However, the proper 1-mm (.040-in.) clearance corresponds to a pedal freeplay of 5 to 7 mm (³⁄₁₆ to ⁹⁄₃₂ in.) as indicated in Fig. 4-3.

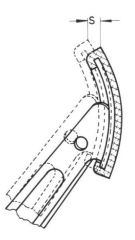

Fig. 4-3. Brake pedal freeplay. Dimension **S** is the distance the brake pedal should travel before the pushrod contacts the master cylinder piston.

The pushrod's length is factory-adjusted. It need not and should not be altered. To adjust pushrod clearance, simply loosen the clamp bolt and move the brake pedal's stop bracket as indicated in Fig. 4-4. Tighten the clamp bolt when freeplay is correct.

Fig. 4-4. Pedal stop bracket movement (double arrow).

6 Brakes and Wheels

Detecting and Correcting Altered Pushrod Length

The pushrod length is factory-adjusted to produce adequate pedal travel when pedal freeplay is correct. However, the factory setting may have been altered. After adjusting pedal freeplay to the specifications given earlier, measure from the back of the pedal to the frame cross member. This distance should be at least 200 mm (7⅞ in.). If clearance is less, the pedal will not have sufficient travel to provide braking if one of the brake circuits should fail. In such cases, the pushrod length must be increased so that there is 200 mm (7⅞ in.) of available pedal travel when pedal freeplay is 5 to 7 mm (³⁄₁₆ to ⁹⁄₃₂ in.).

To adjust pushrod length, push the rubber boot toward the master cylinder to uncover the locknut on the pushrod. Loosen the locknut. Then, using your fingertips, screw the pushrod farther out of its clevis. Readjust the pedal freeplay and the pushrod length alternately until both are within specifications. Then tighten the locknut and pull the boot back to cover the end of the clevis.

4.2 Removing and Installing Master Cylinder

All brake fluid should be drained from the reservoir before the master cylinder is removed from the car.

> **WARNING** ——
> Do not start a siphon with your mouth or spill fluid onto the car. Brake fluid is both poisonous and damaging to paint.

To remove:

1. Pull the elbows out of the top of the master cylinder and detach the wires. Disconnect the pressure lines and seal them with dust caps.

2. Unbolt the master cylinder from the frame. Do not lose the spacers (Fig. 4-5).

Fig. 4-5. Spacer (arrow) under master cylinder bolt.

3. Take the master cylinder out from under the car.

To install:

1. Bolt the master cylinder to the frame.

> **NOTE** ——
> Do not forget the spacers and rubber boot. The vent hole in the boot goes downward. Make sure the pushrod enters the boot and the master cylinder piston properly.

2. Remove the dust caps from the pressure lines and install the lines on the master cylinder.

3. Lubricate the elbows with brake fluid and snap them into the top of the master cylinder. Then attach all wires.

4. Adjust pushrod clearance as described in **4.1 Pushrod Adjustment.**

5. Fill the reservoir with fresh brake fluid and bleed the system as described in **8. Bleeding Brakes.**

> **CAUTION** ——
> Use only new, unused brake fluid that meets SAE recommendation J 1703 and conforms to Motor Vehicle Safety Standard 116.

6. Check the reservoir cap to see that the vent is open. Check the brake operation and the brake lights before road-testing the car.

4.3 Repairing Master Cylinder

Faulty master cylinders can be replaced with new or rebuilt units, or the existing master cylinder can be rebuilt. Replacement is fairly easy; rebuilding demands know-how and some special tools. Repair kits are available for the master cylinder and contain all the seals, cups, pistons, valves, and springs needed for rebuilding.

Except for 1973 and later models, all cars with four-wheel drum brakes have residual pressure valves between the master cylinder outlets and the pressure lines. These valves are shown in Fig. 4-6.

Fig. 4-6. Master cylinder with hexagonal residual pressure valves screwed into pressure outlets.

BRAKES AND WHEELS 7

Cars with front disc brakes have a master cylinder with restriction drillings instead of residual pressure valves. These drillings (See Fig. 4-7) are 3.50 mm (.138 in.) in diameter.

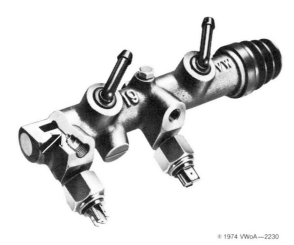

Fig. 4-7. Master cylinder with restriction drillings. The arrow indicates one of the two drillings.

The master cylinders for 1973 and later models with four-wheel drum brakes also have restriction drillings rather than residual pressure valves. These master cylinders must not be interchanged with those used on cars with disc brakes. To prevent confusion, the drum brake type master cylinder is marked by a countersink in the mounting flange, adjacent to the boss for the supply line from the fluid reservoir. Disc brake type master cylinders are marked with a V-notch on the side of the mounting flange, near the mounting bolt hole.

Fig. 4-8 is an exploded view of a master cylinder. Although the exterior of the master cylinder shown in the illustration may not completely resemble the unit on the car, the internal parts of all master cylinders used on the cars covered by this Manual are identical in number and position.

Rebuilding the master cylinder requires a tool called a brake cylinder hone. Do not install new parts in a master cylinder that has a worn or scored bore. If wear cannot be corrected by honing—without oversizing the bore—a new master cylinder should be obtained.

To disassemble master cylinder:

1. Remove the rubber boot.
2. Remove the lockring over the stop washer. Then remove the stop screw from the center of the master cylinder.
3. Remove the internal parts from the cylinder body.
4. Unscrew the brake light/warning light switches and the residual pressure valves (where fitted).
5. Remove the elbows and (where fitted) the sealing plugs.

Fig. 4-8. Exploded view of master cylinder. Parts are installed in the same positions on all types.

1. Brake light/warning light switch (2)
2. Elbow (2)
3. Sealing plug (where fitted) (2)
4. Stop screw
5. Stop screw seal
6. Cylinder body
7. Front brake piston spring
8. Spring support ring (2)
9. Spring seat (integral with spring support ring on some late master cylinders) (2)
10. Cup (4)
11. Primary cup washer (2)
12. Front brake piston
13. Stroke limiting screw
14. Stop sleeve
15. Rear brake piston spring
16. Rear brake piston
17. Secondary cup
18. Stop washer
19. Lockring
20. Rubber boot

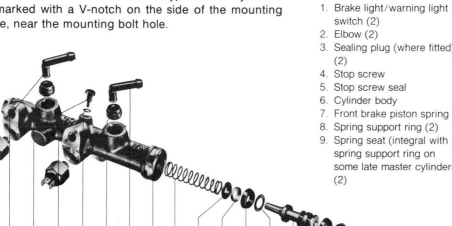

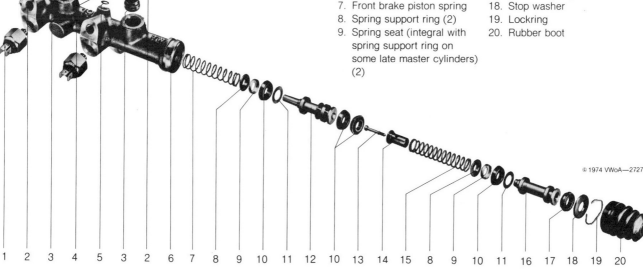

8 Brakes and Wheels

Inspect the moving parts. If any are worn, replace them all from a rebuilding kit. Replace the entire master cylinder if the cylinder bore is deeply scored. If the bore shows only moderate irregularities from normal wear, remove them with a brake cylinder hone.

When using a brake cylinder hone, lubricate the stones with brake fluid only, never with mineral oil or kerosene. Move the hone in and out of the bore rapidly while it is spinning to achieve an even polish over the entire cylinder wall.

After honing, check the clearance between the new pistons and the cylinder bore. If the clearance exceeds 0.10 mm (.004 in.), the entire master cylinder should be replaced. The bore can be measured by inserting a feeler gauge between the piston and the cylinder wall. However, a brake cylinder bore measuring tool or a snap (telescope) gauge and micrometer are preferable tools.

> **WARNING** ———
> After honing the master cylinder, clean any burrs from the compensating ports. Make certain the ports are not blocked. Otherwise, brake lockup or failure could result.

To assemble:

1. Clean all parts in brake fluid. Make certain the compensating ports are clear.

   > **WARNING** ———
   > Never use kerosene, gasoline, or other petroleum-based solvents for cleaning. These substances are highly damaging to rubber brake parts.

2. Using a conical cup sleeve as shown in Fig. 4-9, install the cups on the pistons.

   > **CAUTION** ———
   > If the cups are put on the pistons without a conical cup sleeve, extreme care is required to prevent damage to the cups.

Fig. 4-9. Installing cup on piston. The cup has been slipped over the conical sleeve from the small end. The sleeve is then placed over the piston, and the cup slipped off the sleeve into its groove.

3. Hold the master cylinder with its closed end down. Lubricate all internal parts with VW brake cylinder paste (see **8.5 Brake Cylinder Paste**) or with fresh brake fluid. Install the parts in the order shown earlier in Fig 4-8.

4. Push the pistons into the cylinder against spring tension, then install the stop washer and lockring.

5. Install the stop screw and its seal. Torque the screw to 0.5 to 1.0 mkg (3.6 to 7 ft. lb.).

   > **CAUTION** ———
   > Be sure the stop screw hole is clear. If blocked by the piston, damage will result when the screw is installed.

6. Screw the brake light/warning light switches and residual pressure valves (where fitted) into the master cylinder. Torque switches and valves to 2.0 mkg (14 ft. lb.).

7. Lubricate the elbows with brake fluid and install them in the top of the master cylinder.

8. Install the rubber boot with the vent hole down.

4.4 Testing and Replacing Brake Light/Warning Light Switches

The brake light/warning light electrical circuit for 1972 and later cars is shown in Fig. 4-10. Illustrated are all the connections found on earlier cars plus several additions and modifications. In this version, the functional check no longer requires that the light lens be pushed in by hand. The warning lamp lights up when the ignition is switched on and goes out in the same manner as the generator and oil pressure warning lamps once the engine has started.

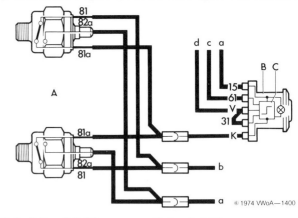

A. Brake light switch (2)
B. Dual circuit brake warning lamp
C. Electronic switch

a. To terminal 15
b. To brake lights
c. From regulator switch terminal 61
d. To ground

Fig. 4-10. Electrical diagram of brake light/brake warning light switches. Terminal **V** was discontinued in early 1973.

BRAKES AND WHEELS 9

To test brake light switch contacts:

1. Check the brake light bulbs. Replace if necessary.
2. Disconnect the front brake wires from the front brake light/warning light switch (81 and 81a, black-red and black wires; see Fig. 4-10).
3. Switch the ignition on and depress the brake pedal. The brake lights should go on.
4. If the lights work, reconnect the wires to the front switch and remove the rear brake wires (81 and 81a) from the other switch. Repeat the test. The brake light should go on.
5. If the brake lights do not work in one of the tests, replace the defective switch.

To replace switch:

1. Disconnect all wires from the defective switch.
2. Unscrew the switch from the master cylinder. Keep the sealing washer.
3. Install the sealing washer and the new switch. Torque the switch to 2.0 mkg (14 ft. lb.).
4. Connect the wires as indicated in Fig. 4-10.

   NOTE ——
   For additional information about the electrical circuits, consult the wiring diagrams in **ELECTRICAL SYSTEM**.

To test brake warning light contact:

1. Check the socket and the light bulb. If necessary, replace them.
2. Switch the ignition on. The warning light should come on.

   NOTE ——
   On pre-1972 models, it is necessary to push the light lens to test the light.

3. Open a bleeder valve in the front brake circuit. (See **8. Bleeding Brakes**).
4. Start the engine and depress the brake pedal. The brake warning light should come on.
5. Close the bleeder valve in the front brake circuit and open a bleeder valve in the rear brake circuit. Repeat the test described in step 4.
6. Check the fluid level in the brake fluid reservoir. If necessary, add fluid.

   CAUTION ——
   Use only new, unused brake fluid that meets SAE recommendation J 1703 and conforms to Motor Vehicle Safety Standard 116.

5. BRAKE LINES AND HOSES

The brake lines are steel tubes mounted on the car's frame. They carry brake fluid from the master cylinder to the flexible brake hoses that serve the wheel cylinders.

The lines are so routed that they are not exposed to moisture and to the hazard of flying stones. The steel clips that secure the lines to the frame at short intervals prevent vibration and chafing that might weaken the tubing.

5.1 Removing and Installing Brake Lines

The brake lines should be inspected regularly, certainly whenever there is brake trouble or the brakes are being serviced. Look for signs of corrosion, leaks around the unions, leaks in the lines themselves, and dents or cracks that may soon cause trouble.

Replacement lines can be obtained from your Authorized VW Dealer. The unions are factory-installed on the replacement lines, and the lines themselves are preformed to the correct shape for immediate installation.

To remove brake line:

1. Unscrew the unions on the line ends (Fig. 5-1).
2. Remove the steel spring clips that hold the line to the frame.
3. Remove the brake line from the car.

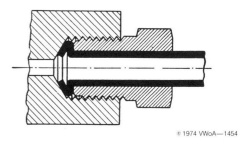

Fig. 5-1. Cross section of brake line union. Notice the double flare that holds the union on the tubing.

To install brake lines:

1. Route the new line so that it follows the routing of the old line.
2. Lubricate the flared ends of the brake lines with brake fluid, then insert the unions and torque them to 1.5 to 2.0 mkg (11 to 14 ft. lb.).

   NOTE ——
   Use a properly fitting wrench to avoid rounding off the union.

10 Brakes and Wheels

3. Carefully install the clips to hold the new line.
4. Bleed the brakes as described in **8. Bleeding Brakes.**

> **WARNING —**
> When installing brake lines, be very careful not to dent, flatten, or bend the tubing enough to collapse it. The resulting restriction can upset brake balance and will create stress points in the tubing that may later cause the tubing to crack. Never attempt to straighten a bent or dented brake line.

5.2 Removing and Installing Brake Hoses

Being flexible, brake hoses are much more subject to wear than brake lines. The hoses should, therefore, be inspected very carefully every time routine maintenance is being carried out.

To remove brake hose:

1. Remove the road wheel.
2. Unscrew the union that holds the hose to the brake line.
3. Remove the steel spring clip from the hose bracket on the frame or axle.
4. Pull the hose off the line and plug the line with a brake bleeder dust cap.
5. Unscrew the hose from the brake caliper or wheel cylinder.

To install brake hose:

1. Obtain a new hose of the correct length.

> **WARNING —**
> If the hose is too long, it may rub the wheel or moving suspension parts. If too short, it could break when drawn tight by wheel travel or steering movements. In either case, partial brake failure could result. Hoses must never be painted, and they can be damaged by grease, oil, gasoline, or kerosene. Brake hoses that bulge or appear oil-soaked or cracked must be replaced immediately.

2. Install the hose, following the removal steps in reverse. The hose must hang down and be free of twists.
3. Torque the hose ends to 1.5 mkg (11 ft. lb.) in disc brake calipers or to 1.5 to 2.0 mkg (11 to 14 ft. lb.) in other locations.
4. Check the hose position and the routing in all steering and suspension travel positions.

6. Disc Brakes

(Karmann Ghia only)

The principal parts of a disc brake assembly are shown in Fig. 6-1.

1. Splash shield   2. Brake disc   3. Brake caliper

Fig. 6-1. Disc brake assembly. Hydraulic pistons in the caliper cause the linings—called pads—to "squeeze" the disc during braking.

The brake caliper is mounted behind the axle. It consists of inner and outer housings, each holding a hydraulic piston for one side of the brake disc. Four bolts join the two halves of the housing; two additional bolts secure the caliper assembly to the steering knuckle.

Three different brake caliper designs have been used on Karmann Ghias during the model years covered by this Manual. The 1970 and 1971 models have calipers with only one bleeder valve and one pad retaining pin. Beginning with the 1972 models, a larger caliper with two bleeder valves and two pad retaining pins has been installed. Beginning with the 1973 models, another caliper with two bleeder valves and a U-shaped pad retaining pin has been gradually introduced. This caliper has been used exclusively from chassis No. 113 2013 627. All three calipers are shown in Fig. 6-2.

Brakes and Wheels 11

1970 and 1971 (ATE)

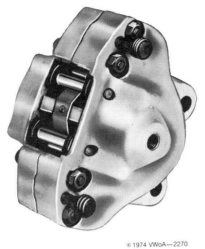
1972 and 1973 (ATE)

1973 and later (Girling)

Fig. 6-2. The three different disc brake calipers used on Karmann Ghias from 1970 to date.

6.1 Removing and Installing Brake Pads

The procedure for replacing pads on both the early and late ATE calipers is similar, despite the greater brake pad area of the later unit. The procedure for the Girling caliper is different and will be described separately. On all types, the pads should be replaced when the friction material has worn to a thickness of 2.00 mm (.080 in.).

To remove pads (ATE):

1. Remove the front wheel.
2. Drive out the friction pad retaining pin(s) and take out the pad spreader spring (Fig. 6-3).

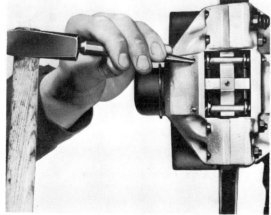

Fig. 6-3. Driving out pad retaining pin. The spreader spring is the cross-shaped piece behind the pins.

3. Pull the pads out of the caliper. The extractor shown in Fig. 6-4 makes the job easier.

Fig. 6-4. Pad being pulled out of caliper with extractor.

NOTE —

If the discs are deeply scored, it may be necessary to press the pistons slightly into the caliper to free the pad.

WARNING —

If the friction pads are to be reused, mark each pad and its original position in the caliper. Changing the location of used pads will result in uneven braking.

It is important that certain preparatory steps be carried out prior to actual installation of the pads. First, make sure there are no hard accumulations of dirt and rust on

12 Brakes and Wheels

the pad sliding surfaces inside the caliper. If there are, the pads may stick and cause dragging brakes or a tendency for the car to pull to one side. It is also necessary to note the following items:

1. If the pads are oily, have deep cracks, or are detached from their metal plate, they must be replaced.
2. If the pads are acceptable for reuse, remove any dirt from the radial grooves.
3. Always replace all four pads at the same time, even if only one is faulty.

To install pads (ATE):

1. Push both pistons into the caliper with a piston retaining device as shown in Fig. 6-5.

Fig. 6-5. Pistons being pushed into caliper bores with piston retaining pliers.

NOTE ——
As you push in the pistons, brake fluid will be forced back into the reservoir. Therefore, remove some fluid first, to prevent the reservoir from overflowing.

WARNING ——
Do not start a siphon with your mouth or spill fluid on the car. Brake fluid is both poisonous and damaging to paint.

2. Remove the piston retaining plates. Scrape clean the pad seating and sliding surfaces in the caliper, then blow out the dirt with compressed air.
3. Check the rubber seal shown in Fig. 6-6. It must not be cracked, hard, or swollen. If necessary, remove the caliper from the car as described in **6.2 Removing and Installing Brake Caliper**. Then replace the seals as described in **6.3 Brake Caliper Repair**.

Fig. 6-6. Rubber seal that should be checked before installing the pads.

4. Make sure the piston is positioned to accept the retaining plate as shown in Fig. 6-7. Then install the retaining plate by firmly pressing the circular part **(a)** into the piston crown. The plate will lie below the relieved part of the piston **(b)** when installed correctly.

Fig. 6-7. Piston retaining plate inserted in piston.

NOTE ——
A gauge, flat along one side so that it fits against the lower pad sliding surface and angled at 20° to match the recess in the piston, is used in VW shops to check piston position. This is faster than making checks using the retaining plate as a guide.

5. If necessary, correct the position of the piston using piston rotating pliers as shown in Fig. 6-8.

BRAKES AND WHEELS 13

Fig. 6-8. Piston being rotated in its bore to correct its position so that the retaining plate will fit.

6. Check the brake disc for wear as described in **6.4 Checking Brake Disc.** If necessary, replace the disc or machine it as described in **6.7 Reconditioning Brake Disc.**

7. Insert the pads in the brake caliper (Fig. 6-9).

    WARNING ——
    Install used pads in the positions marked for them during removal. If the pads are in the wrong location, or bind in the caliper, uneven braking will result.

Fig. 6-9. Inserting pad in brake caliper.

8. On 1970 and 1971 cars, install a new pad spreader spring. Then, while depressing the spreader spring with your fingertip, insert the pad retaining pin.

NOTE ——
The split clamping bushing must move freely on the pin(s). Replace corroded pins.

9. On 1972 and 1973 cars, insert the lower pad retaining pin. Install a new pad spreader spring. Then, while depressing the top of the spreader spring with your thumb, insert the upper pad retaining pin.

    WARNING ——
    Do not grease the retaining pins. Heat produced by braking can melt the lubricant and cause it to flow onto the pads or disc.

10. Using a hammer as shown in Fig. 6-10, drive the pins completely into the caliper.

Fig. 6-10. Hammer being used to drive in a pad retaining pin.

CAUTION ——
Do not drive the retaining pin into the caliper with a punch. Doing so can cause the pin shoulder to be sheared off by the clamping bushing.

11. To ensure that the pads are seated against the disc, depress the brake pedal several times while the car is stationary.

12. Check the level of the brake fluid in the reservoir. If necessary, add fresh fluid.

    CAUTION ——
    Use only new, unused brake fluid that meets SAE recommendation J 1703 and conforms to Motor Vehicle Safety Standard 116.

14 Brakes and Wheels

To remove pads (Girling):

1. Remove the locking clip that is in a hole near the tip of one side of the U-shaped retaining pin. Then remove the retaining pin and the pad spreader spring as shown in Fig. 6-11.

Fig. 6-11. The U-shaped pad retaining pin and the pad spreader spring being removed.

2. Bend down the lockplate tabs on the caliper mounting bolts. Remove the bolts, then take off the caliper and suspend it from the steering tie rod with a stiff wire hook. (See **6.2 Removing and Installing Brake Caliper**).

3. Rotate the pads 90° in the caliper and take them out as illustrated in Fig. 6-12.

Fig. 6-12. Pad being removed from Girling caliper.

4. Push the noise-damping plates (shims) that are between the pads and the pistons toward the center of the caliper and remove them.

WARNING —
If the friction pads are to be reused, mark each pad and its original position in the caliper. Changing the location of used pads will result in uneven braking.

It is important that certain preparatory steps be carried out prior to actual installation of the pads. First, make sure there are no hard accumulations of dirt and rust on the pad sliding surfaces inside the caliper. If there are, the pads may stick and cause dragging brakes or a tendency for the car to pull to one side. Scrape the surfaces clean, then blow out the dirt with compressed air. Inspect the rubber seals around the pistons. If they are cracked, hard, or swollen, replace the seals as described in **6.3 Brake Caliper Repair**. Note the following concerning the pads:

1. If the pads are oily, or have deep cracks, they must be replaced.
2. If the pads are acceptable for reuse, remove any dirt from the radial grooves.
3. Always replace all four pads at the same time, even if only one is faulty.

To install pads (Girling):

1. Install the noise-damping plates (shims) in the caliper with their arrows pointing in the direction of forward wheel rotation. See Fig. 6-13. (The longer of the two bleeder valves marks the top of the caliper.)

2. Check the brake disc for wear as described in **6.4 Checking Brake Disc**. If necessary, replace the disc or machine it as described in **6.7 Reconditioning Brake Disc**.

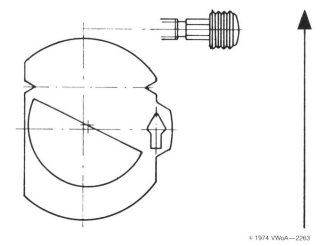

Fig. 6-13. Noise-damping plate installation. Tall arrow at right indicates forward wheel rotation. Long bleeder screw (shown) must be on top when caliper is installed on car.

3. Reversing the removal procedure, install the pads so that the grooved surface of each will be against the disc. (Push the pistons back into the caliper for clearance.)

   WARNING
   Install used pads in the positions you marked for them during removal. If the pads are in the wrong locations, or bind in the caliper, uneven braking will result.

4. Using a new lockplate, install the caliper on the steering knuckle. (The long bleeder valve goes at the top.) Torque the bolts to 4.0 mkg (29 ft. lb.).

   NOTE
   The Girling calipers are symmetrical. As replacement parts, they are supplied without pads, bleeder valves, or noise-damping plates. With the noise-damping plates and the long and short bleeder valves installed in the appropriate positions (long bleeder valve at top), the caliper can be installed on either side of the car.

5. Install a new pad spreader spring, the pad retaining pin, and a new locking clip. Then bend the straight side of the locking clip about 45° as indicated by the arrow in Fig. 6-14.

6. To ensure that the pads are seated against the disc, depress the brake pedal several times while the car is stationary.

7. Check the level of the brake fluid in the reservoir. If necessary, add fresh fluid.

   CAUTION
   Use only new, unused brake fluid that meets SAE recommendation J 1703 and conforms to Motor Vehicle Safety Standard 116.

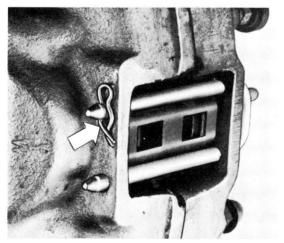

Fig. 6-14. Locking clip properly installed.

6.2 Removing and Installing Brake Caliper

Never attempt to remove a brake caliper until it has cooled. The tabs on the lockplates must be bent away from the bolt heads before the two mounting bolts that hold the caliper on the steering knuckle can be removed.

If the brake caliper is to be completely removed from the car, unscrew the brake hose from the caliper and seal it with a clean bleeder valve dust cap before unbolting the caliper from the steering knuckle. Support the caliper as it is being unbolted.

If the caliper is only being partially removed, for example, in order to obtain clearance for removal of the brake disc, the brake line need not be disconnected. Simply hang the caliper by a stiff wire hook from one of the steering tie rods. Doing so will eliminate the need for bleeding the brakes, which is necessary whenever the hose is removed. Never allow the caliper to hang by its hose.

Use a new lockplate when installing the caliper. Bend the tabs to lock the bolts after they have been torqued to 4.0 mkg (30 ft. lb.). To ensure that the pads are seated against the disc, depress the brake pedal several times while the car is stationary. If the hose has been removed and installed, bleed the brakes as described in **8. Bleeding Brakes.**

NOTE
*If a replacement caliper is being installed, install the pads and related parts as described in **6.1 Removing and Installing Brake Pads.** Note especially the information given concerning Girling replacement calipers.*

6.3 Brake Caliper Repair

The pistons can be taken out of the caliper without separating the two halves of the caliper housing. Although the procedures are illustrated with the ATE caliper, the procedures for replacing piston seals on the Girling caliper are identical. However, the two halves of the Girling caliper must not be separated. O-rings for the fluid channels between the housing halves are not available as spare parts. The ATE calipers' housings can safely be separated as described.

CAUTION
If you lack the skills, tools, or a clean workshop for servicing the brake calipers, we suggest you leave such repairs to an Authorized VW Dealer or other qualified shop. We especially urge you to consult your Authorized VW Dealer before attempting repairs on a car still covered by the new-car warranty.

16 Brakes and Wheels

Fig. 6-15. Exploded view of disc brake caliper.

WARNING —
Do not re-use any fasteners that are worn or deformed in normal use. Many fasteners are designed to be used only once and become unreliable and may fail when used a second time. This includes, but is not limited to, nuts, bolts, washers, self-locking nuts or bolts, circlips and cotter pins. Always replace these fasteners with new parts.

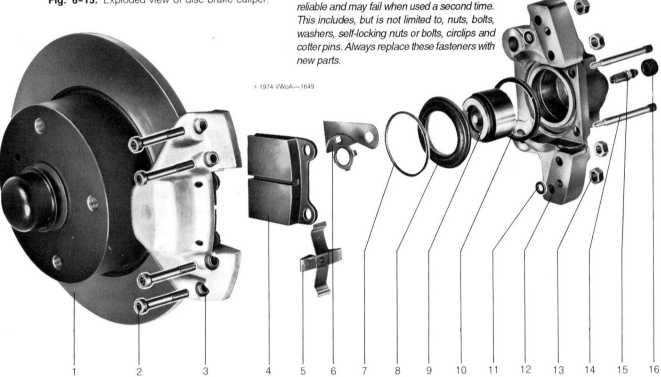

1. Brake disc
2. Socket head screw (4)
3. Caliper outer housing
4. Friction pad (2)
5. Pad spreader spring
6. Piston retaining plate (2)
7. Clamp ring (2)
8. Seal (2)
9. Piston (2)
10. Rubber seal (2)
11. Seal (2)
12. Caliper inner housing
13. Nut (4)
14. Pad retaining pin (2)
15. Bleeder valve
16. Dust cap

To replace piston seals:

1. Remove the brake caliper from the car and clamp its flange in a vise.

   CAUTION —
   Install soft jaws in the vise or clamp the flange between two hardwood blocks. Otherwise, damage to the mounting flange could result that would upset the caliper's alignment with the brake disc.

2. Remove the pads and the piston retaining plates (ATE) or noise-damping plates (Girling). Mark the original positions of the pads if they are to be reused.

3. Using a screwdriver, pry out the clamp ring as shown in Fig. 6-16. Unless the seal is to be replaced, take care not to damage it.

   NOTE —
   The seal, seal clamp ring, and retaining plate should be replaced whenever the pistons are removed. All necessary parts are contained in the VW rebuilding kit.

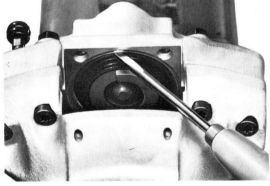

Fig. 6-16. Clamp ring being removed. Use a small enough screwdriver for the tip to be easily inserted behind the ring.

4. Remove the seal as shown in Fig. 6-17.

   NOTE —
   If the seal is simply being lifted to check for fluid leakage, use a plastic rod to avoid puncturing the seal. Use a sharp tool only if the seal is to be replaced.

Brakes and Wheels 17

Fig. 6-17. Seal being removed with a plastic rod.

5. Using compressed air as shown in Fig. 6-18, remove one piston from the brake caliper while the second piston is held by the piston retaining device.

Fig. 6-18. Removing piston with compressed air. Place a 5- to 10-mm (¼- to ⅜-in.) hardwood board in the housing to catch the piston. Otherwise the piston may be damaged.

NOTE—

The cylinders must be serviced one at a time because, with one piston removed, no air pressure can be built up in the brake caliper to expel the other piston.

6. Remove the rubber seal from the cylinder with a plastic rod as shown in Fig. 6-19.

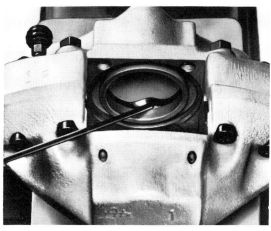

Fig. 6-19. Rubber seal being removed from the groove in the cylinder bore.

To install pistons:

1. Clean all parts with brake fluid only.

2. Check the parts for wear. If the cylinder is damaged, replace the entire brake caliper. Do not hone.

3. Apply a thin coat of VW brake cylinder paste (see **8.5 Brake Cylinder Paste**) to the piston and the new rubber seal.

CAUTION—

To install the pistons, use the installing clamp shown in Fig. 6-20 and not the piston retaining device. The piston retaining device could permit the piston to tilt during installation. Tilting of the piston could result in damage to the seal, piston, or caliper housing.

Fig. 6-20. Piston installing clamp in position.

18 Brakes and Wheels

4. Install a new seal and new clamp ring.

5. Check the 20° angle of the piston recess (ATE only) so that the piston retaining plate can be installed as shown in Fig. 6-21. The plate is at the lower part of the caliper opening.

   NOTE ─
   The VW checking gauge and VW piston rotating pliers can be used if available.

Fig. 6-21. Retaining plate correctly positioned. The plate lies along the lower pad guide surface. On Girling calipers, the arrow on the noise-damping plate should point in the direction of forward wheel rotation.

6. Install the piston retaining plate (ATE) or noise-damping plate (Girling).

7. Repeat the procedure on the other piston.

On ATE calipers, the two halves of the brake caliper housing should be separated only if there are leaks in the joint between them. In this case it is necessary to disassemble the caliper and replace the O-rings surrounding the fluid channels.

CAUTION ─
Do not separate the halves of the Girling caliper housing. Replacement O-rings are not available.

To Disassemble (ATE only):

1. Remove the four socket head screws joining the two halves of the housing.

2. Remove the outer housing half.

3. Remove the old O-rings from the chamfers around the fluid channel.

To Assemble:

1. Clean the mating surfaces with brake fluid only.

2. Install the two new fluid channel O-rings.

3. Using new socket head screws and nuts, loosely join the two housing halves. Check their alignment.

   NOTE ─
   There are two long screws and two short screws. The shorter screws go into the two outermost holes in the housings.

4. Tighten the screws in the sequence illustrated in Fig. 6-22, torquing them to 2.0 to 2.5 mkg (14 to 18 ft. lb.).

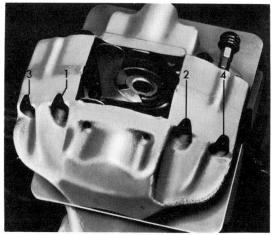

Fig. 6-22. Tightening sequence. Use only the proper driving tool for torquing the screws.

5. Again check the alignment of the housings. If incorrect, loosen the screws and correct it. Then torque the screws to specifications.

6. If the alignment is correct, repeat the tightening sequence, checking that each screw is at the prescribed torque.

6.4 Checking Brake Disc

The brake discs should be checked for wear each time repair work is done on the front brakes. Replace the discs if they are worn, scored with sharp ridges, or cracked. Also replace brake discs that have worn down, or that have been machined, to a thickness of 8.00 mm (.315 in.) or less.

Check the discs for excessive runout. If a low-speed front end shimmy goes away when you release the brakes, excessive brake disc runout is probably the cause of the shimmy.

Brakes and Wheels 19

To measure runout:

1. Remove the front wheel.
2. Adjust the front wheel bearing axial play to 0.03 to 0.12 mm (.001 to .005 in.) as described in **FRONT AXLE**.
3. Install the measuring appliance on the pad retaining pins in the caliper as shown in Fig. 6-23. Tighten the wing nut to hold it solidly in position.
4. Install the dial indicator on the appliance with its measuring pin against the disc surface. Then zero the gauge.
5. Slowly hand-turn the brake disc to check the runout. The maximum allowable runout is 0.02 mm (.0008 in.).
6. If the brake disc runout exceeds specifications, replace the brake disc.

Fig. 6-23. Brake disc runout being measured. A dial indicator on a magnetic base can also be used.

6.5 Removing and Installing Brake Disc

The caliper must first be removed from the steering knuckle. Hang the caliper from the steering tie rod with a stiff wire. Leave the brake hose attached. It will save you the job of bleeding the brakes. The disc is removed the way a brake drum would be removed. See the procedures for wheel bearing repair and adjustment in **FRONT AXLE**. Before driving, depress the brake pedal several times to ensure that the pads are seated against the disc.

WARNING —
Do not re-use fasteners such as nuts, bolts, washers, self-locking nuts, circlips, cotter pins that are worn or deformed in normal use. Always replace.

6.6 Removing and Installing Brake Disc Splash Shield

The brake disc splash shield must be removed whenever the steering knuckle is replaced or if the splash shield itself is damaged. Over the years, minor changes have been made in the shape of the splash shield to accommodate different brake calipers. If the splash shield is to be replaced, replace it with one that is suitable to the brake caliper on the car.

To remove splash shield:

1. Remove the front wheel.
2. Detach the brake caliper from the steering knuckle and suspend it from the tie rod with a wire hook.
3. Remove the brake disc. For details, see the discussion of front wheel bearings in **FRONT AXLE**.
4. Remove the three bolts that hold the splash shield on the steering knuckle (Fig. 6-24).

Fig. 6-24. Removing bolts that hold the splash shield.

To install splash shield:

1. Clean the mounting surface for the splash shield on the steering knuckle. Replace damaged splash shields; do not attempt to repair them.
2. Install the splash shield. Torque the bolts to 1.0 mkg (7 ft. lb.).
3. Check the brake disc for wear, and install the disc if it is still serviceable.
4. Adjust the wheel bearings (see **FRONT AXLE**). Axial play should be 0.03 to 0.12 mm (.001 to .005 in.).
5. Using a new lockplate, install the brake caliper. Torque the bolts to 4.0 mkg (29 ft. lb.).
6. To seat the pads against the disc, depress the brake pedal several times while the car is stationary.

20 BRAKES AND WHEELS

6.7 Reconditioning Brake Disc

The brake discs should be checked for wear whenever brake repairs are made. Replace discs that have worn below the minimum thickness, are scored with sharp ridges, or are cracked. If a disc's thickness is less than 8 mm (.315 in.), the disc must be replaced. Brake discs can be reconditioned if the following restrictions are observed:

1. The minimum allowable thickness after rework is 8.5 mm (.335 in.). New brake discs are between 9.45 and 9.50 mm (.372 and .374 in.) thick.

CAUTION —
Under no circumstances should the brake discs be reworked to a lesser thickness. Although the disc itself is not significantly weakened by such work, bear in mind that a thinner disc causes the pistons to travel farther out in their cylinders. This may severely damage the calipers and pistons.

2. After reworking a brake disc, its thickness should not vary by more than 0.02 mm (.0008 in.) measured at several locations on the disc.

3. The brake disc must be reworked equally on both sides to prevent squeaking, chattering, or brake pedal pulsation.

4. Brake discs must be reworked in a special machine designed for the task (Fig. 6-25).

Fig. 6-25. Reworking brake disc in precision machine.

5. The maximum allowable runout of reworked brake discs is 0.20 mm (.008 in.).

7. DRUM BRAKES

Fig. 7-1 shows the parts of a rear brake. Front drum brakes are similar (Fig. 7-2) but do not have the mechanism for the parking brake.

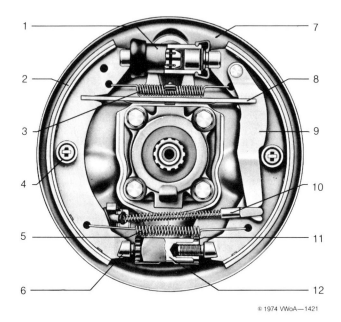

1. Wheel cylinder
2. Brake shoe with lining
3. Upper return spring
4. Spring with cup and pin
5. Lower return spring (2)
6. Adjuster screw
7. Backing plate
8. Connecting link
9. Lever
10. Brake cable
11. Adjuster
12. Anchor block

Fig. 7-1. Components of rear brake (drum removed).

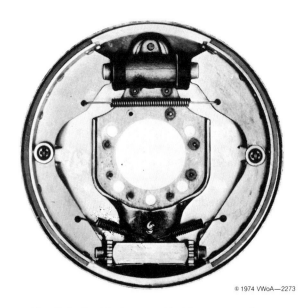

Fig. 7-2. Front brake assembly (drum removed).

Brakes and Wheels 21

7.1 Adjusting Drum Brakes

The clearance between the brake linings and the drum gradually increases due to normal wear. This change is indicated by increased pedal travel in applying the brakes. When pedal travel becomes excessive, the brake shoes must be adjusted to position the linings nearer the drums. These adjustments are made at the individual wheels. The disc brakes at the front of Karmann Ghia models require no adjustment.

To adjust:

1. Raise the car and fully release the parking brake.
2. Depress the brake pedal several times as far as it will go. This centers the brake shoes in the drums.
3. Remove the rubber plugs from the holes in the backing plate.
4. Using an adjusting lever or screwdriver as shown in Fig. 7-3, turn the adjuster until a slight drag is noted when the wheel is turned by hand. Then back off the adjuster three or four clicks so that the wheel turns freely.

   NOTE —
   If the brakes are far out of adjustment, it may be necessary to center the shoes once or twice more during the course of adjustments.

5. Repeat step 4 on the other adjuster on the same wheel's brake assembly.
6. Repeat the entire procedure on the other three wheels.
7. Road-test the car to check the pedal travel.

Fig. 7-3. Lever in position to adjust drum brake.

7.2 Removing and Installing Brake Drums

Removal and installation of the front brake drums is described in conjunction with front wheel bearings in **FRONT AXLE.** The procedure given here is for the rear brake drums only.

To remove:

1. Fully release the parking brake. Then back off the adjusters slightly.
2. Remove the cotter pin from the castellated nut on the rear wheel shaft. Remove the nut, then raise the car.

   WARNING —
   Loosen the wheel shaft nuts while the car is on the ground. The leverage needed for this job is enough to topple a car off the lift.

3. Using a puller, as shown in Fig. 7-4, remove the brake drum from the splines in the rear wheel shaft.

Fig. 7-4. Puller, held by wheel lug bolts, being used to remove the rear brake drum.

To install:

1. Slide the brake drum onto the rear wheel shaft.
2. Using a 36-mm wrench, install the castellated nut. Then, with the car on the ground, torque the nut to 35 mkg (253 ft. lb.) and install a new cotter pin.

   CAUTION —
   The nut must be torqued to specifications to avoid destructive spline wear.

22 Brakes and Wheels

7.3 Removing and Installing Brake Shoes

Absolute cleanliness is necessary in working with brake linings. Any brake lining that has become oil-soaked or saturated with brake fluid must be replaced. A lining must also be replaced if it is worn at any point to a thickness of 2.50 mm (.100 in.). The lining thickness can be checked by removing the rubber plugs from the brake backing plate.

> **CAUTION** ——
> If you lack the skills, tools, or a clean workshop for servicing the brakes, we suggest you leave such repairs to an Authorized VW Dealer or other qualified shop. We especially urge you to consult your Authorized VW Dealer before attempting repairs on a car still covered by the new-car warranty.

Fig. 7–5 is an exploded view of the front drum brake assembly used on Super Beetles. The rear brakes on all models are similar to the front drum brake assembly but with the addition of parts for the parking brake. The front brakes for Beetles are also similar, but with the brake assembly rotated 90° clockwise.

Although the relative positions of the brake components are the same on all models, individual parts are different. So make certain that the correct parts are obtained for the particular car model and year when making replacements. Bonded brake linings are now commonly used on VW cars instead of the riveted type seen in some of the illustrations.

Fig. 7–5. Exploded view of a typical drum brake assembly. A Super Beetle front wheel brake is illustrated, but the relative position of the parts is the same on all drum brakes.

1. E-clip (left side only)
2. Dust cap
3. Clamp nut
4. Thrust washer
5. Outer tapered-roller bearing
6. Brake drum
7. Spring cup
8. Shoe retaining spring
9. Upper return spring
10. Lower return springs
11. Brake shoe with lining
12. Adjuster screw
13. Adjuster
14. Brake mounting bolt
15. Lock washer
16. Backing plate
17. Rubber plugs
18. Steering knuckle
19. Wheel cylinder
20. Lock washer
21. Cylinder mounting bolt
22. Shoe retaining pin

> **WARNING** ——
> Do not re-use nuts, bolts, washers, circlips, cotter pins, etc. that are worn or deformed in normal use. Always use new parts.

To remove brake shoes:

1. Remove the road wheel and brake drum.
2. Check the wheel cylinder for sticking pistons by having someone slowly depress the brake pedal while you watch to see whether the brake shoes move out the same distance at a uniform rate.

   > **NOTE** ——
   > Insert two screwdrivers behind the backing plate flange. Press the screwdrivers against the shoes to limit shoe travel. Also lift the wheel cylinder boot to check for fluid leakage. If the cylinder is sticking or leaking, or if the bleeder valve is rusted tight, rebuild or replace the cylinder.

3. Remove the spring cups and shoe retaining springs as shown in Fig. 7-6. Press the cup in, rotate it 90°, then release it. If necessary, reach behind the backing plate and hold in the shoe retaining pin.

Fig. 7-6. Spring cup and retaining spring being removed.

4. Remove the shoe retaining pins.

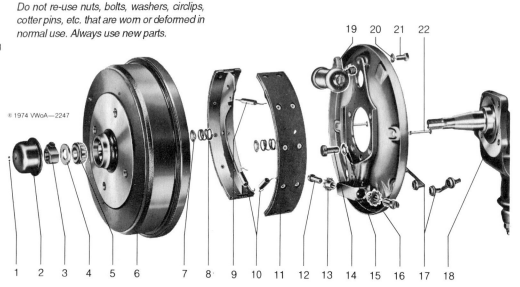

5. Using brake spring pliers, remove the return springs.

6. On rear brakes only, unhook the parking brake cable from the lever at the point indicated in Fig. 7-7. Push the lever forward, then pull the cable eye down to unhook it.

Fig. 7-7. Parking brake cable unhooked from lever.

7. Remove the brake shoes. Then install a wire clip over the wheel cylinder pistons to hold them in while the shoes are off.

To install:

1. Inspect the brake drum and compare it to the specifications given in **7.7 Reconditioning Brake Drums.** Make sure that the same type of linings are used at both wheels on the axle. Fit oversize linings if the drums have been reconditioned.

   WARNING —
   Using linings of different size or composition on opposite sides of the car can cause dangerously uneven braking.

2. Disassemble the adjusters and adjuster screws. Lightly coat the threads and the bearing surface on the adjuster with multipurpose grease.

3. Install the adjusters and screws. Fully back them off if new brake shoes are being installed.

4. On rear brakes only, install the lever for the parking brake cable as shown in Fig. 7-8.

5. On rear brakes only, install both the connecting link and the upper return spring. On front brakes, install only the upper return spring.

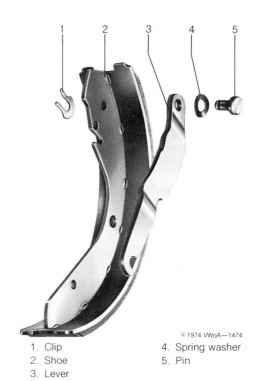

1. Clip
2. Shoe
3. Lever
4. Spring washer
5. Pin

Fig. 7-8. Brake lever installation on rear brake shoe.

6. Remove the wire clip from the wheel cylinder pistons. Then install the brake shoes on the pistons and adjuster screws.

7. Install the lower return springs.

8. Install the shoe retaining pins, the shoe retaining springs, and the spring cups.

9. On rear brakes only, hook the parking brake cable eye over the lever.

10. Install the brake drum over the brake shoes.

11. On front wheel brakes, adjust the front wheel bearings as described in **FRONT AXLE.** Then loosely install the road wheel.

12. On rear wheel brakes, install the castellated nut, but do not torque it. Loosely install the road wheel.

13. Adjust the brakes as described in **7.1 Adjusting Drum Brakes.**

14. Lower the car to the ground. Torque the wheel lug bolts to 13 mkg (94 ft. lb.). On rear wheels only, torque the castellated nut to 35 mkg (253 ft. lb.), then install a new cotter pin.

   WARNING —
   Tighten wheel lug bolts and rear wheel shaft nuts with the car on the ground. The leverage needed for these jobs is enough to topple a car off the lift.

7.4 Removing and Installing Wheel Cylinder

To remove a wheel cylinder, first carry out the procedure for removing the brake shoes as given in **7.3 Removing and Installing Brake Shoes.** Disconnect the brake hose from the wheel cylinder and seal it with a clean bleeder valve dust cap. Then, working behind the backing plate, remove the wheel cylinder mounting bolt and take off the wheel cylinder.

Installation is the reverse of removal. Torque the mounting bolt for front wheel cylinders to 2.5 mkg (18 ft. lb.), and torque the mounting bolt for rear wheel cylinders to 2.0 to 3.0 mkg (14 to 22 ft. lb.). Torque the brake hose to 1.5 to 2.5 mkg (11 to 18 ft. lb.). Carry out the procedure for installing brake shoes as given in **7.3 Removing and Installing Brake Shoes.** Then bleed the brakes as described in **8. Bleeding Brakes.**

7.5 Wheel Cylinder Repair

Because replacement wheel cylinders are inexpensive, it is usually more economical to replace them as a unit. However, repair kits are available. Other than a very light honing to remove tarnish or gummy deposits, no machine work should be done.

Many different wheel cylinders have been used on VW cars. So when you buy replacement parts, take the old cylinder with you for comparison. Fig. 7-9 is an exploded view that shows the position of the working parts in the wheel cylinder.

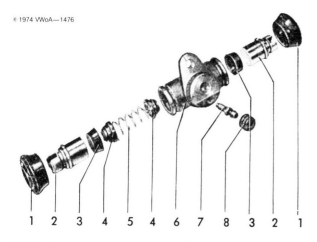

1. Boot (2)
2. Piston (2)
3. Cup (2)
4. Cup expander (2)
5. Spring
6. Housing
7. Bleeder valve
8. Dust cap

Fig. 7-9. Exploded view of wheel cylinder.

The internal parts can be hand-pressed out of the housing once the boots are removed. Prior to assembly, clean all parts with brake fluid only.

Check the cylinder for wear. Do not machine or hone metal from the cylinder bore. A new, lubricated piston must be an airtight fit. If not, replace the cylinder.

NOTE —
Vacuum should keep a new piston (lubricated with brake fluid) from falling out of the cylinder when the bleeder valve and brake hose holes are sealed and you cover the opposite end of the cylinder with your thumb.

Lubricate the cups with brake fluid during installation. Coat the pistons with VW brake cylinder paste (see **8.5 Brake Cylinder Paste**) and insert them in the cylinder, then install the remaining parts.

7.6 Removing and Installing Backing Plate

If the backing plate is bent, or if the raised areas that the brake shoes ride against are badly worn, replace the plate. Otherwise, the brake linings will not line up properly with the drum and will wear to a taper.

To remove the backing plate, remove the brake shoes as described in **7.3 Removing and Installing Brake Shoes.** Remove the wheel cylinder as described in **7.4 Removing and Installing Wheel Cylinder.** On rear wheel brakes, unbolt the parking brake cable from the backing plate.

CAUTION —
If the bolt that holds the parking brake cable is locked in place by corrosion, treat it with rust solvent or penetrating oil before attempting to force it loose. If the bracket is allowed to turn with the bolt, the flexible cable housing may be damaged.

The backing plate can be taken off the steering knuckle or the rear wheel bearing housing after the backing plate mounting bolts have been removed. On rear wheel brakes, the rear wheel bearing cover must come off along with the backing plate.

Installation of the front brake backing plate is simply the reverse of removal. Prior to installation, however, make sure that the mating surfaces on the backing plate and steering knuckle are completely clean. Torque the backing plate mounting bolts to 5.0 mkg (36 ft. lb.). Then install the wheel cylinder, brake shoes, and brake drum as described in **7.3 Removing and Installing Brake Shoes** and **7.4 Removing and Installing Wheel Cylinder.**

To install rear brake backing plate:

1. Check the condition and position of the oil seal in the bearing cover. Replace the seal if it is worn, cracked, or distorted.

2. Clean all mating surfaces on the backing plate, bearing cover, and bearing housing flange.

3. Position the backing plate on the bearing housing.

4. Using a new gasket, install the bearing cover with the vent hole down as indicated in Fig. 7-10.

Fig. 7-10. Bearing cover being installed. The hole and lip (arrow) must face downward.

5. Torque the bearing cover/backing plate mounting bolts to 6.0 mkg (43 ft. lb.).

6. Install the wheel cylinder, brake shoes, and brake drum as described in **7.4 Removing and Installing Wheel Cylinder** and **7.3 Removing and Installing Brake Shoes.**

   WARNING —
   Tighten the wheel lug bolts and rear wheel shaft nuts with the car on the ground. The leverage needed for these jobs is enough to topple the car off a lift.

7.7 Reconditioning Brake Drums

Check the brake drums whenever new linings are installed. Taper, scoring, or other wear must, if possible, be corrected on a machine similar to the one shown in **6.7 Reconditioning Brake Disc.** Both drums on an axle must be machined to the same dimensions and oversized linings must be used on both wheels. Unless oversized linings radiused to fit the reconditioned drums are used, uneven or ineffective braking will result.

The machine work should be carried out by an Authorized VW Dealer or a qualified automotive machine shop. The specified brake drum dimensions, wear limits, and lining dimensions are given in **Table a**.

Table a. Brake Drum Specifications

Brake drum inside diameters mm (in.)	
1970 cars	
New diameter	
Front	230.10 + 0.20 (9.059 + .008)
wear limit	231.50 (9.114)
Rear	230.00 + 0.20 (9.055 + .008)
wear limit	231.50 (9.114)
Permissible turned diameter	
Front (except Ghia)	231.10 + 0.20 (9.098 + .008)
wear limit	231.50 (9.114)
Rear	231.00 + 0.20 (9.094 + .008)
wear limit	231.50 (9.114)
113/115 (convertible) 1971 on	
New diameter	
Front	248.10 + 0.20 (9.768 + .008)
wear limit	249.50 (9.823)
Rear	230.00 + 0.20 (9.055 + .008)
wear limit	231.50 (9.114)
Permissible turned diameter	
Front	249.00 + 0.20 (9.803 + .008)
wear limit	249.50 (9.823)
Rear	231.00 + 0.20 (9.094 + .008)
wear limit	231.50 (9.114)
111/Karmann Ghia, 1971 on	
New diameter	
Front (except Ghia)	230.10 + 0.20 (9.059 + .008)
wear limit	231.50 (9.114)
Rear	230.00 + 0.20 (9.055 + .008)
wear limit	231.50 (9.114)
Permissible turned diameter	
Front (except Ghia)	231.10 + 0.20 (9.098 + .008)
wear limit	231.50 (9.114)
Rear	231.00 + 0.20 (9.094 + .008)
wear limit	231.50 (9.114)
Brake drum irregularity tolerances	
Taper	0.10 (.004) max.
Out-of-round	0.10 (.004) max.
Lateral runout	0.25 (.010) max.
Radial runout	0.15 (.006) max.
Brake lining dimensions	
Width	
All 1970 cars	
Front (except Ghia)	40.00 (1.575)
Rear	40.00 (1.575)
Sedan 111/Ghia, 1971 and on	
Front (except Ghia)	40.00 (1.575)
Rear	40.00 (1.575)
Sedan 113 and Convertible, 1971 and on	
Front	45.00 (1.772)
Rear	40.00 (1.575)
Lining thickness (new)	4.0-3.8 (.157-.150)
wear limit	2.50 (.100) remaining
Lining thickness (oversize)	4.5-4.3 (.177-.169)
wear limit	2.50 (.100) remaining

26 BRAKES AND WHEELS

8. BLEEDING BRAKES

Bleeding the brakes removes air from the hydraulic system. This task must be performed whenever the brake lines have been disconnected or after a brake cylinder has been replaced or repaired. If the brake pedal begins to feel spongy when you apply the brakes, it is an indication that air has entered the system. If bleeding fails to correct the problem, there are probably leaks to be fixed.

Brake Fluids

Additional fluid must be added to the system when it is bled. The quality of the new brake fluid is important. All VW brake fluids have similar chemical and performance characteristics and may be mixed regardless of differences in color. Using a brake fluid that does not conform with SAE recommendation J 1703 and Motor Vehicle Safety Standard 116 can cause brake failure, premature wear, or erratic operation.

8.1 Changing Brake Fluid

Change the brake fluid in your VW every two years. Brake fluid tends to absorb moisture from the air, and water can initiate corrosion. Water can also cause the fluid to boil when the brakes are used very hard.

The brake calipers surround the friction linings, so they pick up a great deal of heat from the linings. Therefore, it is particularly important that the brake systems of cars with disc brakes use a fluid with a high boiling point.

On cars with drum brakes at all four wheels, it is possible to remove all old brake fluid through the single bleeder valve provided at each wheel. All newer VW disc brake calipers have a second bleeder valve on the lower part of their housing. This makes it possible to completely remove all fluid when it has to be changed.

To change fluid:

1. Attach a suitable hose to the bleeder valves for draining the fluid into containers.

2. Open the rear wheel cylinder bleeder valves and pump the pedal until fluid ceases to flow out of the valves.

3. At the front wheels, open the lower bleeder valves on late cars with front disc brakes, the single bleeder valve on early cars with front disc brakes, or the bleeder valve on the front wheel cylinders of cars with four-wheel drum brakes. Pump the pedal until fluid ceases to flow out of the valves.

4. On late cars with front disc brakes, open the upper bleeder valves. Leave them open until the fluid stops draining from the lower bleeder valves.

5. Close all bleeder valves.

6. Fill the fluid reservoir with new unused brake fluid that meets SAE recommendation J 1703 and conforms to Motor Vehicle Safety Standard 116.

7. Bleed the system by either of the following methods.

8.2 Bleeding with Pressure Bleeder

Whenever possible, brake bleeding should be done with a pneumatic pressure bleeder similar to that shown in Fig. 8-1. This device is connected to the brake fluid reservoir. It fills the system with fluid under pressure and will complete the job in a very short time. Simply open the bleeder valve, quickly depress and slowly release the pedal several times, and then close the bleeder valve. A fluid receptacle supplied with the bleeding device must be fitted to the wheel being bled.

Fig. 8-1. Pressure bleeder for bleeding brake system.

8.3 Bleeding by Pumping

For car owners, the pumping method of bleeding the brakes is usually more practical, even though it requires two persons. Have your helper sit in the car to pump the brake pedal. You will then be free to move from wheel to wheel to perform the actual bleeding.

To bleed:

1. Fully fill the fluid reservoir with brake fluid that meets SAE recommendation J 1703 and conforms to Motor Vehicle Safety Standard 116.

2. Take the dust cap off the bleeder valve at the right-hand front wheel. Slip a 4-mm (5/32-in.) I. D. hose over the bleeder valve and submerge the other end in a clear glass jar partially filled with clean brake fluid.

   NOTE —
   A clear glass jar must be used so the bubbles of air can be seen coming out of the hose.

3. Open the bleeder valve a half turn. Have your helper slowly depress the brake pedal until it reaches the floor and keep it there while you close the bleeder valve.

4. Have your helper slowly release the pedal until it is completely up. Repeat the preceding step until no more air bubbles emerge from the hose.

5. Repeat the entire bleeding procedure on the other three wheels in the sequence left front, right rear, left rear.

NOTE —

Refill the reservoir after bleeding each wheel cylinder. If the system contains a great deal of air or if the brake fluid is being changed, it will be necessary to add more fluid once or twice while bleeding each wheel cylinder. Never let the reservoir be emptied completely, or you will have to start bleeding the brakes all over again.

CAUTION —

Do not allow brake fluid to come in contact with painted surfaces. Brake fluid contains a solvent damaging to most finishes.

WARNING —

Do not use soft drink bottles or other food containers to store brake fluid or to bleed the brakes. Brake fluid is poisonous.

8.4 Flushing Brake System

Never use anything but brake fluid to flush the brake system. Alcohol must not be used since it will destroy residual lubrication and will encourage the accumulation of water in the system.

NOTE —

Do not rely on flushing alone to clean a brake system contaminated by dirt or rust. To remove all foreign matter, you must disassemble the system and clean the parts individually.

8.5 Brake Cylinder Paste

Brake cylinders will give longer service when lubricated with VW brake cylinder paste than when lubricated by brake fluid alone. When cleaning or rebuilding a brake cylinder, wash all parts thoroughly in brake fluid. Coat the pistons and cylinder walls with VW brake cylinder paste, then assemble the cylinder.

VW brake cylinder paste is available from Authorized VW Dealers and should always be the preferred form of lubrication when brake cylinders are repaired. Brake fluid should be considered a satisfactory substitute only if VW brake cylinder paste is unavailable.

9. Parking Brake

The parking brake operates only on the rear wheels. It is completely mechanical and independent of the hydraulic brake system.

Two cables extend from the parking brake lever to the left and right rear wheels. Raising the parking brake lever tightens the cables and moves the rear brake shoes into contact with the rear brake drums. The cable operates on a lever attached to the rear brake shoe of each rear brake. The movement of the lever is transmitted to the front shoe of each rear brake by a flat steel bar called the connecting link.

Once the parking brake lever has been raised, it is held in position by a ratchet segment and pawl inside the lever. The lever will remain in the same position until the ratchet is released by pressing in a thumb button on the end of the lever.

The parts that make up the parking brake lever are shown in Fig. 9-1.

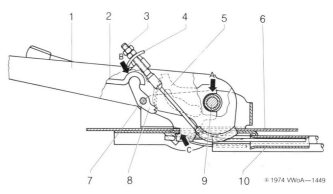

1. Parking brake lever
2. Pawl rod
3. Brake cable
4. Compensating lever
5. Ratchet segment
6. Frame
7. Pawl pin
8. Pawl
9. Lever pin
10. Cable guide tube

Fig. 9-1. Parking brake lever assembly. The lever pin is retained by a circlip (arrow **A**). A hook on the pawl rod engages the pawl at arrow **B**. The ratchet segment is held by a lug (arrow **C**).

9.1 Adjusting Parking Brake

The parking brake should be adjusted whenever the rear brake linings have worn enough so that it is possible to raise the lever five clicks without noticeable braking action. Although the compensating lever atop the parking brake lever makes up for some inequality of adjustment

28 Brakes and Wheels

between the left and right cables, it is important that the cables be adjusted to keep the compensating lever horizontal when the parking brake is applied.

To adjust:

1. Raise and support the vehicle properly.

    WARNING —
    If a lift is not available, support the car on safety stands designed for the purpose. Cement blocks, which may crumble, or other makeshift supports must never be used.

2. Adjust the rear brakes as described in **7.1 Adjusting Drum Brakes**.

3. For access to the cable adjusting nuts, pull back the rubber boot on the hand lever to open the slots in the sides of the boot.

4. Insert a screwdriver in the slotted cable ends so that the cables will not be damaged by twisting. Then loosen the locknuts.

5. Raise the parking brake lever two clicks on 1970 through 1972 models and three clicks on later models. Hold the cable ends with a screwdriver and tighten the adjusting nuts until each rear wheel can just be turned by hand. Braking must be equal on both sides of the car.

6. Check whether the compensating lever is horizontal. If not, see **9.3 Removing and Installing Brake Cable**.

7. If the compensating lever is horizontal, tighten the locknuts and refit the rubber boot, pushing the cover flaps for the adjusting slots forward.

9.2 Removing and Installing Parking Brake Lever

To remove the parking brake lever, remove the rubber boot, locknuts, adjusting nuts, and compensating lever. These parts were illustrated earlier in Fig. 9-1, which can also be used as a guide for assembling and installing the lever. Remove the circlip from the lever pin as shown in Fig. 9-2. Take out the pin and remove the lever. Do not press the release button or the ratchet segment will fall into the frame tunnel.

Before installation, disassemble the lever and clean the internal parts. Grease the parts before assembling them in the lever. Installation is the reverse of removal. Grease the lever pin before inserting it and refitting the circlip. Adjust the parking brake during installation of the cables in the compensating lever.

Fig. 9-2. Circlip being removed from lever pin.

9.3 Removing and Installing Brake Cable

If the rear wheel brake shoes are correctly adjusted, but the compensating lever is not horizontal after adjusting the parking brake, one cable may be stretching. Left in service, the cable will eventually break.

To replace a parking brake cable, take off the locknut and adjusting nut that hold the cable in the compensating lever. Remove the rear brake drum. Then unhook the cable from the lever on the rear brake shoe. Unbolt the bracket that holds the cable in the backing plate and take the cable and flexible housing out of the backing plate.

CAUTION —
If the bolt that holds the parking brake cable is locked in place by corrosion, treat it with rust solvent or penetrating oil before attempting to force it loose. If the bracket is allowed to turn with the bolt, the flexible cable housing may be damaged.

Pull the old cable out to the rear. Then grease the new cable and insert it. Use a long screwdriver or a stiff wire hook to loop the cable around the curved guide channel on the parking brake lever. Adjust the parking brakes as you attach the cable. Check the adjustment again after about 300 mi. (500 km).

10. Brake Troubleshooting

Table b lists brake problems, probable causes, and suggested remedies. The numbers in bold type in the Remedy Column refer to the numbered headings in this section of the Manual under which the suggested repairs are described.

Table b. Brake Troubleshooting

Problem	Probable Cause	Remedy
1. Pedal goes all the way to floor in braking	a. Linings worn b. Insufficient fluid	a. Adjust brake shoes (never adjust at pedal). See **7.1**. b. Find and repair leaks. Fill and bleed system. See **8**.
2. Low pedal even after adjustment and bleeding	Master cylinder defective	Replace or rebuild master cylinder. See **4.2**, **4.3**.
3. Pedal spongy or brakes work only after pedal is pumped	a. Insufficient fluid in reservoir b. Air in system c. Spring weak in master cylinder	a. Top up fluid and bleed system. See **8**. b. Check for leaks and bleed system. See **8**. c. Replace or rebuild master cylinder. See **4.2**, **4.3**.
4. Decreased braking action following shoe adjustment	a. Brake lines leaking b. Defective master or wheel cylinders	a. Tighten connections or fit new lines and hoses. See **5**. b. Replace or rebuild faulty cylinder. See **4.2**, **4.3**, **6.2**, **6.3**, **7.4**, **7.5**.
5. Brakes overheat	a. Compensating port blocked b. Inadequate pushrod clearance c. Brake shoe return springs weak d. Rubber parts swollen	a. Clean master cylinder. See **4.3**. b. Adjust pushrod clearance at pedal stop. See **4.1**. c. Fit new return springs. See **7.3**. d. Flush system. Replace fluid and all rubber parts. See **4.3**, **5.2**, **6.3**, **7.5**, **8.4**.
6. Brakes inefficient despite high pedal pressure	a. Linings oiled up b. Unsuitable brake linings	a. Clean drums. Replace linings and oil seals. See **7.3**. b. Fit new linings. See **7.3**.
7. Brakes bind while car is in motion	a. Compensating port blocked b. Brake fluid unsuitable c. Inadequate pushrod clearance	a. Disassemble master cylinder and clear port. See **4.3**. b. Flush system and refill. See **8.4**. c. Adjust pushrod clearance at pedal stop. See **4.1**.
8. Brakes chatter and tend to grab	a. Linings worn b. Drums out-of-round	a. Fit new linings. See **7.3**. b. Recondition or replace drums. See **7.2**, **7.7**.
9. Drum brakes squeak	a. Unsuitable or badly fitted linings b. Brake linings dirty c. Backing plates distorted d. Brake shoe return springs weak e. Poor lining contact pattern due to shoe distortion	a. Fit new linings properly. See **7.3**. b. Clean brakes. See **7.2**. c. Check backing plates for distortion and fit new parts if necessary. See **7.6**. d. Fit new return springs. See **7.3**. e. Align shoes with backing plate with 0.20 mm (.008 in.) clearance at lining ends and contact across full width. See **7.3**.
10. Disc brakes squeak	a. Unsuitable pads b. Spreader spring faulty or missing c. Pad guide surfaces dirty or rusted d. Pads dirty or glazed e. Lining loose on pad	a. Fit new pads. See **6.1**. b. Install new spreader spring. See **6.1**. c. Clean pads and calipers. See **6.1**. d. Clean and replace pads. See **6.1**. e. Replace pads. See **6.1**.
11. Brakes give uneven braking action	a. Oil or grease on linings b. Poor contact between lining and drum due to brake shoe distortion c. Brake shoes too tight in the adjusting screw or piston slots d. Different types of linings on same axle e. Incorrect tire pressures or unevenly worn tires f. Drums or discs out-of-round or scored g. Disc brake pads sticking in caliper h. Disc brake pads reinstalled in wrong location i. Brake shoes not in contact with backing plate j. Pistons tight in wheel cylinders k. Dirt in brake lines or hoses	a. Clean drums, fit new linings and seals or wheel cylinders if necessary. See **7.2**, **7.3**, **7.4**, **7.5**. b. Shape shoes to leave 0.20 mm (.008 in.) clearance at lining ends. See **7.3**. c. Free up shoes. See **7.3**. d. Fit new shoes or pads. See **6.1**, **7.3**. e. Correct pressures or replace worn tires. See **12.2**, **12.6**. f. Recondition or replace discs or drums. See **6.5**, **6.7**, **7.2**, **7.7**. g. Clean caliper and pads. See **6.1**. h. Install new pads. See **6.1**. i. Reposition shoes or align or replace backing plate. See **7.3**, **7.6**. j. Free up pistons. See **7.5**. k. Clean system and replace defective parts. See **8.4**.
12. Brakes pulsate	a. Drums out-of-round (drum brakes only) b. Excessive disc runout or thickness variations (disc brakes only) c. Dirt trapped between drum and hub	a. Recondition or replace drums. See **7.2**, **7.7**. b. Recondition or replace discs. See **6.5**, **6.7**. c. Clean surfaces, reinstall drum. See **7.2**.

30 Brakes and Wheels

11. Removing, Repairing, and Installing Pedal Cluster

The pedal cluster contains the clutch pedal, brake pedal, and accelerator pedal on cars with manual transmissions. On cars with Automatic Stick Shift transmissions, the pedal cluster contains only the brake pedal and accelerator pedal.

To remove:

1. Detach the clutch cable from the clutch lever on the transmission.

2. Using a screwdriver, press aside the accelerator return spring. Then withdraw the accelerator pedal pivot pin and take off the rubber covered part of the accelerator pedal.

3. Unhook the accelerator cable from the connecting lever.

4. Remove the brake master cylinder pushrod clip, but do not disconnect or remove the pushrod.

5. Using pliers, unhook the brake pedal return spring from the brake pedal.

6. Remove the pedal cluster mounting bolts as shown in Fig. 11-1.

7. Carefully pull out the pedal cluster, together with the clutch cable, so that the cable is not detached. At the same time, take the master cylinder pushrod off its pin. Then detach the clutch cable.

Fig. 11-1. Removing pedal cluster mounting bolts.

Installation is the reverse of removal. After attaching the clutch cable, hold the clutch pedal as nearly vertical as possible to keep the cable from coming off. Make certain that the brake pedal return spring is positioned correctly and that the accelerator cable is attached.

Following installation, adjust the brake pedal freeplay and travel as described in **4.1 Pushrod Adjustment.** Also adjust the clutch pedal freeplay as described in **ENGINE.** Fig. 11-2 shows both pedal clusters completely disassembled. If disassembly is undertaken, be sure to lubricate all moving parts with universal grease during reassembly.

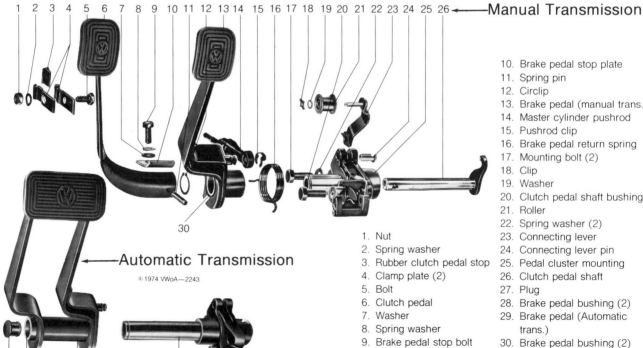

1. Nut
2. Spring washer
3. Rubber clutch pedal stop
4. Clamp plate (2)
5. Bolt
6. Clutch pedal
7. Washer
8. Spring washer
9. Brake pedal stop bolt
10. Brake pedal stop plate
11. Spring pin
12. Circlip
13. Brake pedal (manual trans.)
14. Master cylinder pushrod
15. Pushrod clip
16. Brake pedal return spring
17. Mounting bolt (2)
18. Clip
19. Washer
20. Clutch pedal shaft bushing
21. Roller
22. Spring washer (2)
23. Connecting lever
24. Connecting lever pin
25. Pedal cluster mounting
26. Clutch pedal shaft
27. Plug
28. Brake pedal bushing (2)
29. Brake pedal (Automatic trans.)
30. Brake pedal bushing (2)

Fig. 11-2. Main pedal clusters disassembled.

12. Wheels and Tires

Tires are subject to many stresses. If they are to perform as intended, they must be inflated to specifications and correctly balanced. Properly maintained, the factory-installed tires will provide long service with comfort and safety. But they must never be kept in service when worn out or damaged by accidents or careless driving.

12.1 Wheels

Karmann Ghias have 4½ J × 15 wheels. As of March, 1972, this rim size has also been used on the other models covered by this Manual. Offset is the difference between the center of rim width and the mounting face that bolts to the brake drum or brake disc hub. Karmann Ghia wheels have a 46-mm (1.811-in.) offset, 1974 Super Beetle wheels have a 41-mm (1.614-in.) offset, and the wheels for other late models have a 34-mm (1.388-in.) offset. The 4 J × 15 wheels used on cars other than the Karmann Ghia prior to March 1972 have a 40-mm (1.575-in.) offset. The 1973 Sports Bug has 5½ J × 15 wheels with a 26-mm (1.024-in.) offset. The 1974 model has the same rim size with a 34-mm (1.339-in.) offset. Wheels for 1974 Super Beetles and Sports Bugs cannot be installed on earlier cars that have strut front suspension.

The use of wheels other than those standard for a given year and model is discouraged. Many of the wider wheels sold by accessory companies do not have the correct offset dimension. An incorrect offset dimension may impose excessive stress on the wheel bearings.

The wheels on all models covered by this Manual are suitable for use with tube or tubeless tires of the correct size, including radials. Tires of non-standard size may be used only if the tire manufacturer specifies them for your particular make and model VW.

12.2 Tire Types and Pressures

Conventional (bias ply) tires are factory-installed on all cars covered by this Manual, with the one exception of the limited-production Sports Bug which has 175/70 × 15 radial ply tires. All models built before March 1972, including Karmann Ghias, are equipped with 5.60 × 15 tires. Those built after that date have 6.00 × 15 4 tires.

Conventional Tires

Conventional bias ply tires offer very good riding characteristics and are well suited to both highways and rough roads. One of their principal advantages is versatility of design. Bias-constructed tires can be made extremely strong through the use of an almost infinite number of plies. This basic simplicity results in lower construction costs. The factory-installed tires are all of the tubeless type, which further reduces the overall cost to the purchaser.

Radial Tires

There is less friction between the fabric layers in radial ply tires, so they generate less heat. This makes them especially suitable for long-distance, high-speed driving. Their rigid tread improves wet-weather traction, but tends to produce a harsher ride and increased road noise on some kinds of surfaces.

Winter Tires

Although inferior to regular tires for dry-road wear and handling, winter (mud and snow) tires can greatly improve operation on snowy or slushy roads. Studded winter tires improve traction on icy surfaces, but can be damaged by fast driving on dry roads and may damage some road surfaces. They should be used only if icy conditions predominate throughout the winter months. Also, check your local motor vehicle laws. The use of studded tires may be restricted in your area.

Tire Pressures

For prolonged high speeds, we recommend that you increase the pressure in conventional bias ply tires by 3 psi over the specifications given here. Winter tires will also work better when inflated an additional 3 psi. However, cold-tire pressure must never exceed the maximum inflated pressure marked on the tire. The following inflation pressures are prescribed:

Conventional (bias ply) Tires				
	With 1-2 persons		Fully loaded	
	Front	Rear	Front	Rear
Beetle, Ghia, to Dec. 1972, plus 1970 Convertible	16 psi	24 psi	17 psi	26 psi
Super Beetle, Convertible, from 1971 up to Dec. 1972	16 psi	27 psi	18 psi	27 psi
All models, from Jan. 1973	18 psi	29 psi	18 psi	29 psi

Radial (radial ply) Tires				
	With 1-2 persons		Fully loaded	
	Front	Rear	Front	Rear
Beetle, Ghia, to Dec. 1972, plus 1970 Convertible	18 psi	27 psi	18 psi	27 psi
Super Beetle, Convertible, from 1971 up to Dec. 1972	18 psi	27 psi	18 psi	27 psi
All models, from Jan. 1973	18 psi	29 psi	18 psi	29 psi

Because steel and textile cord radial tires have different traction characteristics, it is important that all four tires on the car have the same cord material. Tire effectiveness under various road conditions is shown in the following chart:

Operating conditions	Dry	Wet	Snow	Ice
Snow with studs radial ply	0	–	X	X
Snow with studs bias ply	0	–	X	X
Snow radial ply	X	X	X	0
Snow bias ply	X	X	X	0
Standard radial ply	X	X	0	0
Standard bias ply	X	X	0	–

X = Effective   0 = Restricted effectiveness   – = Noneffective

12.3 Normal Tire Wear

The original equipment tires on your VW have built-in tread wear indicators. These indicators are molded into the bottom of the tire tread grooves. The indicators eventually appear as the result of normal wear. They are about 13 mm (½ in.) wide in visible bands when the tire tread depth gets down to 1.5 mm (1/16 in.).

When these indicators appear in two or more adjacent grooves of a tire tread, as shown in Fig. 12-1, the tire must be replaced. It is recommended that tires be replaced well before the indicators are as evident as those shown. Worn tires cannot grip even a dry road surface properly and are almost completely ineffective on a wet road surface.

WARNING —
Do not assume that a tire is sound merely because the tread wear indicators have not yet appeared. Always check for cuts, cracks, rubber separation, and internal damage. Normal wear is only one factor in determining tire serviceability.

Fig. 12-1. Indicator showing on worn out tire.

For best all-round handling, always replace all four tires at the same time. If this is not possible, replace both tires on one axle. Do not combine tires of different ply construction, size, or tread pattern.

WARNING —
Break in new tires by driving at moderate speeds for the first 60 to 100 mi. (100 to 160 km). New tires do not have full traction when first installed.

Normal tire wear is accelerated by higher speeds. Wear at a constant 35 mph (56 kph) is only about a third of that incurred at a constant 70 mph (112 kph)—not half as might seem logical.

Weather also affects normal tire wear. Hot weather is the most damaging, and when heat is combined with high speeds and underinflation, tire structure is seriously endangered. Cold weather prolongs tire life; so does wet weather, which reduces the friction between tire and roadway.

12.4 Removing and Installing Wheels

Wheel removal is a simple task but one that can be dangerous if the proper precautions are not observed. The VW Owner's Manual supplied with the car lists the proper procedures for this job. The following points, however, deserve mention to mechanics not familiar with the VW Owner's Manual or with VW cars.

First, use the hubcap remover. This is a wire hook that can be slipped into holes along the edges of the hubcaps. It prevents scratches on the painted wheel and makes the job quicker and easier.

Next, use only the VW jack supplied with the car. Never attempt to lift a VW with an ordinary bumper jack. The cross section of VW bumpers is not contoured for the lifting hook on such jacks, and the vehicle cannot be lifted safely by them. If a hydraulic-type floor jack is used, be certain to position it carefully. Place blocks ahead and behind the wheels that remain on the ground to prevent the car from rolling.

CAUTION —
Under no circumstances must the car be lifted by placing a jack under the engine, transmission, or floor pan. Serious damage can result from these practices.

All wheel lug bolts are removed by turning them counterclockwise. When tightening the wheel bolts, torque them 12.0 to 13.0 mkg (87 to 94 ft. lb.). Use a torque wrench. Pneumatic tools are seldom capable of attaining the prescribed torque with accuracy.

12.5 Wheel Rotation

Although the tires will develop a normal wear pattern under most conditions, abnormal road surfaces or variations in driving technique may produce unequal wear of the four tires on a car. If, after a period of service, the tires on your car show uneven wear, all four wheels can be rotated as shown in Fig. 12-2.

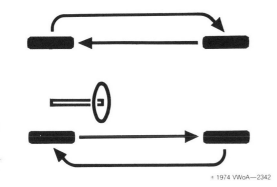

Fig. 12-2. Pattern of recommended wheel rotation. Wheel rotation is not required but may be desirable if tires are wearing unevenly.

12.6 Changing Tires

Dismounting or mounting tires on the rims requires a tire appliance with power enough to force the tire bead over the inner hump on the rim. When carrying out these operations, be sure that the rubber lining on the inner wall of the tire and tire beads is not damaged.

To dismount tire:

1. Take off the valve cap, carefully unscrew the valve core, and let the air out of the tire.

2. Press the tire bead off the rim as shown in Fig. 12-3.

Fig. 12-3. Pressing tire bead off rim.

3. Pry the tire sidewalls, one after the other, over the rim edges as shown in Fig. 12-4.

Fig. 12-4. Prying tire sidewalls over rim edges.

4. Check the airtight lining inside the tire for damage and bruises that have developed between the lining and casing. Carefully inspect the outside of the tire for embedded stones, cuts, grease, and signs of uneven wear.

5. Check the rubber part of the valve for cracks and damage. If the valve is faulty, remove it from the rim. Lubricate a new valve with soapy water. Then install the new valve as shown in Fig. 12-5.

Fig. 12-5. A valve tool being used to install a new valve in the rim.

34 Brakes and Wheels

To mount tire:

1. Check the rim for damage. Runout must not exceed 1.50 mm (.060 in.).

    NOTE
    Check the runout both laterally and radially (wobble and out-of-round).

    WARNING
    Never attempt to straighten bent, dented, or otherwise distorted rims. Doing so could weaken them. Make certain that all wheels have J-type rims, as installed in the factory, before installing tubeless radial tires. Other rim types are unsuitable for use with radial tires unless tubes are installed.

2. Using a wire brush, remove any dirt from the rim shoulders and flanges. Smooth any sharp edges before mounting the tire on the wheel.

3. Insert the valve with the valve installing tool.

    NOTE
    Use inner tubes with radial tires if the car will be used on rough roads or off the road.

4. Mount the tire on the rim as shown in Fig. 12-6. If there is a red dot in the sidewall, position it toward the valve.

Fig. 12-6. Mounting tire on rim. Use soft soap or a special rubber lubricant on the rim for safe, easy mounting.

5. Remove the valve core, if not already removed.

6. Inflate the tire to at least 43 psi (3 kg/cm^2).

    NOTE
    Use a bead expander strap, if necessary, to obtain an airtight seal. When inflating the tire you should hear the bead snap over the inner hump on the rim.

7. Install the valve core and inflate the tire to the correct running pressure, as listed earlier in **12.2 Tire Types and Pressures.**

8. Immerse the wheel in water. Check for leaks.

9. Balance the wheel and install it on the car.

12.7 Abnormal Wear

Following are the six most common causes of abnormal tire wear (extra wear):

1. Underinflation or overinflation
2. Hard driving (high speed driving, violent braking, sudden cornering)
3. Rough or abrasive road surfaces, high crown roads, and very uneven road surfaces
4. Poorly aligned wheels (front or rear)
5. Poorly balanced wheels (front or rear)
6. Vehicle overloading or excessive weight for tire capacity.

Improper Inflation

Tire life depends greatly on correct inflation. Unfortunately, there are many ways for a tire to lose air. Every tire normally loses some air pressure due to the diffusion of air molecules, through the rubber. Although tubeless tires hold air pressure longer than tires with inner tubes, it is recommended that you nevertheless check the pressure of tubeless tires once a week.

Tire pressure should be checked before driving, when the tires are still cold. If tire pressure is checked after driving, the pressure will have increased from the heat of road friction and internal flexing. If air is bled from a warm tire to obtain the pressure recommended for a cold tire, the tire will actually be underinflated. A tire that is driven while underinflated will overheat because of increased tire flexing and will rapidly lose its road-holding ability as tire strength begins to diminish.

There is always a reason for the loss of a significant amount of air from a tire in a short period of time. Aside from a hole in the tire, the possible causes are a leaky rim or valve, loose-fitting tire beads, foreign matter on the rim, or an uneven surface between the rim shoulder and the tire.

Fig. 12-7 shows three tire inflation conditions. The shape of the tire is changed by the degree of inflation. This can cause abnormal wear. An underinflated tire wears at its edges. An overinflated tire wears at its center. The profile of an underinflated tire is similar to that of an overloaded tire. The same kind of wear will result from overloading, with the added possibility of severe heat damage and possible structural failure.

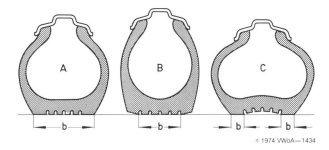

Fig. 12-7. Tire inflation. Tire **A** is normally inflated. Tire **B** is overinflated and Tire **C** is underinflated. Dimension **b** is the width of the tire tread in contact with the road.

Fig. 12-8 shows typical tire wear due to overinflation. Because a narrower portion of the tread is in contact with the road, wear occurs faster than normal. This tire is worn out despite the good tread remaining at the edges.

Fig. 12-8. Overinflation wear. Tire worn out by running it with excessive air pressure.

Fig. 12-9 shows the tread wear pattern of an underinflated tire. This worn-out tire still has deep tread in its center, although the edges are completely bald. Such tires should be removed from service immediately. As with tires worn unevenly by overinflation, tires worn unevenly by underinflation provide very limited adhesion in wet weather and almost no traction in deep snow. In addition, side grip for cornering is seriously impaired.

Fig. 12-9. Underinflation wear. Tire worn out by running it at too low an air pressure.

Hard Driving

Tires may wear abnormally because of excessive speed, heavy braking (see Fig. 12-10), high-speed cornering, and similar violent or abrupt maneuvers. Conservative driving preserves tire life.

Fig. 12-10. Abnormal tire wear caused by locking the brakes. Uneven braking can also cause heavy tire wear.

Road Surfaces

Some road surfaces tend to increase tire wear. Rough, anti-skid surfacing materials abrade the tread and accelerate wear. Roadways with high crowns necessitate constant steering correction to keep the car headed parallel to the highway. This will eventually produce the kind of abnormal wear usually associated with improper front wheel alignment. Obstructions and road hazards can cause severe tire damage and can even break the casing internally, as shown in Fig. 12-11.

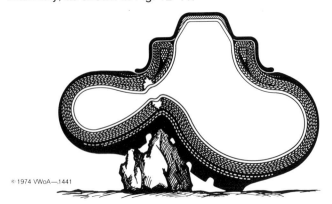

Fig. 12-11. Internal damage. Striking an obstacle in the road may tear the ply layers. Such damage renders the tire unfit for use. The tire may have to be taken off its rim for such defects to be seen.

Faulty Wheel Alignment

Excessive tire wear will result from misalignment of any of the four wheels on the car. If irregular tire wear appears, check the front wheel track, the front wheel angles at full steering lock, the ride height and rear wheel track, the position of the axles relative to each other, the wheelbase on both sides of the car, the front and rear wheel camber, the setting of the rear spring plates, and the condition of the shock absorbers. The abnormal wear caused by misalignment usually takes the form of greater wear at one edge of the tread than on the other. Alignment specification are given in **TRANSMISSION AND REAR AXLE, FRONT AXLE,** and **STRUT FRONT SUSPENSION.**

Wheels Out of Balance

Unbalanced wheels bounce, tramp, and wobble. You can correct one kind of imbalance with static balancing equipment, that shown in Fig. 12-12, which checks the balance of the wheel and tire without spinning them. Notice that the heavier part of the tire is located on the vertical axis of the wheel.

Sometimes simple static balancing will not completely cure the problem. In such cases the tire and wheel must be checked and balanced with dynamic balancing equipment that spins the wheel while imbalance is being measured. Fig. 12-13 shows the type of imbalance that only dynamic balancing can detect and cure. Although in perfect static balance, the concentrations of mass are not in line with one another or with the tire centerline. The wheel wobbles sideways when it spins, as shown on the right-hand side of the illustration. This causes deep wear in the form of cupped areas on the tread. Check that wheel runout does not exceed 1.5 mm (.060 in.) either radially or laterally before dynamic balancing the wheel. Replace the wheel if it does.

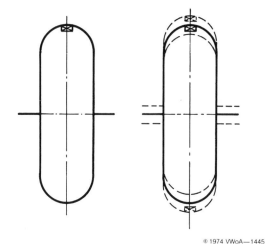

Fig. 12-12. Imbalance that can be cured by static balancing. Vibration of the wheel while driving is shown at the right.

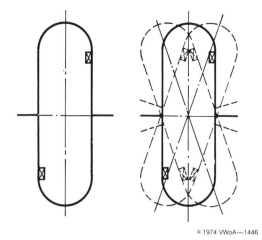

Fig. 12-13. Imbalance that can be cured only by dynamic balancing. Vibration of the spinning wheel shown at the right.

Overloading

Overloading the tires causes damage similar to that produced by underinflation. The excessive heat generated by an overloaded tire weakens the tire and causes difficult vehicle handling. The maximum load capacity (given in the Owner's Manual), plus the weight of the vehicle, must never exceed the total capacity of the four tires (marked on their sidewalls).

13. Technical Data

The following tables give all specifications related to the brake system, tires, and wheels. Wheel alignment specifications are given in **TRANSMISSION AND REAR AXLE, FRONT AXLE,** and **STRUT FRONT SUSPENSION.**

Brakes and Wheels

I. Tightening Torques

Location	Part	Thread	mkg	ft. lb.
Master cylinder				
Stop bolt in housing	bolt	M 6	0.5-1.0	3.5-7
Residual pressure valve on housing	—	M 12 x 1	2.0	14
Brake light switch to housing	—	M 10 x 1	2.0	14
Master cylinder to frame	bolt	M 8	2.5	18
Brake line to master cylinder	union nut	M 10 x 1	1.5-2.0	11-14
Front wheel brakes				
Backing plate on steering knuckle	bolt	M 10	5.0	36
Splash shield to steering knuckle (KG)	bolt	M 7	1.0	7
Wheel cylinder on backing plate	bolt	M 8	2.5	18
Caliper housing	socket head screw	M 7	2.0-2.5	14-18
Caliper to steering knuckle,	bolt	M 10	4.0	29
Bleeder valve in cylinder/caliper	—	M 6 M 7	max. 0.5	max. 3.5
Hose to cylinder/caliper	—	M 10	1.5-2.0	11-14
Screw for clamp nut	socket head screw	M 7	1.0 max. 1.3	7 max. 9
Rear wheel brakes				
Cylinder to backing plate	bolt	M 8	2.0-3.0	14-22
Cover to bearing housing	bolt	M 10	6.0	43
Brake drum to shaft	slotted nut	M 24	30.0	217
Wheels				
Wheel to drum/disc	bolt	M 14 x 1.5	12.0-13.0	87-94
Pedals				
Pedal cluster to frame	bolt	M 10	4.0-4.5	29-33
Pedal stop plate to frame	bolt	M 8	2.0-2.5	14-18

WARNING —
Do not re-use any fasteners that are worn or deformed in normal use. Many fasteners are designed to be used only once and become unreliable and may fail when used a second time. This includes, but is not limited to, nuts, bolts, washers, self-locking nuts or bolts, circlips and cotter pins. Always replace these fasteners with new parts.

II. Tire Data

Original equipment, tires	After March 1972	Prior to March 1972
Tire (tubeless)	6.00 x 15 4 PR	5.60 x 15 4 PR
Rim x wheel size	4½ J x 15	4 J x 15
Radial runout Lateral runout	Maximum 1.5 mm (.060 in.)	Maximum 1.5 mm (.060 in.)

Inflation pressures in psi
(Under no circumstance must the cold tire pressure exceed maximum inflated pressure marked on the tire)

Car Model and Year	bias, front	bias, rear	radial, front	radial, rear
Beetle, Ghia, to Dec. 1972, plus 1970 Convertible				
a. Up to 2 occupants	16 psi	24 psi	18 psi	27 psi
b. Fully loaded	17 psi	26 psi	18 psi	27 psi
Super Beetle, Convertible, from 1971 up to Dec. 1972				
a. Up to 2 occupants	16 psi	27 psi	18 psi	27 psi
b. Fully loaded	18 psi	27 psi	18 psi	27 psi
All models, from Jan. 1973				
a. Up to 2 occupants	18 psi	29 psi	18 psi	29 psi
b. Fully loaded	18 psi	29 psi	18 psi	29 psi
Prolonged high speeds (bias ply tires only)				
Add to the above specifications	3	3		

38 BRAKES AND WHEELS

III. Tolerances, Wear Limits, and Settings

Parts	All 1970 except Ghia mm (in.)	113 & Convertible, 1971 on mm (in.)	Karmann Ghia mm (in.)	Sedan 111, 1971 on mm (in.)
a. Master cylinder				
Front wheel circuit stroke	15.5 (.610)	17.5 (.689)	14.0 (.551)	15.5 (.610)
Rear wheel circuit stroke	12.5 (.492)	11.5 (.453)	14.0 (.551)	12.5 (.492)
Diameter	19.05 (.750)	19.05 (.750)	19.05 (.750)	19.05 (.750)
b. Front wheel cylinder			(in caliper)	
Diameter	22.20 (.874)	23.81 (.937)	ATE: 40.00 (1.575) Girling: 40.40 (1.591)	22.20 (.874)
c. Rear wheel cylinder				
Diameter	17.46 (.687)	17.46 (.687)	17.46 (.687)	17.46 (.687)
d. Brake drums				
Front diameter	230.10 + 0.20 (9.059 + .008)	248.10 + 0.20 (9.768 + .008)	—	230.10 + 0.20 (9.059 + .008)
wear limit	231.50 (9.114)	249.50 (9.823)	—	231.50 (9.114)
Rear diameter	230.00 + 0.20 (9.055 + .008)	230.00 + 0.20 (9.055 + .008)	230.00 + 0.20 (9.055 + .008)	230.00 + 0.20 (9.055 + .008)
wear limit	231.50 (9.114)	231.50 (9.114)	231.50 (9.114)	231.50 (9.114)
Permissible turned diameter, front	231.10 + 0.20 (9.098 + .008)	249.00 (9.803)	—	231.10 + 0.20 (9.098 + .008)
wear limit	231.50 (9.114)	249.50 (9.823)	—	231.50 (9.114)
Permissible turned diameter, rear	231.00 + 0.20 (9.094 + .008)	231.00 + 0.20 (9.094 + .008)	231.00 + 0.20 (9.094 + .008)	231.00 + 0.20 (9.094 + .008)
wear limit	231.50 (9.114)	231.50 (9.114)	231.50 (9.114)	231.50 (9.114)

	New/turned	Wear limit
Front and rear wall thickness	—	4.00 (.157) min.
taper	0.10 (.004)	—
out-of-round	0.10 (.004)	—
lateral runout	0.25 (.010)	—
radial runout (measured at friction surface)	0.15 (.006)	—
e. Brake discs (Ghia only)		
Thickness	9.50 − 0.10 (.374 − .004)	8.00 (.315)
Permissible turned thickness	8.50 (.335) min.	8.00 (.315)
Machining limit per side	0.50 (.020) max.	—
Thickness tolerance	0.02 (.001) max.	—
Disc runout	0.20 (.008) max.	—

	New parts	Wear limit
f. Brake linings		
Front width, Sedan 113 & convertible	45.00 (1.772)	—
Front width, Sedan 111, all 1970 cars	40.00 (1.575)	—
Rear width, all cars	40.00 (1.575)	—
Thickness	4.00 − 3.80 (.157 − .150)	2.50 (.100) min.
Thickness, oversize	4.50 − 4.30 (.177 − .169)	2.50 (.100) min.
Friction pad thickness, Ghia .. ATE	10.00 (.394)	2.00 (.079) min.
Girling	11.50 (.453)	2.00 (.079) min.

Section 6

FRONT AXLE

Contents

Introduction 2	6.2	Checking Steering Knuckle 12

1. General Description 3
    Axle Beam 3
    Suspension 4
    Steering 4
    Wheel Bearings 4

2. Maintenance 4

3. Front Wheel Alignment 4
   3.1 Checking and Adjusting Camber 4
   3.2 Checking and Adjusting Toe 5
   3.3 Checking and Adjusting Steering
       Lock 6

4. Shock Absorbers 6
   4.1 Checking Shock Absorbers 7
   4.2 Replacing Shock Absorbers 7

5. Front Wheel Bearings 8
   5.1 Removing and Installing
       Front Wheel Bearings 8
   5.2 Checking and Adjusting
       Front Wheel Bearings 9

6. Steering Knuckles 9
   6.1 Removing and Installing
       Steering Knuckle 9

**7. Torsion Arms, Suspension Ball Joints,
and Torsion Bars** 13
   7.1 Checking Ball Joints 13
   7.2 Removing and Installing Torsion Arm . . 13
   7.3 Checking Torsion Arms 13
   7.4 Repairing and Replacing Ball Joints . . . 15
   7.5 Removing and Installing Torsion Bars . . 17
   7.6 Replacing Needle Bearings
       and Metal Bushings 17

8. Removing and Installing Axle Beam . . . 19

9. Steering 21
   9.1 Adjusting Steering 21
   9.2 Removing and Checking Steering
       Damper 24
   9.3 Removing and Checking Tie Rods 24
   9.4 Removing and Installing
       Steering Column 25
   9.5 Removing and Installing
       Steering Gearbox 27
   9.6 Disassembling, Assembling, and
       Adjusting Steering Gearbox 28

10. Front Axle Technical Data 31
    I. Tolerances, Wear Limits, and Settings . . 32
   II. Tightening Torques 32
  III. Wheel Alignment Specifications 33

© 1974 VWoA

Front Axle

Only Beetles and Karmann Ghias are equipped with the front axle described in this section. The strut front suspension used on 1971 and later Super Beetles and Convertibles is covered separately in **STRUT FRONT SUSPENSION**.

The front axle has three functions: springing (or suspension), steering, and wheel alignment. Tire wear and vehicle handling are good indicators of how well these functions are being fulfilled. However, it is always best to maintain the front axle in serviceable condition so that abnormal tire wear and poor handling never affect the car.

On cars with the front axle, the spring action of the independent front wheel suspension is accomplished by trailing links and transverse torsion bars. This system not only prevents the transferal of road shocks from one front wheel to the other, but also provides excellent resistance to the transferal of road shocks to the suspension parts, chassis, and passengers. Vertical travel of the front suspension is limited by progressively acting rubber stops. The greater the impact of the torsion arms (the trailing links) against the rubber stops, the greater their springing and energy-absorbing reaction. But because the trailing link system minimizes the suspension deflection caused by severe bumps, the stops are seldom called upon during normal highway driving.

Ball joints at the ends of the torsion arms provide a flexible mounting for the steering knuckle. These joints not only permit the free vertical movement of the front wheels during bump and rebound, but also allow the wheels to be turned around a vertical axis for steering. Front wheel camber adjustments are incorporated in the upper ball joint mountings. And, since the joints are lubricated at the factory, they rarely require further lubrication.

Many repair procedures require special equipment that car owners and small repair shops may not have—a hydraulic press and precision measuring jigs, for example. If you lack the skills, tools, or a clean workshop for servicing the front axle and steering, we suggest you leave such repairs to an Authorized VW Dealer or other qualified shop. We especially urge you to consult your Authorized VW Dealer before attempting repairs on a car still covered by the new-car warranty.

FRONT AXLE 3

1. General Description

The front axle is a rigid beam with pivoting members that provide suspension movement, steering movement, and rotational movement of the wheels. The steering gearbox and steering linkage are mounted on the front axle. Therefore, all steering adjustments and repairs are covered in this section of the Manual. The names and positions of the front axle's parts are shown in Fig. 1-1.

Axle Beam

The axle beam itself is an electrically welded assembly consisting of two large parallel steel tubes and a number of heavy-gauge steel stampings. It is bolted solidly to the frame head and includes two mounting points for the car body. The axle beam tubes contain the torsion bars and their bearings and mounts. The shock absorbers are mounted on welded, stamped steel uprights.

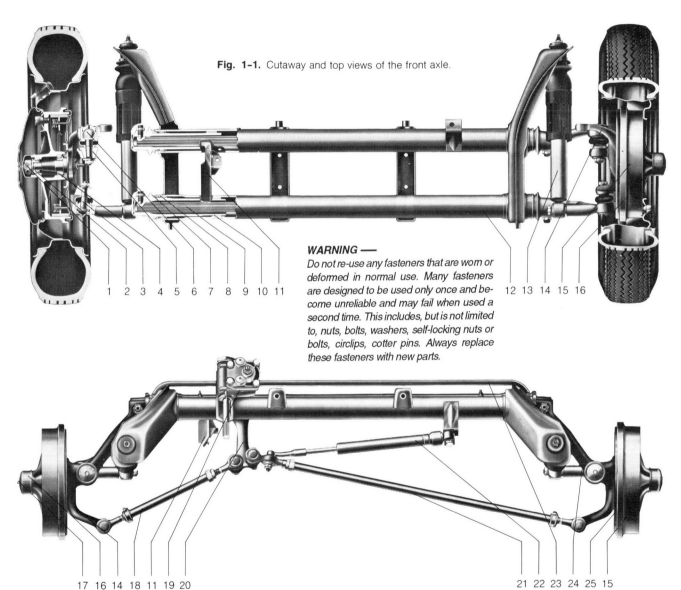

Fig. 1-1. Cutaway and top views of the front axle.

WARNING —
Do not re-use any fasteners that are worn or deformed in normal use. Many fasteners are designed to be used only once and become unreliable and may fail when used a second time. This includes, but is not limited to, nuts, bolts, washers, self-locking nuts or bolts, circlips, cotter pins. Always replace these fasteners with new parts.

1. Wheel bearing clamp nut
2. Outer wheel bearing
3. Lower ball joint
4. Inner wheel bearing
5. Eccentric camber adjusting bushing
6. Upper ball joint
7. Torsion arm seal
8. Torsion arm needle bearing
9. Torsion bar
10. Plastic seat and metal bushing
11. Steering lock stop
12. Front axle beam
13. Shock absorber
14. Steering knuckle
15. Brake drum
16. Dust cap
17. Speedometer cable
18. Left tie rod
19. Steering gearbox
20. Drop arm
21. Right tie rod
22. Steering damper
23. Stabilizer bar
24. Torsion arm
25. Brake backing plate

4 Front Axle

Suspension

Multileaf torsion bars provide the springing. There are two bars: one in the upper axle tube and one in the lower. Setscrews hold the bars stationary in their center bushings. Each half of each torsion bar has a torsion arm mounted at its outer end. In this way, all four torsion arms are sprung by only two torsion bars. The lower torsion arms are joined by a stabilizer bar that increases the front suspension's resistance to roll during cornering.

Steering

The worm and roller steering gearbox used with the front axle is completely different from the worm and roller steering gearbox used with strut front suspension. The steering gearbox has two threaded adjustments that make it possible to compensate for minor gearset wear.

The ball-joint tie rod ends do not require lubrication during their service life. And because the tie rods have threaded ends, you can adjust front wheel toe simply by loosening the locknuts and turning the tie rods on their threaded ends. A hydraulic steering damper is linked to the right-hand tie rod to minimize the road shock transmitted to the steering wheel.

Wheel Bearings

Tapered-roller wheel bearings are used on all cars covered in this Manual. The inner race for each of the four bearing assemblies consists of a cone surrounded by a number of caged tapered rollers. The outer races are a press-fit in the wheel hub bore. The hubs are an integral part of the front wheel brake drum. Tapered-roller bearing adjustment is possible by turning the clamp nut on the steering knuckle's stub axle.

2. Maintenance

The diagnosis and maintenance steps that must be performed at regular mileage intervals are listed here. Lubrication and checking procedures are described fully in **LUBRICATION AND MAINTENANCE** or under the listed headings in this section of the Manual.

1. Lubricating the front axle
2. Lubricating and adjusting the front wheel bearings
3. Checking the dust seals and plugs on the suspension ball joints and the tie rod ends (see **7. Torsion Arms, Suspension Ball Joints, and Torsion Bars** and **9. Steering**)
4. Checking the ball joint play (see **7. Torsion Arms, Suspension Ball Joints, and Torsion Bars**)
5. Checking the steering play (see **9. Steering**)
6. Checking the front wheel camber and toe (see **3. Front Wheel Alignment**).

3. Front Wheel Alignment

Only camber and toe are adjustable. Caster angle and kingpin inclination are determined by the manufactured dimension of the suspension parts, so damaged parts must be replaced to correct these alignment factors.

The following preparatory steps are essential to accurate alignment measurements:

1. Have the car on a level surface.
2. Inflate the tires to specifications, and unload the car except for the spare wheel and a full fuel tank. Then jounce the car several times and let it settle into its normal position.
3. Check the adjustment of the steering gear and the front wheel bearings. Adjust them if necessary. See **9.1 Adjusting Steering** and **5.2 Checking and Adjusting Front Wheel Bearings**.
4. Make sure there is no play in the tie rod ends, drop arm, or other parts of the steering linkage.

Measuring wheel camber and toe requires suitable gauges. Professional-grade instruments may cost several hundred dollars, but the modestly priced gauges available from mail order houses are adequate for home use. Instructions are supplied by the manufacturer.

3.1 Checking and Adjusting Camber

Camber is the angle at which wheels depart from the true vertical, when viewed from directly in front of the car. If the tops of the wheels lean outward slightly, the wheels are said to have positive camber; if the tops of the wheels lean inward, the wheels have negative camber.

To check:

1. After placing the car on a level surface with its front wheels pointing straight ahead, apply a bubble protractor as shown in Fig. 3–1.

Fig. 3–1. Bubble protractor applied against front wheel.

2. Using chalk, mark the wheel at the points where it is contacted by the protractor.

3. Turn the spirit level carrier on the protractor until the bubble is centered, then read the camber angle on the scale.

   NOTE ——
   If you are using a different type of gauge, follow the manufacturer's instructions.

4. Roll the car forward a half-turn of the wheels, and then repeat the measurement at the chalk-marked points.

5. Take the new reading and average it with the one you obtained earlier. The result is the camber angle, corrected for wheel run out.

6. Repeat the entire procedure on the other front wheel.

On cars with the front axle, the front wheels should have 0° 30′ ± 20′ of positive camber. Also, the difference in camber between the wheels should not vary more than 30′. If not within specifications, the camber of each wheel should be adjusted to as near 0° 30′ as possible.

To adjust camber:

1. Loosen the self-locking nut on the upper ball joint stud as shown in Fig. 3-2.

   NOTE ——
   The car must be standing on its wheels while adjustments are being made.

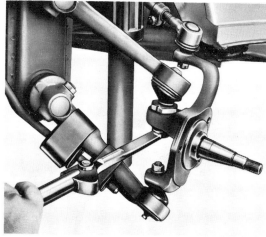

Fig. 3-2. Nut on ball joint stud being loosened. (Wheel and brake assembly removed for clarity.)

2. Set the spirit level carrier on the protractor to the specified angle of 0° 30′ positive camber.

3. Turn the eccentric camber adjusting bushing (Fig. 3-3) until the bubble in the protractor is centered when the protractor is applied to the chalk-marked points.

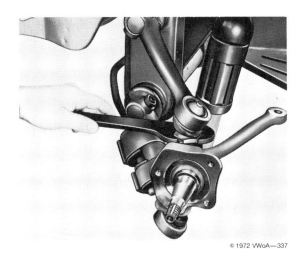

Fig. 3-3. Eccentric camber adjusting bushing being turned. (Wheel and brake removed for clarity.)

4. Install a new self-locking nut on the ball joint stud. Torque it to 5.0 to 7.0 mkg (36 to 50 ft. lb.). Again check the camber; if necessary, repeat the adjustments to bring the camber within specifications.

5. Check toe and adjust it if necessary.

3.2 Checking and Adjusting Toe

Cars with the front axle are designed to operate with a small amount of toe-in. This means that the front edges of the tires are slightly closer together than the rear edges. Shops with optical aligning devices should follow the equipment manufacturer's instructions to obtain a toe-in angle of 30′ ± 15′—or 5′ ± 15′ after adding 8 to 12 kg (18 to 26 lb.) of extra weight above the wheel. The maximum toe change should not exceed 25′.

Most small shops and individual car owners check the toe with a track gauge. This device is used to measure first the distance between two marked points at the front edges of the rims, then the distance between the same two points after the car has been rolled forward until the marks are at the rear. The measurement made at the rear should be 1.8 to 5.4 mm (.071 to .212 in. or approximately ¹⁄₁₆ to ⁷⁄₃₂ in.) greater. This should decrease to 1.2 to 2.4 mm (.047 to .094 in. or approximately ³⁄₆₄ to ³⁄₃₂ in.) when the weight is added above the wheels.

NOTE ——
These specifications apply only when the wheels are in their straight-ahead position.

6 Front Axle

To check wheel toe:

1. Turn the steering to its center position.

2. Check the toe. If a track gauge is used, mark the measuring points with chalk. This allows you to make measurements between the same two points after you roll the car forward a half-turn of the wheels.

3. Adjust the toe if it is not within specifications.

To adjust:

1. Bend up the lockplate on the inner end of the right-hand tie rod. Then loosen the locknut and clamp bolts on the tie rods.

2. Knock the hexagonal tapered rings off the inner end of the right-hand tie rod.

3. Install a new lockplate.

4. Correct the toe-in by turning both tie rod tubes in the same direction. Check toe measurements frequently.

   NOTE ——
   Rotating the tops of the tie rods toward the front of the car increases the toe-in; rotating them to the rear decreases toe-in.

5. When the toe-in is correct, position the tie rods so that the ball joints are not angled. Torque the clamp bolts to 1.5 mkg (11 ft. lb.) and the locknut to 2.5 mkg (18 ft. lb.). Then bend the lockplate over.

If the steering wheel is not centered while you are driving straight ahead on a level surface, center it as illustrated in Fig. 3-4. This procedure will move the steering wheel without moving the front wheels or changing the toe-in.

Fig. 3-4. Steering wheel being centered. Large arrows point to front of car. To move the steering wheel as shown by the curved arrow, turn tie rods in directions indicated by arrows around rods.

CAUTION ——
Never center the steering wheel either by altering its position on the column, or by moving the column on the steering gear. Doing so leaves the real fault uncorrected.

3.3 Checking and Adjusting Steering Lock

To check the steering lock, turn the wheels either full left or full right. The gap between the front tire's inner sidewall and the upper torsion arm pivot should not be less than about 10 mm (3/8 in.). Check the gap at the opposite lock too. If the gap on either side exceeds specifications, loosen the locknuts on the adjusting bolts in the steering lock stop. (The steering lock stop is a bracket to the left of the steering gearbox on the front axle.) Turn the adjusting bolts until the gap on both sides is correct, then torque the locknuts to 2.0 mkg (14 ft. lb.).

4. Shock Absorbers

Make sure you install only shock absorbers intended for the front of the car. Mismatched shock absorbers will impair handling and ride. It is not necessary, however, to replace both shock absorbers if only one is defective. Also, front axle shock absorbers of different manufacture can be combined as long as their damping characteristics are identical. If the car is subjected to heavy loads, rough roads, or extremely high temperatures, it may be advisable to install heavy-duty shock absorbers in the interest of longer service—despite their adverse effect on the riding qualities of the car.

CAUTION ——
Install heavy-duty shock absorbers on all four wheels at the same time. Otherwise, handling will be adversely affected.

4.1 Checking Shock Absorbers

You can quick-check the shock absorbers by grasping the front bumper and rocking the car vigorously. When you let go, the car should rebound only once and then settle into its normal attitude. If the car continues to rock or bob, the shock absorbers are worn. Excessive bobbing on the highway also signals defective shock absorbers. Badly worn shock absorbers often make knocking noises when the car is driven.

You can hand-check a shock absorber by extending and compressing it while holding it in its installed position. It should operate smoothly and with uniform resistance throughout its entire stroke. If possible, compare the used shock absorber with a new one. New shock absorbers that have been in storage may have to be pumped several times before they reach full efficiency.

An adequate supply of fluid is placed in the shock absorbers during manufacture to compensate for small leaks. Minor traces of fluid are acceptable if the shock absorber still functions efficiently.

4.2 Replacing Shock Absorbers

Shock absorbers cannot be repaired or serviced and should be replaced if faulty.

To replace:

1. Raise the car and remove the front wheel.
2. Remove the M 10 nut from the top of the buffer pin. If necessary, slide the buffer and outer tube down and hold the buffer pin as shown in Fig. 4-1.

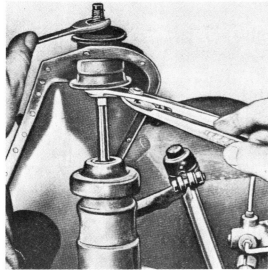

Fig. 4-1. M 10 nut being removed with open end wrench while buffer pin is held with pliers.

3. Push the piston rod down into the shock absorber until the buffer pin is out of the mounting upright.
4. Remove the nut from the lower shock absorber mounting pin, then remove the shock absorber.
5. Remove the buffer pin from the piston rod as shown in Fig. 4-2.

Fig. 4-2. Buffer pin being gripped with pliers so that it can be unscrewed from piston rod. Grip flat sides on piston rod with an open end wrench.

Installation is the reverse of removal. The rubber bushings, buffer pin, and buffer should be replaced or reused depending on their condition. Clean the corrosion from the pin on the torsion arm. If the old pins are damaged, press in new ones. The rubber bushing goes on the buffer pin with the shoulder upward, as shown in Fig. 4-3. Torque the nut on the torsion arm pin to 3.0 to 3.5 mkg (22 to 25 ft. lb.) and the nut atop the buffer pin to 2.0 mkg (14 ft. lb.).

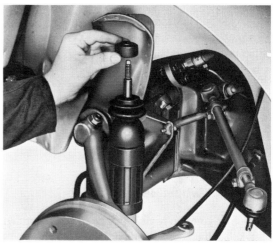

Fig. 4-3. Rubber bushing being installed.

8 Front Axle

5. Front Wheel Bearings

There are two tapered-roller bearing assemblies at each front wheel. Each bearing has a solid steel outer race and an inner race that consists of a steel cone and a number of caged tapered rollers. The bearings and related parts are shown in Fig. 5-1.

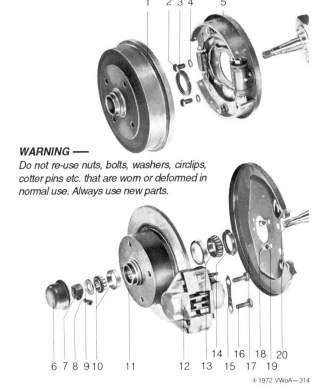

WARNING —
Do not re-use nuts, bolts, washers, circlips, cotter pins etc. that are worn or deformed in normal use. Always use new parts.

1. Brake drum (Beetle)
2. M 10 bolt
3. Drum grease seal
4. Lock washer
5. Brake assembly
6. Dust cap
7. Clamp nut
8. Socket head screw for clamp nut
9. Thrust washer
10. Outer tapered-roller bearing
11. Brake disc (Ghia)
12. Brake caliper assembly
13. Inner tapered-roller bearing
14. Disc grease seal
15. Lockplate
16. M 10 bolt
17. Splash shield
18. M 10 bolt
19. Lock washer
20. Steering knuckle

Fig. 5-1. Wheel bearings, brakes, and steering knuckles. Parts at top are for Beetle; those at bottom are for Karmann Ghia models.

5.1 Removing and Installing Front Wheel Bearings

Before you can remove the brake disc and wheel bearings from Karmann Ghia models, the brake caliper must be removed as described in **6.1 Removing and Installing Steering Knuckle.** You can remove the inner races with common hand tools. The outer races, however, are pressed into the brake disc or drum and should be removed with a hydraulic press and appropriate mandrels.

To remove inner races:

1. Remove the road wheel. On Karmann Ghia models, remove the brake caliper.

2. Pry off the dust cover.

3. Using a 6-mm ($\frac{15}{64}$-in.) hex key, loosen the socket head screw in the clamp nut. Then unscrew the clamp nut from the stub axle on the steering knuckle.

   NOTE —
   The clamp nut for the left front wheel has a left-hand thread.

4. Pull off the brake drum or disc, being careful not to let the thrust washer and the outer tapered-roller bearing inner race fall out and onto the floor.

5. Place the brake drum or disc on the workbench, then carefully remove the thrust washer and the outer bearing's inner race. Store them in a clean, dust-free place.

6. Pry the grease seal out of its recess in the rear of the drum or disc. Then lift out the inner tapered-roller bearing's inner race. Store it with the outer bearing's inner race.

To install inner races:

1. Carefully clean the inner bearing races with solvent, then dry them with compressed air.

   CAUTION —
   Do not use solvents such as gasoline because they remove all lubrication. Also, do not let blasts of compressed air spin the races. Unlubricated bearings can be damaged by rapid movement.

2. Inspect the inner bearing races. Replace them if they are worn, burred, rough, or heat-blued.

3. Clean the brake drum or disc hub and inspect the outer bearing races. Replace them if they are worn, burred, rough, or heat-blued.

4. Pack the inner bearing's inner race with multipurpose grease as described in **LUBRICATION AND MAINTENANCE.** Carefully place it inside the hub.

5. Press a new grease seal into place.

6. Inspect the stub axle for burrs or blued areas. Check the bearing seat dimensions as described in **6.2 Checking Steering Knuckle.** If satisfactory, lightly coat the stub axle with multipurpose grease.

7. Carefully slide the brake disc or drum onto the stub axle so that the grease seal or bearing races are not accidentally damaged by the sharp threads.

8. Pack the outer bearing's inner race with multipurpose grease. Then carefully slide it onto the stub axle and into the hub.

9. Install the thrust washer and the clamp nut. Then tighten the clamp nut on the stub axle until the bearings just contact their outer races.

10. Adjust the bearings as described in **5.2 Checking and Adjusting Front Wheel Bearings**. Install the dust cap, the brake caliper (where fitted), and the road wheel.

To remove outer races:

1. Support the brake disc or drum, outer surface down, on the press bed. Using a suitable driving mandrel, press out the outer bearing's outer race.

2. Turn over the brake disc or drum.

3. Using a suitable driving mandrel, press out the inner bearing's outer race.

Installation is the reverse of removal. The inner bearing's outer race, being the larger, is pressed out last, but the sequence is unimportant to installation. Make certain, however, that the hub recesses are clean and free of burrs or pressure marks that could prevent the races from seating completely.

5.2 Checking and Adjusting Front Wheel Bearings

Wheel bearings should turn smoothly and not have excessive axial play. If the bearings feel gritty, have tight spots, or make noises when the wheel turns, they probably need to be replaced. Excess axial play, though, can be corrected by adjusting.

To adjust bearings:

1. Raise the wheel, then pry off the dust cap.

2. If the bearings have just been installed, torque the clamp nut to about 1.0 mkg (7 ft. lb.) while you hand-turn the brake disc or drum.

   CAUTION
   Never torque the clamp nut to more than 1.3 mkg (9.5 ft. lb.). Doing so will damage the bearing races.

3. To measure the bearing axial play, install a dial indicator on one of the wheel lugs (or use a dial indicator with a magnetic base).

4. Position the dial indicator pin against the end of the stub axle as shown in Fig. 5-2.

5. Move the brake disc or drum in and out by hand. Turn the clamp nut one way or the other until the axial play is between 0.03 and 0.12 mm (.001 and .005 in.).

Fig. 5-2. Dial indicator being used to measure wheel bearing axial play.

NOTE
Turn the brake drum or disc and repeat the measurement at several different points. The readings should not vary greatly and their average should fall within the prescribed range. Replace bearings that will not adjust properly.

6. Torque the socket head screw to 1.0 to 1.3 mkg (7.0 to 9.5 ft. lb.).

7. Install the dust cap and lower the wheel to the ground.

6. STEERING KNUCKLES

The steering knuckles, with their integral steering arms, are held onto the torsion arms by ball joints. For safety as well as driveability, it is important that the steering knuckles are not bent. Check them carefully after an accident or other severe impact.

6.1 Removing and Installing Steering Knuckle

When removing the steering knuckle, take off the brake assembly only if you are planning to replace the steering knuckle itself. The brake assembly can be left in place if only the ball joints require attention.

10 Front Axle

To remove steering knuckle:

1. On cars with drum brakes, remove the brake drum and the wheel bearings as described in **5.1 Removing and Installing Front Wheel Bearings**.

2. On cars with disc brakes, remove the brake caliper from the steering knuckle as shown in Fig. 6-1.

   CAUTION ——
   The caliper must have cooled to room temperature before you remove it. Otherwise, it may be damaged by heat distortion.

Fig. 6-1. Removing disc brake caliper. First bend down the lockplate tabs, then remove the two M 10 bolts as shown.

3. Using a stiff wire hook, suspend the brake caliper from the car body so that it is not being supported by the brake hose.

   WARNING ——
   Never let a brake caliper or brake backing plate assembly hang by the brake hose. Doing so could weaken the hose and cause subsequent brake failure.

4. On cars with drum brakes, remove the three M 10 bolts that hold the brake backing plate assembly on the steering knuckle. Remove the brake backing plate assembly and suspend it from the car body so that its weight is not supported by the brake hose.

5. Remove the cotter pin and the castellated nut from the tie rod end stud. Then press out the tie rod end as shown in Fig. 6-2.

CAUTION ——
Do not hammer out the tie rod end. Doing so will ruin the threads and make reinstallation impossible.

Fig. 6-2. Tie rod end being pressed out.

6. On cars with disc brakes, remove the three M 10 bolts that hold the splash shield on the steering knuckle. Then remove the splash shield.

7. Remove the M 12 self-locking nut from the lower suspension ball joint stud. Then install an M 12 × 1.5 cap nut, and press the ball joint stud loose as shown in Fig. 6-3.

Fig. 6-3. Ball joint being pressed loose in steering knuckle (arrow).

FRONT AXLE 11

8. Remove the M 12 self-locking nut from the upper ball joint stud. Then turn the eccentric camber adjusting bushing to free it in the steering knuckle.

9. Lift the upper torsion arm as shown in Fig. 6-4 so that the steering knuckle can be removed.

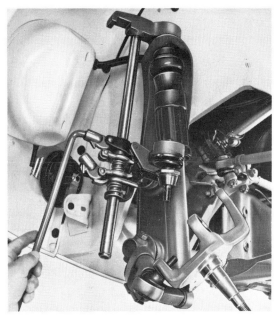

Fig. 6-4. Special tool used to lift torsion arm.

NOTE —
If the steering knuckle is being removed so that the upper ball joint can be repaired or replaced, press the ball joint stud out of the eccentric bushing as shown in Fig. 6-5. We recommend that the lower ball joint be replaced at the same time.

Fig. 6-5. Upper ball joint stud being pressed out of eccentric camber adjusting bushing. Note the cap nut on the stud to protect the threads.

To install:

1. Raise the upper torsion arm using the tool shown in Fig. 6-4.

2. Loosely install the steering knuckle on the lower ball joint stud. Then gradually relieve tension from the upper torsion arm while guiding the eccentric camber adjusting bushing into the steering knuckle.

3. Position the eccentric camber adjusting bushing so that its notch points forward as shown in Fig. 6-6.

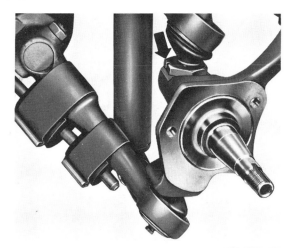

Fig. 6-6. Notch (arrow) in eccentric camber adjusting bushing pointing toward front of car.

4. Install new self-locking nuts on the ball joint studs. Torque the nut on the lower ball joint stud to 5.0 to 7.0 mkg (36 to 50 ft. lb.).

5. Install the tie rod in the steering knuckle. Torque the castellated nut to 3.0 mkg (22 ft. lb.). Advance it, if necessary, to uncover the cotter pin hole, then install a new cotter pin.

6. On cars with drum brakes, install the brake backing plate assembly. Torque the M 10 bolts to 5.0 mkg (36 ft. lb.).

7. On cars with disc brakes, install the splash shield. Torque the M 10 bolts to 1.0 mkg (7 ft. lb.).

8. Install the brake drum or disc. Then adjust the wheel bearings as described in **5.2 Checking and Adjusting Front Wheel Bearings**.

9. On cars with disc brakes, install the brake caliper. Use a new lockplate and torque the bolts to 4.0 mkg (29 ft. lb.).

10. Adjust the camber and toe as described in **3. Front Wheel Alignment**. Torque the self-locking nut on the upper ball joint stud to 5.0 to 7.0 mkg (36 to 50 ft. lb.).

12 Front Axle

6.2 Checking Steering Knuckle

The steering knuckle can be checked either on or off the car. The stub axle should be measured at the two points indicated in Fig. 6-7. The diameter of the inner tapered-roller bearing seat **A** should be 28.98 to 29.00 mm (1.141 to 1.142 in.). The diameter of the outer tapered-roller bearing seat **B** should be 17.45 to 17.46 mm (.687 to .688 in.). The dimensions should be checked with a micrometer or a vernier caliper.

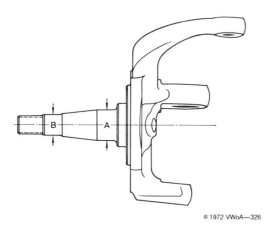

Fig. 6-7. Stub axle measuring points. Dimensions are given in preceding text.

You can check the steering knuckle's stub axle for distortion using a vernier caliper and a machinist's square as shown in Fig. 6-8. Make your measurements at a minimum of three points around the stub axle. The difference between any two measurements must not exceed 0.25 mm (.010 in.).

Fig. 6-8. Vernier caliper and machinist's square being used to check stub axle for distortion.

You can also use a vernier caliper in conjunction with a straightedge to check the steering arm for distortion. Use a C-clamp to hold the straightedge, as shown in Fig. 6-9. The distance from the straightedge to the outer edge of the tie rod hole should be 92.70 to 93.30 mm (3.649 to 3.673 in.).

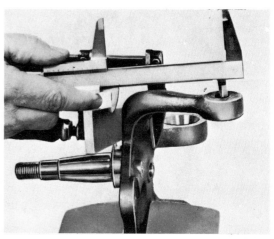

Fig. 6-9. Steering arm being checked for distortion.

The caliper mounting flange on steering knuckles used with disc brakes can be checked for distortion as shown in Fig. 6-10. The dial indicator should not show a variation of more than ± 0.05 mm (.002 in.) from one mounting hole to the other.

Fig. 6-10. Caliper mounting flange being checked. Special gauge holder rotates on stub axle.

CAUTION —
Bent steering knuckles must be replaced and not straightened. Bending them back to their original shape will seriously weaken them structurally.

7. Torsion Arms, Suspension Ball Joints, and Torsion Bars

The torsion bars that provide springing for the front wheels are housed inside the axle beam. Socket head setscrews hold the torsion arms on the outer ends of the bars. The suspension ball joints are a press fit in the torsion arms.

7.1 Checking Ball Joints

A special lever and a vernier caliper should be used to check the ball joint play. If the steering knuckle has been removed, check the ball joints as described in **7.4 Repairing and Replacing Ball Joints**.

To check:

1. Lift the car. Then turn the steering to one side.
2. Install the special lever as shown in Fig. 7-1.
3. Place a vernier caliper over the ball joint with one jaw on the steering knuckle and the other on the ball joint stud. Note the reading.
4. Pull down the lever, as indicated by the curved arrow in Fig. 7-1, so that the ball joint is expanded. Note the new reading on the vernier caliper.

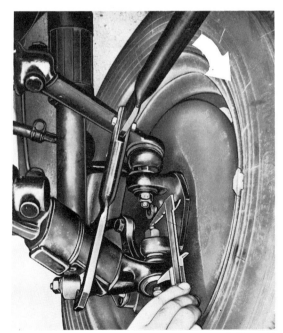

Fig. 7-1. Lever and vernier caliper in position. Lever should pry torsion arms away from each other.

5. Compare the two readings. The difference between them is the ball joint play.

NOTE
New ball joints must not exceed 0.50 mm (.020 in.) play; used lower ball joints must not exceed 1.00 mm (.040 in.) play; and used upper ball joints must not exceed 2.00 mm (.080 in.) play. Joints that exceed specifications must be replaced.

7.2 Removing and Installing Torsion Arm

Each torsion arm is held on its torsion bar by a socket head setscrew and locknut.

To remove:

1. Remove the steering knuckle complete with brake assembly.
2. If the lower torsion arm is to be removed, disconnect the stabilizer bar by driving the retainers off the two stabilizer bar rubber mounting clamps.

NOTE
It is necessary to bend down the locking tabs on the retainers before driving them off. Obtain new retainers for use during assembly.

3. Loosen the locknut on the socket head setscrew that holds the torsion arm on the torsion bar. Then remove the setscrew.
4. Remove the torsion arm from the end of the torsion bar.

Installation is the reverse of removal. Inspect the rubber seal for the torsion arm. Replace the seal if it is worn, cracked, or loose-fitting.

7.3 Checking Torsion Arms

The torsion arms can be checked for bending only after they have been removed from the car. A special measuring jig, VW 282d, is required for this job. Otherwise, accurate checks are impossible.

Suspension ball joints used prior to the 1973 model year have screw-in plastic plugs in their end plates. The latest type ball joints do not have these plugs. Instead, there is a small depression in the end plate with a 6-mm (.236-in.) diameter test surface. This modification has necessitated a minor change in the test jig. When checking torsion arms with the late-type ball joints, the test plates on the test jig must be changed. The ball joint without a plastic plug is shown in Fig. 7-2.

14 Front Axle

Fig. 7-2. Late-type ball joint with test surface (arrow) instead of plastic plug.

To check torsion arms:

1. Carefully clean the torsion arm and ball joint.

2. Inspect the bearing surfaces on the torsion arm. If they are worn, replace the torsion arm complete with ball joint, as well as the needle bearing and metal bushing in the axle tube.

3. On early-type ball joints, remove the plastic plug from the end plate and install the test point in its place as shown in Fig. 7-3.

Fig. 7-3. Test point being installed. Screw the point into the hole for the plastic plug.

4. With the late-type ball joints, install the proper measuring plates on the test jig, and screw the measuring pin into the appropriate hole in the measuring plate.

5. Using the correct bushings from the set supplied with VW 282d, install the torsion arm in the test jig.

6. See whether the test point contacts the test jig measuring plate (or the point on the measuring plate contacts the test surface on the ball joint) as shown in Fig. 7-4 or Fig. 7-5.

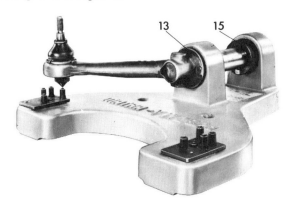

Fig. 7-4. Early-type ball joint. Test point should contact the boss on test plate.

NOTE

The numbers in Fig. 7-4 indicate the correct bushings for an upper torsion arm. Substitute bushing 14 for bushing 13 when checking a lower torsion arm. Upper torsion arms should contact the boss closest to the torsion arm pivot. Lower torsion arms should contact the boss furthest away from the torsion arm pivot, but on the side of the test plate closest to the torsion arm's axis.

Fig. 7-5. Late-type ball joint. Point on jig should contact test surface on ball joint.

CAUTION

Bent torsion arms must be replaced complete with ball joints and not straightened. Bending them back to their original shape will seriously weaken them structurally.

7. On early-type ball joints, remove the test point and screw in a new plastic plug.

Front Axle 15

7.4 Repairing and Replacing Ball Joints

To check the ball joint play after the torsion arm has been removed, place a vernier caliper over the ball joint, as shown in Fig. 7-6. Press the stud all the way in and note the measurement. Then pull the stud out and make another measurement. Compare the two readings. The difference between them is the ball joint play.

Fig. 7-6. Ball joint play being checked. Make measurements with stud all the way in and all the way out, as indicated by double-ended arrow.

NOTE ─
New ball joints must not exceed 0.50 mm (.020 in.) play. Used lower ball joints must not exceed 1.00 mm (.040 in.) play. Used upper ball joints must not exceed 2.00 mm (.080 in.) play. Joints that exceed specifications must be replaced.

Damaged ball joint seals can be replaced. However, the late-type ball joints without plastic plugs must be replaced if dirt has entered them. The early-type ball joints should be cleaned thoroughly in solvent after the damaged seal has been removed. Then remove the plastic plug from the tapped hole and install a grease fitting. Force grease through the joint until all traces of dirt are expelled.

Install the new seal on the ball joint. Then place a conical sleeve over the seal as shown in Fig. 7-7. Slide the new retaining ring over the conical sleeve and off its large end onto the ball joint seal. The conical sleeve will prevent the retaining ring from accidentally puncturing the new seal.

Install a grease fitting as described earlier, then pump multipurpose grease into the ball joint until the seal begins to expand. Being careful to avoid damaging the seal, lift its lower edge away from the ball joint stud slightly.

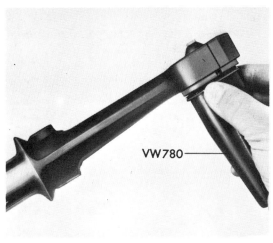

Fig. 7-7. Retaining ring being installed with the help of a special conical sleeve.

Move the ball joint stud through its operating angle at least twice to permit any excess grease to escape. Using a smaller-diameter conical sleeve, slip the plastic retaining ring over the seal to hold it on the ball joint stud.

Depending on the tolerance range of the torsion arm, either a standard size ball joint or an oversize ball joint is installed. Oversize ball joints are 0.40 mm (.016 in.) larger in diameter and are identified by two notches as indicated in Fig. 7-8. When replacing ball joints it is essential to check for these marks on the old ball joints, since the torsion arm itself has no marks to show which ball joint must be used.

Fig. 7-8. Standard (left) and oversize (right) ball joints. Two notches (as at arrow) 45° from installation-position grooves identify oversize ball joints.

The ball joints are a press fit in the torsion arms. They must fit properly and, once pressed out, must never be reinstalled in a torsion arm.

16 Front Axle

To replace ball joint:

1. Press the old ball joint out of the torsion arm as shown in Fig. 7-9.

Fig. 7-9. Hydraulic repair press being used to press ball joint out of torsion arm.

2. If necessary, press the eccentric camber adjusting bushing off the ball joint for an upper torsion arm as shown in Fig. 7-10.

Fig. 7-10. Ball joint being pressed loose from eccentric camber adjusting bushing. Note the nut installed on the ball joint stud. This nut will keep the ball joint from flying out suddenly.

3. Align the installation-position groove with the notch in the torsion arm as indicated in Fig. 7-11.

   WARNING —
   Never reinstall used ball joints. They will not fit tightly and could come out of the torsion arm while the car is being driven.

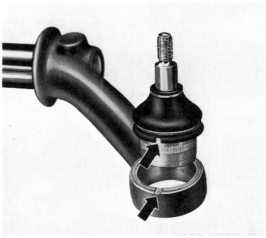

Fig. 7-11. Groove in ball joint (top arrow). When you install the ball joint, align the groove with the notch in the torsion arm (bottom arrow).

4. Being careful to keep the groove and the notch aligned, press the new ball joint into the torsion arm as shown in Fig. 7-12.

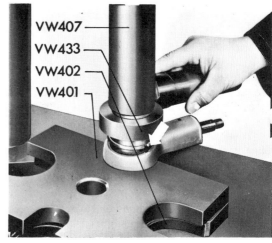

Fig. 7-12. Lower ball joint being pressed in. Arrow indicates the aligning notch and groove.

5. Check to see that the seal has not been damaged or displaced during installation and that the ball joint stud moves freely throughout its operating range.

FRONT AXLE   17

7.5 Removing and Installing Torsion Bars

The torsion bars are anchored at their centers in bushings inside the axle tubes and secured with socket head setscrews.

To remove:

1. Remove both steering knuckles, as described in **6.1 Removing and Installing Steering Knuckle**.

   NOTE
   It is not necessary to dismount the brake assembly from the steering knuckle. However, the brake hose will have to be disconnected. Bleed the brakes as described in **BRAKES AND WHEELS** after the hose is reinstalled.

2. Remove the torsion arm from one end of the torsion bar, as described in **7.2 Removing and Installing Torsion Arm**.

3. Loosen the locknut on the socket head setscrew. Then remove the setscrew as shown in Fig. 7-13.

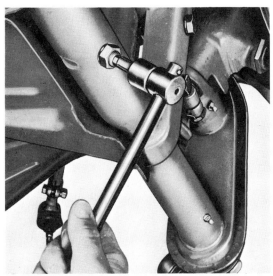

Fig. 7-13. Torsion bar setscrew being removed.

4. Gripping the torsion arm that is still installed on the torsion bar, pull the torsion bar out of the axle beam tube.

Installation is the reverse of removal. Clean the torsion bar, inspect it for cracks and breakage, and then coat it liberally with multipurpose grease before you install it. Align the countersunk mark in the center of the bar with the setscrew hole. Torque first the setscrew, and then the locknut to 4.0 to 5.0 mkg (29 to 36 ft. lb.). Then install the torsion arm and both steering knuckles.

7.6 Replacing Needle Bearings and Metal Bushings

The needle bearings are more likely to require replacement than the metal bushings, which are subject to very little wear. However, if wear is noted on the torsion arm bearing surface, replace the metal bushing as well as the worn torsion arm.

To remove:

1. Remove both steering knuckles complete with their brake assemblies. Then take off the torsion arm(s) on one side of the car and remove the torsion bar(s).

2. Using an expansion tool or a toggle washer behind the needle bearing, pull out the needle bearing with a slide hammer as shown in Fig. 7-14.

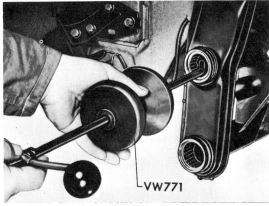

Fig. 7-14. Needle bearing being pulled out of axle tube with slide hammer and toggle washer.

3. Measure the metal bushing for wear as shown in Fig. 7-15.

Fig. 7-15. Internal measuring gauge being used to check inside diameter of metal bushing.

18 Front Axle

4. If the metal bushing has worn to a diameter greater than 37.38 mm (1.472 in.), pull it out with a slide hammer and the tools shown in Fig. 7-16.

CAUTION

Do not pull out the plastic sleeve, as it is not subject to wear. Replacement sleeves are not supplied, so if you remove or damage them the entire axle beam must be replaced.

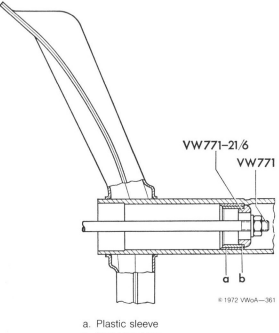

a. Plastic sleeve
b. Metal bushing

Fig. 7-16. Toggle washer in position to pull out metal bushing without removing plastic sleeve.

5. Clean the needle bearing seats. Then, using the internal measuring gauge illustrated in Fig. 7-15, check the diameter of the needle bearing seats.

NOTE

This measurement is essential since either standard or oversize needle bearings might be used.

Install standard bearings in upper axle tubes that measure 45.97 to 45.99 mm (1.8098 to 1.8106 in.); install oversize bearings in upper tubes that measure 46.17 to 46.19 mm (1.8177 to 1.8185 in.). Install standard bearings in lower axle tubes that measure 49.97 to 49.99 mm (1.9673 to 1.9681 in.); install oversize bearings in lower tubes that measure 50.17 to 50.19 mm (1.9752 to 1.9760 in.). Standard upper-tube bearings are 46.00 mm (1.8110 in.). in diameter; standard lower-tube bearings are 50.00 mm (1.9685 in.). Oversize upper-tube bearings are 46.20 mm (1.8189 in.) in diameter; oversize lower-tube bearings are 50.20 mm (1.9764 in.).

Clean the axle tubes, particularly at the needle bearing and metal bushing seats. Then drive in the needle bearing and the metal bushing until the shoulder on the special drift contacts the axle tube as shown in Fig. 7-17.

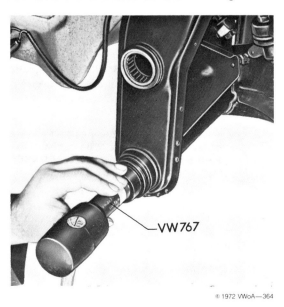

Fig. 7-17. Drift used to drive in lower needle bearing and metal bushing. Use VW 768 for upper tube.

If the special drifts are unavailable, position the needle bearing and the metal bushing as shown in Fig. 7-18. Dimension **a** is 122 + 2 mm (4 $\frac{13}{16}$ + $\frac{5}{64}$ in.) for the upper metal bushing and 132 + 2 mm (5 $\frac{9}{64}$ + $\frac{5}{64}$ in.) for the lower metal bushing. Dimension **b** is 3.5 + 0.2 mm ($\frac{9}{64}$ + $\frac{1}{128}$ in.) for the upper needle bearing and 5.0 + 0.2 mm ($\frac{13}{64}$ + $\frac{1}{128}$ in.) for the lower needle bearing.

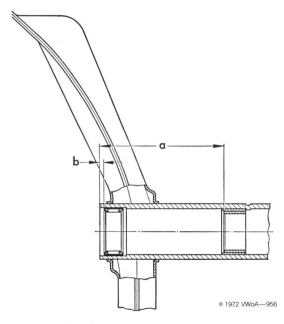

Fig. 7-18. Needle bearing and metal bushing positions.

Front Axle

8. Removing and Installing Axle Beam

Although most front axle repairs can be carried out while the axle beam is mounted on the car, it may be necessary to remove the axle beam when correcting body or frame damage, or to facilitate a complete front axle rebuild. On 1974 and later cars with energy absorbing bumpers, you must saw off the bumper brace ahead of the front axle before you can remove the axle from the car. Weld the brace back together after you have installed the axle. If you suspect that the axle beam has been bent by accident damage, you can check the axle tubes with a straightedge. Bent axle beams must be replaced and not straightened. Bending them back to their original shape will seriously weaken them structurally.

To remove:

1. Raise the car and remove the front road wheels.
2. Working through the left front wheel opening, pull the fuel hose off the tube in the frame head and plug the hose.

    WARNING —
    Disconnect the battery ground strap. Do not smoke or work near heaters or other fire hazards. Have a fire extinguisher handy.

3. Remove the fuel tank as described in **FUEL SYSTEM**.
4. Loosen the clamp on the steering column, and disconnect the horn wire at the flexible coupling.
5. Remove the bolt indicated in Fig. 8-1, then detach the steering damper from the bracket on the upper axle beam tube.

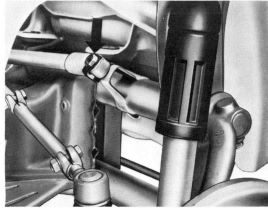

Fig. 8-1. Bolt (arrow) that holds steering damper on its bracket.

6. Remove the E-clip from the speedometer cable where it passes through the dust cap on the left front wheel hub. Then, working behind the brake assembly, pull the speedometer cable out of the steering knuckle.
7. Unscrew the brake hoses from the front brake assemblies and plug them with the dust covers from the brake bleeder valves.
8. Remove the cotter pins from both tie rod end studs on the long (right-hand) tie rod. Then remove the castellated nuts.
9. Using a tie rod removal tool, press the tie rod ends out of the drop arm and the right-hand steering knuckle. Then remove the long tie rod together with the steering damper.

    CAUTION —
    Do not hammer out the tie rod ends. Doing so will ruin the threads and make reinstallation impossible.

10. Working in the front luggage compartment through the fuel tank opening, remove the two body mounting bolts as shown in Fig. 8-2.

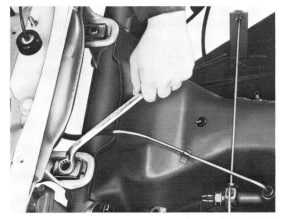

Fig. 8-2. Body mounting bolt being removed.

11. Position a floor jack with the VW 610 front axle supporting adapter under the front axle.
12. Raise the jack until the adapter is in firm contact with the axle beam.

    WARNING —
    If you do not have a suitable jack and adapter, have at least two helpers support the front axle while you are unbolting it. Trying to handle this job by yourself could lead to serious injury owing to the weight of the axle.

20 Front Axle

13. Remove the four axle beam mounting bolts as shown in Fig. 8-3.

Fig. 8-3. Axle beam mounting bolts being removed.

14. While carefully guiding the steering coupling out of the steering column, lower the axle beam with the floor jack.

To install:

1. Place the axle in the floor jack front axle adapter VW 610. Adjust the axle to the correct installation angle with a piece of chain as, shown in Fig. 8-4.

Fig. 8-4. Chain loop being used to adjust the angle of the axle on VW 610 adapter.

2. If you removed them earlier, place the two rubber pads over the threaded bushings for the body mounting bolts.

3. Using new lock washers under the four mounting bolts, install the axle beam on the frame head. Torque the mounting bolts to 5.0 mkg (36 ft. lb.).

4. Make certain the rubber pads and washers are positioned as shown in Fig. 8-5. Then torque the body mounting bolts to 1.5 to 2.0 mkg (11 to 14 ft. lb.).

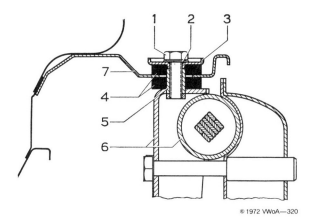

1. Body mounting bolt
2. Lock washer
3. Spacer
4. Rubber pads
5. Threaded bushing
6. Front axle beam
7. Body

Fig. 8-5. Body mounting point on front axle beam.

5. Center the steering gear as described in **9.4 Removing and Installing Steering Column**. Position the steering wheel with its spokes horizontal, then torque the steering column clamp bolt to 2.0 to 2.5 mkg (14 to 18 ft. lb.).

6. Install the long tie rod and the steering damper piston rod end. Torque the castellated nuts to 3.0 mkg (22 ft. lb.). Then advance them to uncover the cotter pin hole, if necessary, and install a new cotter pin.

7. Using a new lockplate, install the steering damper on its bracket on the upper axle beam tube. Torque the mounting bolt to 4.0 to 4.5 mkg (29 to 32 ft. lb.).

NOTE
Install the lockplate so that the open end of the "U" faces the front of the car and the narrow angled part contacts the bracket on the axle.

8. Install the brake hoses so that they hang downward and are not twisted. Torque them to 1.5 to 2.0 mkg (11 to 14 ft. lb.), then make sure that they do not contact any part of the car while the steering is being turned from one lock to the other.

9. Bleed the brakes as described in **BRAKES AND WHEELS**. Then adjust the front wheel alignment as described in **3. Front Wheel Alignment**.

FRONT AXLE 21

9. STEERING

The worm and roller steering gearbox used on Beetles and Karmann Ghias is totally different from the worm and roller steering used on 1971 and later Super Beetles and Convertibles. The steering system normally requires no maintenance. However, the steering gearbox may be damaged if the car is operated for a long time with improperly adjusted steering. The steering linkage parts seldom require replacement unless damaged by a collision. The components of the steering gearbox are shown in Fig. 9-1.

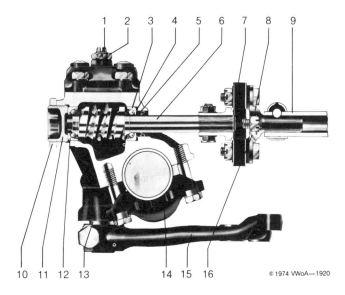

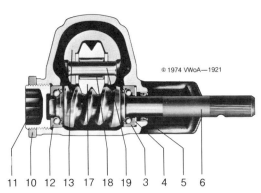

1. Roller shaft adjusting screw
2. Roller shaft adjusting screw locknut
3. Upper worm bearing
4. Worm adjusting shim
5. Worm spindle seal
6. Worm spindle
7. Coupling disk
8. Coupling disk flange
9. Steering column
10. Worm adjuster locknut
11. Worm adjuster
12. Lower worm bearing
13. Roller shaft
14. Mounting clamp
15. Drop arm
16. Ground connection terminal
17. Steering roller
18. Roller needle bearings
19. Roller support pin

Fig. 9-1. Cutaway view of worm and roller steering gearbox. Note position of two adjustments.

9.1 Adjusting Steering

You should not raise the car when the steering adjustment is being checked. With the front wheels in their straight-ahead position, grip the steering wheel with your fingers as shown in Fig. 9-2. Turn the steering wheel lightly in both directions. The freeplay should not exceed 25 mm (1 in.) measured at the wheel rim.

Fig. 9-2. Checking steering freeplay. Freeplay is measured at steering wheel rim as indicated by **a**.

If the steering freeplay is excessive, check to see that the looseness is not caused by worn tie rod ends or a loose drop arm. Make sure that the steering gearbox is mounted firmly and that its cover bolts are torqued to 2.0 to 2.5 mkg (14 to 18 ft. lb.). If no faults are found, correct excessive centerpoint freeplay by making adjustments at the steering gearbox. Check the three factors given in Fig. 9-3 in the order in which they are listed.

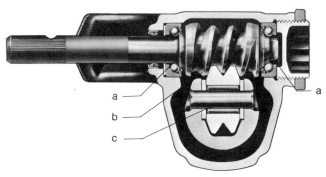

Fig. 9-3. Possible sources of excessive steering gear play: **a**, axial play of worm spindle; **b**, play between roller and worm; **c**, axial play of roller on its support pin.

22 Front Axle

To check and adjust worm:

1. Check the worm spindle for axial play as indicated in Fig. 9-4.

   NOTE —
   Although some of the illustrations show the steering gearbox installed on a Type 3 VW front axle, the gearbox is the same as that in the Beetle and the Karmann Ghia.

Fig. 9-4. Worm play being checked. Grasp the coupling and turn the worm shaft as indicated by the left double arrow. Look for play in the shaft (right double arrow). If there is play, adjust the worm.

2. Turn the steering fully left to right. Loosen the worm adjuster locknut as shown in Fig. 9-5.

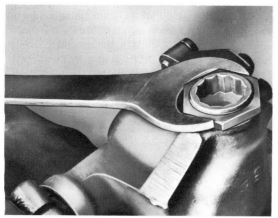

Fig. 9-5. Worm adjuster locknut being loosened.

3. Turn the worm spindle, slowly tightening the adjuster as shown in Fig. 9-6. Continue tightening up to—but not beyond—the point at which play disappears from the worm shaft.

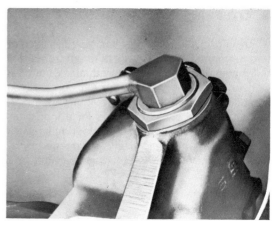

Fig. 9-6. Special wrench inserted into worm adjuster so that adjuster can be turned.

4. Hold the adjuster in its optimum position while tightening the locknut as shown in Fig. 9-7. Then torque the locknut to 5.0 to 6.0 mkg (36 to 43 ft. lb.).

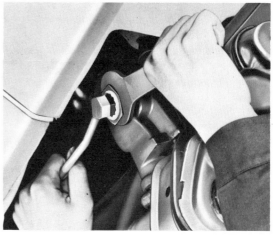

Fig. 9-7. Locknut being tightened after worm adjustment.

5. Turn the worm spindle from lock to lock. There should be no tight spots.

   CAUTION —
   Tight spots indicate that the adjuster has been advanced too far. Bearing damage will result if the adjustment is not loosened.

If you cannot eliminate play by adjusting the worm, adjust the roller-to-worm setting. With the steering in its centered position, there should be no play between the roller and the worm. You can make adjustments with the car lifted, but they must be checked when the vehicle is on the ground.

FRONT AXLE 23

To adjust roller:

1. Turn the steering wheel 90° to either the left or the right of its centered position. Then raise the front luggage compartment hood and remove the spare wheel from the car.

2. Remove the Phillips head screw and take off the access panel for the steering gearbox.

3. Loosen the roller shaft adjusting screw locknut. Then turn the roller shaft adjusting screw (Fig. 9-8) out about one full turn.

Fig. 9-8. Roller shaft adjusting screw (arrow).

4. Turn in the adjusting screw until you feel the roller just contact the worm. Lightly tighten the locknut as shown in Fig. 9-9, then torque it to 2.5 mkg (18 ft. lb.).

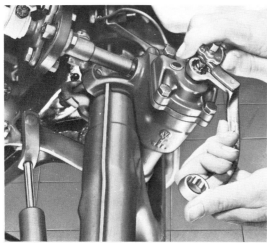

Fig. 9-9. Adjusting screw being held in its optimum position while locknut is lightly tightened.

5. With the car standing on its wheels, turn the steering wheel 90° to each side and check the adjustment. The freeplay at this point must not exceed 25 mm (1 in.) measured at the rim of the steering wheel.

NOTE
If there is more play on one side, repeat the roller shaft adjustment at 90° to that side.

6. Check the toe-in and, if necessary, adjust it.

7. Road-test the car. Take a corner at 10 to 12 mph (15 to 20 km). If the steering wheel does not return to about 45° from its centered position, the roller is too tight. The adjustment must then be repeated to prevent damage to the worm and roller.

If excessive freeplay still exists after carrying out the adjustments described so far, remove and disassemble the steering gearbox. (See **9.6 Disassembling, Assembling, and Adjusting Steering Gearbox**.) Once apart, check the axial play of the roller on its support pin as shown in Fig. 9-10.

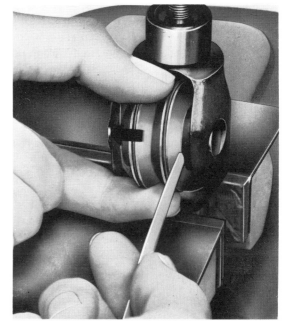

Fig. 9-10. Axial play of roller being checked. Measure the clearance between the roller and washer (arrow) with a 0.05-mm (.002-in.) feeler gauge. If you can push the feeler gauge between the roller and washer, replace the complete roller shaft assembly.

After you have assembled and reinstalled the steering gearbox, check the toe of the front wheels and, if necessary, adjust the toe, as described in **3.2 Checking and Adjusting Toe**.

9.2 Removing and Checking Steering Damper

The steering damper is a hydraulic cylinder mounted between an eye in the long (right-hand) tie rod and a bracket on the axle beam. To remove the steering damper, bend up the lockplate and unscrew the M 10 bolt that holds the damper on the axle-mounted bracket. Then remove the self-locking nut from the damper's piston rod end.

Hand-check the steering damper by extending and compressing it while holding it in its installed position. It must operate smoothly and with uniform resistance throughout its entire stroke. If necessary, compare the used unit with a new steering damper. Minor fluid leakage does not make replacement necessary as long as efficiency is not impaired.

To prevent steering trouble due to premature failure of the steering damper, make certain that the damper you install is the correct one for the car. Check the rubber bushings and sleeves in the removed damper before reinstallation. If they are worn or damaged, replace them.

Install a new lockplate so that the open end of the "U" faces the front of the car and the narrow angled part contacts the bracket on the axle. When installing the piston rod in the long tie rod, use a new self-locking nut.

9.3 Removing and Checking Tie Rods

To remove the tie rods, take out the cotter pins in the tie rod end studs and remove the castellated nuts. Use a tool such as the one shown in Fig. 9-11 to press the tie rod ends out of the drop arm and the steering arms.

CAUTION
Do not hammer out the tie rod ends. Doing so will ruin the threads and make reinstallation impossible.

Fig. 9-11. Tool used to press out tie rod ends.

Installation is the reverse of removal. Carefully inspect the tie rods for cracks. Distortion can be detected by rolling the tie rods over a flat surface. Check the tie rod ends for play and replace any that are worn.

NOTE
Do not confuse the tie rod ends for the strut front suspension with those for the front axle. The tie rod ends for the strut front suspension are identified by either an indentation or a protrusion at the point indicated in Fig. 9-12.

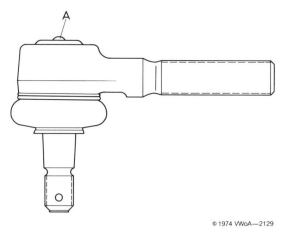

Fig. 9-12. Projection (**A**) that identifies tie rod ends for use with strut front suspension.

NOTE
If the tie rod end boots are torn or cracked, they can be replaced. However, the entire tie rod end should be replaced if dirt has entered the ball socket.

Install a new lockplate between the tapered ring and the locknut when you install the long tie rod on the inner tie rod end. Replace worn or damaged self-locking nuts on the bolts for the three tie rod end clamps.

After installing the tie rod ends in the drop arm and the steering arms, torque the castellated nuts to 3.0 mkg (22 ft. lb.). Then advance them, if necessary, to uncover the cotter pin holes. Install new cotter pins in all four tie rod end studs.

Check front wheel toe and adjust it to the specifications given in **3.2 Checking and Adjusting Toe**. Then torque the locknut to 2.5 mkg (18 ft. lb.) and bend the lockplate tabs in opposite directions. (That is, one tab should lie along a flat face on the tapered ring and the other against one of the locknut flats.) Twist the tie rod as far as it will go so that the tie rod ends are parallel. Then torque the clamp bolts to 1.5 mkg (11 ft. lb.).

FRONT AXLE 25

9.4 Removing and Installing Steering Column

Fig. 9-13 shows the energy-absorbing steering column installed. The energy-absorbing section is at arrow **A**. Plastic rivets at arrow **B** are sheared when forces cause the steering column to collapse.

The energy-absorbing steering wheel can be service-installed on earlier cars if the horn control is modified as well. The energy-absorbing section of the steering wheel and of the steering column should be inspected carefully for cracks and distortion any time the car has been involved in an accident. Even if only slightly damaged, the complete column or steering wheel should be replaced.

There have also been several changes to the steering column switch and horn control during the model years covered by this Manual. These modifications and detailed instructions for adjusting the turn signal switch and testing the horn circuit are described in **ELECTRICAL SYSTEM**

To remove steering column and tube:

1. Remove the fuel tank as described in **FUEL SYSTEM**.

   WARNING —
   Disconnect the battery ground strap. Do not smoke or work near heaters or other fire hazards. Have a fire extinguisher handy.

2. Remove the nut indicated at arrow **A** in Fig. 9-15 from the bolt in the steering column clamp. Then bend up the tab on the support ring (arrow **B**) and remove the support ring.

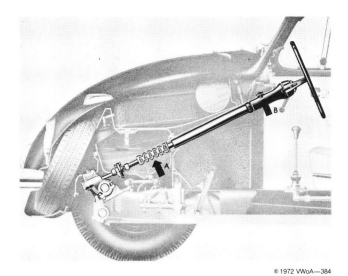

Fig. 9-13. Steering column installed in car.

Beginning with 1972 models, the plastic rivets were eliminated in favor of the energy-absorbing steering wheel shown in Fig. 9-14.

1. Padded cap
2. Steering wheel
3. Trim
4. Contact ring

Fig. 9-14. Energy-absorbing steering wheel.

Fig. 9-15. Nut (arrow **A**) and tab (arrow **B**).

26 Front Axle

3. Remove the steering wheel. Then remove the circlip from the groove below the steering wheel mounting splines on the steering column.

   NOTE ——
   To remove the steering wheel, carefully pry out the cap for the horn ring or, on 1972 and later cars, pry off the padded cap. Then unscrew the large nut on the end of the steering column. Pull the wheel upward and off the splines.

4. Turn the ignition key to the "on" position.
5. On cars with a column-mounted windshield wiper/washer control, release the air pressure from the windshield washer reservoir.
6. On 1971 and later cars, disconnect all wiring and hoses from the steering column switch.
7. On 1971 and later cars, remove the two bolts indicated in Fig. 9-16.

Fig. 9-16. Column tube mounting bolts.

8. On 1970 models, remove the four Phillips head screws in the bottom of the column switch housing. Then pull the column switch assembly upward and off the steering column. Suspend the assembly so that no strain is placed on the wires.
9. Pull the steering column out into the passenger compartment.
10. On 1971 and later cars, remove the bolt that holds the column switch assembly to the column tube and slide the switch assembly off upward.
11. Pull the column tube off the steering column.

To install:

1. On 1971 and later cars, install the column switch assembly on the column tube and align it.

2. Torque the socket head screw at arrow **A** in Fig. 9-17 to 0.5 to 1.0 mkg (3.6 to 7 ft. lb.) and install the sealing ring at arrow **B**.

Fig. 9-17. Column switch and sealing ring installation on column tube.

3. Install the contact ring on the steering column. Then insert the steering column into the column tube from below and secure it with the circlip shown in Fig. 9-18.

1. Contact ring   2. Circlip

Fig. 9-18. Relative positions of circlip and contact ring on steering column.

4. Install the steering column assembly in the car. Do not forget the steering column clamp.
5. With the steering gear in its straight-ahead position, slide the steering column over the end on the flexible coupling.
6. Align the steering column and column clamp with the clamp bolt recess in the side of the flexible coupling end. Then install the clamp bolt with a new self-locking nut and torque it to 2.5 mkg (18 ft. lb.).
7. Install the support ring above the energy-absorbing section and secure it by bending down the tab.
8. On 1970 cars, install the column switch assembly.

Front Axle 27

9. Loosely install the column tube mounting bolts.

10. With the steering gear centered and the turn signal switch in its central position, install the steering wheel so that its spokes are horizontal. Then install the M 18 nut and torque it to 5.0 mkg (36 ft. lb.).

11. Adjust the gap (Fig. 9-19) between the steering wheel hub or hub trim and the column switch to 2 to 4 mm (1/16 to 5/32 in.) by moving the column tube (or the column switch on 1970 cars) up or down. Then torque the column tube mounting bolts to 1.5 mkg (11 ft. lb.). On 1970 cars tighten the column switch screws.

12. Reconnect all wires and hoses to the column switch.

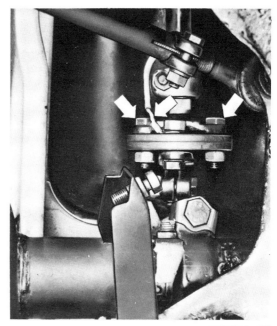

Fig. 9-20. Bolts and wire that must be removed from flexible coupling.

Fig. 9-19. Gap (**a**) between wheel and switch.

9.5 Removing and Installing Steering Gearbox

Unless the drop arm will later be needed to adjust the steering gearbox, as described in **9.6 Disassembling, Assembling, and Adjusting Steering Gearbox**, it is easier to remove the drop arm from the gearbox than to remove the tie rods from the drop arm.

To remove steering gearbox:

1. Remove the fuel tank as described in **FUEL SYSTEM**.

   WARNING —
   Disconnect the battery ground strap. Do not smoke or work near heaters or other fire hazards. Have a fire extinguisher handy.

2. Remove the bolts and the horn ground wire indicated by the arrows in Fig. 9-20.

3. Bend the lockplate away from the clamp bolt head on the drop arm, then remove the bolt (Fig. 9-21).

Fig. 9-21. Clamp bolt in drop arm.

4. Turn the front wheels to an angle that will make it easy to remove the drop arm, then pull the drop arm off the steering gearbox roller shaft.

5. Bend the lockplate tabs away from the bolt heads for the steering gearbox mounting clamp.

6. Remove the mounting clamp bolts and the mounting clamp. Then remove the steering gearbox.

28 Front Axle

Installation is the reverse of removal. Use new lockplates on the mounting clamp bolts. On Beetles, make certain that the cutout in the side of the clamp marked **13** engages the lug on the axle tube; on Karmann Ghias, make certain that the side of the clamp marked **14** engages the lug on the axle tube. Also, be sure that both the steering gear and the steering wheel are both centered while you are bolting together the flexible coupling. Torque the gearbox mounting bolts to 2.5 to 3.0 mkg (18 to 22 ft. lb.). Torque the drop arm clamp bolt to 7.0 mkg (50 ft. lb.). Install all remaining parts. Then check the toe and, if necessary, adjust it.

9.6 Disassembling, Assembling, and Adjusting Steering Gearbox

The steering gearbox can be disassembled in order to replace worn parts or damaged bearings. A special fixture is needed to properly reassemble the unit.

> **CAUTION** ———
> If you lack the skills or the special fixture for repairing the steering gearbox, we suggest you leave such repairs to an Authorized VW Dealer or other qualified shop.

To disassemble:

1. Remove the locknut from the roller shaft adjusting screw. Then remove the four bolts that secure the gearbox housing cover.

2. Remove the housing cover by turning the roller shaft adjusting screw counterclockwise until the cover is free.

3. Remove the lubricant from the gearbox housing.

4. Turn the steering worm to its centered position. Using a drift, push out the roller shaft out and upward.

   > **NOTE** ———
   > A special protective sleeve (VW 649) can be slipped over the splined end of the shaft and into the seal to prevent the splines from damaging the rubber lip. If this sleeve is not available, replace the seal during assembly.

5. Remove the circlip that holds the adjusting screw in the roller shaft. Then remove the adjusting screw and shim.

   > **NOTE** ———
   > No further disassembly of the roller shaft is possible. If the shaft is worn or if the bearings are faulty, the entire roller shaft must be replaced as a unit.

6. Loosen the worm adjuster locknut. Then remove the worm adjuster as shown in Fig. 9-22.

Fig. 9-22. Worm adjuster being removed.

7. Pry off the ring (marking ring) that fits around the worm shaft where it projects from the gearbox.

8. Remove the worm and the lower ball bearing from the gearbox by tapping the worm spindle lightly with a rubber hammer.

9. Drive out the oil seal, shim, and upper bearing as shown in Fig. 9-23.

Fig. 9-23. Oil seal, shim, and upper bearing being driven out toward the inside of the gearbox.

10. Check all parts for wear and damage. Replace as necessary.

    > **NOTE** ———
    > Obtain a new worm spindle seal and marking ring for use during reassembly. If the roller shaft seal has been damaged by removal of the roller shaft, obtain a replacement for it also.

To assemble:

1. Clean all parts thoroughly and replace any that are worn or damaged.

2. Insert the worm spindle, as shown in Fig. 9-24, along with the upper bearing and the shim.

FRONT AXLE

NOTE ━
The oil seal must not be installed until after you have adjusted the worm.

Fig. 9-24. Worm spindle being inserted into gearbox.

3. Install the lower bearing.
4. Install the worm adjuster after applying sealing compound to its threads. Then tighten it lightly in order to press the bearings into their seats.
5. Loosen the worm adjuster, then tighten it again until the worm feels slightly rough as you turn it.
6. Install a torque gauge on the worm spindle. Then check the turning torque as shown in Fig. 9-25.

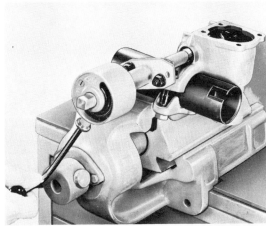

Fig. 9-25. Worm spindle turning torque being measured.

7. If the torque gauge does not indicate 1.5 to 2.5 cmkg (1.3 to 2.2 in. lb.), turn the worm adjuster until you obtain a reading in this range. Then torque the locknut to 5.0 to 6.0 mkg (36 to 43 ft. lb.).

8. Check the axial play of the roller as shown in Fig. 9-26. If it exceeds 0.04 mm (.0016 in.) replace the roller shaft as a unit.

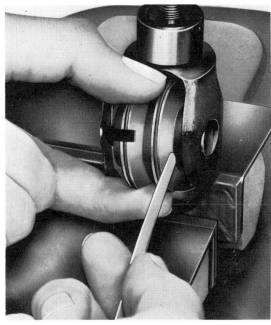

Fig. 9-26. Roller axial play being checked. The play should be such that a 0.05-mm (.002-in.) feeler gauge will not enter between the roller and the spacer washer.

9. Check the adjusting screw to see that the threads are not damaged. If undamaged, install the adjusting screw as illustrated in Fig. 9-27.

NOTE ━
Adjusting washers are available in thicknesses from 2.00 to 2.50 mm (.078 to .098 in.) in increments of 0.05 mm (.002 in.). Select a washer that will barely permit the adjusting screw to be finger-turned.

Fig. 9-27. Installation of adjusting screw.

30 Front Axle

10. Attach the roller shaft to the housing cover by screwing the adjusting screw through the cover as far as it will go.

11. Install the roller shaft as shown in Fig. 9-28. Mesh the roller with the worm at a right angle.

Fig. 9-28. Roller shaft being installed in gearbox. Note that the oil seal protection sleeve is in place.

NOTE —
Do not put lubricant in the gearbox until you have adjusted the steering gear and have installed both oil seals.

12. Torque the cover bolts to 2.0 to 2.5 mkg (14 to 18 ft. lb.) as illustrated in Fig. 9-29.

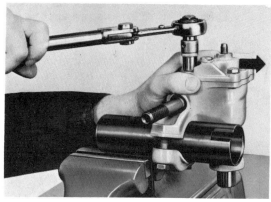

Fig. 9-29. Cover bolts being torqued. Press the cover (as indicated by arrow) to keep it from shifting, which could cause premature steering wear.

13. Using a new lockplate, install the drop arm on the roller shaft.

NOTE —
The lower face of the drop arm should be flush with the top of the chamfer on the shaft end as indicated in Fig. 9-30.

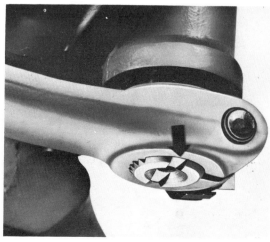

Fig. 9-30. Lower face of drop arm installed in line with top of chamfer (arrow) on roller shaft.

14. Torque the drop arm clamp bolt to 7.0 mkg (50 ft. lb.), then bend over the lockplate tab.

15. Install measuring fixture VW 280 as shown in Fig. 9-31.

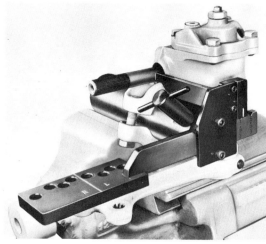

Fig. 9-31. Measuring fixture installed on gearbox. There are a number of holes in the fixture that match the drop arms of other VW steering gearboxes. Make certain you use the correct hole.

16. Insert a pilot pin through the hole in the fixture and into the drop arm as shown in Fig. 9-32.

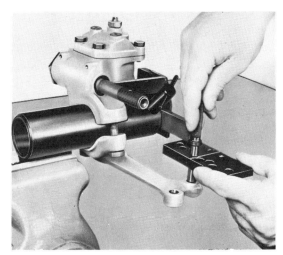

Fig. 9-32. Pilot pin being installed through hole in measuring fixture plate, into hole in drop arm.

17. Install VW 279 as shown in Fig. 9-33 and adjust its pointer to zero. Then pull the pilot pin out of the drop arm and the test fixture.

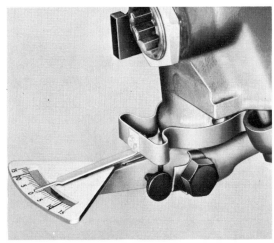

Fig. 9-33. Drop arm movement gauge VW 279 installed on drop arm.

18. Turn the drop arm 5° (11° if you are installing a new roller shaft or worm) to the left or the right.

19. Turn in the roller shaft adjusting screw until you feel no play when the drop arm is moved and the worm spindle is held stationary (Fig. 9-34).

Front Axle 31

Fig. 9-34. Roller adjustment. Tighten the adjuster (arrow) until there is no play in the drop arm (as indicated by the double arrow) when the worm spindle is held.

20. Torque the locknut to 2.5 mkg (18 ft. lb.). Check the no-play range with the drop arm moved to the opposite side. If the range is 5° (11° if you are installing a new roller shaft or worm), drive in the new oil seal with a 21.5-mm ($^{27}/_{32}$-in.) tube.

NOTE —

If the no-play range extends beyond 5° (or 11°) from center, a thicker shim is needed between the upper worm spindle bearing and the gearbox housing; if the no-play range is less than 5° (or 11°) from center, a thinner shim is needed. Repeat the adjustment with different shims until you get the correct range. Shims are available in thicknesses from 0.20 mm (.008 in.) to 0.50 mm (.020 in.) in increments of 0.05 mm (.002 in.).

21. Remove the cover. Fill the gearbox with about 160 cc (5.4 oz.) of liquid transmission grease.

22. Fit new plastic plugs in the cover. Then, using a gasket, reinstall the cover as described earlier.

23. Again adjust the roller shaft as described in steps 19 and 20. Mark the center position of the steering gear with a new marking ring and secure the ring to the shaft with paint. The notch in the ring should point upward.

10. Front Axle Technical Data

The technical data on the following pages contains all the dimensions and adjustment specifications needed to service, repair, or rebuild the front axle and steering. Further data on adjusting the alignment of the rear suspension can be found in **TRANSMISSION AND REAR AXLE**.

32 Front Axle

I. Tolerances, Wear Limits, and Settings

Designation	New Installation mm (in.)	Wear Limit mm (in.)
A. Front torsion bars .... number of leaves	10	—
length	954 (37.559)	—
Setting .... top angle	44° ± 30′	—
bottom angle	35° 30′ ± 30′	—
B. Axle beam and torsion arms		
1. Torsion arm bearings in axle beam		
a. Standard seat for upper needle bearing...... inside diameter	45.97–45.99 (1.8098–1.8106)	—
b. Standard upper needle bearing .... outside diameter	46.00 (1.8110)	—
c. Oversize seat for upper needle bearing ...... inside diameter	46.17–46.19 (1.8177–1.8185)	—
d. Oversize upper needle bearing............ outside diameter	46.20 (1.8189)	—
e. Standard seat for lower needle bearing...... inside diameter	49.97–49.99 (1.9673–1.9681)	—
f. Standard lower needle bearing............ outside diameter	50.00 (1.9685)	—
g. Oversize seat for lower needle bearing ...... inside diameter	50.17–50.19 (1.9752–1.9760)	—
h. Oversize lower needle bearing........... outside diameter	50.20 (1.9764)	—
2. Bushing for		
a. Upper torsion arm	37.20–37.25 (1.4646–1.465)	37.38 (1.4716)
b. Lower torsion arm	37.20–37.25 (1.4646–1.465)	37.38 (1.4716)
C. Ball joints, steering knuckles		
1. Upper ball joints .... play	0.50 (.020)	2.00 (.080) using VW 281a
2. Lower ball joints .... play	max. 0.50 (.020)	1.00 (.040)
3. Steering knuckle/stub axle .... distortion	0.15 (.006) using VW 258 k/p	—
4. Steering knuckle/caliper flange .... distortion	± 0.05 (±.002) using VW 258 k/p	—
5. Inner wheel bearing............... inside diameter	29.00–29.01 (1.1417–1.1421)	—
outside diameter	50.29–50.32 (1.9799–1.9811)	—
6. Outer wheel bearing .... inside diameter	17.46–17.48 (.6874–.6882)	—
outside diameter	39.88–39.90 (1.5700–1.5709)	—
7. Seat for inner wheel bearing on stub axle	28.98–29.00 (1.1409–1.1417)	—
in brake drum or disc	50.25–50.28 (1.9783–1.9795)	—
8. Seat for outer wheel bearing on stub axle	17.45–17.46 (.6870–.6874)	—
in brake drum or disc	39.84–39.87 (1.5685–1.5697)	—
9. Wheel bearing play	0.03–0.12 (.001–.005)	—
D. Steering		
1. Steering wheel turns from lock to lock..................	2½	—
2. Steering gear ratio	19.4	—
3. Overall ratio............................Beetle	14.34	—
Karmann Ghia	14.14	—
4. Worm turning torque (for axial adjustment)...... without oil seal	1.5–2.5 cmkg (1.3–2.2 in. lb.)	—
with oil seal	2.0–3.0 cmkg (1.7–2.6 in. lb.)	—
5. Total turning torque (steering gear assembled)................	9.0 cmkg (7.8–10.4 in. lb.)	—
6. Roller shaft seat .... diameter	23.98–23.99 (.9441–.9445)	—
7. Bore for roller shaft .... inside diameter	24.00–24.02 (.9449–.9457)	—

II. Tightening Torques

Location	Designation	mkg	ft. lb.
Front axle to frame....................	bolt	5.0	36
Shock absorber to side plate	nut	2.0	14
Shock absorber to lower torsion arm	nut	3.0–3.5	22–25
Steering ball joints	self-locking nut M12	5.0–7.0	36–50
	self-locking nut M10	4.0–5.0	29–36
Wheel bearing clamp nut (prior to adjusting axial play)...	split nut	1.0–max. 1.3 (while turning wheel)	7–max. 9.5 (while turning wheel)
Clamp nut screw	socket head screw	1.0–1.3	7–9.5
Tie rods....................	slotted nut	3.0	22
Steering damper to tie rod.............	self-locking nut	2.5	18
Steering damper to axle tube	bolt	4.0–4.5	29–32
Setscrew for torsion bar	setscrew	4.0–5.0	29–36
Locknut for setscrew	locknut	4.0–5.0	29–36

Front Axle 33

II. Tightening Torques (cont'd)

Location	Designation	mkg	ft. lb.
Brake caliper on steering knuckle	bolt		29
Body mounting on front axle beam	bolt	1.5–2.0	11–14
Steering column on worm spindle	clamp bolt	2.0–2.5	14–18
Steering column tube on dashboard	grooved bolt	0.5–1.0	3.6–7
Brake hose in brake wheel cylinder	hose end	1.5–2.0	11–14
Steering gearbox cover	bolt	2.0–2.5	14–18
Steering gearbox worm adjuster	locknut	5.0–6.0	36–43
Roller shaft adjusting screw	locknut	2.5	18
Tie rod end retaining clamp	nut	1.5	11
Steering wheel on steering column	nut	5.0	36
Steering gearbox on axle beam	bolt	2.5–3.0	18–22
Drop arm on roller shaft	clamp bolt	7.0	50

III. Wheel Alignment Specifications

Designation	Value
1. Wheelbase	2400 mm (94.5 in.)
2. Turning circle	11 m (36 ft.)
3. Track at front	1322 mm (52 in.)
4. Total toe with vehicle unladen	+ 30' ± 15' or + 1.8 to + 5.4 mm (+ 1/16 to + 7/32 in.)
5. Total toe with pressure applied to wheels	+ 5' ± 15' or + 1.2 to + 2.4 mm (+ 3/64 to + 3/32 in.)
Pressure applied to wheels	10 ± 2 kg (22 ± 4 lb.)
6. Maximum permissible difference between toe with wheels pressed and not pressed	25'
7. Front wheel camber in straight-ahead position	0° 30' ± 20'
Maximum permissible difference between sides	30'
8. Toe-out at a 20° lock to left and right (wheels not pressed) to left	– 1° 20' ± 30'
to right	– 2° 10' ± 30'
9. Offset between stub axles	max. 8 mm (5/16 in.)
10. Caster angle of wheel	3° 20' ± 1°
(corresponds to the camber difference of a wheel on a lock from 20° left to 20° right)	2° 15' ± 40'
11. Rear wheel camber with spring plates properly set (after at least 300 mi. or 480 km) all models	– 1° ± 40'
Maximum permissible difference between sides	45'
12. Rear wheel toe with correct camber	0° ± 15'
13. Maximum permissible deviation in wheel alignment	max. 10'

WARNING —
Do not re-use any fasteners that are worn or deformed in normal use. Many fasteners are designed to be used only once and become unreliable and may fail when used a second time. This includes, but is not limited to, nuts, bolts, washers, self-locking nuts or bolts, circlips, cotter pins. Always replace these fasteners with new parts.

Section 7

ELECTRICAL SYSTEM

Contents

Introduction ... 3	**5. Starting System** ... 9
1. General Description ... 4	5.1 Starting System Troubleshooting ... 9
System Voltage and Polarity ... 4	5.2 Removing and Installing Starter ... 10
Battery ... 4	5.3 Removing, Adjusting, and Installing Solenoid ... 10
Starting System ... 4	VW Starter ... 10
Charging System ... 4	5.4 Disassembling and Assembling Starter ... 11
Ignition System ... 4	Starter Drive Servicing ... 12
Wiring ... 4	5.5 Starter Bench Tests ... 14
Lights ... 4	**6. Charging System** ... 14
VW Computer Analysis ... 4	6.1 Generator Troubleshooting ... 15
Fresh Air Fan ... 4	6.2 In-car Testing of Generator and Regulator ... 15
Windshield Wipers ... 4	6.3 Removing and Installing Regulator or Alternator ... 17
Instruments ... 4	6.4 Removing and Installing Generator ... 17
2. Maintenance ... 4	6.5 Replacing Generator Brushes and Servicing Commutator ... 18
3. Fuses ... 5	6.6 Disassembling and Assembling Generator ... 18
Replacing In-line Fuses ... 6	6.7 Testing Disassembled Generator ... 20
4. Battery ... 6	Testing Field Windings ... 20
Discharging ... 6	6.8 Testing and Repairing Alternator and Regulator ... 21
Charging ... 6	Noisy Alternator ... 22
Temperature Effects ... 6	Replacement Engines ... 22
4.1 Servicing and Testing Battery ... 6	**7. Removing and Installing Ignition Switch** ... 23
Hydrometer Testing ... 7	**8. Lights and Switches** ... 25
Voltage Testing ... 7	8.1 Replacing Sealed Beam Units and Aiming Headlights ... 25
4.2 Removing and Installing Battery ... 8	8.2 Removing and Installing Switches in Dashboard ... 25
4.3 Charging ... 8	
Quick Charging ... 8	
Storing a Battery ... 8	

©1978 VWoA

2 Electrical System

Removing Door Contact Switches 26	14.3 Removing and Installing Instruments 38
8.3 Dimmer Relay 27	

9. Turn Signals and Emergency Flashers 27
    9.1 Testing Turn Signals and Emergency Flashers 27
        Testing Turn Signal/Emergency Flasher Relay 27
        Replacing Turn Signal/Emergency Flasher Relay 27
    9.2 Removing and Installing Steering Column Switches 27
        Adjusting Turn Signal Switch 28

10. Windshield Wipers and Washer 28
    10.1 Windshield Wiper Motor Troubleshooting 28
        In-car Testing 29
    10.2 Removing and Installing Wiper Motor and Frame 29
        Removing and Installing Wiper Motor ... 30
    10.3 Replacing Wiper Shaft and Wiper Shaft Bearing 30
    10.4 Disassembling and Assembling Wiper Motor 31
        Removing and Installing Wiper Switch ... 32
    10.5 Removing, Servicing, and Installing Wiper Blades and Arms 32
        Removing, Installing, and Adjusting Wiper Arms 33
    10.6 Windshield Washers 34
    10.7 Troubleshooting Windshield Wipers 34

11. Heated Rear Window 34

12. Fresh Air Fan 36

13. Horns and Buzzers 36
        Troubleshooting and Adjusting Horn 36
        Warning Buzzer 37

14. Instruments 37
    14.1 Replacing Speedometer Cable 37
    14.2 Fuel Gauge Troubleshooting 38
        Removing and Testing Fuel Gauge Sending Unit 38

15. Wiring Diagrams 39
    A. How to read current flow diagrams .... A
    B. Automatic Stick Shift and Rear Window Defogger Circuit for Sedan 111 (1970 and 1971 Models) .... B
    C. Sedan 111 (1970 and 1971 Models) C
    D. Sedan 113 (1971 Models) .... D
    E. Automatic Stick Shift and Rear Window Defogger Circuit for Sedan 113 (1971 Models) ....E
    F. Automatic Stick Shift Circuit for Sedan 111 and 113 (1972 Models) .... F
    G. Sedan 111 and 113 (1972 Models) ....G
    H. Sedan 111 (1973 and later Models) .... H
    I. Automatic Stick Shift Circuit for Sedan 113 (1973 Models) I
    J. Sedan 113 (1973 Models) ....J
    K. Sedan 113 (1974 and later Models) .... K
    L. Sedan 111—from 1976, Models with Fuel Injection L
    M. Sedan 113—from 1976, Models with Fuel Injection ....M
    N. Karmann Ghia (1970 and 1971 Models) ....N
    O. Karmann Ghia (1972 Models) ....O
    P. Karmann Ghia (1973 and 1974 Models) .... P
    Q. Ignition Switch/Safety Belt Interlock ....Q
    R. Circuits for Alternator with Integral Regulator and EGR (exhaust gas recirculation) System— 1975 and later Models .... R
    S. Circuit that Eliminates Seatbelt Interlock (shown in Diagrams K and Q) .... S

TABLES

a. Fuse Locations in Fuse Box 5
b. Starter Types 9
c. Starting System Troubleshooting 9
d. Starter Data 14
e. Solenoid Switch Test Data 14
f. Generator Types 14
g. Generator Troubleshooting 15
h. Generator Technical Data 15
i. Windshield Wiper Motor Troubleshooting 29
j. Wiper Arm Adjustment Dimensions 34
k. Windshield Wiping System Troubleshooting 35

Electrical System

The electrical system is basically an efficient means for transmitting power from the engine to remote parts of the car. It does this with the help of a generator or alternator that converts some of the engine's mechanical energy into electrical energy. The electrical energy is carried over wires to motors that convert it back into mechanical energy or to bulbs that convert it into heat and light. The battery in the system supplies electrical power mainly when the engine is not running.

Every terminal in the electrical system is numbered. The terminal numbers for all major electrical connections are given in the color wiring diagrams that appear at the end of this section. The terminal number is usually stamped on the component itself as an aid to proper installation.

Though most of the electrical terminal numbers are used only once to denote a particular terminal on a particular component, there are several numbers that do not designate specific terminals and which appear in numerous locations throughout the electrical system. Such numbers identify main sources of electrical current. All terminals numbered **15** originate at the ignition switch and supply current only when the ignition switch is in its on position. Terminals numbered **30** supply positive polarity current directly from the battery with no intervening switch that can be used to turn it off. Terminals numbered **31** are ground connections between a switch and another electrical component. Terminals numbered **85** receive a ground wire that is connected directly to the chassis of the car. Terminals identified by the number **50** receive current only when the ignition switch is in its start position. A letter suffix is sometimes added to the terminal number to distinguish separate parts of the same circuit or to prevent the confusion of two circuits that have similar functions.

All electrical circuits other than those required for starting and operating the engine are protected by fuses. To prevent accidental shorts that might cause a blown fuse or damage wires and electrical components that are not protected by fuses, you should always disconnect the ground strap from the negative pole of the battery before working on the electrical system of your car. If you lack the skills or the equipment needed for testing and repairing the electrical system we suggest that you leave such work to an Authorized VW Dealer or other qualified shop. We especially urge you to consult your Authorized VW Dealer before attempting repairs on a car still covered by the new-car warranty.

4 Electrical System

1. General Description

The components of the VW electrical system are discussed in detail in later parts of this section. However, a brief description of the principal components is presented here for purposes of familiarization.

System Voltage and Polarity

The VW has a 12-volt, negative-ground electrical system. In other words, the voltage regulator keeps voltage in the system at approximately the 12-volt rating of the battery and the negative pole of the battery is connected directly to the car frame.

Battery

The six-cell, 12-volt lead-acid battery is located under the right-hand side of the rear seat on the Beetle and Super Beetle and in the left-hand side of the engine compartment on the Karmann Ghia. The battery is rated at 45 ampere-hours on all models covered by this Manual.

Starting System

The starter is series-wound and has an overrunning clutch. Output is 0.7 horsepower on cars with manual transmissions and 0.8 horsepower on cars with Automatic Stick Shift transmissions. The starter and its attached solenoid are located near the front of the engine at the right-hand side of the flywheel bellhousing.

Charging System

A belt-driven direct current (DC) generator is installed on all 1970, 1971, 1972, and many 1973 cars. Some 1973 cars built after January 1973 have an alternator with a separate regulator. All 1974 and later cars have alternators but, beginning with the late 1974 models, the alternator has an integral regulator.

Ignition System

The ignition is a conventional coil and battery, distributor-controlled system. Ignition troubleshooting, maintenance, and repair are covered in **ENGINE** and in **LUBRICATION AND MAINTENANCE**. Radio suppression by resistance is built into the spark plug connectors and distributor rotor.

Wiring

All components of the VW electrical system (except for the heavy battery cables) have push-on connectors. A system of fuses prevents short circuits or excessive current from damaging the electrical system and wiring.

Lights

The lighting system includes the parking lights, side marker lights, turn signals, back-up lights, interior lighting, and sealed beam headlights. The headlight beams are dimmed or raised by pulling the turn signal lever toward the steering wheel. Actual switching is carried out by a dimmer relay mounted behind the instrument panel.

VW Computer Analysis

Since June 1971, VW cars have been equipped with a separate wiring harness that serves the VW Computer Analysis system. This harness leads to a central socket in the engine compartment that receives individual wires connected to various measuring points on the car. These connections are marked by encircled numbers in the wiring diagrams that appear in this section.

Although they are not vital to car operation, all such connections must be kept intact if the computer diagnosis system is to work properly. Never connect any device other than the test plug of the VW Computer Analysis system to the test network central socket in the engine compartment. Incorrect equipment could damage the test sensors or the vehicle components containing them.

Fresh Air Fan

All 1971 and later Super Beetles and Convertibles have a two-speed fan that delivers fresh air through the windshield and dashboard vents. The fan became standard on all 1976 models. Removal, replacement, and installation of the fresh air fan are covered in **BODY AND FRAME**.

Windshield Wipers

The blades of the two-speed windshield wiper system automatically return to their parked position when they are switched off. The wiper switch includes a windshield washer control. The washers work by compressed air from the spare tire.

Instruments

A flexible cable driven by the left-hand front wheel operates the speedometer. Karmann Ghia models are equipped with an electrically wound clock in addition to the speedometer. The electric fuel gauge is integral with the speedometer on Beetle and Super Beetle models and integral with the clock on Karmann Ghia models.

2. Maintenance

No routine lubrication of the generator, starter, or other motors is required. However, the following checks are included in **LUBRICATION AND MAINTENANCE**.

1. Checking the lights and switches
2. Checking the windshield wipers and washers
3. Checking the battery
4. Testing the charging and starting systems.

© 1976 VWoA

ELECTRICAL SYSTEM

3. FUSES

Most of the fuses in the electrical system are located in a fuse box under the dashboard next to the steering column. Fig. 3-1 is a diagram of a typical VW fuse box, although the fuse boxes of some cars covered in this Manual have only ten fuses rather than the twelve shown.

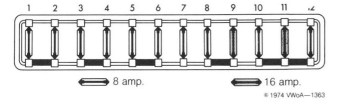

Fig. 3-1. Typical fuse box. The location of the 16-amp fuses varies among models and model years.

Table a lists the fuse number for each electrical component on each model covered in this Manual. Find the component along the column headings at the top. The fuse number is given where this column intersects the row for your particular VW. The Convertible, model 151, has the same fuses as the model 113. (In 1970, the model 151 had the same fuses as the model 111.)

Always replace red (16-amp) fuses with other red fuses and white (8-amp) fuses with other white fuses. If in doubt, consult the Owner's Manual supplied with the car.

CAUTION —
Never patch a blown fuse with aluminum foil or replace it with wire or a fuse of greater capacity. Such practices can damage the electrical system.

Table a. Fuse Locations in Fuse Box

	Accessory	Buzzer	Control Valve	Clock	Emergency Flashers	Fresh Air Fan	Fuel Gauge	High Beam Indicator	High Beam, Left	High Beam, Right	Horn	Interior Light	License Lights	Low Beam, Left	Low Beam, Right
1970 113	N/A	no fuse	in-line	N/A	9	N/A	1	3	3	4	1	9	8	5	6
1970 111	N/A	no fuse	in-line	N/A	9	N/A	1	3	3	4	1	9	8	5	6
1970 Ghia	1	no fuse	in-line	9	9	N/A	2	3	3	4	2	9	8	5	6
1971 113	5	6	in-line	N/A	4	in-line	12	7	7	8	3	6	12	9	10
1971 111	N/A	9	in-line	N/A	10	N/A	12	3	3	4	1	9	8	5	6
1971 Ghia	N/A	9	in-line	9	10	N/A	2	3	3	4	2	9	8	5	6
1972 113	7	9	11	N/A	8	10	12	5	5	6	11	9	7	3	4
1972 111	7	9	11	N/A	8	N/A	12	5	5	6	11	9	7	3	4
1972 Ghia	7	9	12	no fuse	8	N/A	12	5	5	6	11	9	1	3	4
1973 113	N/A	no fuse	11	N/A	8	10	12	6	5	6	11	9	7	3	4
1973 111	N/A	9	11	N/A	8	N/A	12	6	5	6	11	9	7	3	4
1973 Ghia	7	9	11	no fuse	8	N/A	12	5	5	6	11	9	1	3	4
1974-8 113	7	8	11	N/A	8	10	12	5	5	6	11	9	2	3	4
1974-7 111	7	9	11	N/A	8	N/A	12	6	5	6	11	9	1	3	4
1974 Ghia	7	9	11	no fuse	8	N/A	12	5	5	6	11	9	2	3	4

	Parking Light, Left	Parking Light, Right	Rear Window Defogger	Side Marker Light, Left	Side Marker Light, Right	Stop Lights	Tail Light, Left	Tail Light, Right	Turn Signals	Turn Signal Indicator	Warning Light, ATF	Warning Light, Brake	Warning Light, Generator	Warning Light, Oil Pressure	Windshield Wipers
1970 111	7	8	7	N/A	N/A	1	7	8	1	1	1	1	2	2	2
1970 Ghia	7	8	7	N/A	8	2	7	8	1	1	1	1	2	2	1
1971 113	11	12	2	12	12	3	11	12	1	1	2	2	12	12	2
1971 111	7	8	1	7	8	2	7	8	1	1	1	2	12	12	2
1971 Ghia	7	8	1	7	8	2	7	8	1	1	1	1	2	2	1
1972 113	1	1	10	1	1	11	1	1	12	12	11	12	12	12	10
1972 111	1	1	10	1	1	11	1	1	12	12	11	12	12	12	10
1972 Ghia	1	1	10	1	1	11	2	1	12	12	11	12	12	12	10
1973 113	1	2	10	1	2	11	1	1	12	12	11	12	12	12	10
1973 111	1	1	10	1	1	11	1	1	12	12	11	12	12	12	10
1973 Ghia	1	1	10	1	1	11	2	1	12	12	11	12	12	12	10
1974-8 113	1	1	10	1	1	11	1	2	12	12	11	12	12	12	10
1974-7 111	1	1	10	1	1	11	2	1	12	12	11	12	12	12	10
1974 Ghia	1	1	10	1	1	11	2	1	12	12	11	12	12	12	10

6 Electrical System

Replacing In-line Fuses

In addition to the fuses located in the fuse box, one or more fuses are located in in-line fuse holders. The in-line fuse holder for the back-up lights is mounted on the engine, just above the ignition coil. The in-line fuse holder for the control valve on 1971 cars with Automatic Stick Shift transmissions is also located at this point. On 1970 Beetles, this fuse holder is mounted in a similar position at the opposite side of the engine.

On 1971 and later Karmann Ghia models the fuse for main current to the rear window defogger is also in an in-line holder above the ignition coil. On Beetles and Super Beetles built during the same period this fuse holder is located under the rear seat, near the voltage regulator. On 1971 Beetles and Super Beetles an in-line fuse for the fresh air fan can be found in a separate holder near the fuse box behind the dashboard. All in-line fuses are of 8-amp capacity.

4. Battery

Each of the six battery cells contains a set of brown lead oxide positive plates and gray sponge lead negative plates. The cells are connected in series by heavy lead bars and are enclosed in a plastic case having six compartments. The battery case also serves as a tank for the electrolyte—a solution of sulfuric acid diluted with water to a specific gravity of 1.285, which means that the electrolyte weighs 1.285 times as much as an equal volume of water. The battery plates that make up the cells are completely immersed in the electrolyte.

The terminal posts are labeled + and − and are further identified by having a positive post that is the thicker of the two. The ground strap connected to the negative post bolts to the frame of the car. The cable attached to the positive post extends to the starter solenoid where it is joined by the main line serving the rest of the electrical system. The batteries in VW cars built since June 1971 have an additional central terminal with a small gauge wire attached to it. This is the Computer Diagnosis connection used to check the electrolyte level.

Discharging

The battery does not store electricity. Rather, it produces electrical current by means of a reversible electrochemical reaction. When a circuit is completed between the two battery posts, sulfuric acid from the electrolyte combines with the lead in the plates to produce lead sulfate, releasing a great many electrons in the process.

Charging

The electrochemical reaction by which the battery produces electrical current is reversed when direct current is sent back into the cells. The charging system of the car supplies this current. When the discharged battery plates are charged with direct current from an outside source, the lead sulfate in the plates is converted back to its original state, returning sulfuric acid to the solution in the electrolyte.

A battery can never be charged to a voltage level in excess of the voltage it is capable of producing electrochemically. As charging proceeds, the battery's voltage builds to a peak called terminal voltage. If charging is continued beyond the terminal voltage, the water in the electrolyte begins to decompose into hydrogen and oxygen. This condition is called gassing.

Temperature Effects

Temperature changes modify the efficiency of the battery as well as alter the specific gravity of its electrolyte. Low outside temperatures can create slow starting by thickening the engine and transmission oils and simultaneously reducing the battery power available for running the starter motor. The current-producing capacity of a battery chilled to −15°C (5°F) is only half its capacity at 20°C (68°F).

In addition, there is danger of partly-discharged batteries freezing in cold weather due to the higher proportion of water in their electrolyte. A frozen battery will produce no current, but can usually be restored to service if thawed out slowly. The following list shows the safe low temperature limits for batteries in various states of charge.

Specific gravity	Freezing point
1.285	−68°C (−90°F)
1.200	−27°C (−17°F)
1.120	−11°C (12°F)

4.1 Servicing and Testing Battery

The level of the electrolyte should never be allowed to fall below the tops of the plates in any cell. As water is lost through evaporation and electrolysis, fresh water must be added to maintain the electrolyte's level at the bottoms of the indicator tubes which are built into the battery filler openings. Use only distilled water to replenish the electrolyte. Water that is not chemically pure may have an adverse effect on battery life and efficiency.

The battery will lose more water in summer than in winter. In very hot weather it may be necessary to check the electrolyte level as often as once a week. Never overfill the cells. This could cause the electrolyte to boil over during a long daylight drive when the load on the electrical system is light and generator output is high.

Battery terminals must be tight-fitting and free of corrosion and acid salts. If you notice even a trace of corrosion, remove the positive cable and ground strap from the battery posts and clean the posts and terminal clamps with a battery terminal cleaning tool. After the terminals have been cleaned and the positive cable and ground strap tightly installed on the battery posts, the terminals and posts should be coated lightly with petroleum jelly or sprayed with a commercial battery terminal corrosion inhibitor.

WARNING —
Keep sparks and open flame away from the top of the battery. Hydrogen gas from the battery could explode violently.

The top of the battery should always be kept clean. Evan a thin layer of dust containing conductive acid salts can cause the battery to discharge. Corrosion and acid salt accumulations should be washed away with baking soda solution. Be extremely careful that none of this solution enters the cells through the vent holes. Even a drop or two will seriously impair the efficiency of the battery.

Periodic battery tests should be made to help keep track of battery condition. Such tests can also be made to help pinpoint the source of suspected battery trouble.

WARNING —
Wear goggles when you work with battery electrolyte and do not allow the liquid to contact your skin or clothing. Electrolyte is corrosive and can cause severe burns. If it should spill onto your skin, flush the area of contact immediately with large quantities of water. Spilled electrolyte can be neutralized with a strong baking soda solution.

Hydrometer Testing

The simplest tool for testing the battery is a hydrometer. It consists of a glass cylinder with a freely moving float inside. When electrolyte is drawn into the cylinder by squeezing and releasing a rubber bulb, the level to which the float sinks indicates the specific gravity of the electrolyte. A specific gravity scale on the float is read at the point where it intersects the surface of the electrolyte. The more dense the concentration of sulfuric acid in the electrolyte, the less the float will sink and the higher the reading. Specific gravity values for different states of charge are as follows:

State of charge	Specific gravity
Fully discharged	1.120
Half discharged	1.200
Fully charged	1.285

Voltage Testing

Total battery voltage can be tested with a special voltmeter such as the one shown in Fig. 4-1.

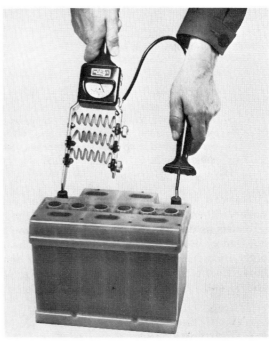

Fig. 4-1. Using voltmeter to test total voltage of battery.

The tester should consist of a voltmeter connected in parallel with a test load of approximately 110 amps. The minimum voltage indicated should not be less than 9.6 volts. If total voltage drops below this value during the 5- to 10-second test, the battery is either discharged or sulfated. A sulfated battery is one in which the plates are covered by a layer of lead sulfate that is difficult to reconvert. Sulfating is visible as a gray coating on the plates.

CAUTION —
A discharged battery should be recharged immediately. Otherwise, sulfating will lead to the loss of active plate materials and to reduced battery capacity.

The voltage of an individual cell should not vary from the others by more than 0.2 volts. This can be determined by applying one prong of the tester to the negative battery post, then dipping the other prong into the electrolyte of successive cells, and finally applying it to the positive post. The readings should be 2, 4, 6, 8, 10, and 12 volts. This test should last for no more than 10 seconds.

8 Electrical System

4.2 Removing and Installing Battery

The battery is fastened to the floor by two clamps. The front clamp is welded to the floor and the rear clamp is held by a bolt. Before removing the bolt, disconnect first the ground strap and then the positive cable from the battery posts. On batteries used since June 1971, also remove the wire from the center terminal that serves the VW Computer Diagnosis system.

CAUTION —
If a new ground strap must be installed on a car equipped for VW Computer Diagnosis, make certain the part number is correct. The wrong ground strap will cause incorrect voltage data to be fed into the computer.

Remove the mounting clamp bolt, then remove the battery. When installing the battery, clean and install the terminals as described earlier in **4.1 Servicing and Testing Battery.** Be sure to install the wire for the VW Computer Diagnosis system. The battery must be mounted firmly to the floor during installation to prevent road shocks and vibration from damaging the plates.

4.3 Charging

Normally, a battery should be charged at no more than 10 percent of its rated capacity. For example, a charging current of 4.5 amperes would be used on a battery having 45 Ah (ampere-hours) capacity. However, a current as low as 5 percent of the rated capacity (2.25 amps for a 45 Ah battery) can be used in normal charging and should always be used the first time a new battery is charged.

In normal charging, the battery is considered fully charged when it is gassing freely and the voltage of the individual cells has risen to 2.5 to 2.7 volts each (about 15 volts for the battery). An hour or so after you have switched off the charging current, use the voltmeter/tester to determine the rest voltage of the battery. This should be 2.1 to 2.2 volts per cell, or approximately 12.5 to 13.0 volts for the battery.

Quick Charging

To save time in an emergency, a higher current can be used to charge batteries in good condition. Only sound batteries that are already in service are suitable for quick charging. Neither factory-new nor sulfated batteries should ever be quick-charged.

WARNING —
Do not boost a sulfated battery at a high charging rate. The battery could explode.

To quick charge:

1. Disconnect the positive cable and the ground strap from the battery.

2. Remove the battery caps, then connect a battery charger and voltmeter to the battery. Quick-charge at 40 amperes for three minutes.

3. Observe the voltmeter reading during charging. If total battery voltage exceeds 15.5 volts, the battery plates are sulfated or worn out and the battery should be replaced.

4. If the total voltage is less than 15.5 volts, test the individual cell voltages. If cell voltages vary by more than 0.1 volt, the battery plates are worn out and the battery should be replaced. If cell voltages are within 0.1 volt, measure the specific gravity and continue quick-charging as follows:

Specific gravity	Period of charge
1.150 or less	1 hour
1.150 to 1.175	45 minutes
1.175 to 1.225	15 minutes
Above 1.225	Slow charge only

WARNING —
Smoking and open flames should not be permitted in a room where batteries are charged. Charging causes excess water in the electrolyte to decompose into hydrogen and oxygen, a dangerously explosive combination of gases.

CAUTION —
Do not store precision tools in a room where batteries are charged. The corrosive fumes generated during charging can severely damage such tools.

Storing a Battery

A battery that is not in use will gradually discharge itself. At room temperature it will lose about one percent of its remaining capacity each day. The rate of discharge increases with higher temperatures. If the battery is allowed to remain in a partly or fully discharged condition for long periods, it will become badly sulfated and may never be serviceable again.

To store battery:

1. Charge the battery. Check the electrolyte level and the specific gravity. If necessary, add distilled water to the electrolyte.

2. Store the battery in a cool, dry place.

3. Every 6 to 8 weeks, discharge the battery and recharge it.

4. Before returning the battery to service, charge it with a very low current (not over 3 amps).

© 1976 VWoA

Electrical System

5. Starting System

Table b lists the starters installed in Type 1 cars built during the years covered by this Manual. Starter 003 911 023A is installed only in cars having Automatic Stick Shift transmissions. Bosch starters 311 911 023B, 311 911 023C, and 311 911 023D are similar in construction and identical in technical specifications. However, the latter two starter types (C and D) have aluminum field coils rather than the copper field coils used in other types. They also have a modified drive mechanism and are equipped with four carbon brushes instead of two. Starter 311 911 023D can also be distinguished from the other Bosch starters by its shorter length—31.8 mm (1¼ in.) less than the other starters.

Table b. Starter Types

Starter	Date Introduced	Output
111 911 023A (VW)	August 1967	.7 bhp
311 911 023B (Bosch)	August 1967	.7 bhp
311 911 023C (Bosch)	August 1969	.7 bhp
311 911 023D (Bosch)	March 1970	.7 bhp
003 911 023A (Bosch)	March 1967	.8 bhp

5.1 Starting System Troubleshooting

Troubleshooting procedures that are applicable to cars with any type starter appear in **Table c.** The bold numbers in the Remedy column refer to the headings in this section under which the prescribed tests and repairs can be found. If more than one probable cause is listed, check them one by one in the order in which they appear.

Table c. Starting System Troubleshooting

Problem	Test and Probable Cause	Remedy
1. Starter does not operate when ignition key is turned to start position—seat belt interlock operating correctly on cars that are so equipped	Turn lights on for test: a. Lights are out. Loose cables or poor ground connection. Battery run down b. Lights go out when key is moved to starting position. Insufficient current due to loose connections or corroded terminals c. Lights go dim when key is moved to starting position. Battery run down d. Connect a jumper cable between starter terminals 30 and 50. If starter runs, cable 50 to ignition switch or cable 30 to lighting switch is faulty, or ignition switch is defective e. Lights stay bright, solenoid operates. Connect jumper cable from starter terminal 30 to connector strip terminal. Solenoid is faulty if starter runs	a. Check battery cable terminals. Test battery. Charge if necessary. See **4.1, 4.3.** b. Clean and tighten all battery cable connections. See **4.1, 4.2.** c. Charge battery. See **4.3.** d. Eliminate open circuits. Replace defective parts. See **7, 15.** e. Replace solenoid. See **5.3.**
2. Starter does not operate when battery cable is directly connected with terminal stud of connector strip	a. Brushes sticking b. Brushes worn c. Weak spring tension. Brushes do not make contact d. Commutator dirty e. Commutator rough, pitted, or burned f. Armature or field coils defective	a. Clean brushes and brush holders. See **5.4.** b. Replace brushes. See **5.4.** c. Replace springs. See **5.4.** d. Clean commutator. See **5.4.** e. Recondition or replace starter motor. See **5.2, 5.4.** f. Recondition or replace starter motor. See **5.2, 5.4.**
3. Starter turns too slowly or fails to turn engine over	a. Battery run down b. Insufficient current flow due to loose or corroded connections c. Brushes sticking d. Brushes worn e. Commutator dirty f. Commutator rough, pitted, or burned g. Armature or field coils defective	a. Charge battery. See **4.3.** b. Clean battery terminals and cable clamps, tighten connections. See **4.1, 4.2.** c. Clean brushes and brush holders. See **5.4.** d. Replace brushes. See **5.4.** e. Clean commutator. See **5.4.** f. Recondition or replace starter motor. See **5.2, 5.4.** g. Recondition or replace starter motor. See **5.2, 5.4.**
4. Starter makes unusual sounds, cranks engine erratically or fails to crank	a. Drive pinion defective b. Flywheel ring gear defective	a. Replace drive pinion. See **5.4.** b. Replace flywheel. See **ENGINE.**
5. Drive pinion does not move out of mesh	a. Drive pinion or armature shaft dirty or damaged b. Solenoid switch defective	a. Recondition or replace starter motor. See **5.2, 5.4.** b. Replace solenoid switch. See **5.3.**

© 1976 VWoA

10 ELECTRICAL SYSTEM

5.2 Removing and Installing Starter

Removing the starter from the car will be made easier if you place the car on a lift and take off the right-hand rear wheel.

To remove:

1. Disconnect the ground strap from the negative post of the battery, then disconnect the wires from terminals 30 and 50 at the starter solenoid.
2. While a helper holds the nut inside the engine compartment, remove the upper starter mounting bolt.
3. Remove the lower starter mounting nut from its stud, then withdraw the starter motor.

    NOTE
    After removing a Bosch starter motor, check the bushing in the transmission case. If the bushing is worn or damaged, replace it.

To install:

1. Lubricate the starter shaft bushing with multipurpose grease, then apply a good sealing compound around the starter opening in the transmission case.
2. Slide the starter onto the stud and secure it loosely with the nut. Then loosely install the bolt and nut.
3. Working alternately, gradually torque both fasteners to 3.0 mkg (22 ft. lb.)
4. Clean the wires and terminals, then tightly install all the electrical connections.

5.3 Removing, Adjusting, and Installing Solenoid

The following procedure applies only to the Bosch starter. The VW unit is covered later.

To remove:

1. Remove the starter from the car.
2. Remove the nut and the connector strap. Then remove the two screws that hold the solenoid to the drive end plate.
3. Lift the pull rod upward and out of the engaging lever. This will be easier if you pull the pinion clockwise and outward at the same time.
4. Withdraw the solenoid.

When installing a new solenoid on a Bosch starter, be sure to adjust the pull rod length as indicated in Fig. 5-1. However, you should never adjust pull rod length to accomodate a defective solenoid. A defective solenoid should be replaced.

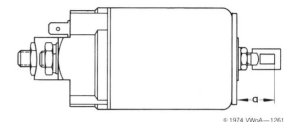

Fig. 5-1. Pull rod adjustment. Loosen locknut, then turn rod clockwise or counterclockwise until dimension **a** is 19 mm ± 0.10 mm (.748 ± .004 in.). Tighten locknut and recheck dimension **a**.

To install solenoid:

1. Seat the molded rubber seal properly on the drive end plate.
2. Place a small strip of plastic sealing compound on the outer edge of the solenoid end face.
3. Withdraw the drive pinion as far as possible and install the solenoid as shown in Fig. 5-2.
4. Install the nut and connector strap.

Fig. 5-2. Solenoid being installed. Hook pull rod eye over engaging lever. Use sealing compound between solenoid end face and drive end plate.

VW Starter

The solenoid is attached to the VW starter by two nuts on studs built into the solenoid housing, a connecting strap, and an electrical connection. The solenoid core is a separate part. To remove the core from the operating lever, take off the spring clips that lock the two pins, push out the pins, remove the insulating plate, and then turn the contact plate 90°. Unlike the Bosch starter, no pull rod adjustment is called for. During assembly, use a good plastic sealing compound to seal the pin holes and the mating surfaces between the solenoid housing and the starter drive end plate.

5.4 Disassembling and Assembling Starter

Three representative starters are shown disassembled in Fig. 5-3, Fig. 5-4, and Fig. 5-5. These exploded views include the VW starter, one Bosch starter having two brushes, and another Bosch starter equipped with four brushes. Basic disassembly requires only that the circlips or C-washers be removed from the armature shaft and the two long through bolts be removed. Further disassembly can be done with a screwdriver.

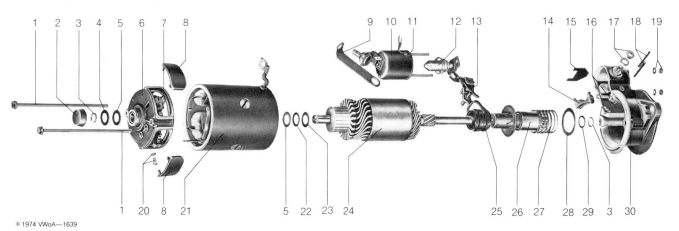

Fig. 5-3. Disassembly of VW starter 111 911 023A

1. Through bolt (2)
2. Cap
3. Circlip (2)
4. Steel washer
5. Bronze washer (2)
6. Commutator end plate
7. Brush holder and brushes (2)
8. Brush inspection cover (2)
9. Connecting strap
10. Solenoid housing and winding
11. Insulating disk
12. Solenoid core
13. Linkage
14. Molded rubber seal
15. Insulating plate
16. Spring clip (2)
17. Nut and lock washer (2)
18. Pins (2)
19. Small nut and lock washer (2)
20. Screw and lock washer
21. Pole housing
22. Dished washer
23. Steel washer
24. Armature
25. Connecting bushing
26. Drive pinion
27. Spring
28. Intermediate washer
29. Dished washer
30. Drive end plate

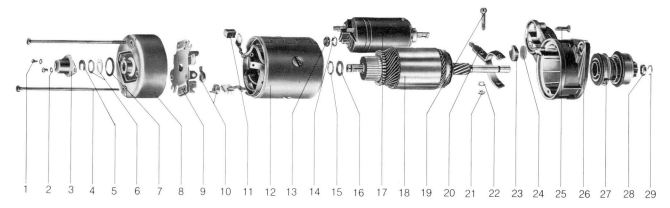

1. Screw (2)
2. Washer (2)
3. End cap
4. Through bolt (2)
5. C-washer
6. Shim
7. Sealing ring
8. End plate
9. Brush holder
10. Spring (2)
11. Rubber grommet
12. Pole housing
13. Nut
14. Lock washer
15. Insulating washer
16. Thrust washer
17. Solenoid
18. Armature
19. Pin
20. Engaging lever
21. Nut
22. Lock washer
23. Molded rubber seal
24. Disk
25. Screw (2)
26. Drive end plate
27. Drive pinion
28. Stop ring
29. Circlip

Fig. 5-4. Disassembly of Bosch starter 311 911 023B.

12 Electrical System

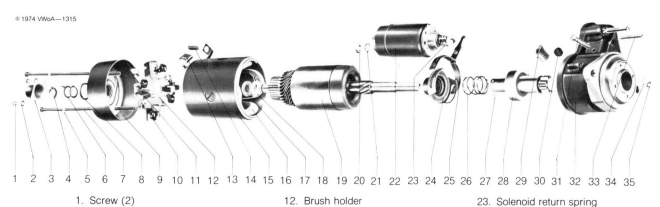

1. Screw (2)
2. Washer (2)
3. End cap
4. C-washer
5. Washer (2)
6. Shim
7. Through bolt (2)
8. Sealing ring
9. End plate
10. Retaining spring (4)
11. Negative brush (2)
12. Brush holder
13. Positive brush (2)
14. Rubber grommet
15. Pole housing
16. Field winding
17. Insulating washer
18. Thrust washer
19. Armature
20. Nut
21. Lock washer
22. Solenoid
23. Solenoid return spring
24. Operating sleeve
25. Engaging lever
26. Engaging spring
27. Detent balls (10)
28. Drive pinion
29. Molded rubber seal
30. Disk
31. Pin
32. Drive end plate
33. Screw (2)
34. Lock washer
35. Nut

Fig. 5-5. Disassembly of Bosch starter 003 911 023A.

Remove the solenoid before disassembling the starter. The cap must also be removed from the commutator end of the starter to gain access to the C-washer or circlip on the armature shaft. After the C-washer or circlip has been removed, the through bolts can be taken out and the starter end plate and housing lifted off the armature. Note the number and position of all spacer washers to ensure proper reassembly. See Fig. 5-6 for additional information on the VW starter.

Starter Drive Servicing

You need not disassemble the starter drive if you intend only to make electrical tests or replace the brushes. If the commutator must be machined or if the starter drive is defective, however, further disassembly is required. Remove the drive pinion from Bosch starters as indicated in Fig. 5-7.

Fig. 5-6. VW starter disassembly. Remove brush inspection covers and lift brushes to position shown before removing the through bolts. Note circlip groove in armature shaft.

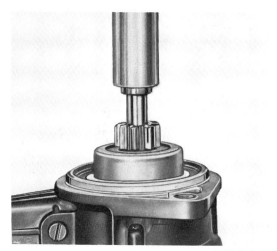

Fig. 5-7. Removing Bosch starter pinion. Remove circlip and stop ring, file burrs from circlip groove, then press off pinion as illustrated.

Electrical System 13

To remove the armature and starter drive from the drive end plate of a VW starter, remove the pins shown in Fig. 5-8. Then remove the solenoid core as described in **5.3 Removing, Adjusting, and Installing Solenoid.**

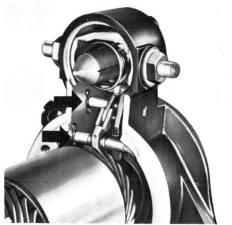

Fig. 5-8. Removing VW starter drive. Remove the two spring clips at arrows, then push pins out of starter drive end casting.

To remove the drive pinion from the VW starter armature, take the circlip off the shaft and withdraw the connecting bushing as indicated in Fig. 5-9. Then turn the whole drive pinion assembly clockwise and jerk it slightly to remove it from the shaft. The connecting bushing and the five steel balls beneath it may then be taken off. During assembly, use multipurpose grease as an adhesive to hold the balls in place.

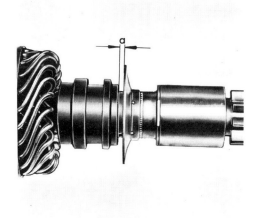

Fig. 5-9. Removing drive pinion assembly from armature shaft of VW starter. The connecting bushing (the grooved sleeve nearest the armature windings) must be pulled back to make gap **a** approximately 3 to 6 mm (⅛ to ¼ in.) wide.

Servicing Brushes and Commutator

To replace brushes, you need only remove the commutator end plate from VW starters or the end plate and brush holder from Bosch starters. However, if the commutator is found to be out-of-round by more than 0.03 mm (.001 in.), the starter has to be completely disassembled so that the commutator can be machined. The diameter of VW starter commutators must not be reduced below 33.00 mm (1.300 in.). The diameter of Bosch starter commutators must not be reduced below 34.50 mm (1.350 in.). After the commutator has been turned on a lathe, the insulation strips should be undercut by about 0.50 mm (.020 in.).

Unsolder the old brush wires and solder the new ones in their place. If the end plate bushing is worn, it can be pressed out and a replacement installed. Heat the end plate in hot oil before pressing in the new bushing.

To assemble starter:

Do not wash the self-oiling end plate bushing or the drive pinion assembly in solvent. This would destroy the lubricant built into these parts at the factory. Lubricate the bushings at both ends of the armature shaft with multipurpose grease. The drive mechanism must not be greased or it may jam in cold weather. During assembly, all points shown in Fig. 5-10 should be made weatherproof with a good sealer.

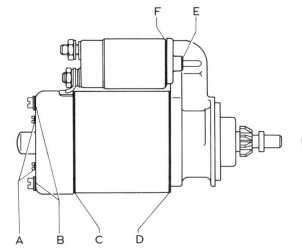

A. Holes for end cap screws
B. Holes for through bolts
C. Joint between pole housing and end plate
D. Joint between pole housing and mounting bracket
E. Holes for solenoid mounting screws
F. Surface between solenoid switch and drive end housing

Fig. 5-10. Sealing locations on Bosch starter. On VW starters, the holes for the engagement linkage pins must also be sealed.

14 ELECTRICAL SYSTEM

5.5 Starter Bench Tests

A battery, or several batteries wired in parallel, with a rated capacity of 135 ampere hours should be available for starter bench tests. This will ensure that decreasing battery power does not influence the test readings. Automotive electrical shops that have a starter motor test stands will find the data in **Tables d** and **e** useful in determining how closely starters conform to factory specifications. Testing should be carried out in the following sequence:

1. No-load test—starter motor running freely.

2. Load test—test stand flywheel braked to limit starter to the rpm given in **Table d**. This test must last no longer than 10 seconds.

3. Stall torque test—test stand flywheel braked to stop the starter under load. This test must last no longer than 5 seconds.

4. Solenoid pull-in voltage under load—starter switched on and off under light load and the pinion checked for proper engagement.

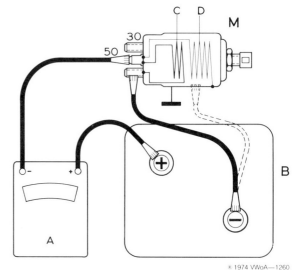

A. Ammeter   C. Pull-in winding   M. Solenoid
B. Battery   D. Holding winding

Fig. 5-11. Solenoid tests. Move negative connection to position shown by dotted lines to check current draw in holding winding. Bosch unit is shown but VW solenoid uses a similar hook-up.

Table d. Starter Data

Starter Type	No-load Test			Load Test			Stall Torque Test		
	Current (amps)	Voltage	Speed (rpm)	Current (amps)	Voltage	Speed (rpm)	Current (amps)	Voltage	Solenoid pull-in voltage
311 911 023B	35-45	12	7400-9100	170-205	9	900-1300	220-260	6	7
111 911 023A	25-40	12	6200-7800	170-195	9	1050-1350	270-290	6	8
003 911 023A	35-50	12	6400-7900	160-200	9	1100-1400	250-300	6	8

Table e lists the normal current draw for the two windings of a solenoid that is in good condition. The pull-in winding is tested as indicated in Fig. 5-11. The holding winding is tested as shown by the dotted lines. Faulty windings cannot be repaired. If readings vary from those listed in **Table e**, the solenoid must be replaced.

Table e. Solenoid Switch Test Data

Solenoid winding	Current draw (amps)	
	Bosch	VW
Pull-in winding, max.	35	30
Holding winding, max.	11	12

6. CHARGING SYSTEM

Table f lists the generators installed in Type 1 cars built during the years covered by this Manual. In addition, two alternators have been installed intermittently on Type 1 vehicles beginning with chassis No. 113 2414 931. One alternator is manufactured by Bosch and the other by Motorola.

Table f. Generator Types

Generator	Years Fitted
113 903 031G	1970 and 1971
113 903 031K	1971 onward
113 903 031P	
113 903 031Q	

Electrical System 15

CAUTION—
Tests and checks suitable for DC generator charging systems must not be used when working with alternator charging systems. Some DC generator tests will damage the diodes in an alternator. With the exception of data appearing in **6.8 Alternator,** *all material in this section applies solely to charging systems equipped with DC generators.*

6.1 Generator Troubleshooting

Troubleshooting procedures applicable to cars with DC generators appear in **Table g.** The bold numbers in the Remedy column refer to the headings in this section under which the prescribed tests and repairs can be found. If more than one probable cause is listed, check them in the order in which they appear.

Some of the cars covered by this Manual are equipped with alternators instead of DC generators. Alternators and alternator regulators can be damaged by some of the tests used in troubleshooting DC generators. Note that **Table g** has been devised with only the DC generator in mind. The table should never be used to troubleshoot an alternator.

6.2 In-car Testing of Generator and Regulator

A number of in-car tests are possible with precision measuring equipment. These tests will accurately determine the condition of the generator and regulator. All DC generators must meet the specifications given in **Table h.** Instructions for testing follow the table.

Table h. Generator Technical Data

Nominal data	
Maximum current, amps	30
Mean regulating voltage	14
Nominal output speed, rpm	2000
Test data	
Cut-in speed, generator rpm (1.9 times engine rpm)	1450
Cut-in voltage	12.4-13.1
Return current, amps	2.0-7.5
No-load regulating voltage	13.5-14.5
Load regulating voltage	12.8-13.8
Load current, amps	45

Table g. Generator Troubleshooting

Problem	Probable Cause	Remedy
1. Igniton is on but warning light does not glow	a. Discharged battery b. Broken battery case or plates c. Bulb burned out d. Corroded and/or loose battery terminals e. Loose connections and/or broken wiring f. Defective ignition/starter switch g. Generator brushes do not make contact with commutator h. Regulator faulty	a. Charge battery. See **4.3.** b. Replace battery. See **4.2.** c. Replace bulb. See **14.3.** d. Clean and/or tighten terminals. See **4.1.** e. Tighten and/or repair wires. See **15.** f. Replace ignition/starter switch. See **7.** g. Free the brushes or replace them. If necessary, replace the brush springs. See **6.5.** h. Replace regulator. See **6.3.**
2. Light stays on or flickers when engine is running	a. Fan belt broken or slips badly b. Regulator faulty c. Loose connections or broken wire d. Generator faulty e. Commutator bars short-circuited	a. Replace or adjust fan belt. See **LUBRICATION AND MAINTENANCE.** b. Replace regulator. See **6.3.** c. Test wires. Repair as necessary. See **15.** d. Test generator. Repair as necessary. See **6.2, 6.7, 6.6.** e. Clean commutator with crocus cloth. See **6.5.**
3. Light goes out only at high speed	a. Generator faulty b. Regulator faulty	a. Test generator. Repair as necessary. See **6.2, 6.7, 6.6.** b. Replace regulator. See **6.3.**
4. Light remains on with the ignition switched off	Regulator contact points sticking (burned)	Replace regulator. See **6.3.**

16 Electrical System

To test no-load voltage:

1. Disconnect the wire from regulator terminal B+ (51), being careful not to ground the lead. Connect a voltmeter as shown in Fig. 6-1.

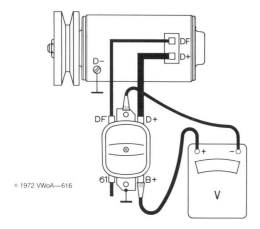

Fig. 6-1. Hook-up for no-load test. Positive voltmeter lead replaces wire normally attached to terminal B+ (51).

2. Start the engine, then slowly increase engine speed to approximately 2000 rpm. The voltmeter should indicate 13.5 to 14.5 volts.

3. When the engine is turned off, the voltmeter reading should drop from 12 to 0 volts just before the engine stops completely. This indicates that the regulator points are not sticking.

4. Check the cut-in speed by temporarily reducing the engine idle to 700 rpm. The voltmeter should suddenly rise from zero as engine speed reaches 760 to 770 rpm.

Load regulating voltage can be tested with the hook-up shown in Fig. 6-2. The test instruments consist of a voltmeter with a range of 0 to 30 volts, an ammeter with a range of −50 to +50 amps, and a rheostat that can be loaded to 50 amps.

With the engine running at approximately 1500 rpm, adjust the rheostat to a load setting of 45 amps. The voltage should be 12.8 to 13.8 volts under these conditions. If it is not, the regulator should be replaced.

Use the hook-up shown in Fig. 6-3 to determine whether the trouble is in the regulator or in the generator. If there is no voltage, the regulator is faulty and must be replaced before you can make the test. Use an 8-gauge or larger jumper wire and limit the test's duration to 15 seconds. At 1500 generator rpm (790 engine rpm), the generator's voltage output should be 12 volts. At double this rpm, the voltage output should be 36 volts. If these specifications are met, the trouble is in the regulator. If not, remove and disassemble the generator for the tests described in **6.7 Testing Disassembled Generator**.

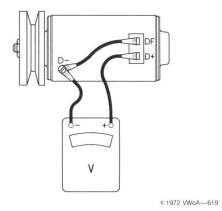

Fig. 6-3. Testing generator without regulator.

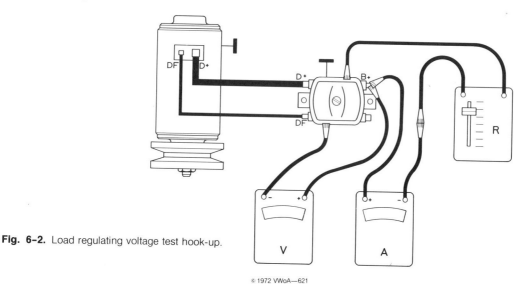

Fig. 6-2. Load regulating voltage test hook-up.

Electrical System

The return current test illustrated in Fig. 6-4 will detect a faulty cutout relay in the regulator unit.

CAUTION —
The D+ terminals on the regulator and generator must not be disconnected while the engine is running. This could burn out the generator's field windings.

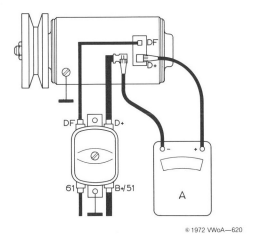

Fig. 6-4. Hook-up used for return current test.

After installing the ammeter as shown, start the engine and reduce the idle speed until the ammeter needle moves into the negative (discharge) range. Turn the engine off. The ammeter needle should jump back to zero before the engine has stopped completely. The maximum discharge reading obtained is the return current, which should be about 2.0 to 7.5 amps. If the ammeter does not return to zero when the engine is completely stopped, the regulator is defective and must be replaced.

6.3 Removing and Installing Regulator

On Beetles, the regulator is under the rear seat at the left-hand side of the car. On Karmann Ghias the regulator is on the left-hand side of the engine compartment near the battery. Always stop the engine and disconnect the battery ground strap before removing the regulator. Attach a tag to each wire as you remove it so that you can later identify the wires and install them on their proper terminals. Be very careful not to interchange wires DF and D+ or B+ and 61. Either error could seriously damage the charging system.

The regulator is held to the car body by two Phillips head sheet metal screws. When installing the regulator, be sure that it makes a good connection with ground. This can be checked with an ohmmeter by testing between the regulator base and an unpainted part of the car. A reading of 0 ohms should be obtained.

6.4 Removing and Installing Generator or Alternator

Before you can remove the generator or alternator, you must partially raise the fan housing as described in conjunction with cooling air system removal in **ENGINE**. With the fan housing raised, remove the four bolts indicated in Fig. 6-5, then remove the alternator or generator and the fan from the fan housing.

Fig. 6-5. Bolts (arrows) that hold fan cover to fan housing.

If necessary, you should polarize generators as shown in Fig. 6-6.

CAUTION —
Polarization of an alternator is both unnecessary and damaging to the alternator.

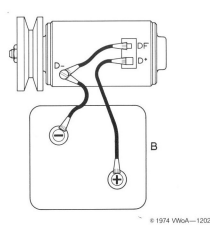

Fig. 6-6. Battery wired to polarize the generator.

You do not need to polarize the generator unless it is a new generator, a rebuilt generator, or a used generator

18 Electrical System

that has been stored for several weeks or more. Also polarize generators that have undergone electrical testing while disassembled. During polarization, the generator should be allowed to run briefly as a motor in order to give the pole shoes residual magnetism of the correct polarity. Without residual magnetism, the generator will not generate enough electricity to close the cut-out relay in the regulator when the engine is started.

If you must disassemble the generator or alternator before you install it, or if you must replace the fan, you can remove the fan from the shaft by taking off the special nut shown in Fig. 6–7. Two nuts hold the fan cover to the generator or alternator.

Always install the fan cover in its original position on the generator. When you install the fan on the generator or alternator shaft, arrange the 0.50-mm (.019-in.) shims so that the clearance between the fan and the fan cover is 2 mm (.080 in.). Make certain that you install the lock washer so that its raised periphery is toward the special nut. Torque the special nut to 5.5 to 6.5 mkg (40 to 47 ft. lb.).

Install the generator or alternator in its original position on the fan housing. After you have installed the generator or alternator on the fan housing, hand-turn the shaft to make sure that the fan is not rubbing against the fan housing. If the shaft turns freely without noise, you have done the job correctly. If the fan rubs, either the fan housing is distorted or the fan shims were incorrectly installed. Install the fan housing and other parts as described in **ENGINE**.

6.5 Replacing Generator Brushes and Servicing Commutator

You can remove the generator brushes through the inspection ports in the generator housing. Lift the brush springs with a wire hook so that the old brushes can be slipped out of their holders. To free their connecting wires, remove the terminal screws. The generator brushes should be replaced whenever they are so worn that they no longer protrude above the tops of the brush holders.

Clean the commutator with trichloroethylene or a similar solvent before installing the new brushes. Bent or rusted brush springs should be replaced. If commutator out-of-round exceeds 0.015 mm (.0006 in.), the generator must be disassembled so that the commutator can be machined. The normal diameter of the commutator is 33.40 mm (1.315 in.). It should not be turned down below 32.80 mm (1.291 in.) when machined. After machining, undercut the insulation between the commutator bars until it is about 0.50 mm (.020 in.) below the surface of the bars themselves.

6.6 Disassembling and Assembling Generator

You may find it necessary to disassemble the generator for electrical tests, commutator servicing, or replacement of bearings or other internal parts. For basic disassembly, simply remove the two nuts that hold the fan cover to the generator and then withdraw the two long through bolts. Further disassembly can be done with only a screwdriver, although it may be necessary to use a press or bearing puller to remove the ball bearing races from the armature shaft. Complete generator disassembly is illustrated in Fig. 6–8.

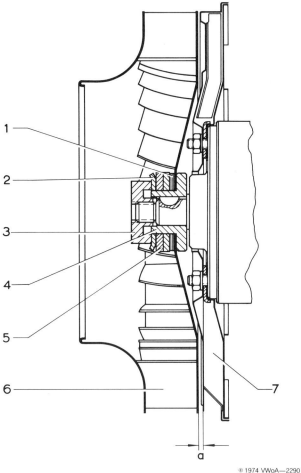

1. 0.50-mm (.019-in.) shims
2. Lock washer
3. Special nut
4. Hub
5. Carrier plate (2)
6. Fan
7. Fan cover assembly

a. Distance between cover and fan cover

Fig. 6–7. Fan installation. During assembly, arrange shims to obtain a 2-mm (.080-in.) clearance at **a**.

Electrical System 19

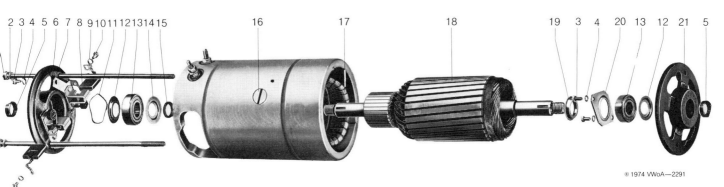

1. Housing through bolt (2)
2. B 6 lock washer (2)
3. AM 4 × 8 fillister head screw (3)
4. B 4 lock washer (3)
5. Spacer ring (2)
6. Brush spring (2)
7. End plate, commutator end
8. Carbon brush (2)
9. B 4 lock washer (2)
10. AM 4 × 6 fillister head screw
11. Lock washer
12. Dished washer (2)
13. Ball bearing (2)
14. Splash shield
15. Thrust washer
16. Pole shoe screw (2)
17. Field coil (2)
18. Armature
19. Splash shield
20. Retaining plate
21. End plate, fan end

Fig. 6-8. Generator disassembly, Remove through bolts to separate armature and end plates from housing.

Using a press and suitable support plates, you can press the armature out of the end plate as shown in Fig. 6-9. The cutout in the thin support plate should have a radius of 11 mm (7/16 in.) so that the bearing will not be damaged. The same support plate can be used for pressing the bearing off the commutator end of the armature.

To assemble:

1. Check the brushes and commutator for wear. If necessary, service them as described in **6.5 Replacing Generator Brushes and Servicing Commutator.**

   NOTE
   The commutator should be smooth, grayish black, and clean. If it is greasy, clean it with trichloroethylene or a similar solvent.

2. Press the bearing onto the commutator end of the armature. Then press the end plate onto the fan end of the armature as shown in Fig. 6-10.

Fig. 6-9. Pressing armature shaft out of end plate at fan end. Same support plate set-up can be used to press bearing off opposite end of armature.

Fig. 6-10. Pressing end plate onto armature.

3. Using a repair press, install the spacer rings.
4. To ensure alignment, fit the end plate notches over the projections on the housing as the parts are assembled. See Fig. 6-11.

Fig. 6-11. Notch (arrow) that must fit over projection on housing. The armature must turn freely following installation of the through bolts.

6.7 Testing Disassembled Generator

If you have tested the charging system, as described in **6.2 In-car Testing of Generator and Regulator** and have determined that trouble lies in the generator, further tests can be made after the generator has been removed and disassembled. Several tests can be made with a simple battery-powered test light, although an ohmmeter is preferable. A resistance-measuring bridge is recommended for testing the field windings for shorts. However, an ammeter used in conjunction with a battery can also be used. Compare the current draw through the two field coils. If one coil consumes 0.5 amperes or more current than the other, it is shorted.

Check the armature for open circuits, shorts, and grounded windings. Burn marks between the commutator segments usually indicate open circuits, which often occur when windings come unsoldered from their commutator bars. Unsoldered windings are usually the result of overheating caused by a faulty regulator. In such cases, an electrical specialist may be able to repair the armature.

Short circuits occur when the insulation between armature windings breaks down. It is usually best to replace shorted armatures. An armature tester called a growler is used to test for shorts. Growlers are available at most automotive supply shops that offer machine shop service. The shop will test an armature for you at low cost.

Armature grounds occur when insulation breaks down and allows the windings to come into electrical contact with the armature laminations or the shaft. It is easy to detect this condition with a battery-powered test light or with an ohmmeter as shown in Fig. 6-12.

Fig. 6-12. Armature ground test. If testing for continuity between the commutator and the armature body or shaft produces an ohmmeter reading or causes a battery-powered test lamp to light, the windings are grounded.

Occasionally, carbon dust from the brushes will short an armature. If arcing current has not permanently damaged the insulation, cleaning with trichloroethylene or a similar solvent will usually cure such shorts.

Testing Field Windings

Shorted field windings result from faulty insulation that allows one strand of wire to come into electrical contact with another. Because shorts reduce the strength of the magnetic field, they limit generator output. Shorts can be detected by the ammeter-and-battery test described earlier, or by the test shown in Fig. 6-13. For this test a resistance measuring bridge is essential.

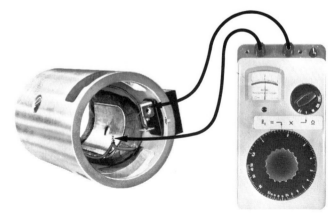

Fig. 6-13. Testing for shorted windings. A resistance measuring bridge should be used to test each field coil. If one coil has lower resistance than the other, it is shorted.

ELECTRICAL SYSTEM 21

In addition to short circuits between strands of wire in the field coils, the generator windings can also become short-circuited to the generator housing. To locate such trouble, test for continuity between the field windings and the housing. These parts should not be in electrical contact. Although Fig. 6-14 shows an ohmmeter making the continuity test, a battery-powered test light will work equally well. The ohmmeter should produce no reading, and a test lamp should not light.

Fig. 6-14. Testing for shorts to ground. If a test between the field coils and generator housing reveals electrical contact, the windings are shorted to ground.

If there is no charging system output, and if the tests described in **6.2 In-car Testing of Generator and Regulator** indicate that the generator is faulty, first check the brushes and commutator. If the brushes and commutator are in good condition, disassemble the generator and test for open field windings as shown in Fig. 6-15. Burned out windings usually result from excessive current loads caused by a faulty regulator or by defective charging system wiring. Repeat the in-car tests after repairing or replacing the generator so that you can detect and repair any remaining charging system faults.

Fig. 6-15. Testing field coils for open circuits. Testing between two terminals should cause the ohmmeter to register or the test lamp to light. If it does not, the windings are open. Note that the positive brush must not touch the housing.

6.8 Testing and Repairing Alternator and Regulator

Beginning with Chassis No. 113 2414 931, some 1973 cars are equipped with alternators that have separate regulators. All 1974 and later cars have alternators. But, since late in the 1974 model year, the alternator has had an integral-type regulator mounted atop it. Determine whether the alternator has a separate regulator or an integral regulator before you test the alternator and regulator. The test specifications are different for each kind of alternator/regulator combination.

Fig. 6-16 shows the hook-up for testing alternator output. The positive battery cable must be disconnected and a battery cutout switch installed as shown, with the cable reconnected to the cutout switch.

The ammeter and the voltmeter used in making the test should be fairly sensitive and accurately calibrated. Ordinary dashboard-type instruments are not suitable. The variable resistance is a standard test device that can be obtained from an automotive supply store. If you use an alternator tester that consists of suitable gauges and a variable resistance built into one unit, please follow the test instructions supplied by the manufacturer of the test equipment. The test hook-up shown in the illustration is applicable to both kinds of alternators.

CAUTION —
An alternator must never be run without a battery. Doing so will severely damage the alternator, the regulator, or both.

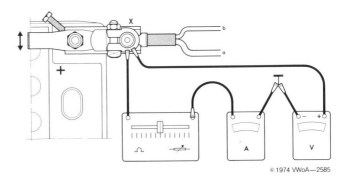

a. To starter
b. To regulator

X. Battery cutout switch (SUN electric No. 7052-003 or similar equipment)

Fig. 6-16. Alternator charging system test. Variable resistance, ammeter, and voltmeter are connected in series as shown. Note that the ammeter lead and the negative voltmeter lead connected to it are grounded to the chassis.

22 Electrical System

To test alternator and regulator:

1. With the engine stopped (alternator stationary), disconnect the battery ground strap. Then make the test connections and reconnect the battery ground strap.

2. Close the cutout switch as shown in Fig. 6-17, then start the engine.

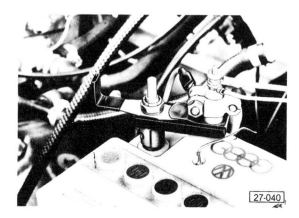

Fig. 6-17. Battery cutout switch in closed position. Notice the test connections.

3. In testing alternators with separate regulators, run the engine at 2500 to 3000 rpm, then adjust the variable resistance so that the ammeter gives a reading of 25 amps. In testing alternators with integral regulators, run the engine at 3000 rpm, then adjust the variable resistance so that the ammeter gives a reading of 45 amps.

4. Move the cutout switch as shown in Fig. 6-18 in order to cut the battery out of the test circuit. The load current is now determined by the variable resistance.

Fig. 6-18. Battery cutout switch in open position. Battery shown here is not installed in one of cars covered by this Manual.

5. Readjust the variable resistance, if necessary, so that the ammeter reading is 25 amps at 2500 to 3000 rpm (separate regulator) or 45 amps at 3000 rpm (integral regulator).

6. Read the voltage indicated on the voltmeter. It should be between 12.5 and 14.5 volts. If not, install a new regulator and repeat the test.

NOTE ——
Remove separate regulators as described in **6.3 Removing and Installing Regulator or Alternator.** To remove an integral regulator, take out the two screws that hold it to the top of the alternator. Then lift out the combined brush holder and regulator and unplug the regulator from the brush holder.

If the charging system output is incorrect even with a new regulator, inspect the brushes to see that they are not worn out, sticking, or dirty. (On alternators with integral-type regulators, the brush holder is located beneath the regulator.) Check that the brush slip rings on the alternator's rotor are not grease-coated. Clean or replace faulty brushes. The brushes should be considered worn out if the remaining length of either brush is less than 10 mm (⅜ in.). This should be measured from the surface of the brush holder to the tip of the brush with the brush holder removed from the alternator.

If the brushes are not faulty, replace the alternator. Further testing of either the regulator or the alternator is of no value because, aside from regulators and brush holder assemblies, individual replacement parts are not available.

Noisy Alternator

Alternator noises are usually mechanical in origin but a high, soft whistling sound may be produced by an alternator that is overcharging because of a faulty regulator or an open diode. The same sound may be heard if there is a shorted diode that is placing abnormal electrical strain on the alternator. Some alternators make this sound when operating normally at maximum output.

Alternator mechanical noises are usually the result of misalignment between the V-belt and the pulley, a loose or broken pulley, worn bearings, or a bent rotor shaft. If any of these faults is found, replace the alternator or, where applicable, correct its installation.

Replacement Engines

Replacement engines with the Motorola alternator have regulator wiring different from that of previous cars with separate regulators. A diagram of the revised wiring used with the integral-regulator Motorola alternator appears as Diagram 0 in **15. Wiring.** When installing a replacement engine with the Motorola alternator in a car

©1978 VWoA

ELECTRICAL SYSTEM 23

that originally had the Bosch alternator with a separate regulator, the wiring must be modified.

To modify wiring:

1. Remove the rear seat. Disconnect the battery ground strap.
2. Remove the old regulator, leaving the connector housing and the four wires in place.
3. In the engine compartment, remove the red D+ wire from the multiple connector on the old alternator.
4. Cut the green DF wire and the brown D− wire close to the plastic insulation tube—insulating the cut wires with electrical tape.
5. Remove and discard the multiple connector, which was attached to the old alternator.
6. Connect the red wire to the new alternator's D+ terminal—when necessary by installing a suitable electrical terminal on the end of the wire.
7. Connect the heavy-gauge B+ wire and, where fitted, the computer analysis wire to terminal B+ on the new alternator.

7. REMOVING AND INSTALLING IGNITION SWITCH

Beginning with the 1972 models, the direct connection of a new-type ignition switch via a plug on the wiring harness makes it possible to remove and install the lock cylinder and the ignition switch without having to remove the steering lock beforehand. The turn signal and wiper switches must, however, first be removed. The new kind of switch is shown in Fig. 7-1.

The procedure for removing and installing the ignition switch lock cylinder is similar for each of the model years covered by this Manual. However, the parts involved have received minor annual modifications.

CAUTION —

Before you begin working on the switch, disconnect the battery ground strap. Otherwise, short circuits could damage the electrical system.

To remove cylinder:

1. Remove the steering wheel as described in **FRONT AXLE** or in **STRUT FRONT SUSPENSION.**
2. Remove the circlip and the plastic bushing from the steering column. Remove the turn signal switch and, where applicable, the wiper switch. Then remove the steering/ignition lock retainer plate.
3. On 1972 and later cars, unplug the wiring harness from the switch; on earlier cars, push the wires through while you withdraw the steering ignition lock far enough to work on it.
4. On 1976 and later cars, inspect the lock assembly's housing. If there is no hole at the point indicated in Fig. 7-2, carefully drill a hole that will allow you to insert a nail or a piece of steel wire.

CAUTION —

Remove the turn signal and windshield wiper switches before you attempt to drill the hole. The sealing plate should not be removed, as doing this may break off the lug on the turn signal switch.

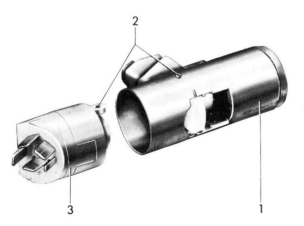

Fig. 7-1. Typical late-type ignition switch. Steering lock is at **1**, setscrew holes at **2**, ignition switch at **3**.

Fig. 7-2. Location for hole in ignition switch housing. Dimension **a** and dimension **b** are both 11 mm.

24 Electrical System

5. Using a nail or a piece of steel wire inserted through the opening, depress the retaining spring on the lock cylinder.

6. Remove the lock cylinder by turning the key and pulling the lock cylinder out of the housing as shown in Fig. 7-3. On 1970 cars it is not necessary to turn the key unless a later-type lock cylinder has been installed.

Fig. 7-3. Removing ignition switch. Wire must be inserted in opening at **A**. Key is withdrawn by pulling in the direction indicated by arrow **C** and turning in the direction indicated by arrow **B**. It may also be necessary to pry outward slightly on the cylinder using a screwdriver.

To install cylinder:

1. Check the slide pin shown at the black arrow in Fig. 7-4. With the pin depressed slightly, you should be able to move the steering column lock's slide to the right until the slide pin is flush with the housing.

Fig. 7-4. Pins in lock housing. Slide pin (black arrow) must move freely. Guide pin (white arrow) must not be bent or broken.

2. If the lock cylinder has a black plastic cap (Fig. 7-5), install the cylinder by pushing it in without the key. Then use the key to turn the cylinder to the left and check the operation. The earlier-type lock cylinder without the black plastic cap must be inserted with the key installed in the lock cylinder. Then turn the cylinder to the left stop position and remove the key.

CAUTION —

Make sure that the switch is in its locked position before you attempt to install the lock cylinder. Do not attempt to install the late-type cylinder with the key installed. By taking these precautions you will avoid damage to the lock. The cylinders with black caps have been used as replacement parts in earlier cars since 1972.

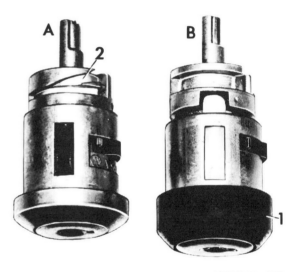

Fig. 7-5. Lock cylinders for cars with locking steering columns. New version (**B**) has black plastic cap (**1**). Early version (**A**) has a helical groove for pin.

3. If you have installed the early-type lock cylinder with the helical groove for the pin, check that the key pulls itself into the lock cylinder housing when the lock cylinder is turned. (This check is unnecessary with the late-type lock cylinder that has the black plastic cap.)

4. Reconnect the electrical wires and, where applicable, reinstall the lock assembly in the housing.

5. Install the plastic bushing and the circlip on the steering column, making sure that the circlip fully engages the groove.

6. Reinstall the steering wheel as described in **FRONT AXLE** or in **STRUT FRONT SUSPENSION**.

ELECTRICAL SYSTEM

8. LIGHTS AND SWITCHES

To replace bulbs in the taillights, license plate light, front turn signals, or parking lights, remove the Phillips head screws in the lens. Then remove the lens and the gasket from the car. This will provide access to the bayonet-base bulbs. Make certain that you install the gasket properly and that it is in serviceable condition. Instrument panel light bulb replacment is discussed in **14.3 Removing and Installing Instruments.**

8.1 Replacing Sealed Beam Units and Aiming Headlights

You can remove the headlight trim ring by loosening screw **a** shown in Fig. 8-1. The screw is permanently installed in the trim ring and can also be used as a handle to pull the trim ring off the car. Remove the sealed beam unit and retaining ring by taking out the three short screws at **b** in Fig. 8-1. You may then pull the sealed beam unit forward and remove the cable connector from its terminals.

CAUTION
Do not alter the position of the long headlight adjustment screws at **c** in Fig. 8-1. If you do, you will have to readjust the headlights.

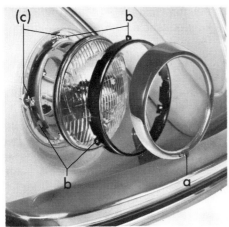

Fig. 8-1. Sealed beam removal. Screw at **a** holds trim ring. Screws at **b** hold retaining ring for sealed beam unit. Screws at **c** are for headlight adjustment only.

When you install a sealed beam unit, be sure its three glass lugs correctly engage the support ring. Install the retaining ring, then hook the trim ring over the upper lug. Make certain that the rubber gasket is properly positioned before you tighten the trim ring mounting screw. It should not be necessary to aim the headlights after changing a sealed beam unit. If you are in doubt, however, the aim should be checked.

Each headlight has two adjusting screws. Adjust vertical aim with the top screw shown in Fig. 8-1. Adjust lateral aim with the screw at the side.

NOTE
Check your state laws to determine whether adjustments must be made by a licensed shop. Your state may also have laws specifying different aiming from that described here.

To adjust:

1. Position the car on a level surface 7.62 m (25 ft.) from a vertical wall. Have the fuel tank about half-filled and make certain that the tire pressures are correct.

2. Roll the car back and forth a few yards to settle the suspension. Then load the driver's seat with one person or a weight of 70 kg (154 lb.)

3. Remove the headlight trim rings. Turn on the low beams and cover one of the headlights. The uncovered light's upper and left edges of high intensity must be in the position shown in Fig. 8-2.

4. Repeat procedure **3** on the opposite headlight and adjust as necessary. The high beams will automatically be in adjustment once the low beams are aimed to the proper specifications.

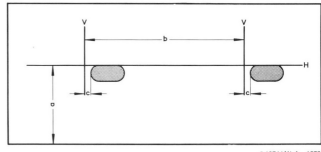

Fig. 8-2. Aiming target on vertical wall. Shaded areas are zones of high light intensity. Vertical lines **(V)** and horizontal line **(H)** intersect at the headlight centers. Dimension **a** is the headlight's height above ground and **b** is the 1260 mm (49 5/8 in.) between the headlights. Dimension **c** equals 50.8 mm (2 in.).

8.2 Removing and Installing Switches in Dashboard

The dashboard-mounted switches installed in most Type 1 vehicles are either push-pull or rotary type. However, 1973 and later Super Beetles and Convertibles are equipped with rocker-type switches.

26 Electrical System

Separate procedures are given here for removing and installing push-pull or rotary-type switches and for removing and installing rocker-type switches.

To remove (push-pull or rotary-type):

1. Disconnect the battery ground strap.
2. Unscrew the knob from the switch.
3. Using a special wrench, remove the escutcheon that holds the switch in the dashboard. See Fig. 8-3.

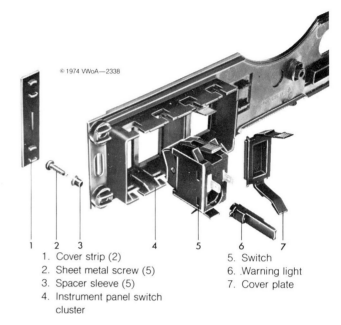

1. Cover strip (2)
2. Sheet metal screw (5)
3. Spacer sleeve (5)
4. Instrument panel switch cluster
5. Switch
6. Warning light
7. Cover plate

Fig. 8-4. Rocker-type switch removal.

1. Light switch
2. Escutcheon
3. Light switch knob
4. Light switch cap

Fig. 8-3. Typical push-pull or rotary-type switch. This is a main lighting switch but other push-pull or rotary switches are removed similarly. Bleed the air from the windshield washer reservoir before removing the hoses from the wiper switch.

4. Open the front hood. Remove the dashboard access panel from the rear of the luggage compartment by removing the two knurled nuts.
5. Pull the switch toward the front of the car and out of the dashboard.
6. Labeling each wire so that you will be able to return it to its correct terminal during installation, disconnect all wires from the switch.

Installation is the reverse of removal. Make sure the notches in the body of the switch engage the projections in the hole in the dashboard.

To remove (rocker-type switch):

1. Disconnect the battery ground strap.
2. Carefully pry off the cover strip at each end of the instrument panel switch cluster. Pry off the cap above the steering column. See Fig. 8-4.

3. Remove the five sheet metal screws that hold the instrument panel switch cluster. There are two screws at each end of the cluster and one in the center.
4. Pull the cluster out of the dashboard. Squeeze the knurled tabs on the switch toward the switch body, then pull the switch out toward the back of the instrument panel switch cluster.
5. Labeling each wire so that you will be able to return it to its correct terminal during installation, disconnect all wires from the switch.

Installation is the reverse of removal. The instrument panel switch cluster must also be removed in order to replace the bulbs in warning lights such as the one shown in Fig. 8-4.

Removing Door Contact Switches

You can remove the door contact switches as illustrated in Fig. 8-5.

Fig. 8-5. Door contact switch removal. Remove screw **3**, withdraw switch **2**, and disconnect wire. Seal **1** comes off along with the switch.

Electrical System 27

8.3. Dimmer Relay

Beginning with the 1971 models, Super Beetles and Convertibles are equipped with a fuse box that has an integral relay console. The dimmer relay for low beam operation is located as shown in Fig. 8-6. To replace it, remove the fuse box attaching screws and lower the fuse box and console. Then unplug the faulty relay and install a new one. All other models have a separate relay mounted on a bracket above the fuse box. The relay can be reached by removing the access panel from the rear of the front luggage compartment.

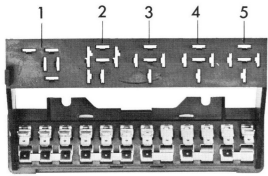

1. Low beam
2. Open
3. Turn signal/emergency flasher
4. Open
5. Buzzer

Fig. 8-6. Relay locations on fuse box console (Super Beetles and 1971 or later Convertibles).

9. Turn Signals and Emergency Flashers

The turn signal switch and relay are often mistakenly blamed for troubles caused by dirty, corroded, or loose-fitting turn signal bulb contacts. Before starting to troubleshoot either the turn signal switch or the turn signal/emergency flasher relay, be sure that all bulb contacts are clean and tight.

9.1 Testing Turn Signals and Emergency Flashers

The turn signals and emergency flashers share the same relay. If the emergency flashers work but the turn signals do not, you can be certain the relay is not faulty. The same is true if the turn signals work but the emergency flashers do not. The following test will help you avoid unnecessarily replacing a good relay.

Testing Turn Signal/Emergency Flasher Relay

With the ignition off, connect terminal + or −49 (depending on the relay manufacturer's designation) of the relay to terminal 30 of the fuse box. Operate the turn signal switch both ways. If the turn signals on both sides light up, the relay and turn signal switch are not defective. If they do not light up, test first the emergency flasher switch and then the turn signal switch.

To test emergency flasher switch:

1. Disconnect the battery ground strap. Then remove the emergency flasher switch and turn it on.
2. With an ohmmeter, test to make sure there is continuity between terminals 30 and + and also between terminal 49a and terminals **R** and **L**.
3. If the resistance is greater than zero ohms, replace the emergency flasher switch.

To test turn signal switch:

1. At the plug guide, disconnect the black/green/white wire to the turn signal switch.
2. With an ohmmeter, test for continuity between terminal 54BL on 1970 cars or terminal 49a on later cars and terminals **R** and **L** of the turn signal switch.
3. If the resistance is greater than zero ohms, replace the turn signal switch.

Replacing Turn Signal/Emergency Flasher Relay

The relay for the turn signals/emergency flashers is located near the headlight dimmer relay. Its position on 1971 and later Super Beetles and Convertibles is shown in Fig. 8-6. All other models have a separate relay mounted on a bracket above the fuse box. The relay can be reached by removing the access panel from the rear of the front luggage compartment.

Always disconnect the battery ground strap before removing the relay. Attach identifying tags to each wire as it is removed so that you will be able to return it to the proper terminal on the new relay. Defective relays cannot be repaired and must be replaced.

9.2 Removing and Installing Steering Column Switches

A number of changes have been made to the steering column-mounted switches during the model years covered by this Manual. A shorter switch housing, bolted to the steering column tube, was introduced on 1971 models. The 1970 models have a longer switch housing that bolts onto the body. In 1972 the windshield wiper switch was moved to the steering column. The turn signal switch and housing were again modified at this time to accommodate the extra lever, the wiring, and the windshield washer water valve and hoses.

28 Electrical System

To remove turn signal switch:

1. Disconnect the battery ground strap.
2. Turn the ignition key to the on position. Place the turn signal lever in its central position.
3. Remove the steering wheel as described in **FRONT AXLE** or **STRUT FRONT SUSPENSION.**
4. Remove the circlip from the groove below the steering wheel splines on the steering column.
5. On 1970 and 1971 cars only, disconnect the wires to the turn signal switch from behind the dashboard. Then push the wires through the opening in the body toward the switch.
6. Remove the four turn signal switch mounting screws shown in Fig. 9-1.
7. Pull the turn signal switch upward and off the steering column.

   NOTE —
   On 1972 and later cars the turn signal and windshield wiper switches must be removed together and then separated after removal. These switches have guide channels that receive the wires from the switches. The guide channels unplug from a connector in the bottom of the steering column switch housing. Bleed the air from the windshield washer reservoir before you remove the water hoses from the valve on the switch.

Fig. 9-1. Screws holding turn signal switch.

During turn signal switch installation, be sure that the contact ring between the ball bearing and the steering column is positioned properly and that the turn signal switch lever is in its central position. Otherwise, the canceling cam will be damaged by the tongue of the contact ring when the steering wheel is installed. On 1972 and later cars, make sure that the wires for the turn signals are not crushed by contact with the windshield wiper switch at the two points indicated in Fig. 9-2.

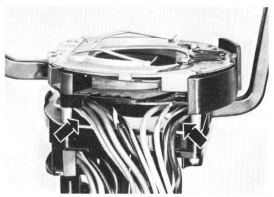

Fig. 9-2. Points where wires must not be caught and crushed between the turn signal and wiper switches.

Following turn signal switch installation, adjust the clearance between the steering column switch housing and the steering wheel hub as described in the procedure that follows.

Adjusting Turn Signal Switch

The distance between the steering wheel hub and the steering column switch housing must be 2 to 3 mm (about $1/16$ to $1/8$ in.). To adjust the distance, loosen the two screws that secure the switch to the car body or steering column tube and then move the switch housing in the slotted holes.

CAUTION —
Move the turn signal switch only when its lever is in the central position. Otherwise, the canceling cams may be damaged.

10. Windshield Wipers and Washer

Operation of the two-speed windshield wipers is controlled by turning a rotary switch on 1970 and 1971 cars, or by depressing a lever on the steering column on later models. The blades park automatically when turned off. A push button in the center of the rotary switch actuates the windshield washers. On later models, lifting the column-mounted lever toward the steering wheel triggers the washers. The washers are not electrical. They operate by compressed air from the spare tire. To replace the water valve, see **9.2 Removing and Installing Steering Column Switches**.

10.1 Windshield Wiper Motor Troubleshooting

Table i is designed to help you determine the cause and remedy for windshield wiper motor malfunctions. The numbers in the Remedy column refer to headings in this section where the suggested repairs are described.

ELECTRICAL SYSTEM 29

Table i. Windshield Wiper Motor Troubleshooting

Problem	Probable Cause	Remedy
1. Windshield wiper motor does not work, operates too slowly, cuts out, or comes to a standstill	a. Brushes worn b. Brush tension spring too weak c. Brushes stuck in their holders d. Commutator dirty e. Moving joints of windshield wiper linkages dry or jammed f. Battery voltage too low g. Armature burned by short circuits h. Switch faulty or wires loose	a. Replace brushes. See **10.4**. b. Replace tension spring. See **10.4**. c. Free brushes. See **10.4**. d. Clean commutator. See **10.4**. e. Thoroughly lubricate all moving joints with universal grease; eliminate jamming. See **10.3**. f. Charge battery; check cables and connections. See **4.1, 4.3**. g. Replace motor or armature. See **10.2, 10.3**. h. Replace switch; repair faulty connections. See **10.4**.
2. Windshield wiper motor continues to run or fails to return blades to parking position after manual switch is turned off	a. Contacts in cover damaged b. Contact mount (insulation plate) broken c. Contacts dirty d. Poor connection from terminal 31b via wiper switch to ground	a. Replace cover. See **10.4**. b. Replace cover. See **10.4**. c. Clean contacts. See **10.4**. d. Check connection; replace parts if necessary. See **10.4**.
3. Wiper linkage squeaks. Motor operates slowly. Armature is burned	a. Moving joints of windshield wiper linkages need grease b. Gear shaft lacks sufficient end clearance c. Drive housing not correctly positioned on motor	a. Thoroughly grease all moving joints with universal grease. See **10.3**. b. Adjust axial play. See **10.4**. c. Install cover properly. See **10.4**.

In-car Testing

The operating condition of the windshield wipers can be evaluated by checking the current draw (at terminal 30) at both low and high speeds. The current draw should be about 2.5 amps at low speed and 3.5 amps at high speed. The test should be made with the wiper blades pulled away from the windshield. Otherwise, the resulting friction will cause your ammeter readings to be inaccurate.

If the wipers make squeaking noises and run slowly with a high current draw, the probable cause is inadequate bearing lubrication. If current draw is high without squeaking noises, the wiper motor's armature is probably shorted.

10.2 Removing and Installing Wiper Motor and Frame

The wiper system introduced along with the larger windshield and new instrument panel on 1973 Super Beetles and Convertibles cannot be interchanged with those of other models. The basic design, however, is the same. Variations in the removal procedure will be pointed out where applicable.

To remove:

1. Disconnect the battery ground strap.
2. Remove the wiper arm cap nuts, then take off the wiper arms.
3. Remove the seals for the wiper shafts. Then remove the nuts, washers, and outer bearing seals.
4. On 1973 and later Super Beetles and Convertibles, remove the cover from the fresh air box by taking out the screws around its edge. (This cover is located at the rear of the front luggage compartment.) On other models, remove the access cover at the rear of the front luggage compartment by taking off the two knurled nuts.
5. Except on 1973 and later Super Beetles and Convertibles, remove the fresh air box as described in **BODY AND FRAME**. Then take out the glove compartment and remove the right-hand fresh air vent.

   NOTE ━━
   No disassembly of the fresh air system is required on 1973 and later Super Beetles and Convertibles. The wiper frame and motor are contained within the fresh air box as shown in Fig. 10-1.

6. Remove the bolt that holds the motor drive gear cover to the rubber mounting and bracket.
7. Unplug the wires to the wiper motor.
8. Pull the wiper bearings downward through the holes in the cowl panel, then lower the frame until it can be removed from the car with the wiper motor still attached.

30 Electrical System

9. On 1973 and later Super Beetles and Convertibles, remove the plastic rain cover as illustrated in Fig. 10-1.

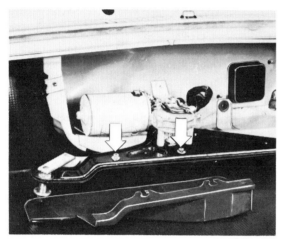

Fig. 10-1. Plastic rain cover removal. Loosen screws indicated by arrows, then lift off cover shown in foreground.

The wiper motor and frame is shown removed in Fig. 10-2. The wire connections are not the same on all models; motors for late models differ from those used in 1970 and 1971.

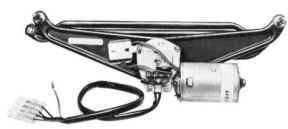

Fig. 10-2. Wiper motor and frame.

Installation is the reverse of removal. Make sure that the wiper shafts are perpendicular to the surface of the windshield and that the bearing washers and seals are installed in their proper order. If in doubt, refer to **10.3 Replacing Wiper Shaft and Wiper Shaft Bearing**. When installing the wiper arms, torque the cap nuts to 42 to 60 cmkg (37 to 52 in. lb.). If necessary, the correct installation of the wires can be determined by studying the wiring diagrams presented in this section.

Removing and Installing Wiper Motor

You can remove the wiper motor from the frame for servicing or replacement, but only after you remove the frame—with the motor still attached—from the car.

Two bolts hold the motor to the frame. But before unbolting the motor, remove the E-clip and the spring washer from the drive crank pin, then detach the connecting link. Lubricate the drive crank pin with multipurpose grease during assembly.

10.3 Replacing Wiper Shaft And Wiper Shaft Bearing

The wiper shaft, with its bearing and related parts, is shown in Fig. 10-3. Note the order in which the nuts and washers are installed on the bearing's threaded exterior.

NOTE

Windshield wiper bearings now have a single seal rather than the inner and outer seals shown in Fig. 10-3. The single seal is a grooved grommet that fits into the car body. Such seals should be lubricated with glycerine or silicone spray so that they will not be pushed out of the body when the wiper frame is installed.

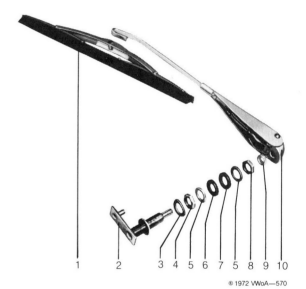

1. Windshield wiper blade
2. Wiper shaft in bearing
3. Spring washer
4. Brass nut
5. Washer
6. Inner bearing seal
7. Outer bearing seal
8. Nut
9. Wiper shaft seal
10. Bracket and arm

Fig. 10-3. Wiper shaft and bearing with related parts.

To remove bearing and shaft:

1. Disconnect the battery ground strap, then remove the wiper frame and motor.

2. Remove the E-clip and the spring washer from the wiper shaft crank pin. Then remove the connecting rod. In the same way, remove the connecting link if the left wiper shaft is being removed.

Electrical System 31

3. Pry the lockring out of the groove in the wiper shaft. Then withdraw the wiper shaft from the bearing complete with its small spring washer.

4. Remove the brass nut, then take the bearing out of the wiper frame.

Installation is the reverse of removal. Make sure that the groove in the bearing engages the projection on the wiper frame. If the connecting rod or connecting link bushings are worn, replace the rod or link.

10.4 Disassembling and Assembling Wiper Motor

The windshield wiper motor is shown disassembled in Fig. 10-4. The drive housing need not be taken apart if only the brushes and commutator require servicing.

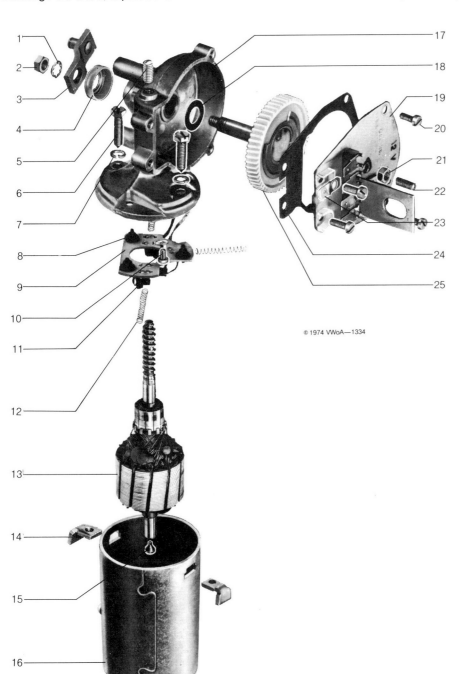

Fig. 10-4. Exploded view of windshield wiper motor.

1. Serrated washer
2. Nut
3. Drive crank
4. Plastic cap
5. Adjusting screw for armature end play
6. Hexagon head screw
7. Spring washer
8. Rubber mounting
9. Brush holder
10. Screw
11. Ground brush
12. Spring
13. Armature
14. Bracket
15. Thrust cone
16. Pole housing and permanent magnet
17. Drive housing
18. Thrust washer
19. Cover
20. Screw
21. Nut
22. Adjusting screw for gearshaft end play
23. Ground plug
24. Gasket
25. Drive gear and driveshaft

To disassemble wiper motor:

1. Remove the wiper motor and frame, then detach the motor from the frame. See **10.2 Removing and Installing Wiper Motor and Frame.**

2. Remove the two hexagon head screws. While slowly turning the drive crank to free the worm gear, lift the drive housing off the motor.

3. Inspect the brushes. If they are oily, clean them with trichloroethylene or a similar solvent. If they are worn, replace them.

   NOTE——
   The brush wires must be cut off at the brush guides before the old brushes can be removed. Place the new brushes in the guides, then solder the new brush wires to their terminals.

4. Lift the armature out of the pole housing. Inspect the commutator.

   NOTE——
   If the commutator is oily, clean it with trichloroethylene or a similar solvent. If the commutator is worn or burned it can be turned down to a minimum diameter of 21.50 mm (.846 in.). Permissible commutator out-of-round is 0.03 mm (.001 in.). After turning, undercut the insulating strips between the commutator bars. Then clean away all metal particles to prevent shorts.

Because the wiper motor uses permanent magnets for its field poles, there are no field coils to test. Armatures can be checked using the same test described in **6.7 Testing Disassembled Generator.** New armatures are available as replacements.

To disassemble drive:

1. Remove the four cover screws, then remove the cover and gasket.

2. Remove the drive crank nut and the serrated washer. Remove the drive crank and the plastic cap.

3. Remove the drive gear and shaft together with the thrust washer.

Assembly of the wiper motor and drive is the reverse of removal. Use universal grease to lubricate the armature and drive gear bearings. Before installing the drive crank, you must adjust the parking position.

To adjust parking position:

1. Connect the wiper motor to the wiper switch according to the terminal designations.

2. Connect the ground terminal of the wiper motor and terminal 31 on the switch to the battery's negative pole.

3. Connect terminal 30 on the switch to the battery's positive pole. Turn on the wiper switch and let the motor run for about 15 seconds.

4. Turn the switch off. The motor should stop in its correct parking position.

5. Install the drive crank. Position it at a 90° angle to the armature shaft when viewed from the end of the driveshaft.

The armature's end play should be 0.20 to 0.30 mm (.008 to .012 in.). The end play can be adjusted with the adjusting screw in the drive housing near the end of the armature. The drive gear shaft's end play should also be 0.20 to 0.30 mm (.008 to .012 in.) and can be adjusted with the screw and locking nut at the end of the drive gear shaft.

Removing and Installing Wiper Switch

Instructions applicable to rotary-type wiper switches are given in **8.2 Removing and Installing Switches in Dashboard.** Lever-type switches (mounted on the steering column) are removed as a unit with the turn signal switch. Follow the procedure given in **9.2 Removing and Installing Turn Signal Switch.**

10.5 Removing, Servicing, and Installing Wiper Blades and Arms

Before removing a wiper blade, first fold the wiper arm away from the windshield. Turn the blade as shown in Fig. 10-5.

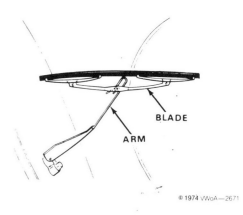

Fig. 10-5. Wiper blade turned to an angle with the wiper arm.

NOTE——
The pivot joint in the wiper arm has a built-in stop that prevents the blade from swinging to an excessive angle when the arm is folded away from the windshield.

With the blade angled until it is against the stop, lift the retaining spring and slide the blade down the wiper arm until the hook of the arm is off the pivot pin. You may then lift the blade off upward. The sequence is illustrated in Fig. 10-6.

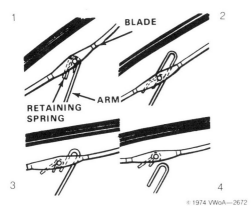

Fig. 10-6. Blade removal sequence.

To install wiper blade:

1. Position the new blade on the arm. The arm must enter through the hole in the blade frame on the side opposite the retaining spring.
2. Slide the blade down the arm until you can slip the hook on the arm over the pivot pin.
3. Pull the blade upward until the retaining spring is fully enclosed in the hook, then place the wiper against the windshield. This installation procedure is the reverse of the sequence given in Fig. 10-6.

The rubber filler can be replaced without replacing the entire wiper blade. See Fig. 10-7.

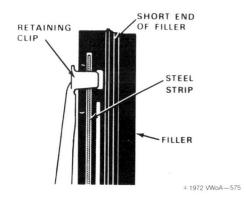

Fig. 10-7. Components of wiper blade filler assembly.

To replace filler:

1. Remove the wiper blade from the wiper arm.
2. Tightly compress the short end of the filler between your thumb and finger, then twist one side of the filler out of the retaining clip.

ELECTRICAL SYSTEM 33

3. With the free side of the retaining clip resting in the groove with the steel strip, repeat the preceding step to free the other side of the filler from the retaining clip.
4. Slide the short end of the filler toward the center of the blade until the short end is completely free of the retaining clip.
5. Shift the filler sideways and unhook the steel strips from the retaining clip as indicated in Fig. 10-8. Then slide the filler out of the other retaining clips.

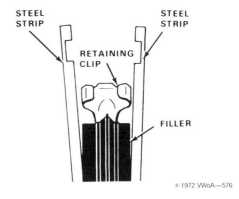

Fig. 10-8. Steel strips unhooked from retaining clip.

6. Place both steel strips in the grooves of the new filler. Make sure the notches in the steel strips face the filler and engage the projections in the filler grooves.
7. Hold the filler so that the strips are kept in the grooves. Starting at the open end of the filler, carefully slide the filler and the steel strips into the retaining clips.
8. When the closed end of the filler has reached the end retaining clip, compress the filler until the retaining clip rides over the raised edge next to the retaining clip recess in the end of the filler.
9. Make sure the retaining clip completely engages the recess in the filler. Then install the wiper blade on the wiper arm.

Removing, Installing, and Adjusting Wiper Arms

On cars built before model year 1973, the wiper arm is held on the wiper shaft by a cap nut. An M 6 hexagon nut is used on 1973 and later models. The nut is covered by a black plastic cap. When removing late-type wiper arms, it is necessary to carefully pry off the cap in order to gain access to the mounting nut.

On all models, proper installation requires that the spring washer be seated under the nut. It is very important that you adjust the angle of the wiper arm to specifications before you torque the mounting nuts. The nuts should be torqued to 42 to 60 cmkg (37 to 52 in. lb.).

34 ELECTRICAL SYSTEM

The wiper arms must be correctly installed after they have been removed for replacement or repairs. Proper wiper operation under all weather conditions is possible only when the arms are adjusted accurately to specifications. The adjustment of the wiper on the wiper arm shafts is measured at the two points indicated in Fig. 10-9. Dimensions **a** and **b** for each model are listed in **Table j**.

Fig. 10-9. Points where wiper arm adjustment is measured. Dimensions **a** and **b** are given in **Table j**.

Table j. Wiper Arm Adjustment Dimensions

Model and Year	Dimension a mm (in.)	Dimension b mm (in.)
111 Beetle, all models and 113 Super Beetle, 1970 to 1972	15 (19/32)	12 (15/32)
113 Super Beetle, 1973 and later	34 (1 11/32)	41 (1 39/64)
Karmann Ghia, all models	23 (29/32)	15 (19/32)

10.6 Windshield Washers

The windshield washer system has no electrical parts. The washers operate by compressed air from the spare tire. To replace the water valve, remove the wiper switch as described in **8.2 Removing and Installing Switches in Dashboard** or, for 1972 and later cars, **9.2 Removing and Installing Steering Column Switches**.

A cutoff valve built into the windshield washer reservoir's cap prevents pressure in the spare tire from dropping below 26 psi (1.8 kg/cm^2). If you replace the cap, be sure to obtain a new cap suitable for use on Type 1 cars. Caps for other models have different minimum pressure settings. Check the spare tire's pressure periodically. Inflate it to a maximum of 42 psi (3.0 kg/cm^2). This high pressure is only for operation of the windshield washer. For road use, adjust the spare tire to the pressure specified in **BRAKES AND WHEELS**.

To clean a windshield washer jet, insert a fine steel wire into its orifice. If the spray does not strike the windshield at a satisfactory angle, adjust the jets as shown in Fig. 10-10.

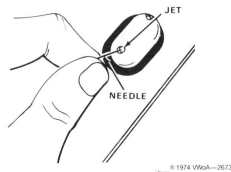

Fig. 10-10. Ball-shaped windshield washer jet being pivoted in its socket. A sewing needle inserted into the orifice is a satisfactory tool.

10.7 Troubleshooting Windshield Wipers

Table k lists the probable causes and suggested remedies for windshield wiping problems. The numbers in bold type in the Remedy column refer to the numbered headings in this section where repairs are described.

11. HEATED REAR WINDOW

A relay beneath the left side of the rear seat controls the temperature of the heated rear window. Disconnect the battery ground strap before removing either the wires or the relay mounting screws from the relay. Make sure there is good ground contact between the relay and the body of the car when a new relay is installed.

Before removing the rear window, disconnect the two wires from the terminals. Removal and installation of the heated rear window are the same as for any other rear window. However, they should only be attempted by someone familiar with automotive glass work who has the special tools required.

If only the conductive grid is damaged, it is unnecessary to replace the window. Repair material is available at VW dealers to patch broken circuits.

To repair:

1. Apply a strip of masking tape along each edge of the broken conductor.

2. Apply the repair material evenly over the break.

3. Allow the repair to dry for one hour at room temperature. Then remove the tape and test the heating effect.

ELECTRICAL SYSTEM 35

Table k. Windshield Wiping System Troubleshooting

Problem	Probable Cause	Remedy
1. Smearing	a. Blade dirty b. Blade lips frayed; rubber damaged or worn out c. Blades old, blade surface cracked	a. Clean blade with hard nylon brush and soap solution or alcohol. b. Install new rubber fillers. See **10.5**. c. Install new rubber fillers. See **10.5**.
2. Traces of water on windshield form small beads	Window soiled by paint, polish, oil, or diesel exhaust deposits	Clean windshield with clean cloth and grease/oil silicone remover.
3. Blade misses parts of windshield	a. Filler torn out of retainer b. Blade not in uniform contact with glass; spring or retainer distorted c. Insufficient wiper pressure	a. Reinstall filler carefully. See **10.5**. b. Install new blade. See **10.5**. c. Lightly lubricate arm linkage and spring or install new arm. See **10.5**.
4. Blade wipes well on one side but badly on other, shudders	a. Filler distorted, no longer flips b. Wiper arm distorted; blade not perpendicular to the windshield	a. Clean blade with hard nylon brush and soap solution or alcohol, or install new filler. b. Carefully twist wiper arm until it is perpendicular to the windshield.

36 ELECTRICAL SYSTEM

12. FRESH AIR FAN

The two-speed fresh air fan is standard equipment on 1971 and later Super Beetles and Convertibles. It became standard on the Beetle in 1976.

To remove fresh air fan:

1. Remove the battery ground strap.
2. Disconnect the wiring harness leading to the fan at the switch, then remove the fresh air box as described in **BODY AND FRAME.**
3. Remove the top of the fresh air box.
4. Remove the four nuts indicated in Fig. 12-1.

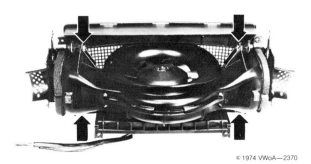

Fig. 12-1. Nuts (arrows) at bonded rubber mountings.

5. Separate the fan and the air duct from the fresh air box, pulling the wiring harness out of the grommet in the fresh air box upper part.
6. Pry off the clips, then separate the two halves of the air duct. Remove the M 5 nuts and the washers that hold the fan. See Fig. 12-2.

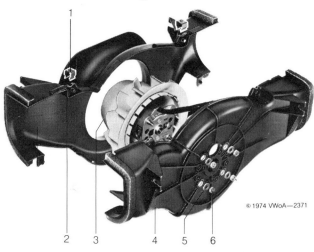

1. Clip
2. Air duct front part
3. Fan
4. Air duct rear part
5. Washer
6. M 5 nut

Fig. 12-2. Components of fresh air fan and duct.

Individual parts are not supplied for the fan and motor. The motor, blower wheel, and wiring harness must be replaced as a unit. Installation is the reverse of removal.

13. HORNS AND BUZZERS

On Beetles and Super Beetles, the horn is mounted under the left front fender. On the Karmann Ghia dual horns are mounted in the front of the spare wheel compartment. Current is supplied to the Karmann Ghia's horns through a horn relay, which is located at the left side of the spare wheel compartment. The horn circuit on other models is controlled directly by the horn lever or button.

Troubleshooting and Adjusting Horn

The best way to troubleshoot the horn circuit is with a voltmeter. Using the wiring diagrams as a guide, you can test whether battery voltage is reaching the horns, the relay, and other parts of the circuit. Once you have isolated the faulty wire or component, you can take steps to repair it.

In particular, check for loose, dirty terminals and worn electrical contacts. If ground (negative polarity) is weak either at the brown wire to the horn or to terminal 85 of the relay on Karmann Ghia models, thoroughly check the horn control circuit shown in Fig. 13-1. The Karmann Ghia's horn relay is located at the left side of the spare wheel compartment.

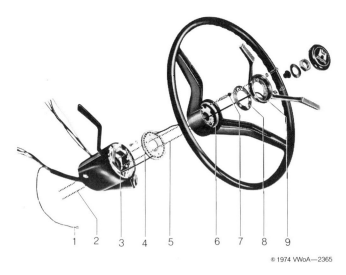

1. Ground wire to horn
2. Ground via steering column
3. Turn signal switch with horn circuit contact spring
4. Slip ring
5. Steering column
6. Contact ring in wheel
7. Insulating bushing
8. Contact disk
9. Horn control

Fig. 13-1. Horn control circuit, 1971 and later models.

On 1970 models, the contact ring in the steering wheel is grounded via a wire from the steering coupling that passes through the steering column. Depressing the horn control grounds the column tube to the contact ring in the steering wheel. The brown ground wire from the horn is connected to the column tube.

On 1971 and later cars, the steering wheel is grounded by contact with the steering column. The steering column is grounded by a short wire that bypasses the steering coupling's flexible disk. When the horn control is depressed, it is grounded against the steering wheel contact ring. A wire from the horn control to the slip ring grounds the horn circuit contact spring that is connected to the horn's brown ground wire.

The horn is burned out if there is no electrical continuity between its two terminals. But if damage has not broken continuity, it may still be possible to disassemble and repair the horn. Look for burned or dirty contact points, a faulty condenser, or water inside the horn caused by faulty sealing.

If the horn is receiving full battery voltage but makes only a soft clicking sound (or has a poor tone), this malfunction can usually be cured by adjustment.

CAUTION
Do not perform needless horn adjustments. Misadjustment can seriously damage the horn.

Chip out the sealing compound over the adjusting screw. Then connect an ammeter in series with the horn. (That is, positive polarity current for the horn must pass through the ammeter.) Make repeated test adjustments, striving for the clearest sound with minimum amperage draw.

CAUTION
Do not turn the adjusting screw while the horn is sounding. It could damage the horn.

If a suitable ammeter is not available, turn the adjusting screw counterclockwise until there is no sound when current is supplied to the horn. Then turn the screw clockwise very gradually (about ¼ turn at a time) until the horn has a clear tone. After adjustment, seal the adjusting screw with a good sealing adhesive.

Warning Buzzer

All cars covered by this Manual are equipped with an ignition key warning buzzer that sounds if the ignition key is in the lock when the driver's door is opened. The buzzer is connected to terminal 30 of the fuse box and, via the left-hand door contact switch, to the warning system contact in the ignition lock.

©1978 VWoA

The seat belt warning buzzer system, installed on cars manufactured since January 1, 1972, shares the same buzzer with the ignition key warning system. The two systems are linked by a wire from terminal 86 on the safety belt warning light to terminal 86 on the buzzer. The 1.5 watt lamp for the illuminated FASTEN SEAT BELTS sign can be replaced after you have pried off the lens.

14. INSTRUMENTS

Beetles and Super Beetles are equipped with a single instrument cluster consisting of a large speedometer dial inset with a fuel gauge. Karmann Ghias built during the 1970 and 1971 model years have a large speedometer that is flanked on the left by a fuel gauge and on the right by a clock. The 1972 and later Karmann Ghias have two large dials in front of the driver. The speedometer is on the left and the clock on the right, its face inset with a fuel gauge.

14.1 Replacing Speedometer Cable

The left front wheel drives the speedometer via a flexible cable.

To replace cable:

1. On 1973 and later Super Beetles and Convertibles, remove the instrument panel switch cluster as described in **8.2 Removing and Installing Switches in Dashboard.** On other models, remove the access panel from the rear of the front luggage compartment by removing the two knurled nuts.

2. Working through the hole in the dashboard (1973 and later Super Beetles and Convertibles) or from behind the dashboard (all other models), unscrew the knurled ferrule nut that holds the cable to the speedometer head.

3. Remove the left front hub cap. Pry off the circlip on the square end of the speedometer cable where it projects from the wheel bearing dust cover.

4. Working from behind the wheel, pull the cable out of the steering knuckle.

5. On 1973 and later Super Beetles and Convertibles, pull the cable out through the hole in the dashboard. On other models, pull the cable up into the luggage compartment.

Installation is the reverse of removal. Pass the new cable through the body grommet without stretching it or bending it sharply. Insert the drive end in the speedometer head and tighten the ferrule nut, then insert the other end into the steering knuckle.

CAUTION
The radius of any bend must be at least 150 mm (6 in.). Otherwise the cable will soon break at the bend.

38 ELECTRICAL SYSTEM

14.2 Fuel Gauge Troubleshooting

The thermal-type fuel gauge is controlled by an electromechanical sending unit in the fuel tank. Inaccurate fuel gauges must be replaced or returned to their manufacturer for calibration (VDO Instruments, Ltd., 116 Victor Ave., Detroit, Mich. 48203).

A fuel gauge that never moves from the 1/1 position has a grounded control circuit. Disconnect the control wire from the sending unit on top of the fuel tank. If the gauge falls from the 1/1 mark, the trouble is in the sending unit. If it does not, the control wire is grounded somewhere between the sending unit and the gauge.

If the gauge fails to register at all, remove the gauge wire from the sending unit and ground it against a clean, unpainted part of the car. If the gauge moves up to 1/1, the sending unit is faulty or not properly grounded. If the gauge still fails to move, the gauge or wire is faulty.

Removing and Testing Fuel Gauge Sending Unit

If an ohmmeter indicates infinite ohms when it is connected between the gauge terminal on the sending unit and ground, the sending unit is burned out and must be replaced.

To remove sending unit:

1. Disconnect the battery ground strap.
2. Remove the liner from the floor of the front luggage compartment. On 1972 Super Beetles and Convertibles, remove the fresh air box.
3. Disconnect the wire (s) from the sending unit.

   NOTE —
   During the 1972 model year a new sending unit was introduced that has a second terminal for a separate ground wire.

4. Take out the five bolts in the top of the sending unit, or unscrew the sending unit as indicated in Fig. 14-1.

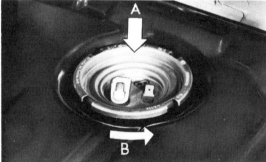

Fig. 14-1. Late-type sending unit removal. Using universal pliers, turn sending unit as at **B,** pressing down as at **A** until bayonet connection is released.

The sending unit can be tested by connecting a battery and voltmeter in series between the sending unit housing (or ground terminal) and the gauge wire terminal.

WARNING —
Do not make tests near the fuel tank. An electrical spark could cause an explosion.

Observe whether the voltmeter reading changes continuously as the sending unit's float is moved by hand through its full range. If the voltmeter needle does not move, or does not move smoothly, the sending unit is defective and should be replaced.

When installing the sending unit, be sure that the rubber gasket has a ground connection clip on its edge. (This clip is not necessary on the late-type unit.) The fuel tank must be clean and bright where the ground connection clip contacts it.

14.3 Removing and Installing Instruments

The instrument cluster introduced on 1973 Super Beetles and Convertibles is held in the dashboard by a ribbed rubber boot. To remove the instrument cluster or to replace light bulbs, remove the instrument panel switch cluster as described in **8.2 Removing and Installing Switches in Dashboard.** Working through the hole, unscrew the knurled ferrule nut that holds the speedometer cable to the speedometer head. The instrument cluster may then be pushed out of the panel as shown in Fig. 14-2.

Fig. 14-2. Late Super Beetle instruments cluster. Notice the ribbed rubber boot that holds the cluster in the dashboard.

ELECTRICAL SYSTEM   39

To remove instruments from other models, take out the access panel at the rear of the front luggage compartment by removing the two knurled nuts. (This step alone will permit you to replace instrument light bulbs.) Take out the mounting screw on each side of the instrument. Press the instrument out toward the car's interior.

15. WIRING

Two types of wiring diagrams are presented in this section. Those for 1970 through 1972 cars show individual components as line drawings. Diagrams for 1973 and later cars are of the current flow type and represent the components schematically. Both types of diagrams show the wires' insulation color. The small numbers in the wire lines indicate the wire's gauge in mm^2. Note that in the wiring diagrams for cars built since May 1971, the test connections for the VW Computer Analysis system are shown. Always reconnect these wires when servicing the electrical system.

CAUTION —
Never connect any device other than the test plug of the VW Computer Analysis system to the test network socket in the engine compartment. Other test equipment will not guarantee accurate readings and could damage the socket, the test sensors, or the car components containing them.

The current flow diagrams for 1973 and later cars have symbols for the electrical components rather than pictorial representations as in earlier diagrams. These symbols are explained in Fig. 15-1. The thin black lines in the current flow diagrams are not actual wires but ground connections via the car's chassis. Along the bottom of each current flow diagram is a yellow band containing numbers which will help you find electrical components easily. Appearing after each component listed in the description are numbers in a yellow legend labeled "current track." These indicate the current track in the diagram that contains the part you are looking for.

Fig. 15-1. Current flow diagram symbols.

A. How to read current flow diagrams

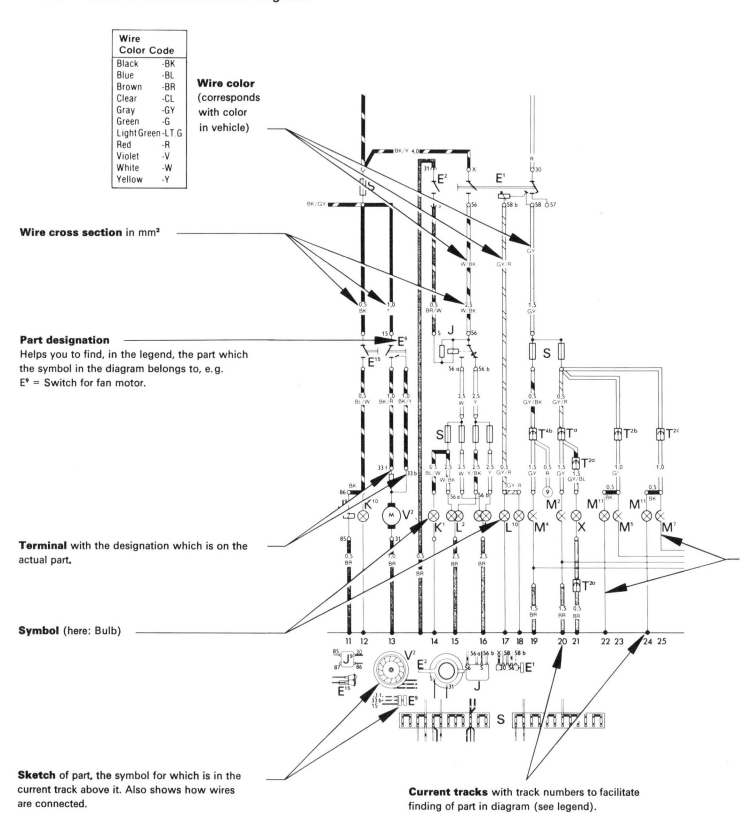

How to read current flow diagrams   **A.**

	Description	current track	
A	– Battery	4	← **Legend,** consisting of:
B	– Starter	5, 6	
C	– Generator	1, 2, 3	
C^1	– Regulator	1, 2, 3	← **Part designation,**
D	– Ignition / starter switch	6, 7, 12	
E	– Windshield wiper switch	8, 9, 10	
E^1	– Light switch	16, 17, 19	
E^2	– Turn signal and headlight dimmer switch	14, 38, 39	← **Current track number,** with the aid of which you can find the part in the diagram, e.g. turn signal switch E^2, current track 14.
E^3	– Emergency flasher switch	38, 39, 42, 43, 44	
E^9	– Fan motor switch	13	
E^{15}	– Rear window defogger switch	12	
T^2	– Wire connector, double a – in engine compartment lid b – in luggage compartment, front, left c – in luggage compartment, front, right		← **Explanation** (where you can find a wire connector, for example, on the vehicle).
T^3	– Wire connector, triple a – in luggage compartment, front, left b – behind the engine compartment insulation, right		

———————— **Internal connections** (thin lines).
These are **not** actual wires but ground connections which go through parts such as bulb holders for example.

Please note:
All switches and contacts are in a mechanically neutral position.
The various contacts in switches are shown in the current track in which they operate.

B. Automatic Stick Shift and Rear Window Defogger Circuit for VW Type 1/Sedan 111—from August 1969 (1970 and 1971 Models)

C. VW Type 1/Sedan 111— from August 1969 (1970 and 1971 Models)

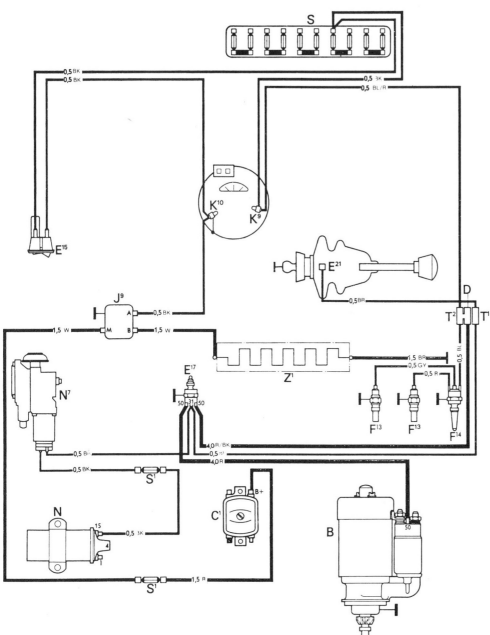

- B — Starter
- C¹ — Regulator
- D — To ignition/starter switch, terminal 50
- E¹⁵ — Switch for rear window defogger
- E¹⁷ — Starter cut-out switch
- E²¹ — Contact at selector lever
- F¹³ — Temperature sensor
- F¹⁴ — ATF temperature sensor selector
- J⁹ — Rear window defogger relay
- K⁹ — ATF temperature warning light
- K¹⁰ — Rear window defogger warning light
- N — Ignition coil
- N⁷ — Automatic Stick Shift control valve
- S — Fuse box
- S¹ — Fuse for rear window defogger, Automatic Stick Shift control valve
- T¹ — Cable connector, single
- T² — Cable connector, double
- Z¹ — Rear window defogger heating element

Wire Color Code	
Black	-BK
Blue	-BL
Brown	-BR
Clear	-CL
Gray	-GY
Green	-G
Light Green	-LT G
Red	-R
Violet	-V
White	-W
Yellow	-Y

- A — Battery
- B — Starter
- C — Generator
- C¹ — Regulator
- D — Ignition/starter switch
- E — Windshield wiper switch
- E¹ — Light switch
- E² — Turn signal and headlight dimmer switch
- E³ — Emergency flasher switch
- F — Brake light switch with warning switch
- F¹ — Oil pressure switch
- F² — Door contact switch, left, with contact for buzzer H 5
- F³ — Door contact switch, right
- F⁴ — Back-up light switch
- G — Fuel gauge sending unit
- G¹ — Fuel gauge
- H — Horn button
- H¹ — Horn
- H⁵ — Ignition key warning buzzer
- J — Dimmer relay
- J² — Emergency flasher relay
- J⁶ — Vibrator for fuel gauge
- K¹ — High beam warning light
- K² — Generator charging warning light
- K³ — Oil pressure warning light
- K⁵ — Turn signal warning light
- K⁶ — Emergency flasher warning light
- K⁷ — Dual circuit brake system warning light
- L¹ — Sealed beam unit, left headlight
- L² — Sealed beam unit, right headlight
- L¹⁰ — Instrument panel light
- M² — Tail and brake light, right
- M⁴ — Tail and brake light, left
- M⁵ — Turn signal and parking light, front, left
- M⁶ — Turn signal, rear, left
- M⁷ — Turn signal and parking light, front, right
- M⁸ — Turn signal, rear, right
- M¹¹ — Side marker light, front
- N — Ignition coil
- N¹ — Automatic choke
- N³ — Electro-magnetic pilot jet
- O — Ignition distributor
- P¹ — Spark plug connector, No. 1 cylinder
- P² — Spark plug connector, No. 2 cylinder
- P³ — Spark plug connector, No. 3 cylinder
- P⁴ — Spark plug connector, No. 4 cylinder
- Q¹ — Spark plug, No. 1 cylinder
- Q² — Spark plug, No. 2 cylinder
- Q³ — Spark plug, No. 3 cylinder
- Q⁴ — Spark plug, No. 4 cylinder
- R — Radio connection
- S — Fuse box
- S¹ — Back-up light fuse
- T — Cable adapter
- T¹ — Cable connector, single
- T² — Cable connector, double
- T³ — Cable connector, triple
- T⁴ — Cable connector (four connections)
- V — Windshield wiper motor
- W — Interior light
- X — License plate light
- X¹ — Back-up light, left
- X² — Back-up light, right

- ① — Battery to frame ground strap
- ② — Transmission to frame ground strap

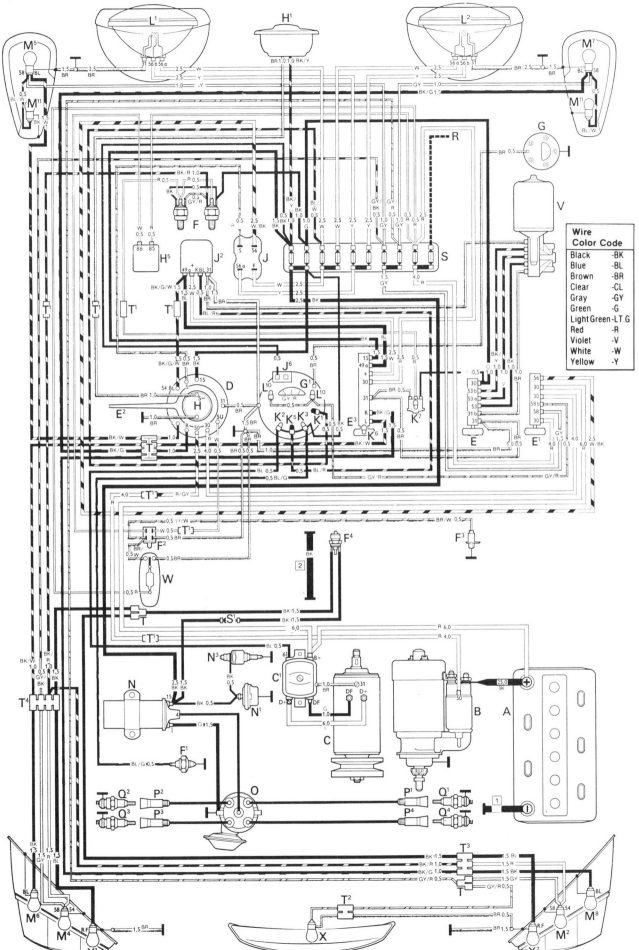

VW Type 1/Sedan 111—from August 1969 (1970 and 1971 Models) C.

D. VW Type 1/Sedan 113—from August 1970 (1971 Models)

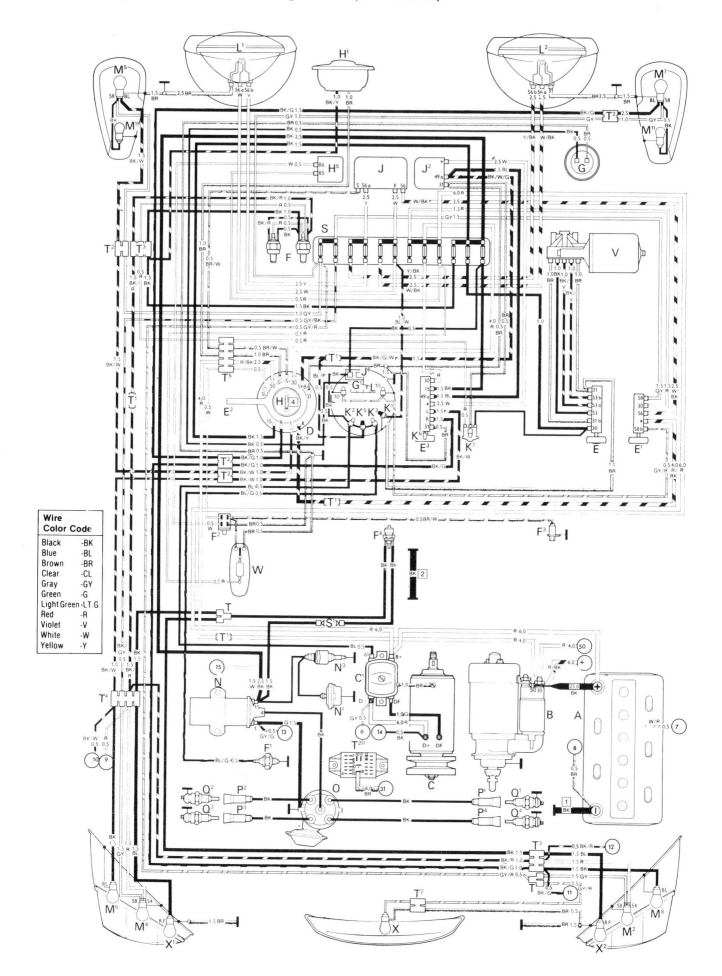

VW Type 1/Sedan 113— from August 1970 (1971 Models) — D.

- A — Battery
- B — Starter
- C — Generator
- C[1] — Regulator
- D — Ignition / starter switch
- E — Windshield wiper switch
- E[1] — Light switch
- E[2] — Turn signal and headlight dimmer switch
- E[3] — Emergency flasher switch
- F — Brake light switch
- F[1] — Oil pressure switch
- F[2] — Door contact and buzzer alarm switch, left
- F[3] — Door contact switch, right
- F[4] — Back-up light switch
- G — Fuel gauge sending unit
- G[1] — Fuel gauge
- H — Horn button
- H[1] — Horn
- H[5] — Ignition key warning buzzer
- J — Dimmer relay
- J[2] — Emergency flasher relay
- J[6] — Fuel gauge vibrator
- K[1] — High beam warning light
- K[2] — Generator charging warning light
- K[3] — Oil pressure warning light
- K[5] — Turn signal warning light
- K[6] — Emergency flasher warning light
- K[7] — Dual circuit brake warning light
- L[1] — Sealed-Beam unit, left headlight
- L[2] — Sealed-Beam unit, right headlight
- L[10] — Instrument panel light
- M[1] — Parking light, left
- M[2] — Tail / brake light, right
- M[4] — Tail / brake light, left
- M[5] — Turn signal and parking light front left
- M[6] — Turn signal, rear, left
- M[7] — Turn signal and parking light front right
- M[8] — Turn signal, rear, right
- M[11] — Side marker light, front
- N — Ignition coil
- N[1] — Automatic choke
- N[3] — Electro-magnetic pilot jet
- O — Distributor
- P[1] — Spark plug connector, No. 1 cylinder
- P[2] — Spark plug connector, No. 2 cylinder
- P[3] — Spark plug connector, No. 3 cylinder
- P[4] — Spark plug connector, No. 4 cylinder
- Q[1] — Spark plug, No. 1 cylinder
- Q[2] — Spark plug, No. 2 cylinder
- Q[3] — Spark plug, No. 3 cylinder
- Q[4] — Spark plug, No. 4 cylinder
- S — Fuse box
- S[1] — Back-up light in-line fuse
- T — Cable adapter
- T[1] — Cable connector, single
- T[2] — Cable connector, double
- T[3] — Cable connector, triple
- T[4] — Cable connector (four connections)
- T[5] — Cable connector (five connections)
- T[20] — Test network, central plug
- V — Windshield wiper motor
- W — Interior light
- X — License plate light
- X[1] — Back-up light, left
- X[2] — Back-up light, right

- [1] — Ground strap from battery to frame
- [2] — Ground strap from transmission to frame
- [4] — Ground cable from front axle to frame

Test network (from June 1971 only)

The numbered circles are the connections in the test network which are wired to the central plug. The numbers correspond to the terminals in the central plug.

Automatic Stick Shift, Rear Window Defogger Circuit, and Fresh Air Fan Circuit for VW Type 1/Sedan 113—from August 1970 (1971 Models) — E.

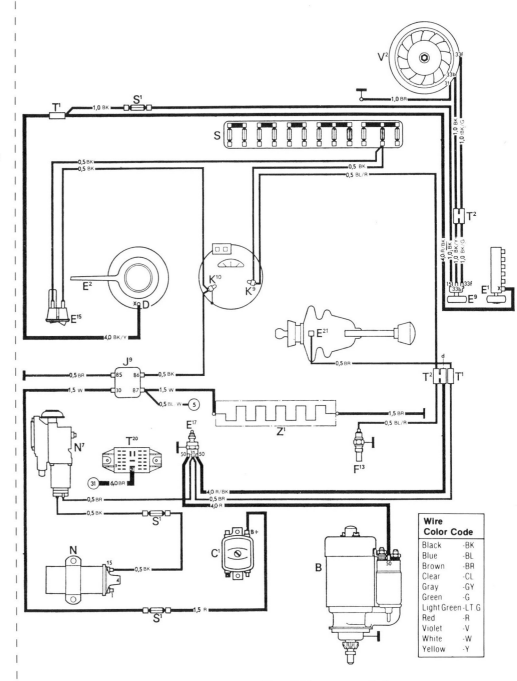

- B — Starter
- C[1] — Regulator
- D — Ignition / starter switch
- d — To ignition / starter switch, terminal 50
- E[1] — Light switch
- E[2] — Turn signal and headlight dimmer switch
- E[9] — Fan motor switch
- E[15] — Rear window defogger switch
- E[17] — Starter cut-out switch
- E[21] — Contact at selector lever
- F[13] — ATF temperature control switch
- J[9] — ATF temperature warning light
- K[10] — Rear window defogger warning light
- N — Ignition coil
- N[7] — Automatic Stick Shift control valve
- S — Fuse box
- S[1] — Fuse for:
  rear window defogger,
  Automatic Stick Shift control valve,
  fan motor
- T[1] — Cable connector, single
- T[2] — Cable connector, double
- T[20] — Test network, central plug
- V[2] — Fan motor
- Z[1] — Rear window defogger heating element

Test network (from June 1971 only)

The circles with the numbers 5 and 31 are connections in the test network which are wired to the central plug.

The numbers in the circles correspond to the terminals in the central plug.

The other cables of the test network can be found in the wiring diagram for the 1971 Sedan 113.

Wire Color Code

Black	-BK
Blue	-BL
Brown	-BR
Clear	-CL
Gray	-GY
Green	-G
Light Green	-LT G
Red	-R
Violet	-V
White	-W
Yellow	-Y

F. **Automatic Stick Shift Circuit for VW Type 1/Sedan 111 and 113—from August 1971 (1972 Models)**

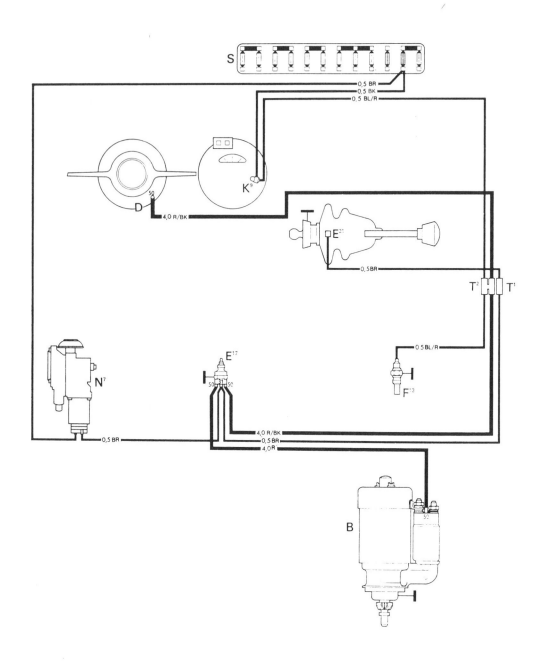

- B — Starter
- D — Ignition / starter switch
- E17 — Starter cut-out switch
- E21 — Contact at selector lever
- F13 — ATF temperature control switch
- K9 — ATF temperature warning light
- N7 — Automatic Stick Shift control valve
- S — Fuse box
- T1 — Cable connector, single
- T2 — Cable connector, double

Wire Color Code	
Black	-BK
Blue	-BL
Brown	-BR
Clear	-CL
Gray	-GY
Green	-G
Light Green	-LT.G
Red	-R
Violet	-V
White	-W
Yellow	-Y

This page intentionally left blank

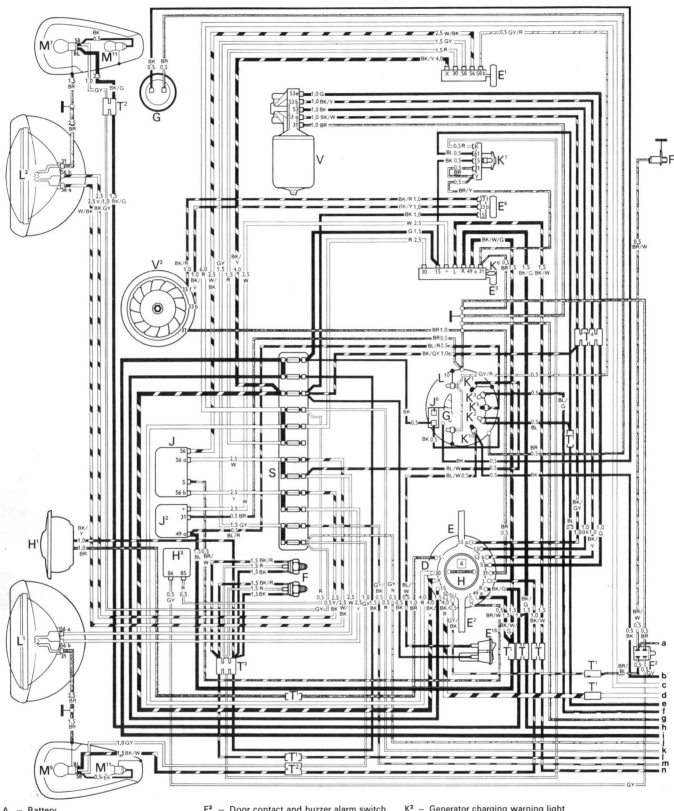

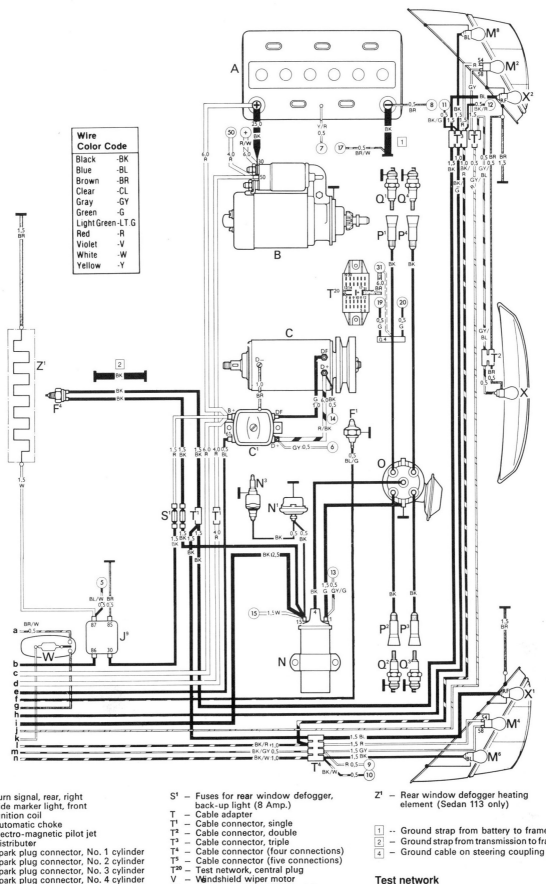

H. VW Type 1/Sedan 111—from August 1972 (1973 and later Models)
page 1 of 3

	Description	current track
A	Battery	4
B	Starter	5, 6
C	Generator	1, 2, 3
C^1	Regulator	1, 2, 3
D	Ignition / starter switch	6, 7, 10, 25
E	Windshield wiper switch	9
E^1	Light switch	12, 14, 15
E^2	Turn signal and headlight dimmer switch	11, 38, 39
E^3	Emergency flasher switch	38, 39, 42, 44, 45
E^{24}	Safety belt lock, left	27
E^{25}	Safety belt lock, right	26
E^{26}	Contact strip in passenger seat	26
F	Brake light switch	30, 31, 32, 33
F^1	Oil pressure switch	36
F^2	Door contact and buzzer alarm switch, left	24, 25
F^3	Door contact switch, right	23
F^4	Backup light switch	51
F^{15}	Transmission switch for safety belt warning system	28
G	Fuel gauge sending unit	34
G^1	Fuel gauge	34
G^4	Ignition timing sensor	47
H	Horn button	29
H^1	Horn	29
H^5	Ignition key warning buzzer	24, 25
H^6	Steering lock contact for ignition key warning system	25
J	Dimmer relay	11, 13
J^2	Emergency flasher relay	39, 40
J^6	Fuel gauge vibrator	34
K^1	High beam warning light	12
K^2	Generator charging warning light	35
K^3	Oil pressure warning light	36
K^5	Turn signal warning light	37
K^6	Emergency flasher warning light	45
K^7	Dual circuit brake warning light	31, 33
K^{19}	Safety belt warning system light	27, 28
L^1	Sealed beam unit, left headlight	11
L^2	Sealed beam unit, right headlight	13
L^{10}	Instrument panel light	14, 15
L^{21}	Light for heater lever illumination	44
M^2	Tail light, right	19
M^4	Tail light, left	21
M^5	Turn signal and parking light front left	15, 38
M^6	Turn signal, rear, left	39
M^7	Turn signal and parking light front right	18, 41
M^8	Turn signal, rear, right	42
M^9	Brake light, left	30
M^{10}	Brake light, right	33
M^{11}	Side marker light, front	16, 17
M^{16}	Backup light, left	51
M^{17}	Backup light, right	52

	Description	current track
N	Ignition coil	48
N^1	Automatic choke	49
N^3	Electromagnetic pilot jet	50
O	Distributor	48
P	Spark plug connectors	48
Q	Spark plugs	48
S^1 to S^{12}	Fuse box	10, 11, 13, 15, 21, 24, 30, 38, 40
S^{13}	Fuse for backup light (8 Amp.)	51
T	Wire connector (close to fuse box)	
T^1	Wire connector, single	
	a – close to fuse box	
	b – below rear seat bench	
	c – behind the engine compartment insulation, front	
T^2	Wire connector, double	
	a – in engine compartment lid	
	b – in luggage compartment, front, left	
	c – in passenger seat	
	d – below rear seat bench	
T^3	Wire connector, triple	
	a – in luggage compartment, front, left	
T^4	Wire connector, four connections	
	a – close to fuse box	
	b – behind engine compartment insulation, right	
	c – behind engine compartment insulation, left	
T^5	Wire connector, double on passenger seat rail	
T^{20}	Test network, test socket	46
V	Windshield wiper motor	8, 9
W	Interior light	23
X	License plate light	20

	Description	current track
①	Ground strap from battery to frame	4
②	Ground strap from transmission to frame	1
④	Ground wire on steering coupling	29
⑩	Ground connector, dashboard	
⑪	Ground connector, speedometer	

Ignition Switch/Safety Belt Interlock—see diagram Q.

Test network
The numbered circles are the connections in the test network which are wired to the central plug (T^{20}). The numbers correspond to the terminals in the central plug.

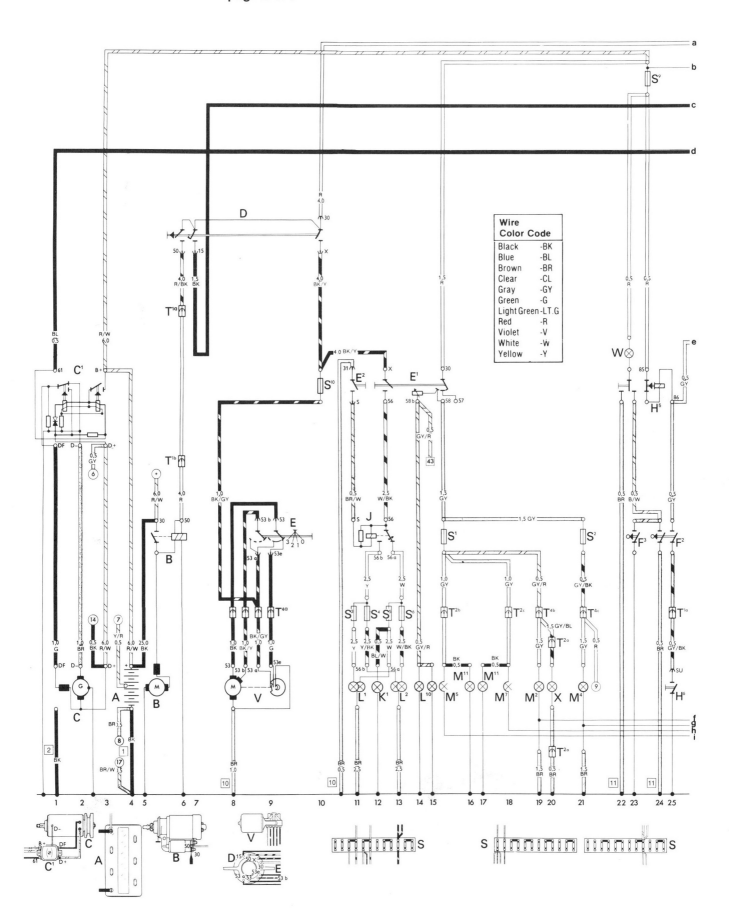

H. VW Type 1/Sedan 111—from August 1972 (1973 and later Models) page 3 of 3

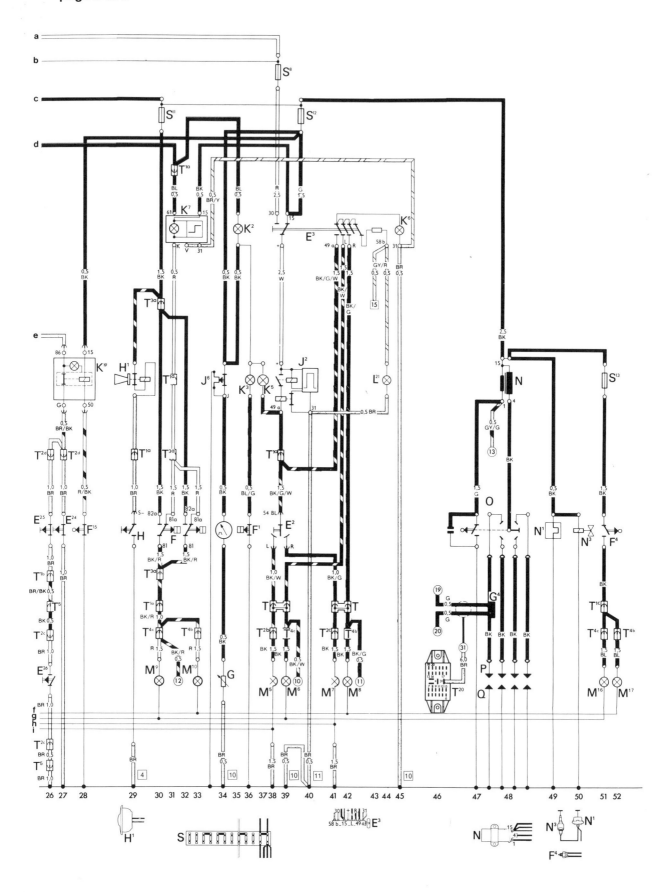

Automatic Stick Shift Circuit for VW Type 1/Sedan 113— from August 1972 (1973 Models)

I.

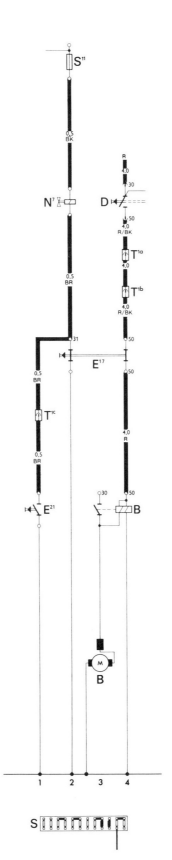

Description

- B — Starter
- D — Ignition / starter switch
- E17 — Starter cutout switch
- E21 — Selector lever contact
- N7 — Control valve
- S11 — Fuse in fuse box
- T1 — Wire connector, single
  - a — below dashboard
  - b — under rear seat bench
  - c — on frame tunnel, right

Wire Color Code	
Black	-BK
Blue	-BL
Brown	-BR
Clear	-CL
Gray	-GY
Green	-G
Light Green	-LT.G
Red	-R
Violet	-V
White	-W
Yellow	-Y

J. VW Type 1/Sedan 113—from August 1972 (1973 Models)
page 1 of 3

Description	current track
A — Battery	4
B — Starter	5, 6
C — Generator	1, 2, 3
C^1 — Regulator	1, 2, 3
D — Ignition / starter switch	6, 7, 12, 29
E — Windshield wiper switch	10, 11
E^1 — Light switch	16, 18, 20
E^2 — Turn signal and headlight dimmer switch	15, 43
E^3 — Emergency flasher switch	41, 42, 43, 45, 47, 48
E^9 — Fan motor switch	14
E^{15} — Rear window defogger switch	12
E^{24} — Safety belt lock, left	31
E^{25} — Safety belt lock, right	30
E^{26} — Contact strip in passenger seat	30
F — Brake light and dual circuit warning light switch	34, 35, 36
F^1 — Oil pressure switch	38
F^2 — Door contact and buzzer alarm switch, left	29
F^3 — Door contact switch, right	27
F^4 — Backup light switch	55
F^{15} — Transmission switch for safety belt warning system (man. transm.)	32
G — Fuel gauge sending unit	40
G^1 — Fuel gauge	40
G^4 — Ignition timing sensor	51
H — Horn button	33
H^1 — Horn	33
H^5 — Ignition key warning buzzer	28
H^6 — Steering lock contact for ignition key warning system	29
J — Dimmer relay	15, 16
J^2 — Emergency flasher relay	41
J^6 — Fuel gauge vibrator	40
J^9 — Rear window defogger relay	8, 12
K^1 — High beam warning light	16
K^2 — Generator charging warning light	37
K^3 — Oil pressure warning light	38
K^5 — Turn signal warning light	39
K^6 — Emergency flasher warning light	48
K^7 — Dual circuit brake warning light	35, 36
K^{10} — Rear window defogger warning light	13
K^{19} — Safety belt warning system light	31, 32
L^1 — Sealed beam unit, left headlight	15
L^2 — Sealed beam unit, right headlight	17
L^{10} — Instrument panel light	18, 19
L^{21} — Light for heater lever illumination	47
M^2 — Tail light, right	24
M^4 — Tail light, left	20
M^5 — Turn signal and parking light front, left	21, 42
M^6 — Turn signal, rear, left	43
M^7 — Turn signal and parking light front, right	26, 45
M^8 — Turn signal, rear, right	46
M^9 — Brake light, left	34
M^{10} — Brake light, right	36
M^{11} — Side marker light, front left + right	22, 25
M^{16} — Backup light, left	55
M^{17} — Backup light, right	56
N — Ignition coil	52
N^1 — Automatic choke	53
N^3 — Electromagnetic pilot jet	54
O — Distributor	50, 52
P — Spark plug connectors	51, 52
Q — Spark plugs	51, 52
S^1 to S^{12} — Fuse box	12, 15, 17, 20, 26, 27, 34, 40, 41

Description	current track
S^{13} — Fuse for backup light (8 Amp.)	55
S^{14} — Fuse for rear window defogger, (8 Amp.)	8
T^1 — Wire connector, single	
a — below rear seat bench	
b — one connector of the eight terminals (strip) behind the dashboard	35
c — behind the engine compartment insulation	
T^2 — Wire connector, double, one of the eight terminals (strip)	33
a — in engine compartment lid	
b — in luggage compartment, front, left two of the eight terminals (strip)	21, 42 42, 45
c — in luggage compartment, front, right	
d — below rear seat bench	
e — two of the eight terminals (strip) behind the dashboard	29, 34
f — in passenger seat	
T^3 — Wire connector, triple, in luggage compartment, front, left	
T^4 — Wire connector, four connections, behind engine compartment insulation, left	
T^5 — Wire connector, fire connections, behind engine compartment insulation, right	
T^6 — Wire connector, six connections, above fuse box, left	
T^7 — Wire connector, double, on passenger seat rail	
T^{20} — Test network, test socket	49
V — Windshield wiper motor	9, 10, 11
V^2 — Fan motor	14
W — Interior light	27
X — License plate light	23
Z^1 — Rear window defogger heating element	8

1	— Ground strap from battery of frame	4
2	— Ground strap from transmission to frame	1
4	— Ground cable on steering coupling	33
10	— Ground connector, dashboard	
11	— Ground connector, speedometer housing	

Test network

The numbered circles are the connections in the test network which are wired to the central plug (T^{20}). The numbers correspond to the terminals in the central plug.

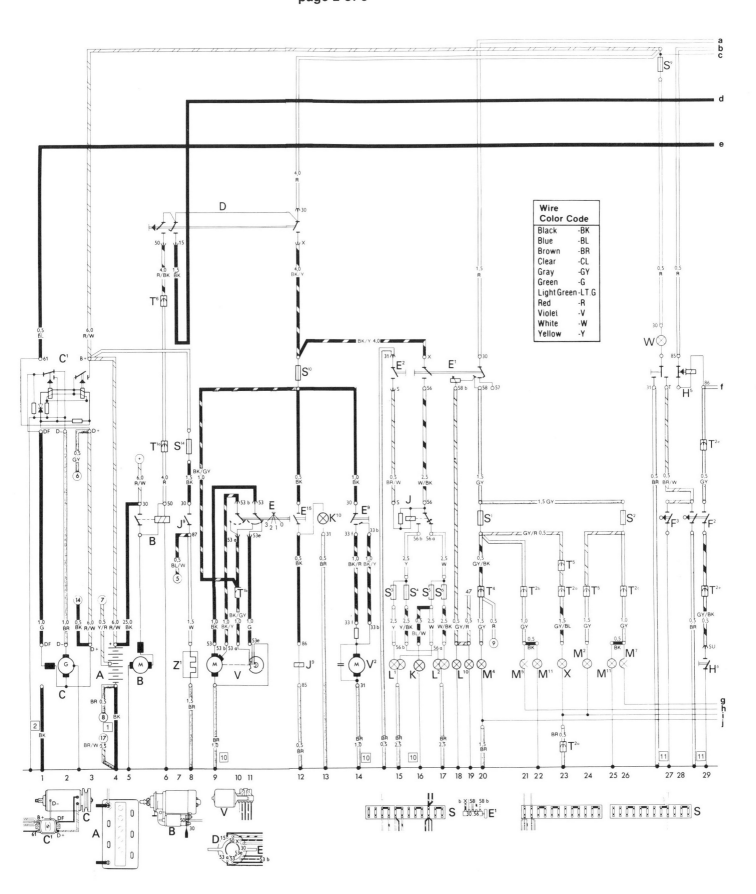

J. **VW Type 1/Sedan 113—from August 1972 (1973 Models)**
page 3 of 3

NOTE—
For cars with alternators, see the charging system as shown in the current flow diagrams for the 1974 Sedan 113.

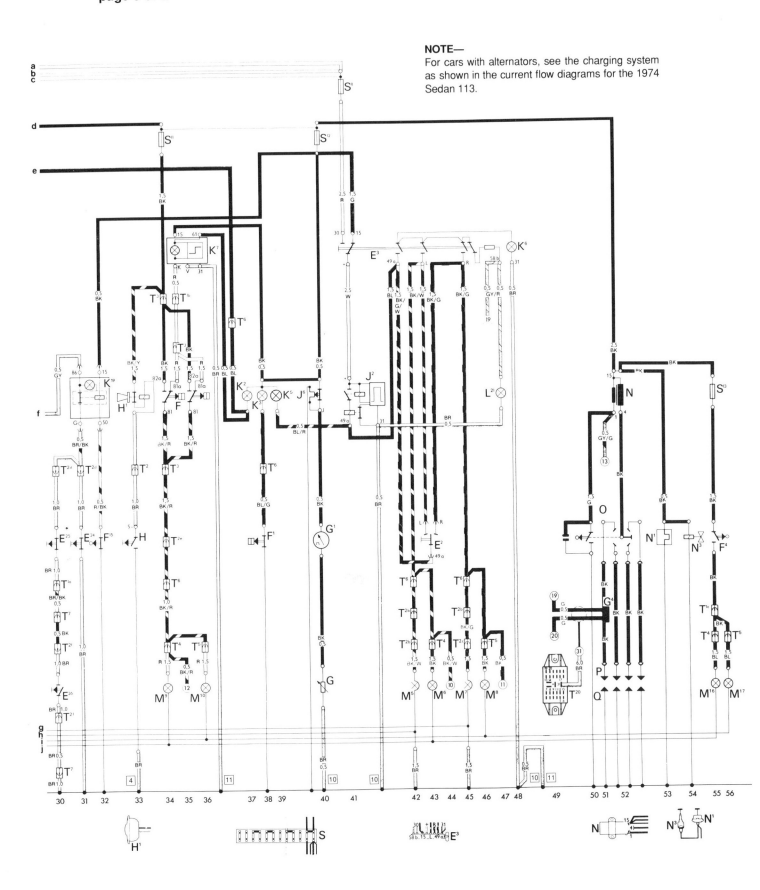

K. VW Type 1/Sedan 113—from August 1973 (1974 and later Models) page 1 of 4

	Description	current track
A	– Battery	26
B	– Starter	27
C	– Alternator	1, 2
C¹	– Regulator	1, 2
D	– Ignition/starter switch	8, 25, 26
E	– Windshield wiper switch	7, 9
E¹	– Light switch	13, 15, 17
E²	– Turn signal switch	46
E³	– Emergency flasher switch	41, 43, 44, 48, 50
E⁴	– Headlight dimmer switch	11
E⁹	– Fresh air fan motor switch	10
E¹⁵	– Rear window defogger switch	4, 5
E²⁴	– Safety belt lock, left	31
E²⁵	– Safety belt lock, right	29
E³¹	– Contact strip in driver seat	30
E³²	– Contact strip in passenger seat	28
F	– Brake light switch	34, 35
F¹	– Oil pressure switch	37
F²	– Door contact and buzzer alarm switch, left	24, 25
F³	– Door contact switch, right	23
F⁴	– Backup light switch	52
F⁹	– Parking brake control light switch	32
G	– Fuel gauge sending unit	40
G¹	– Fuel gauge	40
G⁴	– Ignition timing sensor	56
G⁷	– TDC marker unit	60
H	– Horn button	36
H¹	– Horn	36
H⁶	– Contact in ignition / starter switch for buzzer	25
J	– Dimmer relay	11, 13, 14
J²	– Emergency flasher relay	41, 42
J⁶	– Voltage vibrator	40
J⁹	– Rear window defogger relay	3, 4
J³⁴	– Safety belt warning system relay	25, 26, 27, 28, 29, 30, 31, 32, 33, 34, 35
K¹	– High beam warning light	13
K²	– Alternator charging warning light	39
K³	– Oil pressure warning light	37
K⁵	– Turn signal warning light	38
K⁶	– Emergency flasher warning light	51
K⁷	– Dual circuit brake warning and safety belt interlock warning system	33, 34, 35
L¹	– Sealed beam unit, left headlight	12
L²	– Sealed beam unit, right headlight	14
L⁶	– Speedometer light	15, 16
L²¹	– Light for heater lever illumination	50
M²	– Tail light, right	20
M⁴	– Tail light, left	17
M⁵	– Parking light front, left	18
M⁵	– Turn signal, front, left	44
M⁶	– Turn signal, rear, left	45
M⁷	– Parking light front, right	19
M⁷	– Turn signal, front, right	48
M⁸	– Turn signal, rear, right	47
M⁹	– Brake light, left	34
M¹⁰	– Brake light, right	35
M¹¹	– Sidemarker light front, left and right	18, 19
M¹⁶	– Backup light, left	52
M¹⁷	– Backup light, right	53
N	– Ignition coil	55
N¹	– Automatic choke	59
N³	– Electro-magnetic cutoff valve	58

	Description	current track
O	– Ignition distributor	55, 57
P	– Spark plug connectors	55, 56, 57
Q	– Spark plugs	55, 56, 57
S¹ to S¹²	– Fuses in fuse box	8, 12, 14, 17, 20, 22, 30, 31, 40
S²¹	– Fuse for backup lights (8 Amp.)	52
S²²	– Fuse for rear window defogger (8 Amp.)	3
T	– Cable adapter, behind insulation in engine compartment a – under rear seat bench	
T¹	– Wire connector, single a – behind instrument panel b – under rear seat bench	
T²	– Wire connector, double a – in luggage compartment, left b – in luggage compartment, right c – under passenger seat d – under driver's seat e – in hood of engine compartmnet	
T³	– Wire connector, 3 point a – in luggage compartment, left b – behind insulation in engine compartment, right	
T⁴	– Wire connector, 4 point behind insulation in engine compartment, left	
T⁵	– Wire connector, single a – behind instrument panel b – on passenger seat rail	
T⁶	– Wire connector, double a – under passenger seat b – under driver's seat	
T⁷	– Wire connector, 3 point, in engine compartment	
T⁸	– Wire connector, 4 point, under rear seat bench	
T⁹	– Wire connector, 8 point, behind instrument panel	
T²⁰	– Test network, test socket	54
V	– Windshield wiper motor	6, 7, 8
V²	– Fresh air motor	10
W	– Interior light	22
X	– License plate light	21
Z¹	– Rear window defogger heating element	3

[1]	– Ground strap from battery to frame	
[2]	– Ground strap from transmission to frame	
[10]	– Ground connection on instrument panel	
[11]	– Ground connection on speedometer	

Test network

The numbered circles are the connections in the test network which are wired to the central plug (T²⁰). The numbers correspond to the terminals in the central plug.

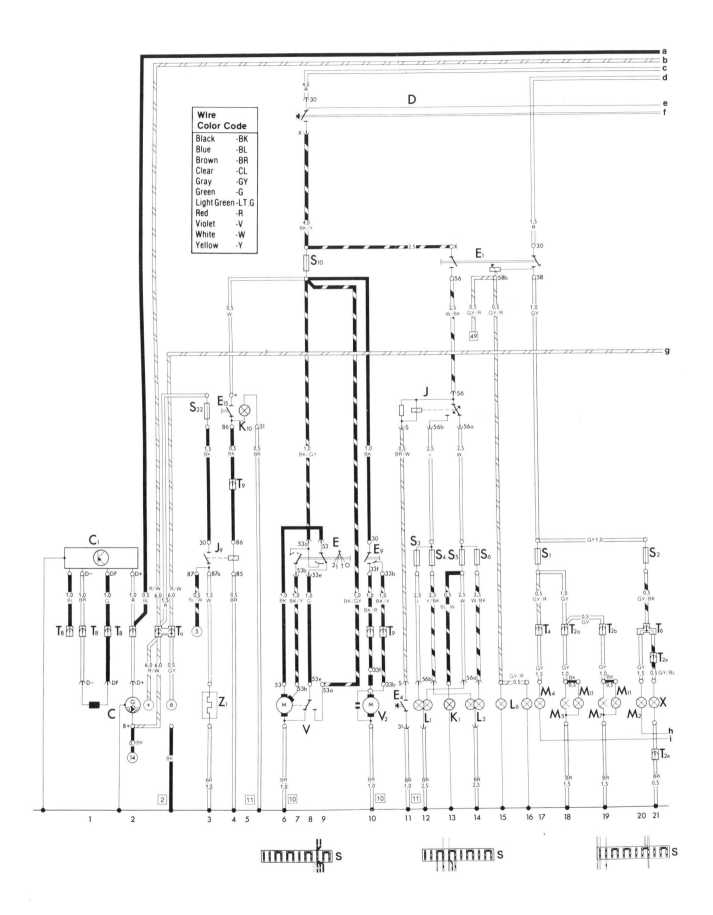

K. VW Type 1/Sedan 113—from August 1973 (1974 and later Models)
page 3 of 4

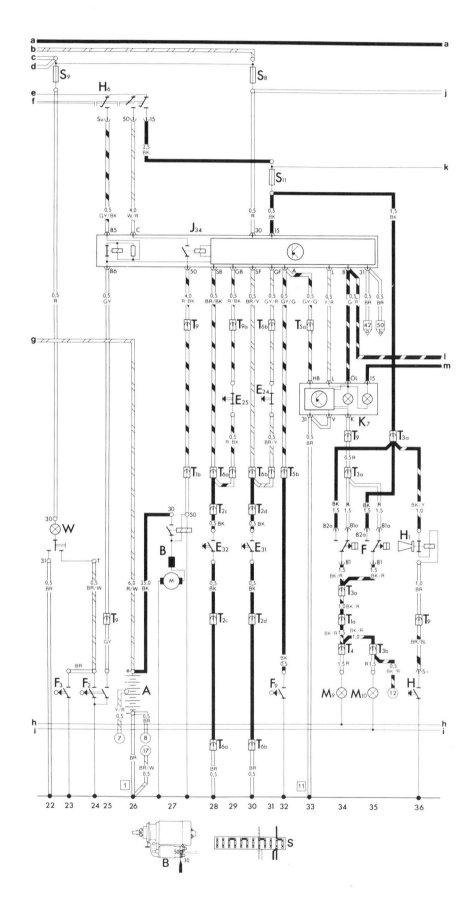

VW Type 1/Sedan 113—from August 1973 (1974 and later Models) page 4 of 4 K.

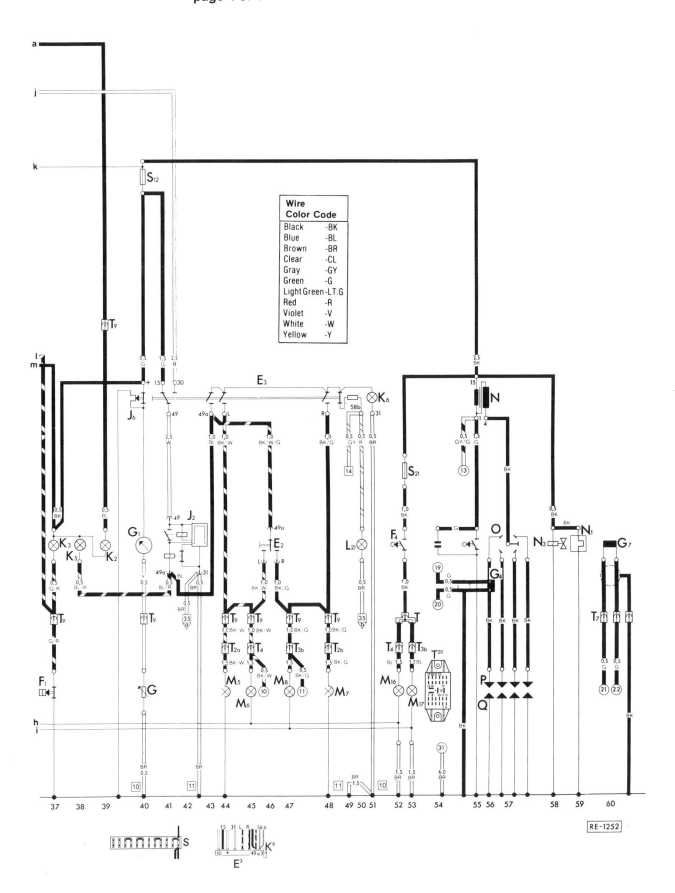

L. VW Type 1/Sedan 111—from 1976, Models with Fuel Injection

	Description	Current track
A	— Battery	2
B	— Starter	3, 4
C	— Alternator	1
C 1	— Regulator	1
D	— Ignition / starter switch	4, 5, 6, 7
E	— Windshield wiper switch	8, 9
E 1	— Light switch	13, 17, 19
E 4	— Headlight dimmer switch	10
E 24	— Safety belt lock with contact, left	30
F	— Dual circuit brake light switch	33, 36
F 2	— Door contact and buzzer alarm switch, front left	28, 29
F 3	— Door contact switch, front right	27
F 9	— Parking brake control light switch	37
H	— Horn button	31
H 1	— Horn	31
J	— Relay for headlight dimmer switch	11, 12, 13
J 34	— Relay for safety belt warning system	27, 29, 30
K 1	— High beam warning light	14
K 7	— Warning light for safety belt warning system, dual circuit brake system and parking brake	34, 35, 36, 37, 38
L 1	— Sealed beam unit, left headlight	11
L 2	— Sealed beam unit, right headlight	16
L 10	— Instrument panel light	17, 18
M 1	— Parking light, left	24
M 2	— Tail light, right	20
M 3	— Parking light, right	22
M 4	— Tail light, left	19
M 9	— Brake light, left	33
M 10	— Brake light, right	36
M 11	— Side marker light, front	23, 25
	— Fuses S 1, S 2, S 3, S 4, S 5, S 6, S 9, S 10, S 11 in fuse box	
Ta	— Wire connector, in engine compartment	
T b	— Wire connector, behind insulation in engine compartment	
T c	— Wire connector, behind insulation in engine compartment	
T d	— Wire connector, behind dashboard	
T e	— Wire connector, behind dashboard	
T 1a	— Wire connector, single; behind dashboard	
T 1b	— Wire connector, single; under rear seat bench	
T 1c	— Wire connector, single; behind dashboard	
T 1d	— Wire connector, single; on frame	
T 1e	— Wire connector, single; behind dashboard	
T 1f	— Wire connector, single; behind dashboard	
T 1g	— Wire connector, single; in luggage compartment	
T 1h	— Wire connector, single; behind dashboard	
T 1i	— Wire connector, single; behind dashboard	
T 1k	— Wire connector, single; behind dashboard	
T 2a	— Wire connector, double; in luggage compartment, left	
T 2b	— Wire connector, double; in luggage compartment, right	
T 2c	— Wire connector, double; in luggage compartment	
T 2d	— Wire connector, double; under driver's seat	
T 2e	— Wire connector, double; in rear hood	
T 2f	— Wire connector, double; behind dashboard	
T 3a	— Wire connector, 3 point; behind insulation in engine compartment, right	
T 3b	— Wire connector, 3 point; in luggage compartment, left	
T 4a	— Wire connector, 4 point; behind dashboard	
T 6	— Wire connector, 6 point; in engine compartment	
V	— Windshield wiper motor	6, 7, 8, 9
W	— Interior light	26
X	— License plate light	21
①	— Ground strap from battery to body	2
⑩	— Ground connection, dashboard	

VW Type 1/Sedan 111—from 1976, Models with Fuel Injection

Description — **Current track**

E 2	Turn signal switch	46
E 3	Emergency flasher switch	46, 48, 49, 50, 51, 52
F 1	Engine oil pressure switch	41
F 4	Backup light switch	55
F 25	Throttle valve switch	71
F 26	Thermo switch for cold starting valve	72, 73
F 27	EGR elapsed mileage odometer	39
G	Fuel gauge sending unit	42
G 1	Fuel gauge	42
G 6	Fuel pump	74
G 7	TDC marker unit	57
G 18	Temperature sensor	66
G 19	Intake air sensor	66, 67, 68
J 2	Emergency flasher relay	46, 47
J 6	Voltage stabilizer	42
J 21	Control unit (electronic)	62, 63, 64, 65, 66, 67, 68, 69, 70, 71
J 40	Double relay	65, 66, 67, 68, 69, 70
K 2	Alternator charging warning light	44
K 3	Engine oil pressure warning light	41
K 5	Turn signal warning light	43
K 6	Emergency flasher warning light	53
K 22	EGR warning light	39
L 21	Heater lever right	51
M 5	Turn signal light, front left	45
M 6	Turn signal light, rear left	47
M 7	Turn signal light, front right	49
M 8	Turn signal light, rear right	50
M 16	Backup light, left	56
M 17	Backup light, right	54
N	Ignition coil	61
N 17	Cold starting valve	72
N 21	Auxiliary air regulator	70
N 30	Injector, cyl. 1	62
N 31	Injector, cyl. 2	63
N 32	Injector, cyl. 3	64
N 33	Injector, cyl. 4	65
N 34	Resistance for injectors	62, 63, 64, 65
O	Ignition distributor	59, 61
P	Spark plug connectors	61
Q	Spark plugs	61
S 8 } S 12 }	Fuses in fuse box	45, 46
S 21	Fuse for backup light	55
T a	Wire connector, in engine compartment	
T b	Wire connector, behind insulation in engine compartment	
T c	Wire connector, behind insulation in engine compartment	
T d	Wire connector, behind dashboard	
T e	Wire connector, behind dashboard	
T 1a	Wire connector, single; behind dashboard	
T 1b	Wire connector, single; under rear seat bench	
T 1c	Wire connector, single; behind dashboard	
T 1d	Wire connector, single; on frame	
T 1e	Wire connector, single; behind dashboard	
T 1f	Wire connector, single; behind dashboard	
T 1g	Wire connector, single; in luggage compartment	
T 1h	Wire connector, single; behind dashboard	
T 1i	Wire connector, single; behind dashboard	
T 1k	Wire connector, single; behind dashboard	
T 2a	Wire connector, double; in luggage compartment, left	
T 2b	Wire connector, double; in luggage compartment, right	
T 2c	Wire connector, double; in luggage compartment	
T 2d	Wire connector, double; under driver's seat	
T 2e	Wire connector, double; in rear hood	
T 2f	Wire connector, double; behind dashboard	
T 3a	Wire connector, 3 point; behind insulation in engine compartment, right	
T 3b	Wire connector, 3 point; in luggage compartment, left	
T 4a	Wire connector, 4 point; behind dashboard	
T 6	Wire connector, 6 point; in engine compartment	
T 20	Diagnosis socket	58
⑩	Ground connection, dashboard	
⑪	Ground connection, speedometer	

L. VW Type 1/Sedan 111—from 1976, Models with Fuel Injection

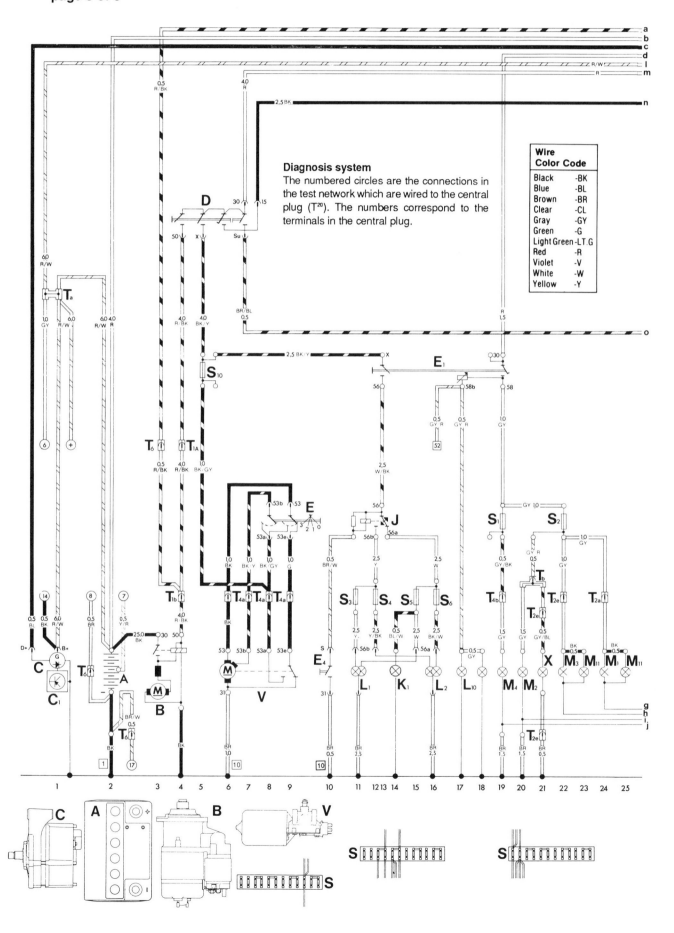

VW Type 1/Sedan 111—from 1976, Models with Fuel Injection
page 4 of 5

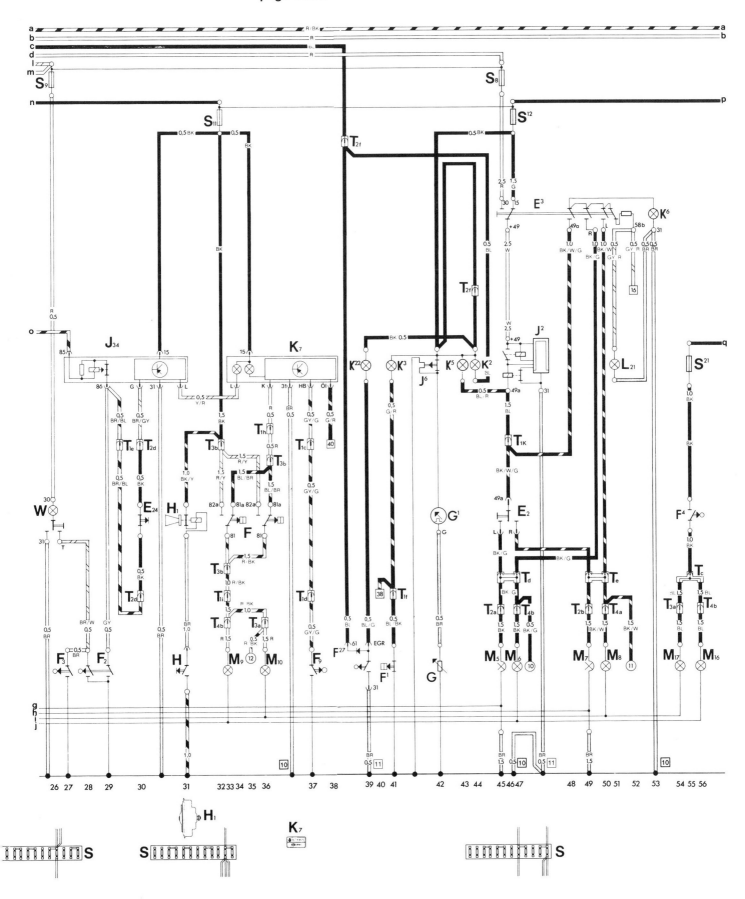

L. **VW Type 1/Sedan 111—from 1976, Models with Fuel Injection**

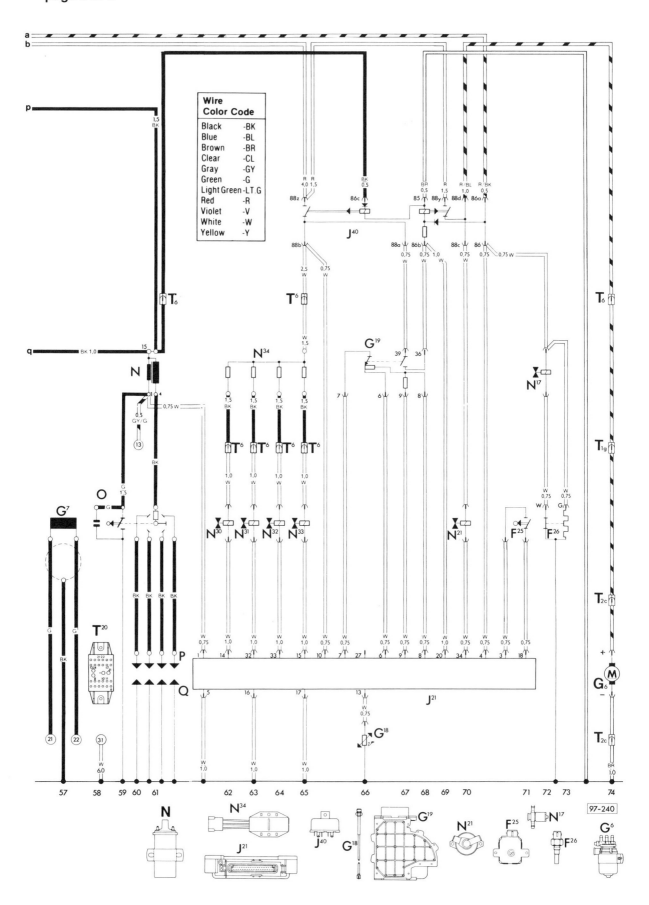

M. VW Type 1/Sedan 113—from 1976, Models with Fuel Injection

Description			Current track
A	—	Battery	5
B	—	Starter	6, 7
C	—	Alternator	1
C 1	—	Regulator	1
D	—	Ignition/starter switch	7, 8, 9, 10
E	—	Windshield wiper switch	11, 12
E 1	—	Light switch	16, 21, 23
E 4	—	Headlight dimmer switch	14
E 9	—	Fresh air fan switch	13
E 15	—	Rear window defogger switch	3
E 24	—	Safety belt lock with contact, left	36
F	—	Dual circuit brake light switch	39, 41
F 2	—	Door contact and buzzer alarm switch, front left	34, 35
F 3	—	Door contact switch, front right	33
F 9	—	Parking brake control light switch	42
H	—	Horn button	37
H 1	—	Horn	37
J	—	Relay for headlight dimmer switch	15, 16, 17
J 9	—	Rear window defogger relay	2, 3
J 34	—	Relay for safety belt warning system	35, 36, 37
K 1	—	High beam warning light	17
K 7	—	Warning light for safety belt warning system, dual circuit brake system and parking brake	39, 40, 41, 42, 43
K 10	—	Rear window defogger control light	4
L 1	—	Sealed beam unit, left headlight	15
L 2	—	Sealed beam unit, right headlight	19
L 6	—	Instrument panel light	21, 22
M 1	—	Parking light, left	28
M 2	—	Tail light, right	24
M 3	—	Parking light, right	26
M 4	—	Tail light, left	23
M 9	—	Brake light, left	38
M 10	—	Brake light, right	41
M 11	—	Side marker light, front	27, 29
	—	Fuses S1, S 2, S 3, S 4, S 5, S 6, S 9, S 10, S 11 in fuse box	
S 22	—	Single fuse for rear window defogger	2
T b	—	Wire connector, behind insulation in engine compartment	25
T c	—	Wire connector, in engine compartment	1
T 1a	—	Wire connector, single; behind dashboard	38
T 1b	—	Wire connector, single; behind dashboard	42
T 1c	—	Wire connector, single; behind dashboard	35
T 1e	—	Wire connector, single; on frame	42
T 1f	—	Wire connector, single; behind dashboard	21
T 1g	—	Wire connector, single; below rear seat	7
T 2a	—	Wire connector, double; in luggage compartment, left	28
T 2b	—	Wire connector, double; in luggage compartment, right	26
T 2e	—	Wire connector, double; in rear hood	25
T 2g	—	Wire connector, double; under driver's seat	36
T 3a	—	Wire connector, 3 point; in lugagge compartment, left	38, 41
T 3b	—	Wire connector, 3 point; behind insulation in engine compartment, right	41
T 4a	—	Wire connector, 4 point; behind engine compartment insulation	23, 28
T 6	—	Wire connector, 6 point; in engine compartment	5, 6
T 8a	—	Wire connector, 8 point; behind dashboard	3, 7
T 8b	—	Wire connector, 8 point; behind dashboard	13, 35, 37, 41
U 1	—	Cigarette lighter	31
V	—	Windshield wiper motor	9, 10, 11, 12
V 2	—	Fresh air fan	13
W	—	Interior light	32
X	—	License plate light	25
Y	—	Clock	30
Z 1	—	Rear window defogger element	2
[1]	—	Ground strap from battery to body	5
[10]	—	Ground connection; dashboard	14, 30, 36, 42
[11]	—	Ground connection; speedometer	4, 9, 13, 31, 41

VW Type 1/Sedan 113—from 1976, Models with Fuel Injection   M.

Description			Current track
E 2	—	Turn signal switch	51
E 3	—	Emergency flasher switch	50, 52, 53, 54, 55, 56
F 1	—	Engine oil pressure switch	46
F 4	—	Backup light switch	60
F 25	—	Throttle valve switch	76, 77
F 26	—	Thermo switch for cold start valve	78, 79
F 27	—	EGR elapsed mileage odometer	44
G	—	Fuel gauge sending unit	47
G 1	—	Fuel gauge	47
G 6	—	Fuel pump	80
G 7	—	TDC marker unit	62
G 18	—	Temperature sensor	71
G 19	—	Intake air sensor	71, 72, 73
J 2	—	Emergency flasher relay	51, 52
J 6	—	Voltage stabilizer	47
J 21	—	Control unit (electronic)	67, 68, 69, 70, 71 73, 74, 75, 76
J 40	—	Double relay	70, 71, 73, 74, 75
K 2	—	Alternator charging warning light	49
K 3	—	Engine oil pressure warning light	46
K 5	—	Turn signal warning light	48
K 6	—	Emergency flasher warning light	58
K 22	—	EGR warning light	44
L 21	—	Heater lever right	57
M 5	—	Turn signal light, front left	48
M 6	—	Turn signal light, rear left	50
M 7	—	Turn signal light, front right	55
M 8	—	Turn signal light, rear right	53
M 16	—	Backup light, left	61
M 17	—	Backup light, right	59
N	—	Ignition coil	66
N 17	—	Cold start valve	78
N 18	—	EGR valve	77
N 21	—	Auxiliary air regulator	75
N 30	—	Injector, cyl. 1	67
N 31	—	Injector, cyl. 2	68
N 32	—	Injector, cyl. 3	69
N 33	—	Injector, cyl. 4	70
N 34	—	Resistance for injectors	67, 68, 59, 70
O	—	Ignition distributor	64, 66
P	—	Spark plug connectors	66
Q	—	Spark plugs	66
S 8 S 12	—	Fuses in fuse box	51, 52
S 21	—	Fuse for backup light	60
T a	—	Wire connector, behind insulation in engine compartment	61
T 1d	—	Wire connector, single; in luggage compartment	80
T 1h	—	Wire connector, single; in engine compartment	71
T 2a	—	Wire connector, double; in luggage compartment, left	48
T 2b	—	Wire connector, double; in luggage compartment, right	55
T 2c	—	Wire connector, double; in luggage compartment	80
T 2f	—	Wire connector, double; in luggage compartment	44
T 3b	—	Wire connector, 3 point; behind insulation in engine compartment, right	53, 59
T 3c	—	Wire connector, 3 point; in engine compartment	62
T 4a	—	Wire connector, 4 point; behind insulation in engine compartment	50, 61
T 6	—	Wire connector, 6 point; in engine compartment	66, 80
T 6b	—	Wire connector, 6 point; in engine compartment	67, 68, 69, 70
T 8a	—	Wire connector, 8 point; in engine compartment	44, 46, 50, 53
T 8b	—	Wire connector, 8 point; in engine compartment	47, 48, 55
T 20	—	Diagnosis socket	63
[10]	—	Ground connection; dashboard	52, 57, 58
[11]	—	Ground connection; speedometer	47
[15]	—	Ground connection; in luggage compartment, left	44, 48, 80
[16]	—	Ground connection; in luggage compartment, right	55

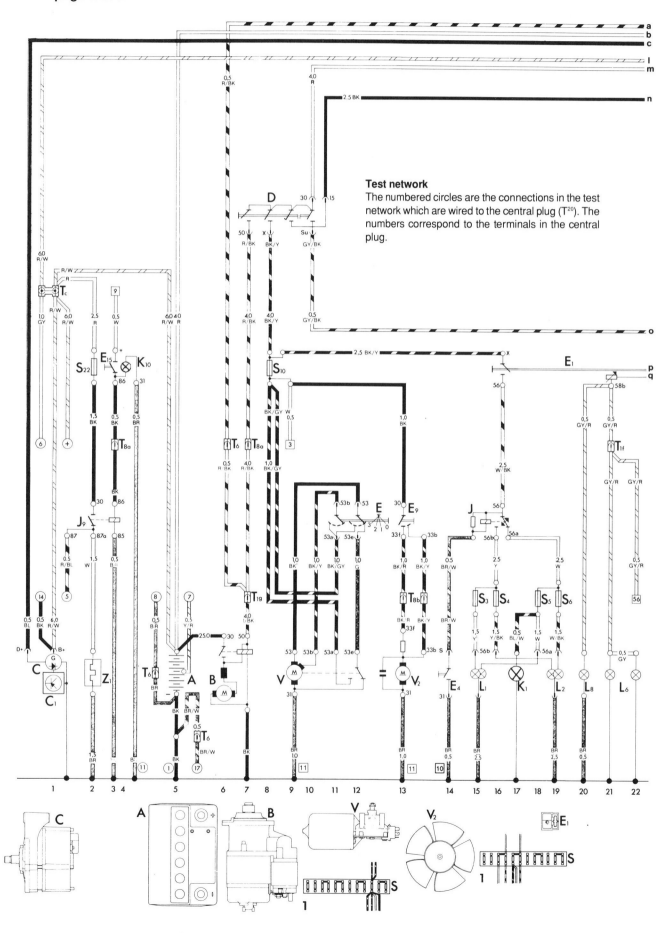

M. VW Type 1/Sedan 113—from 1976, Models with Fuel Injection
page 3 of 6

Test network
The numbered circles are the connections in the test network which are wired to the central plug (T20). The numbers correspond to the terminals in the central plug.

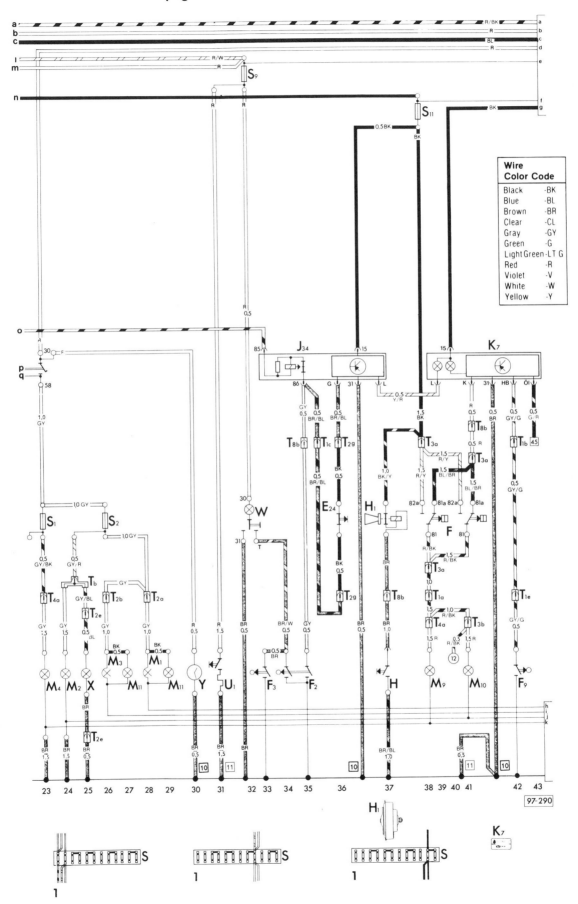

M. VW Type 1/Sedan 113—from 1976, Models with Fuel Injection
page 5 of 6

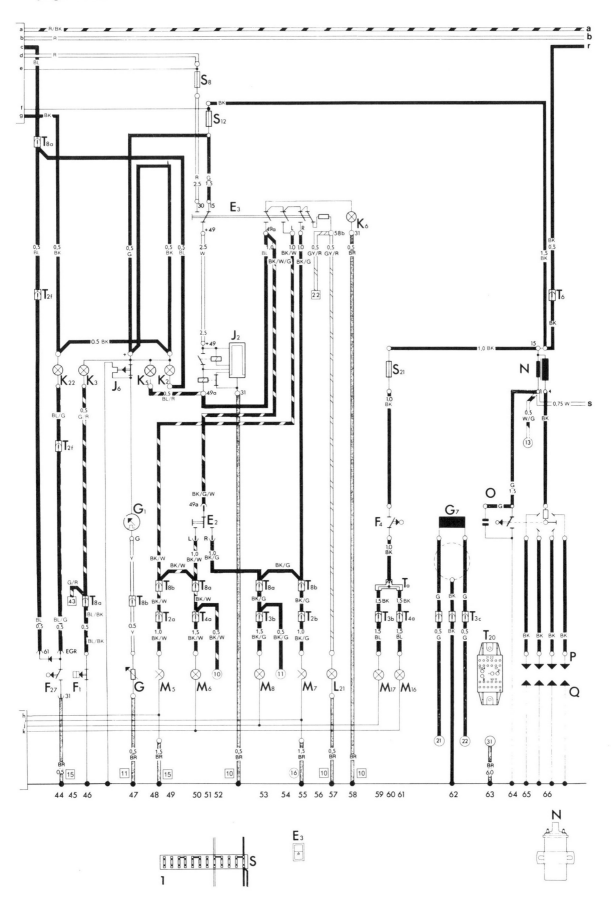

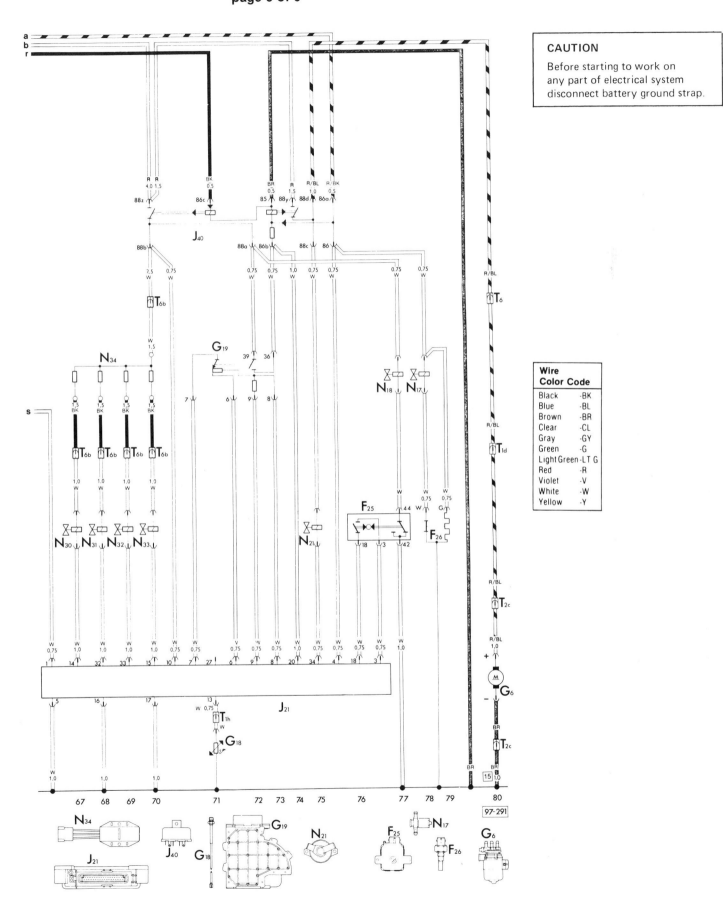

N. VW Karmann Ghia—from June 1969 (1970 and 1971 Models)

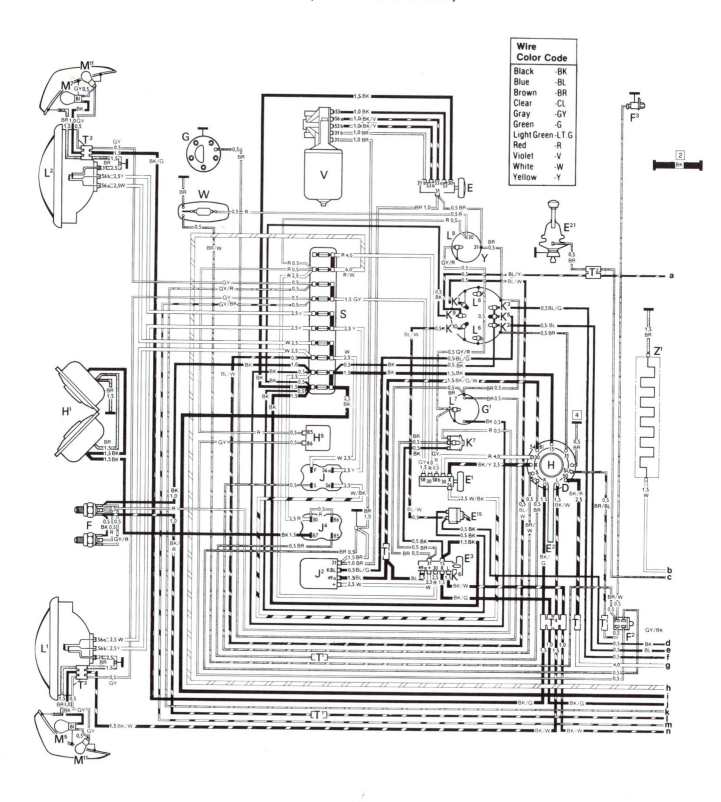

NOTE

On 1970 and early 1971 models (built before June 1971), the mainline from the battery's post to terminal B+ on the regulator is a solid red wire. A solid red wire also joins regulator terminal B+ with a terminal 30 on the light switch—this terminal 30 being in the location shown by terminal X in this diagram. On these earlier models, there is no terminal X on the steering column switch and no black and yellow wire joining it to a terminal X on the light switch.

VW Karmann Ghia—from June 1969 (1970 and 1971 Models) N.

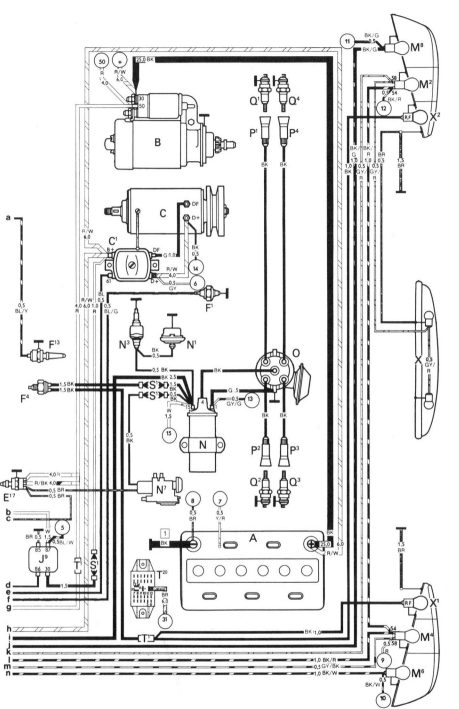

Test network (from June 1971 only)
The numbered circles are the connections in the test network which are wired to the central plug (T20). The numbers correspond to the terminals in the central plug.

- A — Battery
- B — Starter
- C — Generator
- C1 — Regulator
- D — Ignition / starter switch
- E — Windshield wiper switch
- E1 — Light switch
- E2 — Turn signal and headlight dimmer switch
- E3 — Emergency flasher switch
- E15 — Switch for rear window defogger
- E17 — Starter cut-out switch
- E21 — Contact at selector lever
- F — Brake light switch with warning switch
- F1 — Oil pressure switch
- F2 — Door contact switch, left, with contact for buzzer
- F3 — Door contact switch, right
- F4 — Back-up light switch
- F13 — ATF temperature control switch
- G — Fuel gauge sending unit
- G1 — Fuel gauge
- H — Horn button
- H1 — Twin horns
- H5 — Ignition key warning buzzer
- J — Dimmer relay
- J2 — Emergency flasher relay
- J4 — Relay for twin horns
- J9 — Rear window defogger relay
- K1 — High beam warning light
- K2 — Generator charging control light
- K3 — Oil pressure control light
- K5 — Turn signal warning light
- K6 — Emergency flasher warning light
- K7 — Dual circuit brake system warning light
- K9 — ATF temperature warning light
- K10 — Rear window defogger warning light
- L1 — Sealed beam unit, left headlight
- L2 — Sealed beam unit, right headlight
- L6 — Speedometer light
- L7 — Fuel gauge light
- L8 — Clock light
- M2 — Tail and brake light, right
- M4 — Tail and brake light, left
- M5 — Turn signal and parking light, front, left
- M6 — Turn signal, rear, left
- M7 — Turn signal and parking light, front, right
- M8 — Turn signal, rear, right
- M11 — Side marker light, front
- N — Ignition coil
- N1 — Automatic choke
- N3 — Electromagnetic pilot jet
- N7 — Automatic Stick Shift control valve
- O — Ignition distributor
- P1 — Spark plug connector, No. 1 cylinder
- P2 — Spark plug connector, No. 2 cylinder
- P3 — Spark plug connector, No. 3 cylinder
- P4 — Spark plug connector, No. 4 cylinder
- Q1 — Spark plug, No. 1 cylinder
- Q2 — Spark plug, No. 2 cylinder
- Q3 — Spark plug, No. 3 cylinder
- Q4 — Spark plug, No. 4 cylinder
- S — Fuse box
- S1 — Fuse for rear window defogger back-up lights, Automatic Stick Shift control valve
- T1 — Cable connector, single
- T2 — Cable connector, double
- T3 — Cable connector, triple
- T4 — Cable connector, four connections
- T20 — Test network, central plug
- V — Windshield wiper motor
- W — Interior light
- X — License plate light
- X1 — Back-up light, left
- X2 — Back-up light, right
- Y — Clock
- Z1 — Rear window defogger heating element

- ① — Ground strap battery to engine
- ② — Ground strap transmission to frame
- ④ — Ground strap steering column

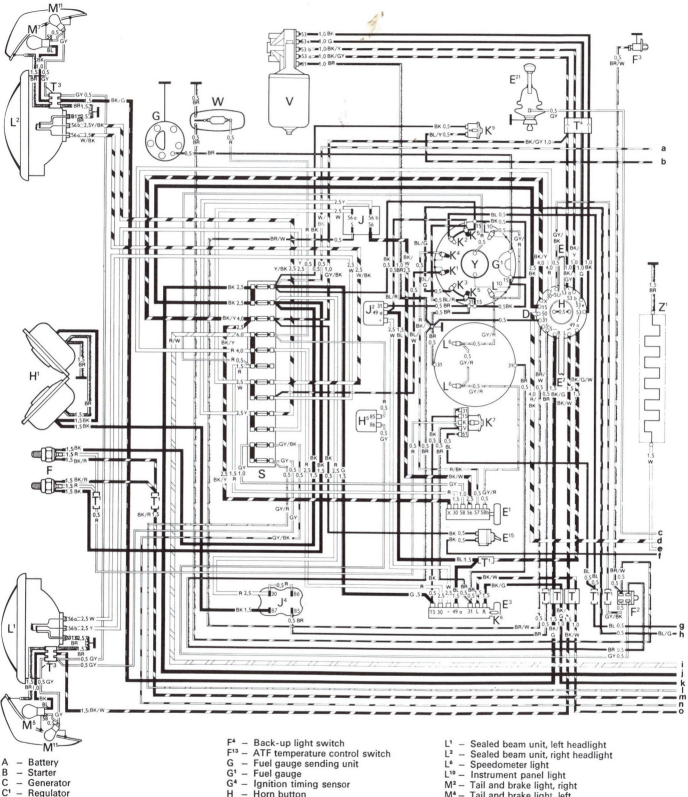

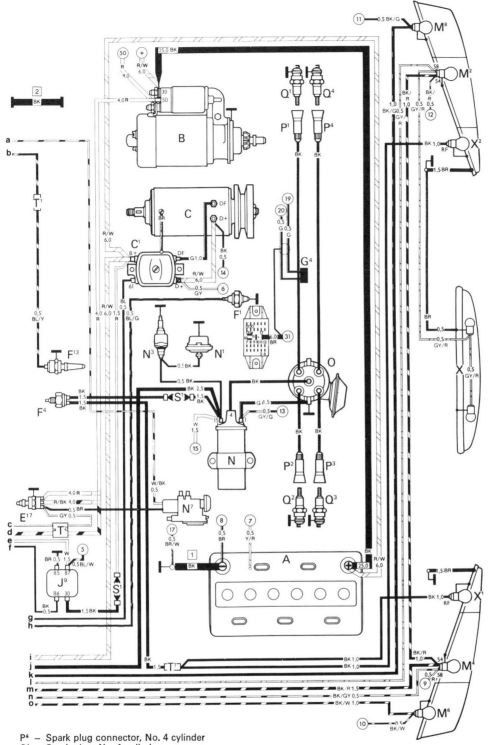

- P4 — Spark plug connector, No. 4 cylinder
- Q1 — Spark plug, No. 1 cylinder
- Q2 — Spark plug, No. 2 cylinder
- Q3 — Spark plug, No. 3 cylinder
- Q4 — Spark plug, No. 4 cylinder
- S — Fuse box
- S1 — Fuse for rear window defogger, (8 Amp) back-up lights, (8 Amp)
- T — Cable adapter
- T1 — Cable connector, single
- T2 — Cable connector, double
- T3 — Cable connector, triple
- T4 — Cable connector, four connections
- T20 — Test network, central plug
- V — Windshield wiper motor
- W — Interior light
- X — License plate light
- X1 — Back-up light, left
- X2 — Back-up light, right
- Y — Clock
- Z1 — Rear window defogger heating element

- ① — Ground strap battery to engine
- ② — Ground strap transmission to frame
- ④ — Ground strap steering column

Test network

The numbered circles are the connections in the test network which are wired to the central plug (T20). The numbers correspond to the terminals in the central plug.

P. VW Karmann Ghia—from August 1972 (1973 and 1974 Models)
page 1 of 4

Description		current track
A	Battery	5
B	Starter	6, 7
C	Generator	1, 2, 3
C^1	Regulator	1, 2, 3
D	Ignition / starter switch	7, 11, 12
E	Windshield wiper switch	9, 10
E^1	Light switch	16, 18, 22
E^2	Turn signal and headlight dimmer switch	13, 50
E^3	Emergency flasher switch	50, 53, 54, 56, 58
E^{15}	Rear window defogger switch	11
E^{24}	Safety belt lock, left	37
E^{25}	Safety belt lock, right	36
E^{26}	Contact strip in passenger seat	36
F	Brake light and dual circuit warning light switch	42, 43
F^1	Oil pressure switch	46
F^2	Door contact and buzzer alarm switch, left	34, 35
F^3	Door contact switch, right	33
F^4	Backup light switch	59
F^{15}	Transmission switch for safety belt warning system	38
G	Fuel gauge sending unit	48
G^1	Fuel gauge	48
G^4	Ignition timing sensor	62
H	Horn button	39
H^1	Twin horns	40, 41
H^5	Ignition key warning buzzer	33, 35
H^6	Steering lock contact for ignition key warning system	35
J	Dimmer relay	13, 16
J^2	Emergency flasher relay	50, 52
J^4	Relay for twin horns	39
J^9	Rear window defogger relay	4, 11
K^1	High beam warning light	15
K^2	Generator charging warning light	45
K^3	Oil pressure warning light	46
K^4	Parking light warning light	22
K^5	Turn signal warning light	47, 49
K^6	Emergency flasher warning light	58
K^7	Dual circuit brake warning light	42, 43
K^{19}	Safety belt warning system light	37, 38
L^1	Sealed beam unit, left headlight	14
L^2	Sealed beam unit, right headlight	16
L^6	Speedometer light	18, 19
L^{10}	Instrument panel light	20, 21
L^{21}	Light for heater lever illumination	56
M^2	Tail/brake light, right	28, 43
M^4	Tail/brake light, left	23, 42
M^5	Turn signal and parking light front, left	24, 50
M^6	Turn signal, rear, left	51
M^7	Turn signal and parking light front, right	27, 53
M^8	Turn signal, rear, right	54
M^{11}	Side marker light, front left + right	25, 26
M^{16}	Backup light, left	59
M^{17}	Backup light, right	60
N	Ignition coil	62
N^1	Automatic choke	64
N^3	Electromagnetic pilot jet	65
O	Distributor	62, 63
P	Spark plug connectors	62, 63
Q	Spark plugs	62, 63
S^1 to S^{12}	Fuse box	11, 14, 16, 25, 28, 33, 40, 50, 51
S^{13}	Fuse for backup light (8 Amp.)	59
S^{14}	Fuse for rear window defogger, (8 Amp.)	4
T	Cable adapter, behind dashboard	
T^1	Wire connector, single a – below rear seat bench b – behind the dashboard c – behind the engine compartment insulation	
T^2	Wire connector, double a – below rear seat bench b – in the passenger seat	
T^3	Wire connector, triple a – in headlight housing, left b – in headlight housing, right	
T^4	Wire connector, four connections, behind the dashboard	
T^7	Wire connector, double on passenger seat rail	
T^{20}	Test network, test socket	61
V	Windshield wiper motor	8, 9, 10
W	Interior light	32
X	License plate light	29, 30
Y	Clock	31
Z^1	Rear window defogger heating element	4

[1]	Ground strap from battery to frame	5
[2]	Ground strap from transmission to frame	1
[4]	Ground cable on steering coupling	39
[10]	Ground connector, dashboard	
[11]	Ground connector, speedometer housing	
[12]	Ground connector, clock	

Test network
The numbered circles are the connections in the test network which are wired to the central plug (T^{20}). The numbers correspond to the terminals in the central plug.

VW Karmann Ghia—from August 1972 (1973 and 1974 Models)

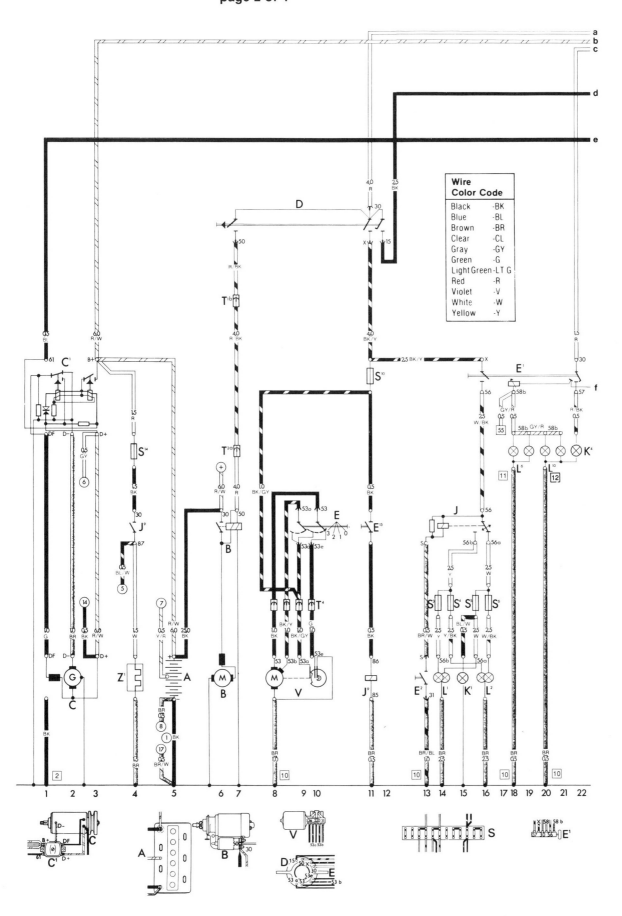

P. VW Karmann Ghia—from August 1972 (1973 and 1974 Models)
page 3 of 4

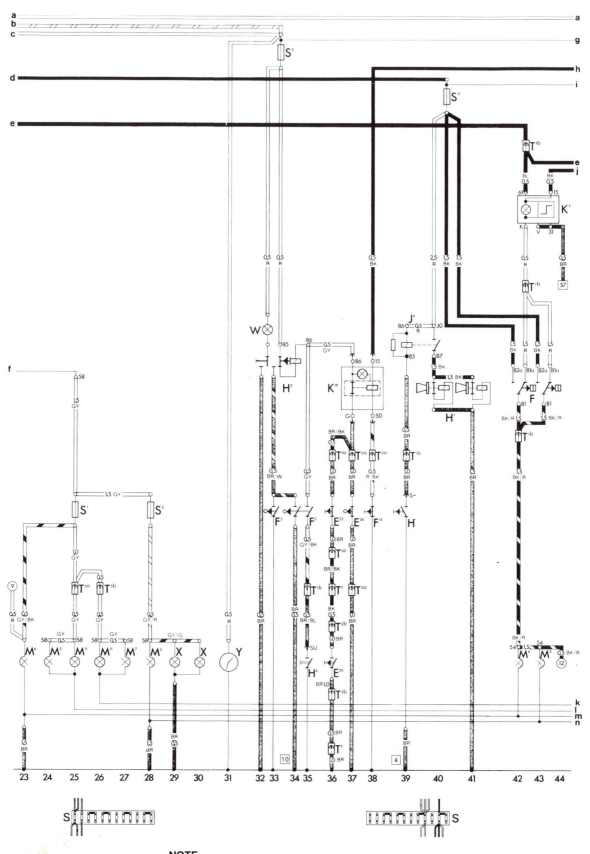

NOTE—
Ignition Switch/Safety Belt Interlock—See diagram Q.

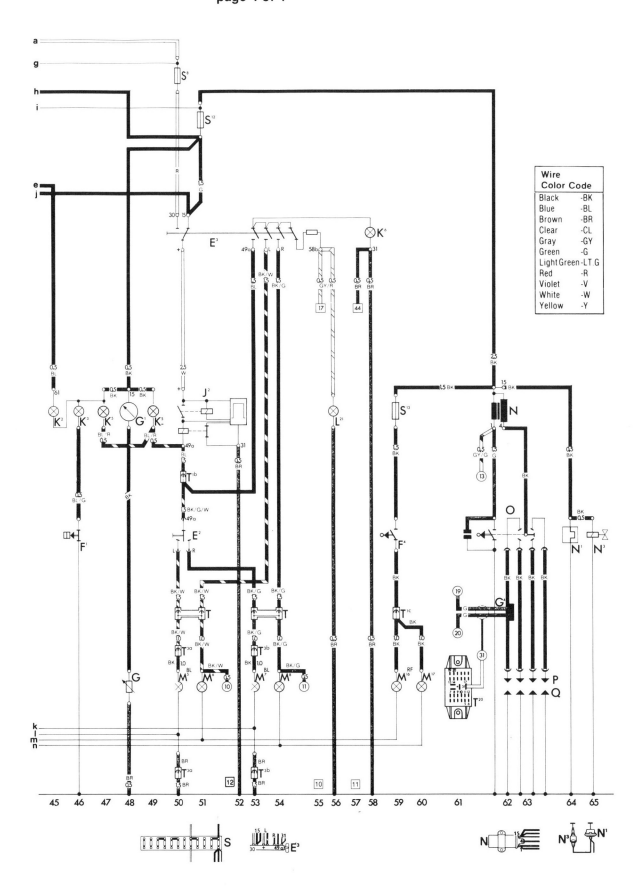

VW Karmann Ghia—from August 1972 (1973 and 1974 Models) page 4 of 4

Q. Ignition Switch/Safety Belt Interlock

© 1973 VWoA—3579

Description		current track
A	—Battery	4
B	—Starter	5
C	—to Alternator	4
D	—Ignition/starter switch	4
E^{24}	—Safety belt lock, with contact	10
E^{25}	—Safety belt lock, right contact	8
E^{31}	—Contact strip in driver's seat	9
E^{32}	—Contact strip in passenger seat	7
F	—Brake light switch	14, 16
F^1	—Oil pressure switch	18
F^2	—Door contact and buzzer alarm switch, front left	3
F^3	—Door contact switch, front right	2
F^9	—Parkingbrake control switch	11
H^6	—Contact in ignition/starter switch for ignition key warning system	3
J^{34}	—Relay for safety belt warning system	3, 6, 7, 9, 11, 14, 15
K^3	—to oil pressure warning light	18
K^7	—Warning light for dual circuit brake system and parkingbrake with warning light for safety belt warning system	12, 14
M^9	—Brake light bulb left	14
M^{10}	—Brake light bulb right	16
S^1	—to fuse box terminal 30 (closed side)	
S^2	—to fuse box terminal 15 (closed side)	
S^3	—to fuse box terminal 30 (open side)	
S^{11}	—Fuse in fuse box	
T^{1a}	—Wire connector, single, behind the dashboard	
T^{1b}	—Wire connector, single, under rear seat	
T^{1c}	—Wire connector, single, behind the dashboard	
T^{1d}	—Wire connector, single, on central tunnel	
T^{2a}	—Wire connector, double, under passenger seat	
T^{2b}	—Wire connector, double, under driver's seat	
T^{2c}	—Wire connector, double, under passenger seat	
T^{2d}	—Wire connector, double, under driver's seat	
T^{3a}	—Wire connector, three-point, in luggage compartment left	
T^{3b}	—Wire connector, three-point, behind engine compartment damping material right	
T^4	—Wire connector, four-point, behind engine compartment damping material left	
T^8	—Wire connector, eight-point, behind instrument panel	
W	—Interior light	1

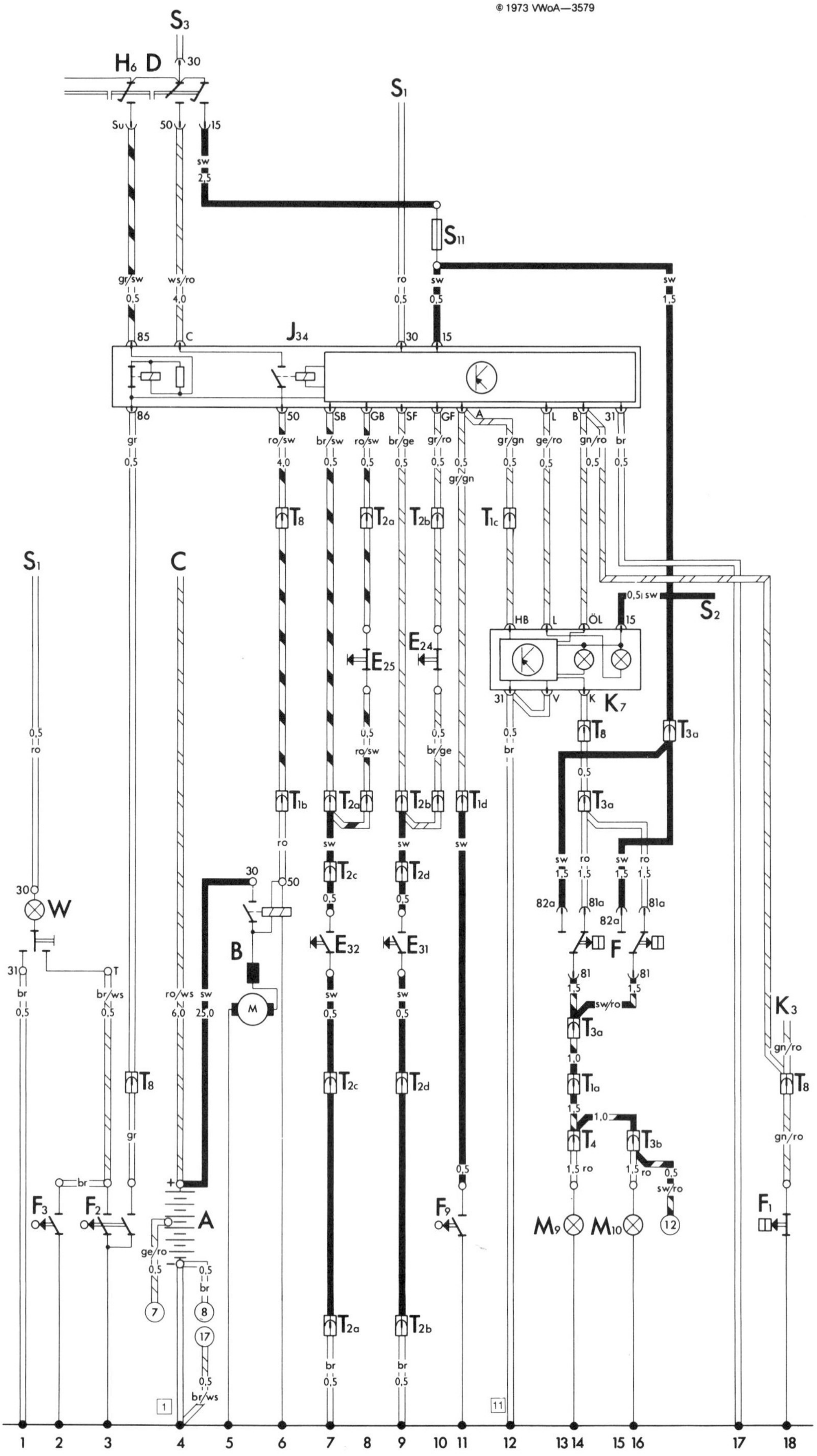

© 1974 VWoA

Circuits for Alternator with Integral Regulator and EGR (exhaust gas recirculation) System (1975 and later Models)

R.

© 1976 VWoA—3840

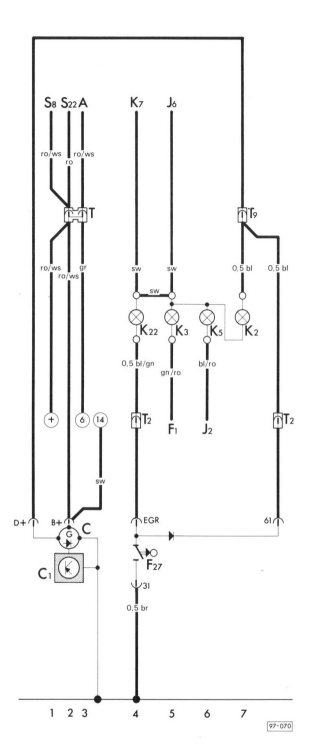

Description		Current track
A	- to battery (positive terminal)	3
C	- to alternator	1, 2, 3
C1	- voltage regulator	2
F1	- to oil pressure switch	5
F27	- switch in EGR elapsed mileage odometer	4, 5, 6
J2	- to turn signal/emergency flasher relay	6
J6	- to voltage vibrator	5
K2	- alternator charging warning light	7
K3	- oil pressure warning light	5
K5	- turn signal warning light	6
K7	- to dual circuit brake warning light, terminal 15	4
K22	- EGR warning light	4
S8	- to fuse 8 in fuse box	
S22	- to single fuse for rear window defogger element	
T	- wire connector in engine compartment	
T2	- wire connector, 2 point behind dashboard	
T9	- wire connector, 9 point behind dashboard	

Color code for wires

bl	- blue	bl/gn	- blue/green
br	- brown	bl/ro	- blue/red
gr	- grey	gn/ro	- green/red
ro	- red	ro/ws	- red/white

NOTE

Where no wire cross section diameters are given, wires are already in current flow diagram.

S. Circuit that Eliminates Seatbelt Interlock (shown in Diagrams K and Q)

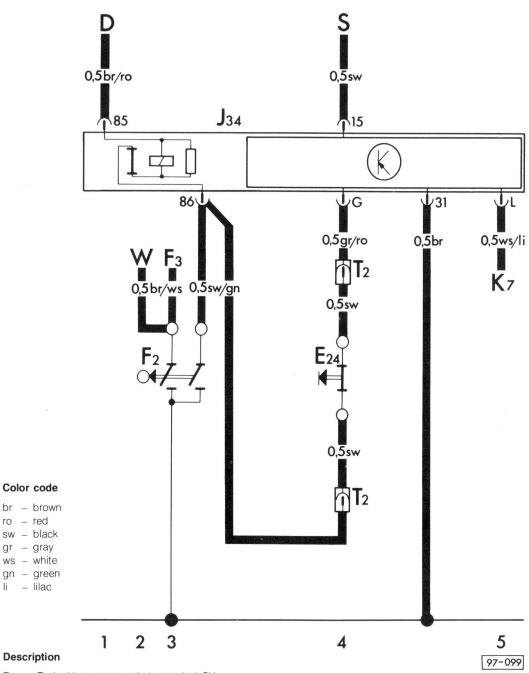

Color code

br – brown
ro – red
sw – black
gr – gray
ws – white
gn – green
li – lilac

Description

- D – To ignition starter switch, terminal SU
- E24 – Contact in safety belt lock, right
- F2 – Door contact switch for buzzer, left
- F3 – Door contact switch, right
- J34 – Safety belt warning system relay
- K7 – To safety belt warning system and dual brake circuit warning light, terminal L
- T2 – Wire connector, double; on frame
- W – Interior light
- S – To fuse box, terminal 15

Section 8

STRUT FRONT SUSPENSION

Contents

Introduction 2	8.3 Removing and Installing Stabilizer Bar . . 17
1. General Description 3	**9. Steering** 17
Suspension Struts 3	9.1 Adjusting Steering 17
Track Control Arms and Stabilizer Bar 4	9.2 Removing, Checking, and Installing Steering Damper (1970 through 1974 only) 18
Steering 4	9.3 Removing, Checking, and Installing Tie Rods 20
2. Maintenance 4	9.4 Removing and Installing Drop Arm 20
3. Front Wheel Alignment 4	9.5 Removing and Installing Idler Arm and Bracket 21
3.1 Checking and Adjusting Camber 4	9.6 Replacing Idler Arm Bracket Bushing (1971 through 1974 cars) 22
3.2 Checking and Adjusting Toe 5	9.7 Removing and Installing Steering Column 22
3.3 Checking and Adjusting Steering Lock 7	Removing Universal Joint Shaft 26
4. Front Wheel Bearings 8	9.8 Removing and Installing Worm and Roller Steering Gearbox 26
5. Steering Knuckles 8	9.9 Disassembling and Assembling Worm and Roller Steering Gearbox 28
5.1 Checking Steering Knuckle 9	Marking the Centerpoint 32
6. Suspension Ball Joints 10	9.10 Removing and Installing Rack and Pinion Steering Gearbox 33
6.1 Removing and Installing Ball Joint . . . 11	
7. Suspension Struts 12	**10. Strut Front Suspension Technical Data** . . . 33
7.1 Removing and Installing Strut 12	I. Tightening Torques 33
7.2 Disassembling, Checking, and Assembling Strut 13	II. Wheel Alignment Specifications 34
Checking Shock Absorber 14	III. Tolerances, Wear Limits, and Settings 34
8. Track Control Arms and Stabilizer Bar . . . 14	
8.1 Removing and Installing Track Control Arm 15	**TABLE**
8.2 Replacing Bushings in Track Control Arm 16	a. Coil Spring Tolerance Groups 14

© 1976 VWoA

Strut Front Suspension

All 1971 and later VW 113 Sedans and Convertibles are equipped with strut-type independent front-wheel suspension. The struts, with their built-in shock absorbing capability, are similar in many respects to the landing wheel struts used on commercial airliners. In addition to providing excellent steering and handling for modern driving conditions, the system has the combined advantages of compact size and comparatively light weight.

A worm and roller steering gearbox is used on 1971 through 1974 Super Beetles with strut front suspension. A rack and pinion steering gearbox was introduced on the 1975 "La Grande Bug." Both of these steering gearboxes are described in this section of the Manual. Neither gearbox shares any components with the worm and roller steering gearbox used on the Beetle Sedan 111 and the Karmann Ghia.

While the spring action of the independent front wheel suspension is accomplished by torsion bars on other Type 1 models, cars with the strut front suspension have coil springs. The springs are installed concentrically over the upper parts of the struts. Frames for cars with strut front suspension have frame heads that are different from those used on Beetle and Karmann Ghia frames. Also, on cars with fuel injection, there is a larger opening in the tunnel part of the frame head in order to provide room for both a fuel feed line and a fuel return line.

Although the suspension struts and the worm and roller steering gearbox can be disassembled for reconditioning, the work requires special tools, experience, and completely clean shop conditions. If you lack the skills, tools, or a suitable workshop for suspension strut or steering gearbox work, we suggest you leave such repairs to an Authorized VW Dealer or other qualified shop. We especially urge you to consult your Authorized VW Dealer before attempting any repairs on a car still covered by the new-car warranty.

Tire wear and vehicle handling qualities are good indicators of the strut front suspension's condition. However, before assuming that the struts or steering require repair or adjustment, make sure that the front tires are properly inflated and that the front wheels have been accurately balanced both statically and dynamically. These checks and adjustments are described fully in **BRAKES AND WHEELS.**

Strut Front Suspension 3

1. General Description

A detailed view of the strut front suspension is presented in Fig. 1-1, along with the nomenclature for the system's major parts. The frame head is a welded assembly. It is completely different from that used on Type 1 models with the front axle, but still containing an opening in its front for removal or installation of the shift rod.

Suspension Struts

Each strut consists of a large hydraulic shock absorber assembly, a concentric coil spring, the steering knuckle, upper and lower strut mountings, and a number of smaller parts. The top of each strut is mounted on the body in a ball thrust bearing. The suspension's upward travel is limited by a hollow rubber buffer, its lower limit of travel by a rubber stop inside the shock absorber.

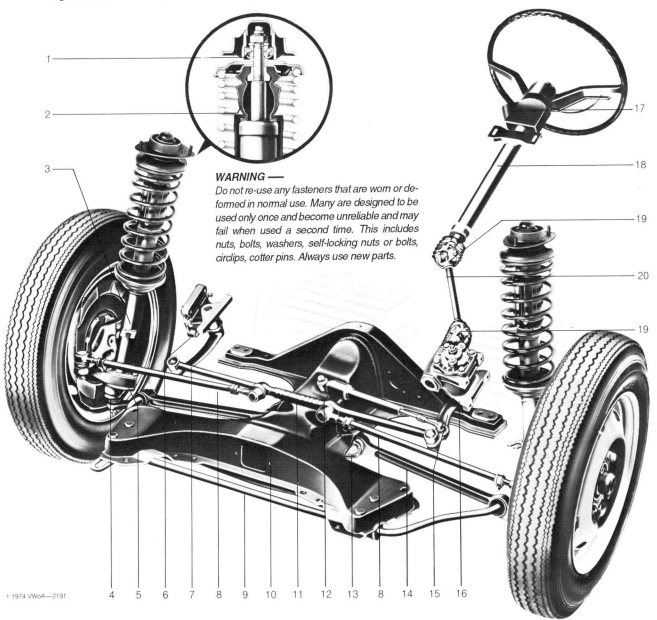

Fig. 1-1. Strut front suspension and steering.

WARNING —
Do not re-use any fasteners that are worn or deformed in normal use. Many are designed to be used only once and become unreliable and may fail when used a second time. This includes nuts, bolts, washers, self-locking nuts or bolts, circlips, cotter pins. Always use new parts.

1. Ball thrust bearing
2. Hollow rubber buffer
3. Suspension strut
4. Ball joint
5. Steering knuckle
6. Idler arm bracket
7. Idler arm
8. Side tie rod
9. Stabilizer bar
10. Frame head
11. Center tie rod
12. Steering damper
13. Eccentric camber adjusting bolt
14. Track control arm
15. Drop arm
16. Steering gearbox
17. Column switch housing
18. Column tube
19. Universal joint
20. Universal joint shaft

4 Strut Front Suspension

Track Control Arms and Stabilizer Bar

The two track control arms locate the suspension struts laterally. The length and adjustment of the track control arms control front wheel camber. Camber can be adjusted by means of the eccentric bolt that forms each track control arm's inner pivot.

Primarily, the stabilizer bar provides anti-roll stabilization, but it also locates the suspension struts longitudinally. Although the trailing length of the stabilizer bar controls the front wheel's caster angle, there is no provision for adjustment. New parts should be installed if wear or accident damage has produced improper caster or altered the wheelbase from specifications.

Steering

The worm and roller steering gearbox used on 1971 through 1974 cars with strut front suspension has two threaded adjustments that can be used to compensate for minor wear. The rack and pinion steering gearbox, introduced on 1975 models, has only one adjustment. Normally, the rack and pinion gearbox requires adjustment only if, after considerable mileage, the unit begins to rattle. Because of its self-damping characteristics, the rack and pinion steering system does not have the hydraulic steering damper used with worm and roller steering. The tie rod ends used with both kinds of steering do not require lubrication during their service life.

Because the front wheel bearings are the same on all cars covered by this Manual, front wheel bearing repair is described in **FRONT AXLE**. Bearing lubrication is discussed in **LUBRICATION AND MAINTENANCE**.

2. Maintenance

The strut front suspension is virtually maintenance-free. The following items only are covered in **LUBRICATION AND MAINTENANCE**:

1. Checking ball joint dust seals and tie rod end dust seals
2. Checking steering play
3. Checking camber and toe
4. Checking front wheel bearing adjustment
5. Lubricating the front wheel bearings.

NOTE —
The lubricant in the worm and roller steering gearbox need be checked only if there are signs of leakage. Normally, lubricant is added only after repairs have been made to correct a leak. The rack and pinion steering gearbox requires neither lubrication nor repair.

3. Front Wheel Alignment

Only camber and toe are adjustable on cars with strut front suspension. Caster angle and king pin inclination are determined by the manufactured dimensions of the suspension parts, so damaged parts must be replaced to correct these alignment factors.

The following preparatory steps are essential to accurate alignment measurements:

1. Have the car on a level surface.
2. Inflate the tires to specifications and unload the car except for the spare wheel and a full fuel tank. Then jounce the car several times and let it settle into its normal position.
3. Check the adjustment of the steering gear and front wheel bearings. Correct these if necessary. See **9.1 Adjusting Steering** and **FRONT AXLE**.
4. Make sure there is no play in the tie rod ends, idler arm, or other parts of the steering linkage.

Measuring wheel camber and toe requires suitable gauges. Although professional-grade instruments may cost several hundred dollars, modestly priced gauges that are adequate for home use are available from mail order houses. Instructions are supplied by the manufacturer.

3.1 Checking and Adjusting Camber

Camber is the angle at which wheels depart from the true vertical when viewed from directly in front of the car. If the tops of the wheels lean slightly outward, they are said to have positive camber. If they lean inward, they are said to have negative camber.

To check:

1. After placing the car on a level surface with the steering centered, apply a bubble protractor as shown in Fig. 3-1.

Fig: 3-1. Bubble protractor against front wheel.

2. Using chalk, mark the wheel at the points where it contacts the protractor.

3. Turn the spirit level carrier on the protractor until the bubble is centered, then read the camber angle on the scale.

   NOTE ▬
   If you are using a different type of gauge, follow the manufacturer's instructions.

4. Roll the car forward a half-turn on the wheels and repeat the measurement at the chalk-marked points.

5. Take the new reading and average it with the one you obtained earlier. The result is the camber angle for the wheel.

6. Repeat the entire procedure on the other front wheel.

The front wheels of all cars with strut front suspension, including 1974 and later models, should have 1° +20′ or −40′ of positive camber. Also, the difference in camber between the wheels should not vary more than 30′. If the camber of each wheel is not within specifications, it should be adjusted to as near 1° as possible.

To adjust camber:

1. Loosen the nut indicated in Fig. 3-2.

   NOTE ▬
   The car must be standing on its wheels while adjustments are being made.

Fig. 3-2. Nut (arrow) on eccentric camber adjusting bolt.

2. Set the spirit level carrier on the protractor to the specified angle of 1° positive camber.

3. Turn the eccentric camber adjusting bolt indicated in Fig. 3-3 until the bubble in the protractor is centered when the protractor is applied to the chalk-marked points on the wheel.

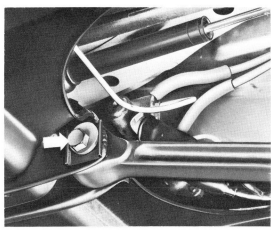

Fig. 3-3. Eccentric camber adjusting bolt.

4. Torque the nut on the eccentric camber adjusting bolt to 4.0 mkg (29 ft. lb.). Again check the camber and repeat the adjustments, if necessary, to bring the camber within specifications.

5. Check the toe and adjust it if necessary.

3.2 Checking and Adjusting Toe

Cars with strut front suspension are designed to operate with a small amount of toe-in. This means that the front edges of the tires are slightly closer together than the rear edges. Shops with optical aligning devices should follow the equipment manufacturer's instructions to obtain a toe-in angle of 30′ ± 15′—or 10′ ± 15′ after adding 8 to 12 kg (18 to 26 lb.) of extra weight above the wheel. The maximum toe change produced by the added weight should not exceed 25′.

Most small shops and individual car owners check toe with a track gauge. This device is used to measure the distance between two points at the front edges of the rims, then the distance between the same two points after the car has been rolled ahead so that these points are at the rear. The measurement made at the rear should be 1.8 to 5.4 mm (.071 to .212 in., or approximately 1/16 to 7/32 in.) greater than the measurement made at the front. This should decrease to −0.6 to +4.2 mm (−.024 to +.165 in., or approximately −1/64 to +11/64 in.) with weight added above the wheels.

NOTE ▬
These specifications apply only with the wheels in their straight-ahead position.

6 Strut Front Suspension

To check wheel toe:

1. Turn the steering to its center position. On cars with rack and pinion steering, you need only have the front wheels pointed straight ahead (steering wheel spokes horizontal).

   NOTE
   On cars with worm and roller steering, the center position is set by using a bolt (VW part No. 113 415 375). The bolt is installed through the drop arm until the point just touches the conical hole in the aluminum plug indicated in Fig. 3-4. (On late models there is no plug. The hole is in the steering gearbox itself.) Tighten the bolt to no more than 0.5 mkg (3.6 ft. lb.).

Fig. 3-4. Aluminum plug (arrow) receives bolt when steering is precisely centered.

   WARNING
   The centering bolt must be removed from the car as soon as adjustments are completed. If left in place, it will prevent the steering wheel from turning.

2. Check the toe. Each wheel should have half the total toe angle specified earlier.

   NOTE
   Since it is not possible to compute the toe angle of an individual wheel when a track gauge is used to check toe, precision adjustments require the use of more sophisticated equipment.

3. Adjust the toe if it is not within specifications.

   NOTE
   If the aluminum centering plug is missing, center the steering by positioning the lug on the ring for the boot in line with the notch indicated by the arrow in Fig. 3-5.

Fig. 3-5. Steering centered by turning worm spindle to position lug on ring in line with notch.

To adjust:

1. On 1971 through 1974 cars, loosen the nut on the clamp (left arrow in Fig. 3-6). Pry up the lockplate, then loosen the locknut for the hexagonal tapered ring (right arrow in Fig. 3-6). On cars with rack and pinion steering, only the right-hand tie rod is adjustable. Loosen the locknuts and hexagonal tapered rings at each end of the right tie rod.

2. Turn the tie rod(s) on the threaded tie rod ends until each wheel is set to half the specified toe angle.

Fig. 3-6. Clamp nut (left arrow) and locknut (right arrow). Drive the tapered ring off the inner end of the tie rod, but leave the clamp in place.

STRUT FRONT SUSPENSION

NOTE —
Rotating the top(s) of the tie rod(s) toward the front of the car increases the toe-in. Rotating the rod(s) to the rear decreases toe-in. If a tie rod is difficult to turn on a car with worm and roller steering, loosen the centering bolt before attempting to free the rod. Otherwise, you may damage the aluminum plug and the bolt.

3.3 Checking and Adjusting Steering Lock

Prior to chassis No. 132 2643 786 or 152 2643 786, built in March 1972, steering lock is determined by the two bolts indicated in Fig. 3-7. Subsequent cars with worm and roller steering have the adjusting bolts located in the drop arm as indicated in Fig. 3-8. Steering lock is built into rack and pinion steering gearboxes and is non-adjustable.

The term "steering lock" describes how far the steering wheel can turn the front wheels in either direction. When the bolts are adjusted properly, they prevent the tires from rubbing the stabilizer bar or the wheel housings when turned to full lock.

To adjust:

1. Lift the front of the car.
2. Loosen the locknuts on the adjusting bolts.
3. Turn the bolts in or out until clearance **a** shown in Fig. 3-9 is 15 mm ($^{19}/_{32}$ in.).

   NOTE —
   Check clearance **a** with the wheels turned full-left and full-right. The same clearance should exist between the rear edge of the wheel and the wheel housing on the opposite side of the car.

Fig. 3-7. Bolts in idler arm bracket (early cars).

Fig. 3-9. Clearance **a** (permissible minimum).

4. Torque the locknuts to 2.0 mkg (14 ft. lb.).
5. Lower the car onto its wheels and recheck clearance **a**. It must not be less than previously specified.

Further alignment checks are in order if you suspect that any part of the strut front suspension system has been damaged in a collision, or if any of the mounting or pivot points in the suspension appear badly worn. These checks can be carried out using the additional specifications given in **10. Technical Data**.

Fig. 3-8. Bolts in drop arm (late cars).

8 Strut Front Suspension

4. Front Wheel Bearings

An exploded view of the front wheel bearing assembly appears in Fig. 4-1. Adjustment procedures are the same as those described in **FRONT AXLE**. Procedures for packing the bearings are given in **LUBRICATION AND MAINTENANCE**.

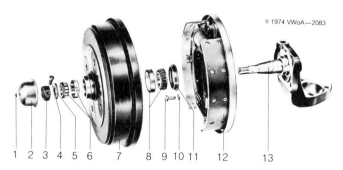

1. Speedometer cable circlip
2. Dust cap
3. Clamp nut
4. Socket head screw for clamp nut
5. Thrust washer
6. Outer tapered-roller bearing
7. Brake drum
8. Inner tapered-roller bearing
9. M 10 bolt
10. Lock washer
11. Grease seal
12. Brake assembly
13. Steering knuckle

Fig. 4-1. Front wheel bearings and related parts.

5. Steering Knuckles

Fig. 5-1 is an exploded view of the steering knuckle mounting used on 1970 through 1973 models. The 1974 steering knuckle mounting is shown in Fig. 5-2.

1. Suspension strut shock absorber
2. M 10 bolt
3. Lockplate
4. Steering knuckle
5. Ball joint
6. Track control arm
7. M 12 self-locking nut

Fig. 5-1. Steering knuckle mounting.

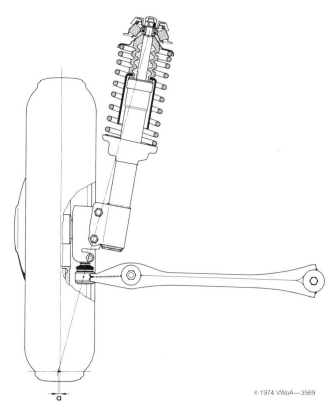

Fig. 5-2. Steering knuckle introduced on 1974 models. The revision of the steering knuckle and its mounting moves the steering axis outside the tire centerline. The revised steering geometry improves handling and stability in the event of a sudden tire blowout.

The steering knuckle used on 1974 and later models bolts to the side of the suspension strut rather than to the bottom of the strut. Also, the suspension ball joint is pressed into the track control arm rather than being bolted to the strut. The ball joint stud points upward, not downward as on previous models, and enters a hole in the bottom of the steering knuckle. It is held in place by a clamp bolt. This modification also changes the procedure used in removing and installing the track arm. However, despite the new track arm and redesigned track arm mountings, the front wheel alignment specifications remain unchanged from those given for the earlier models.

To remove steering knuckle:

1. Remove the road wheel and tire.

2. On the left wheel only, remove the speedometer cable circlip. Then pull the speedometer cable out of the back of the knuckle.

3. Pry off the dust cap. Loosen the socket head screw for the clamp nut and remove the clamp nut.

   NOTE —
   The clamp nut for the left wheel has a left-hand thread.

STRUT FRONT SUSPENSION   9

4. Remove the nut from the tie rod end stud, then press the tie rod end out of the steering arm with a tool such as the one shown in Fig. 5-3.

CAUTION
Do not hammer out the tie rod end. Doing so will ruin the threads and make reinstallation impossible.

Fig. 5-3. Puller in position for tie rod end removal.

5. Pry off the retaining plate shown in Fig. 5-4.

Fig. 5-4. Brake hose retaining plate (arrow).

6. Remove the brake drum together with the wheel bearings.

7. Remove the four bolts holding the brake backing plate on the steering knuckle. Then remove the brake assembly and suspend it from the car with a wire hook to avoid damaging the brake hose.

8. On 1974 and later models, loosen the clamp bolt for the suspension ball joint stud. Then pull the track control arm complete with ball joint down out of the steering knuckle.

9. On 1974 and later models, remove the two bolts that hold the steering knuckle to the strut and remove the steering knuckle from the car.

10. On 1973 and earlier models, bend down the lockplates, remove the three M 10 bolts, and remove the steering knuckle from the car.

Installation is the reverse of removal. On 1970 through 1973 models, install new lockplates under the M 10 bolts, then torque the bolts to 4.0 mkg (29 ft. lb.). On later models, torque the two knuckle mounting bolts to 8.5 mkg (61 ft. lb.) and the clamp bolt for the ball joint stud to 3.5 mkg (25 ft. lb.).

Torque the brake backing plate bolts to 5.0 mkg (36 ft. lb.). Torque the castellated nut on the tie rod end to 3.0 mkg (22 ft. lb.). Advance it, if necessary, to uncover the cotter pin hole, then install a new cotter pin. Adjust the wheel bearings as described in **FRONT AXLE**. Then adjust the camber and toe as described in **3. Front Wheel Alignment.**

5.1 Checking Steering Knuckle

The steering knuckle can be checked either on or off the car. The stub axle should be measured at the three points indicated in Fig. 5-5. The diameter of the outer tapered-roller bearing seat (**A**) should be 17.45 to 17.46 mm (.687 to .688 in.). The diameter of the inner tapered-roller bearing seat (**B**) should be 28.99 to 29.00 mm (1.141 to 1.142 in.). The diameter of the grease seal seat (**C**) should be 40.00 to 40.25 mm (1.575 to 1.585 in.). The dimensions should be checked with a micrometer or vernier caliper.

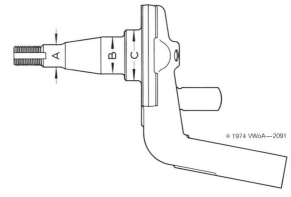

Fig. 5-5. Stub axle measuring points. Dimensions are given in preceding text.

You can check the steering knuckle's stub axle for distortion by using a vernier caliper and a machinist's square

10 STRUT FRONT SUSPENSION

as shown in Fig. 5-6. Make measurements at not less than three points around the stub axle. The difference between any two measurements must not exceed 0.25 mm (.01 in.).

Fig. 5-6. Vernier caliper and machinist's square being used to check stub axle for distortion.

You can also use a vernier caliper in conjunction with a straightedge to check the steering arm for distortion. Use a vise to hold the straightedge, as shown in Fig. 5-7, or a C-clamp if the steering knuckle is on the car. The distance from the straightedge to the outer edge of the tie rod hole should be 54.20 to 54.70 mm (2.130 to 2.150 in.).

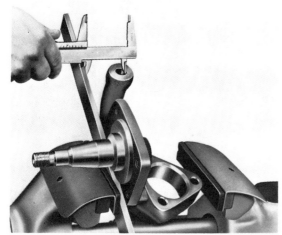

Fig. 5-7. Steering arm being checked for distortion.

CAUTION ——
Bent steering knuckles must be replaced and not straightened. Bending them back to their original shape will seriously weaken them structurally.

6. Suspension Ball Joints

A special lever and a vernier caliper should be used to check ball joint play.

To check:

1. Lift the car. Then turn the steering to one side.

2. Install the special lever on 1970 through 1973 models as shown in Fig. 6-1. Move the lever as indicated by the curved arrow to compress the spring inside the ball joint.

   NOTE ——
   The same procedure is used on 1974 and later models. However, the lever must rest on the nut or bolt head that holds the steering knuckle to the suspension strut.

Fig. 6-1. Lever in position. Hook goes under track control arm on all models; lever goes above steering knuckle flange on 1970 through 1973 models and above knuckle mounting bolt head on 1974 and later models.

3. With the ball joint compressed, place a vernier caliper on the suspension as shown in Fig. 6-2.

Fig. 6-2. Vernier caliper in position over ball joint.

STRUT FRONT SUSPENSION 11

NOTE —
On 1974 and later models, place the upper jaw of the caliper on top of the clamp bolt for the ball joint stud.

4. Note the reading on the caliper. Then, while slowly releasing the lever, note the caliper's travel. The increase in the reading is the ball joint play. On all models, replace the ball joint if play is 2.50 mm (.100 in.) or more.

Fig. 6-3 is a cross section of a typical suspension ball joint. When new, there is 1.00 mm (.040 in.) of play between the upper plastic shell and the housing. This clearance increases with wear, since the spring forces the upper shell downward.

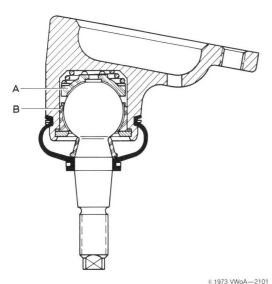

Fig. 6-3. Upper plastic shell **(A)** and lower plastic shell **(B)**. Note the spring above the upper shell.

The gap between the upper and lower plastic shells is 1.50 mm (.060 in.) in a new ball joint. No further wear is permissible when, owing to wear, this gap disappears and the shells meet. If kept in service, the shells will no longer support the ball and the joint will rattle. Maximum permissible ball joint play is therefore 2.50 mm (.100 in.). This figure equals the original 1.00 mm (.040 in.) clearance plus the 1.50 mm (.060 in.) gap between the shells.

6.1 Removing and Installing Ball Joint

If a ball joint exceeds the wear limit specified in **6. Suspension Ball Joints,** it must be replaced. Ball joints should also be replaced if they have become stiff or if their dust seals are cracked or torn and it is suspected that abrasive dirt has entered the joint itself.

The procedures for removing and installing suspension ball joints on 1974 and later models are different from the procedures for earlier models and are given separately.

To remove ball joint (1970 through 1973):

1. Remove the self-locking nut from the ball joint stud where it passes through the track control arm.

2. Press the ball joint stud out of the track control arm as shown in Fig. 6-4.

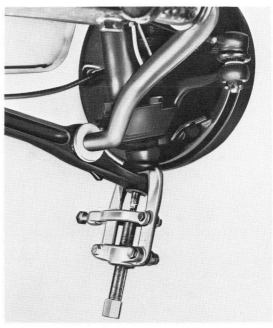

Fig. 6-4. Puller in position to press ball joint stud out of track control arm.

3. Bend down the lockplates, then remove the three M 10 bolts that hold the ball joint and steering knuckle on the strut.

4. Using a loop of wire, suspend the steering knuckle and brake assembly from the brake hose bracket or the strut. Doing so will make it unnecessary to dismount the brake hose.

5. Remove the ball joint from the steering knuckle.

Installation is the reverse of removal. Use new lockplates under the M 10 bolts and a new self-locking nut on the ball joint stud. The ball joint stud must be grease-free when it is inserted into the track control arm. If necessary, grip the flat sides on the lower end of the ball joint stud with an open end wrench while tightening the self-locking nut. Torque the M 10 bolts and the self-locking nut to 4.0 mkg (29 ft. lb.). Check the camber and toe following installation of the ball joint and correct them if necessary.

12 Strut Front Suspension

To replace ball joint (1974 and later models):

1. Remove the track control arm as described in **8.1 Removing and Installing Track Control Arm.**

2. Using a press tool that can be threaded onto or slipped over the ball joint stud, press the ball joint downward out of the track control arm.

3. Using press tools that will apply pressure only to the track control arm eye and the outer housing of the ball joint, press in the new ball joint from the deeper section (bottom) of the control arm (Fig. 6-5).

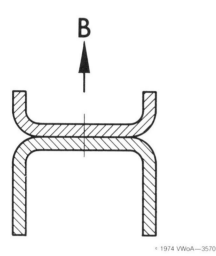

Fig. 6-5. Section of track control arm. Press in the new ball joint in the direction indicated by arrow **B**.

WARNING —
- *Never reinstall used ball joints. They will not fit tightly and could come out of the track arm while the car is being driven.*

- *Do not re-use fasteners that are worn or deformed in normal use. Many are designed to be used only once and become unreliable and may fail when used a second time. This includes, but is not limited to, nuts, bolts, washers, self-locking nuts or bolts, circlips, cotter pins. Always use new parts.*

7. Suspension Struts

Although the steering knuckle mounting is different on the struts introduced in 1974, the working parts are unchanged from the struts used on previous models. Shock absorbers and coil springs can be replaced individually. New rubber buffers, protective sleeves, and new parts for the upper mount are also available.

7.1 Removing and Installing Strut

The steering knuckle and brake assembly can be left on the car when you remove the strut.

To remove:

1. From the left wheel only, pry off the speedometer cable circlip. Then pull the speedometer cable out of the back of the steering knuckle.

2. Pry off the retaining plate indicated in Fig. 7-1.

Fig. 7-1. Brake hose retaining plate (arrow).

3. On 1970 through 1973 models, bend down the lockplates, then remove the three M 10 bolts that hold the ball joint and steering knuckle on the strut. On 1974 and later models, remove the two bolts and nuts that mount the steering knuckle on the strut. See **5. Steering Knuckles**.

4. Pull the strut off the steering knuckle. Then, on 1970 through 1973 models, install one bolt to support the knuckle as shown in Fig. 7-2.

Fig. 7-2. Bolt (arrow) to support knuckle.

Strut Front Suspension

5. Working inside the front luggage compartment, remove two of the three nuts shown in Fig. 7-3.

6. Reach under the fender and support the strut while removing the third nut. Take the strut out downward when the last nut is off.

Fig. 7-3. Nuts that hold upper strut mount. Arrow indicates last nut to be removed.

Installation is the reverse of removal. Torque the three nuts for the upper mount to 2.0 mkg (14 ft. lb.). On 1970 through 1973 models, install new lockplates under the M 10 bolts at the lower end of the strut, then torque the bolts to 4.0 mkg (29 ft. lb.). On 1974 and later models, torque the two knuckle mounting bolts to 8.5 mkg (61 ft. lb.) and the clamp bolt for the ball joint stud to 3.5 mkg (25 ft. lb.). Check the camber and toe following installation of the strut and correct them if necessary.

7.2 Disassembling, Checking, and Assembling Strut

You must have a proper spring compressing appliance to disassemble the strut.

> **WARNING** —
> Do not attempt to tension the spring with makeshift tools. If a compressed spring slips free, it can inflict severe injury.

To disassemble strut:

1. Install the spring compressing appliance in a bench vise, then install the strut in the spring compressing appliance.

2. Pry off the ball thrust bearing dust cap.

3. To compress the spring, tighten the long bolts a few turns at a time, working alternately (Fig. 7-4).

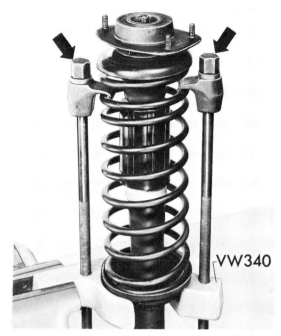

Fig. 7-4. Strut mounted in spring compressing appliance. Arrows indicate the two long bolts.

4. Using an offset box wrench and an Allen wrench as shown in Fig. 7-5, remove the self-locking nut from the shock absorber piston rod.

5. Working alternately, gradually loosen the two long bolts on the compressing appliance until spring tension is relieved. Then take the strut apart.

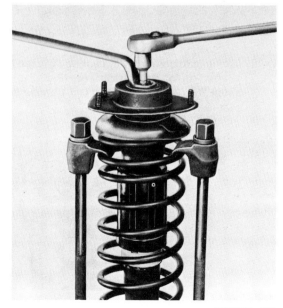

Fig. 7-5. Removing self-locking nut from piston rod.

14 Strut Front Suspension

Checking Shock Absorber

Defective shock absorbers often make knocking noises when the car is driven. You can hand-check a shock absorber by extending and compressing it while holding it in its installed position. It must operate smoothly and with uniform resistance throughout its entire stroke. If possible, compare the used unit with a new shock absorber. New shock absorbers that have been in storage may have to be pumped several times before they reach full efficiency.

Shock absorbers do not require maintenance. An adequate supply of fluid is placed in them during manufacture to compensate for small leaks. Minor traces of fluid are acceptable if the shock absorber still functions efficiently. Defective shock absorbers cannot be serviced and must be replaced.

To assemble strut:

1. Install the rubber buffer and the protective sleeve on the piston rod.

2. Install the shock absorber and the coil spring in the spring compressing appliance and gradually compress the spring.

   NOTE ——
   Replacement springs are supplied in three tolerance groups **(Table a)**. Both springs on the car must have the same color code.

Table a. Coil Spring Tolerance Groups

Group	Color code	Spring pressure
1	1 red paint mark	227-233 kg (500-514 lb.)
2	2 red paint marks	234-240 kg (516-529 lb.)
3	3 red paint marks	241-247 kg (531-545 lb.)

3. Continue to compress the spring until the unthreaded portion of the piston rod (Fig. 7-6) projects about 8 to 10 mm (5/16 to 3/8 in.) above the spring seat.

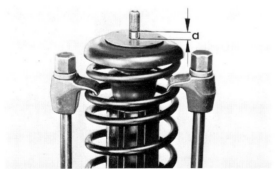

Fig. 7-6. Unthreaded length of piston rod **(a)** projecting above upper spring seat.

4. Install the ball thrust bearing assembly. Then, using the same tools as for removal, install a new self-locking nut on the piston rod. On 1971 through 1973 models, torque the nut to 7.0 to 8.5 mkg (50 to 61 ft. lb.); on 1974 and later models, torque the nut to 6.0 mkg (43 ft. lb.).

5. Fill the spaces indicated in Fig. 7-7 with lithium grease. Then install the ball thrust bearing dust cap.

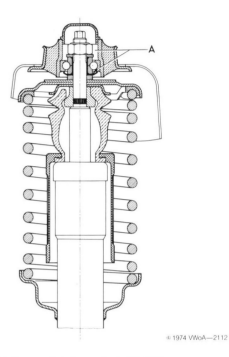

Fig. 7-7. Spaces **A** to be packed with lithium grease.

6. Install the strut on the car.

   NOTE ——
   When installing replacement shock absorbers on cars built prior to chassis No. 111 2810 528 (April 1971), it may be necessary to grind a chamfer in the upper end of the large hole in the steering knuckle and to install longer, 48-mm (1 57/64-in.) M 10 bolts. These modifications are necessary to adapt the early steering knuckle to the latest type replacement struts.

8. Track Control Arms and Stabilizer Bar

The track control arms and stabilizer bar form a flexible support linkage between the frame head and the suspension struts. Their design is such that all wear is confined to easily replaceable rubber bushings. Fig. 8-1 shows the bushings and other mounting parts for the control arms and stabilizer bar.

Strut Front Suspension 15

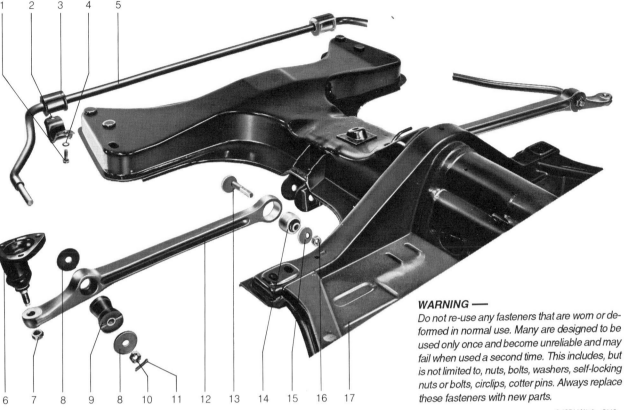

1. M 8 grooved bolt
2. Mounting clamp
3. Rubber stabilizer mounting
4. Spring washer
5. Stabilizer bar
6. Ball joint (1970 through 1973)
7. M 12 self-locking nut
8. Washer for bushing
9. Bonded rubber bushing for stabilizer bar
10. Castellated nut
11. Cotter pin
12. Track control arm
13. Eccentric camber adjusting bolt
14. Bonded rubber bushing for track control arm
15. Eccentric camber adjusting washer
16. M 10 self-locking nut
17. Frame head

WARNING —
Do not re-use any fasteners that are worn or deformed in normal use. Many are designed to be used only once and become unreliable and may fail when used a second time. This includes, but is not limited to, nuts, bolts, washers, self-locking nuts or bolts, circlips, cotter pins. Always replace these fasteners with new parts.

Fig. 8-1. Track control arm and stabilizer bar with mounting parts and frame head.

8.1 Removing and Installing Track Control Arm

It is not necessary to remove the suspension strut or stabilizer bar to replace a track control arm.

To remove:

1. On 1974 and later cars, loosen the clamp bolt for the ball joint stud. On earlier models, remove the self-locking nut from the ball joint stud.

2. On 1974 and later cars, pull the track control arm down and out of the steering knuckle. On earlier models, use a puller, as illustrated earlier in **6.1 Removing and Installing Ball Joint**, to pull the track control arm off the ball joint stud.

3. Remove the cotter pin from the castellated nut on the stabilizer bar. Remove the castellated nut and the self-locking nut at the frame head (Fig. 8-2).

Fig. 8-2. Castellated nut (left arrow) and self-locking nut (right arrow).

16 STRUT FRONT SUSPENSION

4. Pull the track control arm downward and off.

To install:

1. Check the track control arm for cracks and bends.
2. Inspect the bonded rubber bushings for wear. If necessary, replace them as described in **8.2 Replacing Bushings in Track Control Arm.**
3. Insert the stabilizer bar into the track control arm, then install the castellated nut finger-tight.

   NOTE
   The rubber lugs on the end of the bonded rubber bushing for the stabilizer bar must be positioned horizontally.

4. Make sure that the ball joint stud and its hole are grease-free. Then install the ball joint stud.
5. Install the track control arm on the frame head using the eccentric bolt, the eccentric washer, and a new self-locking nut.
6. Torque the castellated nut to 3.0 mkg (22 ft. lb.), advancing it, if necessary, to uncover the cotter pin hole. Then install a new cotter pin.
7. On 1974 or later cars, torque the clamp bolt to 3.5 mkg (25 ft. lb.). On earlier cars, install a new self-locking nut on the ball joint stud and torque it to 4.0 mkg (29 ft. lb.). If necessary, hold the flat sides on the ball joint stud with an open end wrench to keep the stud from turning as the nut is tightened.
8. Adjust the camber, then torque the self-locking nut on the eccentric bolt to 4.0 mkg (29 ft. lb.).
9. Adjust toe to specifications.

8.2 Replacing Bushings in Track Control Arm

Install the bonded rubber bushings in the positions shown in Fig. 8-3.

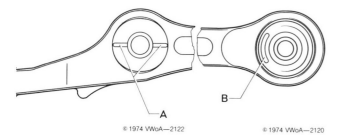

Fig. 8-3. Bushing positions. Rubber lugs on end of stabilizer bar bushing must be horizontal as at **A**. Recess in track control arm bushing must be toward front wheel as at **B**.

To replace bushings:

1. Press the old bushings out of the track control arm.
2. Clean the bushing holes in the track control arm.
3. Lubricate the new bonded rubber bushing for the stabilizer bar with silicone spray.
4. Press the new bonded rubber bushing for the stabilizer bar into the track control arm until the bushing is flush with the top of the arm (Fig. 8-4).

Fig. 8-4. Bonded rubber bushing for stabilizer bar being pressed into track control arm.

5. Press in the new bonded rubber pivot bushing for the track control arm until the bushing is flush with the top of the arm (Fig. 8-5).

Fig. 8-5. Bonded rubber bushing for track control arm being pressed into its hole at inner end of track control arm.

8.3 Removing and Installing Stabilizer Bar

Because the stabilizer bar has a direct influence on front wheel alignment, it is important that a bent stabilizer bar be replaced promptly. For the same reason the car should not be kept in service with worn rubber stabilizer mountings.

If the rubber stabilizer mountings make squeaking noises as the car is driven over undulating pavement, wash them with brake fluid or lubricate them with silicone spray to cure the squeak.

To remove stabilizer bar:

1. Remove the cotter pins from the two castellated nuts, then take the castellated nuts off the stabilizer bar.
2. Remove the four grooved bolts that hold the two mounting clamps to the frame head (Fig. 8-6).

Fig. 8-6. Castellated nut (left arrow) and one of the four grooved bolts (right arrow) that hold the mounting clamps.

3. Take the mounting clamps off the rubber stabilizer mountings.
4. Pull the stabilizer bar out of the bonded rubber bushings in the track control arms.

Installation is the reverse of removal. If the rubber stabilizer mountings are worn, cut them off the stabilizer bar. Clean the stabilizer bar with a wire brush, then lubricate the new rubber stabilizer mountings with silicone spray and slide them into position on the stabilizer bar.

Torque the castellated nuts to 3.0 mkg (22 ft. lb.). If necessary, advance them to uncover the cotter pin holes, then install new cotter pins. On 1971 through 1973 models, torque the grooved bolts for the mounting clamps to 2.0 mkg (14 ft. lb.); on 1974 and later models, torque the bolts to 4.0 mkg (29 ft. lb.).

9. STEERING

The steering normally requires no maintenance. Parts seldom require replacement unless damaged by a collision or, in the case of the steering gearbox, unless the car has been operated for a long time with the steering improperly adjusted.

9.1 Adjusting Steering

If the rack and pinion steering starts to rattle after high mileage, adjust it. To do so, remove the cover from the access hole in the spare wheel well. With the car standing on its wheels, loosen the locknut, then turn in the adjusting screw until it contacts the thrust washer. Hold the screw in this position while you tighten the locknut. The adjusting screw is too tight if the steering is stiff and fails to self-center properly.

To check the worm and roller steering adjustment, raise the front wheels off the ground. With the front wheels in their straight-ahead position, grip the steering wheel with your fingers as shown in Fig. 9-1. Turn the steering wheel lightly in both directions. Freeplay must not exceed 15 mm (about $9/16$ in.) measured at the rim of the steering wheel.

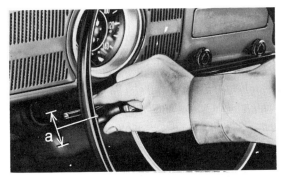

Fig. 9-1. Checking steering freeplay. Freeplay is measured at steering wheel rim as indicated by **a**.

If worm and roller steering freeplay is excessive, check to see that the looseness is not caused by worn tie rod ends or a worn idler arm bracket bushing. Check also to see that the steering gearbox and the idler arm are not loose on their mountings and that the gearbox cover bolts are torqued to 2.0 to 2.5 mkg (14 to 18 ft. lb.). Finally, make certain that the drop arm is not loose and that there is no play in the steering column's universal joints. If none of these faults are found, check and, if necessary, adjust the steering gearbox in the following sequence: (1) the play between the worm and the roller, (2) the axial play of the worm spindle, and (3) the axial play of the roller. If there is any play in a rack and pinion steering gearbox, replace it.

18 Strut Front Suspension

To adjust play between roller and worm:

1. Turn the steering wheel 90° to the left or right.

2. Loosen the adjusting screw locknut indicated in Fig. 9-2. Then turn the adjusting screw out about one full turn.

Fig. 9-2. Locknut for roller shaft adjusting screw. Screw is accessible through a hole in front luggage compartment.

3. Turn the adjusting screw slowly clockwise until you feel the roller contact the worm.

4. While holding the screw in this position, torque the locknut to 2.5 mkg (18 ft. lb.).

5. Turn the steering wheel first 90° to the left and then 90° to the right. Check the freeplay in both these positions as described earlier. It must not exceed 15 mm (about 9/16 in.).

   NOTE ——
   If freeplay is excessive on only one side repeat the adjustment with the steering wheel turned 90° to that side.

6. Road test the car. If after taking a corner at 10 to 12 mph (15 to 20 kph), the steering does not return to about 45° from the center position, the roller is too tight. The adjustment must then be repeated to prevent damage to the worm and roller.

If you have made the preceding adjustment correctly, yet freeplay remains excessive, check the axial play of the worm spindle. To do this, grasp the lower part of the steering column and turn the worm spindle clockwise and counterclockwise. If you can see the worm spindle moving slightly in and out of the steering gearbox, there is excessive play.

Because of the location of the body crossmember, you cannot adjust the worm spindle axial play with the gearbox installed. Adjustment must be carried out with the steering gearbox removed as described in **9.9 Disassembling, Adjusting, and Assembling Steering Gearbox.**

If there is no worm spindle axial play, yet there seems to be play in the gearbox, remove and disassemble the gearbox so the roller can be checked for axial play. This check is described in **9.9 Disassembling, Adjusting, and Assembling Steering Gearbox.** If the roller has excessive axial play, the entire roller shaft must be replaced.

9.2 Removing, Checking, and Installing Steering Damper

(1970 through 1974 only)

The steering damper is a kind of miniature hydraulic shock absorber installed between the drop arm on the steering gearbox and the frame head. It is designed to prevent road shocks from being transmitted to the steering wheel.

To remove:

1. Detach the steering damper piston rod from the drop arm on the steering gearbox by taking out the bolt indicated in Fig. 9-3.

Fig. 9-3. Bolt (arrow) through drop arm that screws into steering damper piston rod end.

2. Remove the spare wheel from the front luggage compartment. They pry the cover off the access hole for the bolt that holds the steering damper on the frame head.

3. Remove first the bolt and then the steering damper from the car.

The steering damper and other parts of the worm and roller steering linkage are shown disassembled in Fig. 9-4.

Strut Front Suspension 19

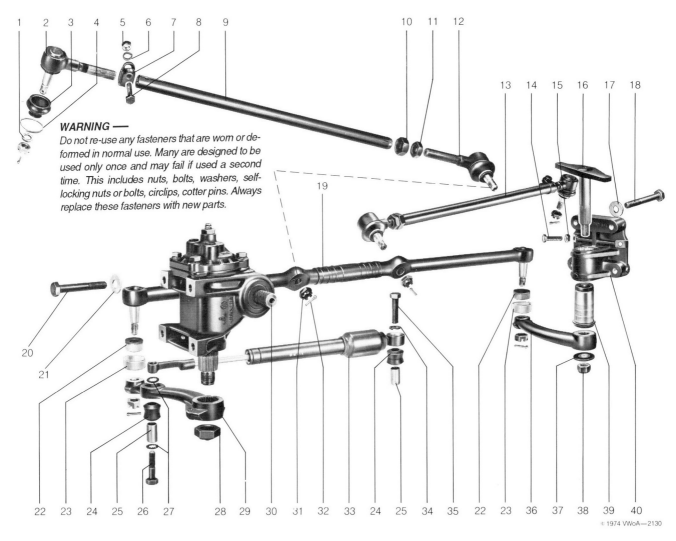

WARNING —
Do not re-use any fasteners that are worn or deformed in normal use. Many are designed to be used only once and may fail if used a second time. This includes nuts, bolts, washers, self-locking nuts or bolts, circlips, cotter pins. Always replace these fasteners with new parts.

1. Sealing ring for boot
2. Tie rod end
3. Boot
4. Retaining ring for boot
5. M 8 nut
6. Spring washer
7. Tie rod clamp
8. M 8 bolt
9. Left tie rod
10. Tapered ring
11. M 14 nut
12. Tie rod end
13. Right tie rod assembly
14. M 8 adjusting bolt
15. Locknut
16. Idler arm shaft
17. Flat washer
18. M 10 bolt
19. Center tie rod
20. M 10 bolt
21. Flat washer
22. Spacer
23. Sealing ring
24. Rubber bushing
25. Bushing sleeve
26. M 10 bolt
27. Lock washer
28. M 20 nut
29. Drop arm
30. Steering gearbox
31. Castellated nut
32. Cotter pin
33. Steering damper
34. Lock washer
35. M 10 bolt
36. Idler arm
37. Washer
38. Self-locking nut
39. Idler arm bracket bushing
40. Idler arm bracket

Fig. 9-4. Exploded view of 1971 through 1974 steering linkage.

Hand-check the steering damper by extending and compressing it while holding it in its installed position. It must operate smoothly and with uniform resistance throughout its entire stroke. If necessary, compare the used unit with a new steering damper. Minor fluid leakage does not make replacement necessary as long as efficiency is not impaired.

Installation is the reverse of removal. Torque the bolt that holds the steering damper to the frame head to 6.0 mkg (43 ft. lb.). Torque the bolt that holds the steering damper's piston rod to the drop arm to 4.0 to 4.5 mkg (29 to 32 ft. lb.). Install the access hole cover and apply caulking compound ("dum-dum") around its edge. Check to see that the steering operates freely before taking the car out on the road.

9.3 Removing, Checking, and Installing Tie Rods

It is best to remove the right, left, and center tie rods as a unit from cars with worm and roller steering. Separate the rods after they are off the car. Cars with rack and pinion steering have left and right tie rods only. Their inner ends are held to the rack by bolts.

To remove:

1. Remove the cotter pins and the castellated nuts. Then press the tie rod ends out of the steering arms with a tool such as the one shown in Fig. 9-5.

   CAUTION ——
   Do not hammer out the tie rod ends. Doing so will ruin the threads and make reinstallation impossible.

Fig. 9-5. Puller in position to press out tie rod end.

2. On cars with worm and roller steering, remove the cotter pins and the castellated nuts from the center tie rod's ends, then press the tie rod ends out of the drop arm and idler arm as illustrated in Fig. 9-5. On cars with rack and pinion steering, bend up the lockplate, then remove the bolts that hold the tie rod inner ends to the steering rack.

3. On cars with rack and pinion steering, remove the tie rod(s).

4. On cars with worm and roller steering, remove the complete tie rod assembly, then clamp the center tie rod in a vise. Remove the cotter pins and castellated nuts from the tie rod ends. Then, using the tool illustrated in Fig. 9-5, press the ends of the outer tie rods out of the center tie rod.

5. Clean the tie rods with a wire brush to make inspections easier.

Installation is the reverse of removal. Carefully inspect the tie rods for cracks. Distortions can be detected by rolling the tie rods over a flat surface. Check the tie rod ends for play and replace any that are worn.

WARNING ——
Do not confuse the tie rod ends used for the strut front suspension with those for the front axle. The tie rod ends for the strut front suspension are identified by either an indentation or a protrusion at the point indicated in Fig. 9-6. The tie rod ends for the front axle do not have an adequate range of angular movement and could cause the steering on cars with strut front suspension to jam or become erratic.

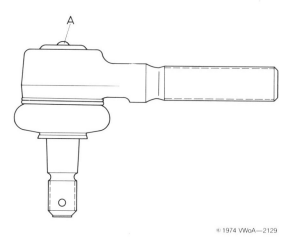

Fig. 9-6. Projection **(A)** that identifies tie rod ends for use with strut front suspension.

NOTE ——
If the tie rod end boots are torn or cracked, they can be replaced. However, the entire tie rod end should be replaced if dirt has entered the ball socket.

On cars with rack and pinion steering, torque the bolts at the inner tie rod ends to 5.5 mkg (40 ft. lb.), then bend over the locking tabs. On all cars, torque the castellated nuts to 3.0 mkg (22 ft. lb.). Then advance them, if necessary, to uncover the cotter pin holes. Install new cotter pins. Check and, if necessary, adjust the wheel alignment as described in **3.2 Checking and Adjusting Toe**.

9.4 Removing and Installing Drop Arm

The drop arm (worm and roller steering only) can be removed without removing the steering gearbox from the car. If you must replace the arm, be sure to get one that has the same spline pattern as the old drop arm.

Strut Front Suspension

To remove:

1. Detach the steering damper's piston rod from the drop arm by taking out the M 10 bolt. See **9.2 Removing, Checking, and Installing Steering Damper.**

2. Remove the cotter pin and the castellated nut, then press the tie rod end out of the drop arm with a tool such as the one shown in Fig. 9-7.

   CAUTION
   Do not hammer out the tie rod end. Doing so would ruin the threads and make reinstallation impossible.

3. Remove the M 20 nut that holds the drop arm on the roller shaft of the steering gearbox. Then pull the drop arm off the shaft as shown in Fig. 9-7.

Fig. 9-7. Puller used for removing center tie rod end and drop arm.

To install:

1. Inspect the rubber bushing. If it is worn, press it out, then lubricate the replacement bushing with silicone spray before pressing it in.

2. Slide the drop arm onto the roller shaft. Install a new M 20 nut finger-tight.

   NOTE
   One or more of the splines on the roller shaft are wider than the others to ensure proper installation of the drop arm. If a replacement drop arm is needed, make sure it has the same size splines as the original.

3. Install the tie rod end in the drop arm. Torque the castellated nut to 3.0 mkg (22 ft. lb.). Then advance it, if necessary, to uncover the cotter pin hole. Install a new cotter pin.

4. Install the steering damper piston rod. Torque the M 10 bolt to 4.0 to 4.5 mkg (29 to 32 ft. lb.).

5. Torque the M 20 nut to 10.0 mkg (72 ft. lb.). Then peen it as indicated in Fig. 9-8.

Fig. 9-8. Nut on roller shaft. Using a blunt cold chisel, peen the lower part of the nut as indicated by the arrows, until it engages the slots in the roller shaft.

9.5 Removing and Installing Idler Arm and Bracket

The idler arm (worm and roller steering only) and bracket should be removed as a unit, then disassembled, if necessary, on the workbench.

To remove:

1. Remove the cotter pin and castellated nut.

2. Press the tie rod end out of the idler arm using a tool such as the one illustrated in Fig. 9-7.

   CAUTION
   Do not hammer out the tie rod end. Doing so would ruin the threads and make reinstallation impossible.

22 Strut Front Suspension

3. To remove the idler arm and bracket from the car, remove the three bolts shown in Fig. 9-9.

Fig. 9-9. Bolts that hold idler arm and bracket on body. Support the idler arm as you remove the last bolt (arrow).

4. If it is necessary to disassemble the idler arm and bracket, remove the self-locking nut and the washer from the end of the idler arm shaft, then pull the idler arm off the splines.

To install:

1. If removed earlier, install the idler arm on its shaft as illustrated in Fig. 9-10.

Fig. 9-10. Idler arm installation. Wide splines (arrows) must be matched as the parts are joined.

2. After installing the idler arm on the shaft, install the washer and a new self-locking nut. Torque the nut to 4.0 mkg (29 ft. lb.).

3. Install the idler arm and bracket on the body. Torque the mounting bolts to 3.0 mkg (22 ft. lb.).

4. Install the tie rod end in the drop arm. Torque the castellated nut to 3.0 mkg (22 ft. lb.). Then advance it, if necessary, to uncover the cotter pin hole. Install a new cotter pin.

9.6 Replacing Idler Arm Bracket Bushing
(1971 through 1974 cars)

The bonded rubber bushing in the idler arm bracket must be replaced if the idler arm has excessive play. To replace the bushing, it is necessary to remove the idler arm and bracket. Then remove the idler arm from its shaft as described in **9.5 Removing and Installing Idler Arm and Bracket**.

To replace:

1. Press the bonded rubber bushing along with its inner metal sleeve out of the idler arm bracket.

2. Press the outer metal sleeve out of the idler arm bracket.

3. Press the new bonded rubber bushing complete with both inner and outer metal sleeves into the idler arm bracket.

9.7 Removing and Installing Steering Column

The universal joints in the steering column used with strut front suspension are shown in Fig. 9-11. The design keeps the steering gearbox and the steering column from being in line with each other. The resulting angle provides the offset necessary to prevent possible collision forces from any direction being transmitted to the steering wheel.

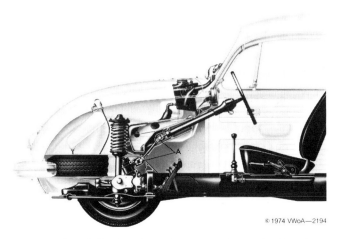

Fig. 9-11. Steering column and universal joint shaft installed. The universal joints are at **A**.

STRUT FRONT SUSPENSION 23

Fig. 9-12 is an exploded view of the steering column assembly. There have been several changes to the steering column switch, horn control, and steering wheel during the model years covered by this Manual. These modifications and detailed instructions for adjusting the turn signal switch and for testing the horn circuit are described in **ELECTRICAL SYSTEM**.

Individual parts for the universal joint shaft are not available. If play is discovered in the universal joints, the shaft must be replaced as a unit. However, trouble is unlikely unless damaged boots have permitted abrasive dirt to enter the universal joint bearings. Torn or cracked boots should always be replaced promptly.

To remove steering column and tube:

1. Remove the fuel tank as described in **FUEL SYSTEM**.

    WARNING —
    - *Disconnect the battery ground strap. Do not smoke or work near heaters or other fire hazards. Have a fire extinguisher handy.*
    - *Do not re-use any fasteners that are worn or deformed in normal use. Many fasteners are designed to be used only once and become unreliable and may fail when used a second time. This includes, but is not limited to, nuts, bolts, washers, self-locking nuts or bolts, circlips, cotter pins. Always replace these fasteners with new parts.*

2. Carefully pull the boot off the support washer above the upper universal joint.

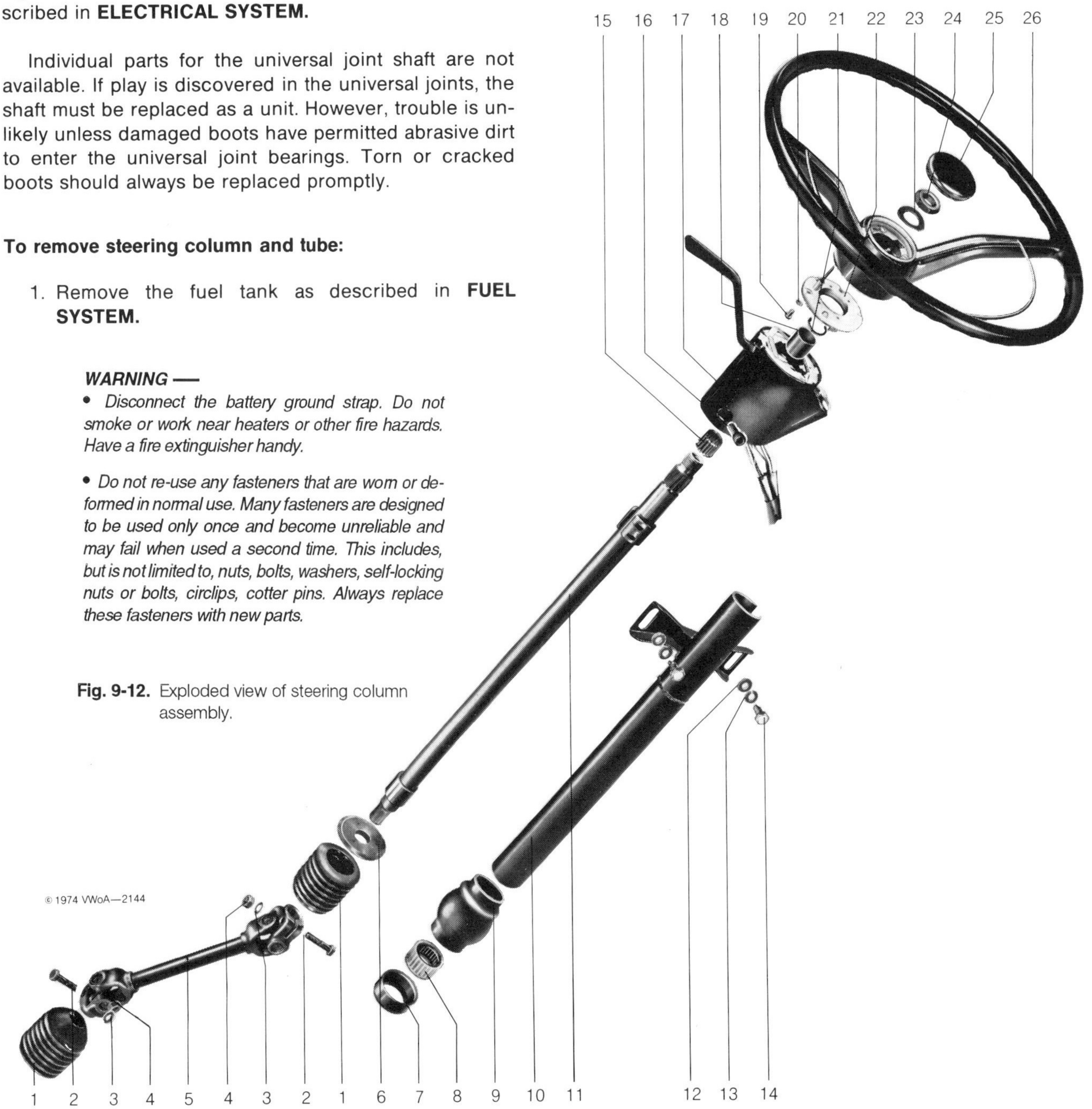

Fig. 9-12. Exploded view of steering column assembly.

1. Boot
2. M 8 bolt
3. Small spring washer
4. Self-locking nut
5. Universal joint shaft
6. Support washer
7. Sealing ring
8. Needle bearing
9. Sealing collar
10. Column tube
11. Steering column
12. Flat washer
13. Lock washer
14. M 8 grooved bolt
15. Contact ring
16. M 8 socket head screw
17. Column switch housing and switches
18. Spacer
19. Screw
20. Serrated lock washer
21. Circlip
22. Contact ring with turn signal cancellation
23. Large spring washer
24. M 18 nut
25. Horn ring cap
26. Steering wheel

24 STRUT FRONT SUSPENSION

3. Completely remove the M 8 bolt that holds the universal joint to the steering column (Fig. 9-13).

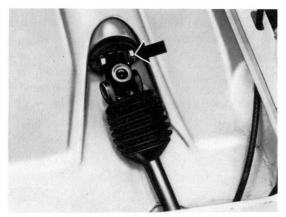

Fig. 9-13. Bolt (arrow) that holds universal joint on steering column.

4. Disconnect all wires to the column switches, then pull the wiring into the passenger compartment.
5. Carefully pry off the horn ring cap (or the steering wheel hub pad on late models). Then remove the M 18 nut and the large spring washer.
6. Take off the steering wheel. Remove the circlip and spacer from the steering column.
7. Remove the two M 8 grooved bolts and the washers that hold the column tube to the dashboard.
8. Pull the column tube out into the passenger compartment complete with the column switches and column.
9. Loosen the M 8 socket head screw that holds the column switch housing and the switches on the column tube. Then pull the column tube downward off the column.

To install:

1. Check the needle bearing for wear and replace it if necessary.

   NOTE—
   The new needle bearing goes in the column tube chamfered-side first. Place the bearing against a level surface. Then lightly tap the column tube, as illustrated in Fig. 9-14, until the needle bearing is flush with the end of the column tube.

2. Place the column in the column tube, being careful not to damage the needle bearing. The bearing seat on the column must contact the bearing needles properly.

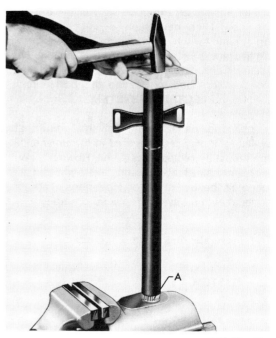

Fig. 9-14. Installing needle bearing (A). Use a hardwood block and light hammer taps to drive the tube down over the bearing.

3. Install the contact ring over the top of the steering column as shown in Fig. 9-15.

Fig. 9-15. Contact ring (A) properly in place. The lower arrow shows the correct position of the steering column's bearing seat in the needle bearing.

4. Push the column switch housing and switches onto the column tube as far as they will go. Then torque the M 8 socket head screw to 0.5 to 1.0 mkg (3.6 to 7 ft. lb.).
5. Slide the sealing collar on the column tube.

6. Coat the sealing ring with talcum powder. Then slide the ring over the lower end of the steering column and onto the column tube.

7. Install the spacer and circlip on the upper end of the steering column.

8. Carefully install the steering column and tube in the car.

9. Install the support washer so that its plastic coated side is toward the universal joint. Then align the groove in the steering column with the bolt hole in the universal joint as indicated in Fig. 9-16.

10. Install the M 8 bolt with a new self-locking nut and torque to 2.5 mkg (18 ft. lb.).

Fig. 9-17. Centering bolt and plug (arrow).

WARNING—
The centering bolt must be removed from the car as soon as adjustments are completed. If left in place, it will prevent the steering wheel from being turned.

13. Slide the steering wheel on the steering column with its spokes in a horizontal position.

14. Make sure the turn signal switch is in its central position. Then install the large spring washer and the M 18 nut. Torque it to 5.0 mkg (36 ft. lb.).

15. Install the cap for the horn ring (or the steering wheel hub pad on late models).

16. Slide the column tube up or down to adjust gap **a** indicated in Fig. 9-18 to 2 to 4 mm (1/16 to 5/32 in.). When the proper gap is obtained, torque the two M 8 grooved bolts to 1.5 to 2.0 mkg (11 to 14 ft. lb.).

Fig. 9-16. Steering column installation. Plastic-coated support washer is at **A**. Groove and bolt hole (arrows) must be aligned.

11. Install the two M 8 groove bolts and the washers finger-tight.

12. Set the steering in its center position. On cars with rack and pinion steering, have the front wheels pointed straight ahead and the steering wheel spokes horizontal.

NOTE—
The center position on cars with worm and roller steering is set by using a bolt—VW part No. 113 415 375. The bolt is installed through the drop arm until the point just touches the conical hole in the aluminum plug indicated in Fig. 9-17. (On late models there is no plug. The hole is in a boss on the steering gearbox itself.) Tighten the bolt to no more than 0.5 mkg (3.6 ft. lb.).

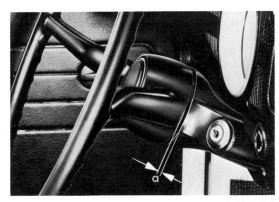

Fig. 9-18. Clearance **a** between steering wheel and column switch housing.

Removing Universal Joint Shaft

The universal joint shaft can be removed without removing the steering column. After freeing the steering column from the universal joint as described in step 3 of the steering column removal procedure, loosen the two M 8 grooved bolts that hold the column tube to the dashboard. Then pull the column and column tube up into the passenger compartment as far as possible so that the universal joint shaft can be taken off the steering column as shown in Fig. 9-19.

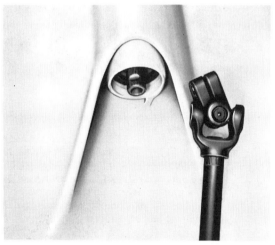

Fig. 9-19. Universal joint removal with steering column remaining in car.

To remove the universal joint shaft at its lower joint, pull off the lower boot and remove the bolt indicated in Fig. 9-20.

Fig. 9-20. M 8 bolt that holds lower universal joint on steering gearbox worm spindle.

The universal joint shaft will have to be pried off the worm spindle with suitable levers as illustrated in Fig. 9-21.

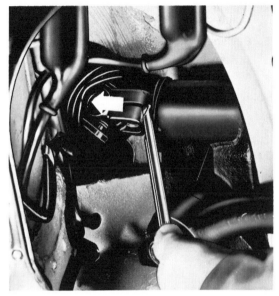

Fig. 9-21. Prying universal joint shaft off steering gearbox worm spindle.

Installation is the reverse of removal. During installation, center the steering and adjust the gap between the steering wheel and the steering column switch housing. These two adjustments are described earlier in the procedure given for installing the steering column.

Align the bolt holes in the universal joints with the grooves in the worm spindle and steering column. Install the M 8 bolts with new self-locking nuts and torque them to 2.5 mkg (18 ft. lb.). Torque the two grooved bolts that hold the column tube to the dashboard to 1.5 to 2.0 mkg (11 to 14 ft. lb.).

9.8 Removing and Installing Worm and Roller Steering Gearbox

The steering gearbox can be removed complete with the drop arm. It is not necessary to remove the steering column or the universal joint shaft when removing the steering gearbox.

To remove:

1. Detach the steering damper's piston rod from the drop arm by taking out the M 10 bolt. See **9.2 Removing, Checking, and Installing Steering Damper.**

2. Push the piston rod completely into the steering damper so that it will be out of your way, or swing the steering damper toward the front of the car.

STRUT FRONT SUSPENSION 27

3. Remove the cotter pin and castellated nut from the drop arm. Then press the tie rod end out of the drop arm with a suitable puller.

   CAUTION —
   Do not hammer out the tie rod end. Doing so would ruin the threads and make reinstallation impossible.

4. Push the lower boot up on the universal joint shaft to uncover the lower universal joint. Then remove the M 8 bolt that holds the lower universal joint on the steering gearbox worm spindle.

5. Remove the three M 10 bolts that hold the steering gearbox on the body side member (Fig. 9-22).

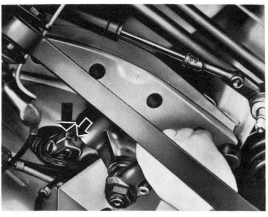

Fig. 9-23. Groove in worm spindle (right arrow) must align with bolt hole in lower universal joint (left arrow).

NOTE —
If a new boot retaining ring is installed on the worm spindle, its lug must be in line with the mark indicated in Fig. 9-24 when the steering is centered as described in **9.7 Removing and Installing Steering Column.**

Fig. 9-22. Steering gearbox mounting bolts. Support the gearbox as the last bolt (arrow) is removed.

Fig. 9-24. Mark (arrow) that should align with lug on retaining ring when steering is centered.

6. Pull down on the steering gearbox to slide the worm spindle out of the universal joint. Then remove the gearbox from the car.

To install:

1. Push the worm spindle into the lower universal joint so that the groove in the spindle is aligned with the bolt hole in the universal joint (Fig. 9-23).

2. Install the three M 10 bolts that hold the steering gearbox on the body side member. Torque the bolts to 4.0 mkg (29 ft. lb.).

3. Install the M 8 bolt in the universal joint with a new self-locking nut and torque to 2.5 mkg (18 ft. lb.). Then pull down the boot and engage it in the groove in the retaining ring.

4. Install the tie rod end in the drop arm. Torque the castellated nut to 3.0 mkg (22 ft. lb.). Then advance it, if necessary, to uncover the cotter pin hole and install a new cotter pin.

28 STRUT FRONT SUSPENSION

9.9 Disassembling and Assembling Worm and Roller Steering Gearbox

Fig. 9-25 is an exploded view of the worm and roller steering gearbox. Assembly and adjustment require special tools and clean working conditions.

CAUTION
If you lack the skills, special tools, or a clean workshop for servicing the steering gearbox, we suggest you leave such repairs to an Authorized VW Dealer or other qualified shop. We especially urge you to consult your Authorized VW Dealer before attempting repairs on a car still covered by the new-car warranty.

1. Worm adjuster locknut
2. Worm adjuster
3. Ball thrust bearing (2)
4. Worm spindle
5. Adjusting shim
6. Bushing sleeve
7. Rubber bushing
8. Drop arm
9. Roller shaft seal
10. Centering bolt (installed only during adjustments)
11. M 20 nut
12. Aluminum plug (early models)
13. Washer (2)
14. M 10 bolt (2)
15. Roller shaft adjusting screw locknut
16. M 8 cover bolt (4)
17. Plastic cover plugs (2)
18. Housing cover
19. Cover gasket
20. Circlip
21. Adjusting washer
22. Roller shaft adjusting screw
23. Roller shaft
24. Worm spindle seal
25. Retaining ring
26. Steering gearbox housing

Fig. 9-25. Worm and roller steering gearbox disassembled.

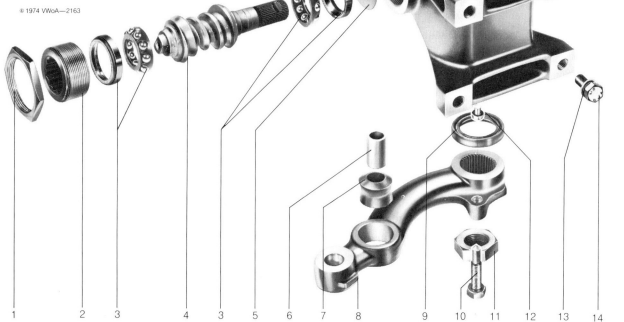

STRUT FRONT SUSPENSION

To disassemble:

1. Clamp the drop arm in a bench vise and remove the M 20 nut from the roller shaft.

2. Using a two-arm puller, pull the drop arm off the roller shaft. Then install the steering gearbox in the repair fixture as shown in Fig. 9-26.

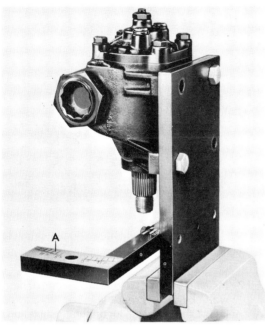

Fig. 9-26. Gearbox installed in repair fixture with two bolts. The hole in the degree scale at **A** is in line with the center of the tie rod hole in the drop arm when the steering is centered.

3. Remove the roller shaft adjusting screw locknut and the four M 8 cover bolts. Then lift off the housing cover by turning the roller shaft adjusting screw as shown in Fig. 9-27.

Fig. 9-27. Removing housing cover from gearbox housing.

4. Place a pan under the gearbox to catch the oil. Then turn the worm spindle until you can lift out the roller shaft as shown in Fig. 9-28.

Fig. 9-28. Roller shaft being lifted out (arrow).

5. Clamp the roller shaft in a vise and remove the circlip. Then lift out the adjusting washer and the roller shaft adjusting screw.

6. Loosen the worm adjuster locknut.

7. Using a hexagonal adapter, remove the worm adjuster as shown in Fig. 9-29.

Fig. 9-29. Wrench in position to turn hexagonal adapter inserted into worm adjuster.

30 STRUT FRONT SUSPENSION

8. Remove the worm spindle as shown in Fig. 9-30.

Fig. 9-30. Worm spindle being withdrawn from steering gearbox housing.

9. Carefully knock the worm spindle seal, adjusting shim, and upper ball thrust bearing out of the steering gearbox housing as shown in Fig. 9-31.

Fig. 9-31. Rubber mallet and tube-type driver being used to knock bearing, shim, and seal out of gearbox housing.

Inspect all parts following disassembly. Replace any that are worn or damaged. A new cover gasket and new seals should be installed during assembly. Be sure to check whether the roller has excessive axial play in the roller shaft. If its axial play exceeds 0.04 mm (.0015 in.), the entire roller shaft should be replaced.

To assemble and adjust:

1. Install the worm spindle together with the upper ball thrust bearing and adjusting shim, as shown in Fig. 9-32, but do not install the worm spindle seal.

   NOTE

   The worm spindle seal must not be installed until after adjustment is complete. It is best to start with an adjusting shim of medium thickness—0.35 mm (.014 in.). The final shim thickness will be determined later.

Fig. 9-32. Worm spindle being installed with bearing and shim in place on it.

2. Install the lower ball thrust bearing. Then coat the worm adjuster threads with sealing compound and install the adjuster just tight enough to press the bearings into their seats.

3. Slightly loosen the worm adjuster. Then, while turning the worm spindle with a torque gauge (Fig. 9-33), slowly tighten the worm adjuster until turning torque is 2.0 to 3.0 cmkg (1.7 to 2.6 in. lb.).

4. Torque the worm adjuster locknut to 6.5 mkg (47 ft. lb.). Recheck the worm spindle's turning torque to make sure it is within 2.0 to 3.0 cmkg (1.7 to 2.6 in. lb.).

Fig. 9-33. Torque gauge installed on worm spindle.

5. Install the roller shaft adjusting screw, adjusting washer, and circlip in the roller shaft (Fig. 9-34).

   NOTE
   Adjusting washers are available in thicknesses from 2.00 to 2.25 mm (.079 to .089 in.) in increments of 0.05 mm (.002 in.). Select a washer that will barely permit the adjusting screw to be finger-turned.

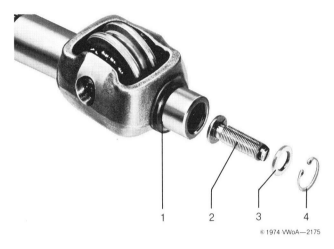

1. Roller shaft
2. Roller shaft adjusting screw
3. Adjusting washer
4. Circlip

Fig. 9-34. Adjusting screw installation.

6. Assemble the roller shaft and the housing cover by threading the roller shaft adjusting screw into the cover as far as it will go.

7. Install a new roller shaft seal. Then, using a new gasket, install the roller shaft and housing cover as shown in Fig. 9-35. The roller must engage the worm at an angle of 90°.

Fig. 9-35. Installing roller shaft and housing cover.

8. Install the four M 8 cover bolts and torque them to 2.0 to 2.5 mkg (14 to 18 ft. lb.).

   NOTE
   Do not put oil into the steering gearbox until the gearbox has been adjusted.

9. Install the drop arm with a new M 20 nut. There are one or more wide splines on the roller shaft which match wide grooves on the drop arm. Therefore, the drop arm can only be installed correctly.

10. Torque the M 20 nut to 10.0 mkg (72 ft. lb.). Then peen the nut into the roller shaft as described in **9.4 Removing and Installing Drop Arm.**

11. Install a pointer in the drop arm. Then turn the drop arm 11° to the left or right.

12. Hold the worm spindle while you move the drop arm back and forth as indicated in Fig. 9-36. Turn in the roller shaft adjusting screw until the drop arm no longer has any play.

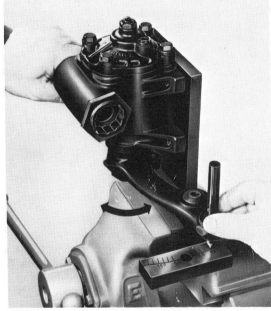

Fig. 9-36. Drop arm being moved through its play range at 11° off center while worm spindle is held stationary.

13. Torque the roller shaft adjusting screw locknut to 2.5 mkg (18 ft. lb.).

14. Install the torque gauge on the roller shaft.

15. Using the torque gauge, turn the worm spindle so that the drop arm moves from 11° on one side of the scale to 11° on the other. Maximum torque through the center point should be 5.0 to 8.0 cmkg (4 to 7 in. lb.). If it is not, readjust the roller shaft.

32 Strut Front Suspension

16. Check the no-play range of the roller shaft and drop arm. It should extend from 11° ± 2° on one side of the scale to 11° ± 2° on the other side.

    NOTE ——
    If the no-play range extends beyond 11° ± 2° to the left side of the scale, disassemble the gearbox and install a thicker adjusting shim on the worm spindle. If the no-play range does not extend to 11° ± 2° on the left side of the scale, install a thinner adjusting shim. Keep repeating the adjustment until it is correct. Adjusting shims are available in thicknesses from 0.20 to 0.50 mm (.008 to .020 in.) in increments of 0.05 mm (.002 in.).

17. After the adjustment is correct, drive in the new worm spindle seal using a tube-type driver.

18. Fill the gearbox with transmission grease.

    NOTE ——
    The fastest way to fill the housing is to remove the housing cover. However, the roller shaft must then be readjusted. If the gearbox is filled by removing both plastic cover plugs, install new plugs to ensure a perfect seal. Turn the gears to help air escape while filling. The housing holds 175 cc (about 5⅞ oz.) of grease.

Marking the Centerpoint

After the steering gearbox has been assembled and adjusted, the centerpoint should be marked. Otherwise, accurate toe adjustments and proper steering installation will not be possible.

If the steering gearbox was manufactured before March 1972, install a new aluminum plug. Later gearboxes have a boss on the housing that accepts the centering mark. On these gearboxes the original mark should be filed off.

Center the steering gears as shown in Fig. 9-37. Then install the centering bolt as indicated in the same illustration until it barely marks the aluminum plug or the boss on the gearbox housing. Remove the gearbox from the repair fixture. Place it in a repair press so the roller shaft adjusting screw is against the press bed and the press ram is against the end of the roller shaft. Apply a pressure of 1.6 tons, being careful never to exceed 2.0 tons. This load will keep the circlip from being forced out of the roller shaft as the centering bolt is turned tighter to make its mark.

Torque the centering bolt to 1.5 mkg (11 ft. lb.) so that it will make a conical depression in the aluminum plug. On gearboxes without the plug, the mark may need to be made deeper with a center punch.

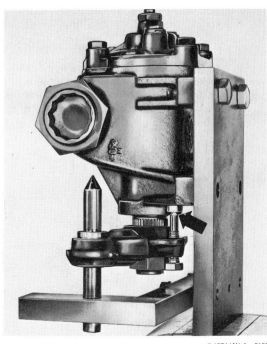

Fig. 9-37. Pin through drop arm to center steering. Centering bolt is contacting the plug at arrow.

Install the retaining ring for the universal joint boot with its lug in line with the notch on the steering gearbox (Fig. 9-38). Then remove the marking bolt from the drop arm. However, if the steering gearbox is to be installed on the car immediately, leave the centering bolt in place until installation is complete and wheel alignment has been checked.

Fig. 9-38. Boot retaining ring correctly installed. Lug (arrow) is in line with the notch in the gearbox housing when the steering is centered.

Strut Front Suspension

WARNING —
The centering bolt must be removed from the car as soon as wheel alignment adjustments are completed. If left in place, it will prevent the steering wheel from turning.

Fill the drop arm hole for the centering bolt with a small amount of grease. This will prevent the formation of rust on the threads. Such rust might interfere with installation of the centering bolt during subsequent wheel alignment and steering repair work.

9.10 Removing and Installing Rack and Pinion Steering Gearbox

A rack and pinion steering gearbox is used on 1975 and later cars with strut front suspension. This gearbox, similar to the one used on the VW Dasher, requires no lubrication—and seldom requires adjustment—during its service lifetime. Adjustment is described in **9.1 Adjusting Steering**.

The installed position of the rack and pinion steering gearbox is shown in Fig. 9-39. New rubber boots are available to replace steering rack boots that have become cracked or otherwise damaged. These boots can be replaced only after you have removed the rack and pinion steering gearbox from the car.

Because of its inherent self-damping qualities, the rack and pinion steering system does not require the hydraulic steering damper that is used on cars that have the worm and roller steering system. The inner ends of the tie rods are held to the rack by bolts. Though Fig. 9-39 shows adjustment locknuts on both tie rods, only the right-hand tie rod is actually adjustable.

To remove gearbox:

1. Raise the car on a hoist or support it on jack stands. Remove the front road wheels.

2. Push the lower boot up on the universal joint shaft to uncover the lower universal joint. Then remove the M 8 bolt that holds the lower universal joint to the steering gearbox pinion spindle.

3. Using suitable levers, as previously shown in Fig. 9-21, pry the universal joint shaft off the steering gearbox pinion spindle.

4. Bend up the lockplate. Remove the bolts that hold the tie rod inner ends to the steering rack. Then disconnect the tie rods at their inner ends.

5. Remove the nuts and bolts that hold the ends of the steering rack housing to the brackets on the body side members. Then remove the rack and pinion steering gearbox from the car.

Installation is the reverse of removal. If you have removed the brackets from the side members, torque the two bolts that hold each bracket to 4.5 mkg (32 ft. lb.) during bracket installation. Torque the bolts that hold the gearbox to the brackets to 2.5 mkg (18 ft. lb.). Torque the bolts that hold the tie rods to the rack to 5.5 mkg (40 ft. lb.), then bend over the lockplate tabs. With the car on the ground, adjust the toe as described in **3.2 Checking and Adjusting Toe**.

Fig. 9-39. Rack and pinion steering gearbox installed in chassis. Body side member is at **1**, bolts that hold bracket are at **2**, bolt that holds gearbox to bracket is at **3**, adjusting screw is at **4**.

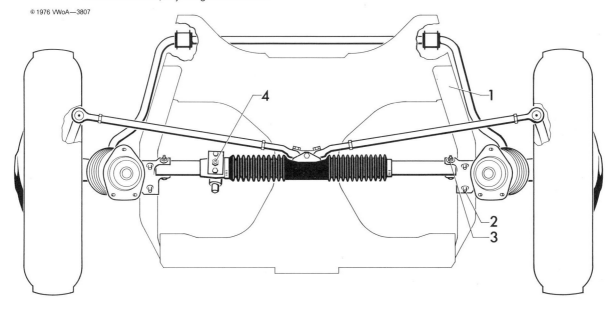

34 Strut Front Suspension

10. Strut Front Suspension Technical Data

Because the strut front suspension used on Super Beetles and Convertibles is completely different from the front axle used on Beetles and Karmann Ghias, it is important that you do not confuse the specifications for the two systems when making adjustments or repairs. The specifications given in this section should never be applied to cars with the front axle.

In addition to the wheel alignment specifications given in **3. FRONT WHEEL ALIGNMENT,** other wheel alignment factors are given here in Table II. This additional data is not a necessary part of routine alignment checks, but is primarily for use in repairing collision damage.

I. Tightening Torques

Location	Designation	mkg	ft. lb.
Track control arm to frame head	nut	4.0	29
Track control arm to ball joint	nut	4.0	29
Ball joint stud in steering knuckle (1974)	clamp nut	3.5	25
Stabilizer to track control arm	slotted nut	3.0 (turn onto cotter pin hole)	22 (turn onto cotter pin hole)
Stabilizer clamp to frame head (1971-1973)	bolt	2.0	14
Stabilizer clamp to frame head (1974)	bolt	4.0	29
Strut mounting to shock absorber (1971-1973)	nut	7.0-8.5	50-61
Strut mounting to shock absorber (1974)	self-locking nut	6.0	43
Steering knuckle axle ball joint to strut	bolt	4.0	29
Steering knuckle to strut (1974)	bolt	8.5	61
Strut to body	nut	2.0	14
Backing plate to steering knuckle	bolt	5.0	36
Wheel bearing	clamp nut	1.0 (1.3 max.) (while turning wheel)	7 (9 max.) (while turning wheel)
Clamp nut screw	socket head screw	1.0-1.3	7-9
Idler arm bracket to body	bolt	3.0	22
Idler arm to bracket	nut	4.0	29
Adjusting bolt in idler arm bracket	nut	2.0	14
Center tie rod to drop arm and idler arm	slotted nut	3.0 (turn onto cotter pin hole)	22 (turn onto cotter pin hole)
Side tie rods to center tie rod and steering knuckle	slotted nut	3.0 (turn onto cotter pin hole)	22 (turn onto cotter pin hole)
Clamps on tie rods	nut	1.5	11
Locknut on tie rod	nut	2.5	18
Steering damper to frame head	bolt	6.0	43
Steering damper to drop arm	bolt	4.0-4.5	29-32
Steering gearbox to body	bolt	4.0	29
Tie rod end	M12 castellated nut	3.0	22
	M10 castellated nut	2.5	18
Drop arm on roller shaft	nut	10.0 (peen after torquing)	72 (peen after torquing)
Universal joint shaft to worm spindle and steering column	bolt and self-locking nut	2.5	18
Cover on steering gearbox	bolt	2.0-2.5	14-18
Roller shaft adjusting screw locknut	nut	2.5	18
Worm adjuster locknut	nut	6.5	47
Column tube to instrument panel	bolt	1.5	11
Column switch to column tube	bolt	0.5-1.0	4-7
Tie rod inner end to steering rack	bolt	5.5	40
Tie rod mounting bracket to steering rack	nut and bolt	2.0	14
Rack and pinion steering to steering mounting brackets	nut and bolt	2.5	18
Rack and pinion steering mounting brackets to side members	bolt	4.5	32

WARNING —

Do not re-use any fasteners that are worn or deformed in normal use. Many fasteners are designed to be used only once and become unreliable and may fail when used a second time. This includes, but is not limited to, nuts, bolts, washers, self-locking nuts or bolts, circlips, cotter pins. Always replace these fasteners with new parts.

II. Wheel Alignment Specifications

Designation	Value
Front wheel alignment	
1. Total toe angle of front wheels, not pressed	+30' ± 15'
2. Total toe angle of front wheels, pressed	10' ± 15'
3. Pressure applied to wheels	10 ± 2 kg (22 ± 4 lb.)
4. Maximum permissible difference between total toe angle with wheels pressed and not pressed	25'
5. Front wheel camber with wheels straight ahead (adjust as close as possible to 1°)	1° {+20' / −40'}
6. Maximum permissible difference between sides	30'
7. Toe-out angle at 20° lock to left and right (not pressed)	−30' ± 30'
8. Stub axle offset	max. 11 mm (7/16 in.)
9. Caster angle of a wheel	2° ± 35'
10. Corresponds to the camber difference of a wheel on a lock from 20° left to 20° right	1° 20' ± 25'
Rear wheel alignment	
1. Rear wheel camber with spring plates properly set (after at least 300 miles, or 480 km)	−1° ± 40'
Maximum permissible difference between sides of the car	45'
2. Rear wheel toe angle with correct camber	0° ± 15'
3. Maximum permissible deviation in wheel alignment	max. 10'

III. Tolerances, Wear Limits, and Settings

Designation		
A. Suspension struts	**Color code**	**Spring Pressure**
1. Coil spring .... group 1 (both springs on one axle must have same color code) group 2 group 3	1 red paint mark 2 red paint marks 3 red paint marks	227-233 kg (500-514 lb.) 234-240 kg (516-529 lb.) 241-247 kg (531-545 lb.)
2. Number of coils	10.5	
3. Number of effective coils	9	
4. Coil diameter	110 mm (4 5/16 inch)	
5. Wire diameter	10.45 mm (.411 inch)	
6. Length unloaded	326 mm (12 27/32 inches)	
7. Shock absorber piston diameter	32 mm (1.260 inches)	
8. Shock absorber maximum stroke	175 mm (6 7/8 inches)	
B. Ball joints, steering knuckles	**New Installation mm (inch)**	**Wear Limit mm (inch)**
1. Steering joints .... play	1.00 (.039)	2.50 (.098) using VW 281 a
2. Steering knuckle/stub axle .... distortion	0.15 (.006) using VW 258 k/p	—
3. Wheel bearing inner .... inside diameter outside diameter	29.00-29.01 (1.1417-1.1421) 50.29-50.32 (1.9799-1.9811)	— —
4. Wheel bearing outer .... inside diameter outside diameter	17.46-17.48 (.6874-.6882) 39.88-39.90 (1.5701-1.5709)	— —
5. Seat for inner wheel bearing on stub axle ...... outside diameter	28.98-29.00 (1.1409-1.1417)	—
6. Seat for inner wheel bearing in brake drum ..... inside diameter	50.25-50.28 (1.9783-1.9795)	—
7. Seat for outer wheel bearing on stub axle ..... outside diameter	17.45-17.46 (.6870-.6874)	—
8. Seat for outer wheel bearing in brake drum ..... inside diameter	39.84-39.87 (1.5685-1.5697)	—
9. Wheel bearing play	0.03-0.12 (.001-.005)	—
C. Worm and roller steering		
1. Steering wheel turns from lock to lock	2 3/4	
2. Steering gear ratio	17.8	
3. Overall steering ratio	16.5	
4. Worm turning torque (for axial adjustment) without oil seal	2.0-3.0 cmkg (1.7-2.6 in. lb.)	
5. Total turning torque (steering gear assembled)	5-9 cmkg (4.3-7.8 in. lb.)	

© 1976 VWoA

Section 9

AUTOMATIC STICK SHIFT

Contents

Introduction	3	
1. General Description	4	
Torque Converter	4	
Pump	4	
Clutch	4	
Control Valve	4	
Temperature Warning Device	5	
Gears	5	
2. Maintenance	6	
3. Troubleshooting	6	
3.1 Road Testing	6	
3.2 Stall Speed Test	6	
3.3 Pressure Test	6	
4. Service Adjustments	8	
4.1 Adjusting Clutch Freeplay	8	
4.2 Adjusting Control Valve	8	
Clutch Engages Too Quickly	9	
Clutch Engages Too Slowly	9	
4.3 Cleaning and Adjusting Selector Lever Contacts	9	
5. Control Valve	9	
5.1 Removing and Installing Control Valve	9	
5.2 Disassembling and Assembling Control Valve	10	
6. Torque Converter	11	

6.1	Removing, Checking, and Installing Torque Converter	11
6.2	Replacing Converter Seal	12
6.3	Replacing Converter Bushing	13
6.4	Cleaning Torque Converter	13
7. Removing and Installing Automatic Stick Shift		14
8. Clutch		15
8.1	Removing and Installing Bellhousing and Clutch	15
8.2	Removing and Installing Clutch Release Shaft	17
8.3	Replacing Main Drive Shaft Seal	17
8.4	Disassembling and Assembling Clutch	17
8.5	Adjusting Clutch Linkage	19
9. Transmission		20
9.1	Disassembling and Assembling Gearshift Housing	20
9.2	Removing and Installing Transmission Gears in Case	20
9.3	Adjusting Selector Forks	23
9.4	Disassembling and Assembling Gear Carrier	24
9.5	Disassembling and Assembling Main Drive Shaft	24
9.6	Disassembling and Assembling Drive Pinion	26
9.7	Removing and Installing Reverse Shaft, Needle Bearings, and Parking Lock	28

2 Automatic Stick Shift

10. Differential . . . 29
   10.1 Removing and Installing Differential . . . 29
   10.2 Disassembling and Assembling Differential . . . 31
   10.3 Replacing Oil Seal for Constant Velocity Joint Flange . . . 32

11. Adjusting Final Drive . . . 32
   11.1 Adjusting Pinion . . . 33
   11.2 Adjusting Ring Gear . . . 35

12. Selector Lever . . . 38
   12.1 Removing and Installing Selector Lever and Shift Rod . . . 39

13. Tanks . . . 40
   13.1 Removing and Installing Vacuum Tank . . . 40
   13.2 Removing and Installing ATF Tank . . . 41

14. Automatic Stick Shift Technical Data . . . 41
   I. Ratios . . . 41
     Rear Suspension and Rear Wheel Alignment Data . . . 42
   II. Tolerances, Wear Limits, and Settings . . . 42
   III. Tightening Torques . . . 42

TABLES

a. Automatic Stick Shift Troubleshooting . . . 7
b. Final Drive Replacement and Adjustments . . . 32
c. Available S_3 Shim Sizes . . . 35
d. S_3 Shim Measurements . . . 35

Automatic Stick Shift

The VW Automatic Stick Shift, optional on cars covered by this Manual, sharply reduces the amount of gear changing required in ordinary driving and does away with the clutch pedal altogether. A torque converter is used between the engine and the clutch. The torque converter acts as a fluid coupling, making it unnecessary for the clutch to disengage when the car is stopped with the engine running.

The clutch disengages only when the selector lever is in neutral or when a new drive range is being selected. Clutch operation is triggered by an electrical signal from the selector lever. Actual disengagement is carried out by a vacuum servo. A vacuum tank installed between the engine's intake manifold and the servo ensures adequate vacuum for operation of the clutch. Evacuation of the vacuum tank is regulated by an electromechanical control valve. The same control valve meters vacuum to the servo for clutch operation.

The transmission gears are similar to those of the four-speed manual transmission that is standard equipment on cars covered by this Manual. However, the Automatic Stick Shift has only three forward gears. The function of the manual transmission's low gear is taken over by the torque converter. Many parts in the gear system of the Automatic Stick Shift are identical, or of similar design, to corresponding parts of the manual transmission.

The ring gear and differential for the Automatic Stick Shift are the same as those for the manual transmission. However, the mounting and adjustment of the pinion are different. For this reason, only final drive repairs applicable to the Automatic Stick Shift are covered in this section. Repairs and adjustments shared with the manual transmission are covered fully in **TRANSMISSION AND REAR AXLE**.

Many of the repairs described in this section require special tools and skills that only a trained VW mechanic or transmission specialist is likely to have. If you lack the tools, experience, or a clean workshop suitable for transmission work, we recommend that you leave such repairs to an Authorized VW Dealer or other qualified shop. We especially urge you to consult your Authorized VW Dealer before attempting repairs on a car still covered by the new-car warranty.

Although many of the repairs described in this section may prove practical only to professional mechanics, an understanding of the Automatic Stick Shift's construction and operating principles can improve your ability to discuss transmission malfunctions with your mechanic or VW Service Advisor. This same basic knowledge is similarly useful when troubleshooting the Automatic Stick Shift transmission.

4 Automatic Stick Shift

1. General Description

Compare the cutaway view of the Automatic Stick Shift shown in Fig. 1-1 with the similar cutaway of the manual transmission given in **TRANSMISSION AND REAR AXLE**.

Torque Converter

During engine operation, the torque converter is filled with automatic transmission fluid (ATF). The converter housing spins with the engine's crankshaft. An impeller, made up of curved vanes welded to the interior of the housing, sets the ATF into rotary motion. Centrifugal force causes the ATF to flow out radially. It is then redirected by the housing into the curved vanes of the turbine, causing the turbine to turn. The stator vanes redirect ATF flowing from the turbine back into the impeller at a favorable angle to aid its rotation, thereby multiplying torque.

Pump

The ATF pump, which is mounted on the engine, is a dual pump. One pump supplies lubrication to the engine; the other supplies ATF pressure to the torque converter.

Clutch

The clutch is a diaphragm-spring, single-disk unit. It works similarly to the clutch used with manual transmissions. However, it is driven by the torque converter turbine and not directly by the engine.

Control Valve

A solenoid on the control valve opens the passage from the vacuum tank line to the clutch servo line when the driver closes the electrical contacts in the selector lever by touching the knob. Also, when the engine's

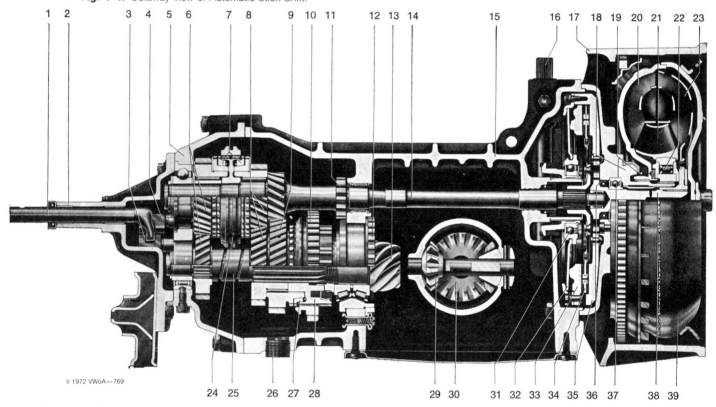

Fig. 1-1. Cutaway view of Automatic Stick Shift.

1. Inner shift lever
2. Gearshift housing
3. Low/reverse shift rod
4. 1st/2nd shift rod
5. Gear carrier
6. 2nd speed gears
7. 1st/2nd gear synchronizer rings
8. 1st speed gears
9. Low speed gears
10. Operating sleeve for low/reverse gear
11. Reverse gear drive
12. Retaining ring for pinion
13. Pinion
14. Main drive shaft
15. Transmission case
16. Clutch release shaft
17. Converter housing
18. Support tube for one-way clutch
19. Seal for converter
20. Impeller
21. Stator
22. One-way clutch
23. Turbine
24. 1st/2nd gear operating sleeve
25. 1st/2nd gear spacer spring
26. Magnetic oil drain plug
27. Low gear synchronizer ring
28. Synchronizer hub low/reverse gear
29. Differential pinion
30. Differential side gear
31. Clutch release bearing
32. Diaphragm spring
33. Pressure plate
34. Clutch plate
35. Carrier plate
36. Oil seal for converter housing
37. Bearing for turbine shaft
38. Turbine shaft
39. Torque converter

AUTOMATIC STICK SHIFT 5

1. Torque converter
2. Clutch
3. Clutch operating lever
4. Servo
5. Control valve
6. Vacuum tank
7. ATF tank
8. ATF pump
9. Carburetor

A. Pressure line
B. Return line
C. Vacuum line—intake manifold/control valve
D. Vacuum line—vacuum tank/control valve
E. Vacuum line—servo/control valve
F. Vacuum line—reduction valve/venturi

Fig. 1-2. Schematic view of torque converter hydraulic system and electropneumatic clutch operating system.

throttle is closed, the valve permits air to be evacuated from the vacuum tank into the engine's intake manifold. See Fig. 1-2. The clutch servo is shown in Fig. 1-3.

Temperature Warning Device

Cars equipped for trailer towing should have a warning light wired to a temperature sensor on the torque converter bellhousing. The light goes on whenever the ATF temperature approaches a level that may damage the converter seals.

Gears

The transmission gears are of the constant-mesh type with balk ring synchronizers. The transfer gears for reverse are housed in the transmission section of the case rather than in the final drive portion as on the manual transmission.

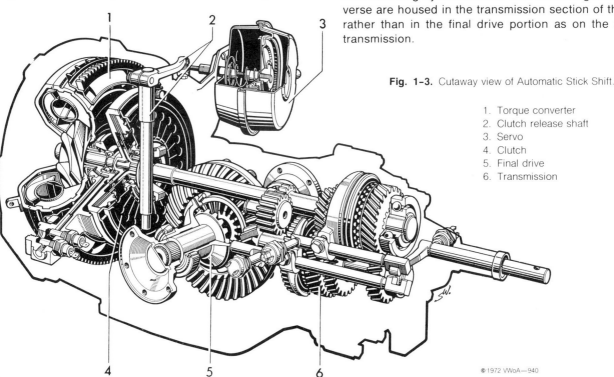

Fig. 1-3. Cutaway view of Automatic Stick Shift.

1. Torque converter
2. Clutch release shaft
3. Servo
4. Clutch
5. Final drive
6. Transmission

6 Automatic Stick Shift

2. Maintenance

There are eight items that require routine inspection or maintenance at a prescribed mileage. These items are listed here and described in **LUBRICATION AND MAINTENANCE**.

1. Checking the selector lever contacts
2. Cleaning the air filter on the control valve
3. Checking and correcting the ATF level
4. Checking the axle boots
5. Checking the constant velocity joint screws
6. Checking and correcting the transmission oil level
7. Checking the clutch freeplay
8. Lubricating the rear wheel bearings

3. Troubleshooting

One of the most important phases of troubleshooting is distinguishing those clutch or torque converter problems that can be corrected without removing the transmission from the car from more serious malfunctions that require transmission removal. If the gears grind when you attempt to change drive ranges, it is seldom the gears' fault. Usually, the clutch is not disengaging. If there is no drive in any range, then the clutch is not engaging. Converter or hydraulic system trouble can be distinguished from clutch trouble by making stall speed and pressure tests. If the clutch is found to be slipping, the adjustment described in **4.1 Adjusting Clutch Freeplay** will usually cure the trouble.

3.1 Road Testing

If a defect is suspected in the Automatic Stick Shift, it can be very helpful to compare its performance and operation with the Automatic Stick Shift on a similar car known to be in good condition. During a road test, check the operation of all four drive ranges. Note especially the speed of clutch engagement when you shift down and accelerate.

If the engine revs up excessively—particularly in the **L**, **1**, and **R** ranges—the cause is likely to be a slipping clutch, faulty torque converter, or poor supply of ATF to the torque converter. Find the results of your road test on **Table a. Automatic Stick Shift Troubleshooting**.

3.2 Stall Speed Test

A stall speed test will determine whether or not the torque converter is functioning properly. It should be performed only if the car cannot reach specified maximum speed or if the acceleration is poor. During this test the ATF in the converter heats up rapidly, so do not extend the test beyond the time it takes to read the gauges.

To test:

1. Connect an electric tachometer according to the instrument manufacturer's instructions.
2. Start the engine. Then set the parking brake and step hard on the foot brake to hold the car stationary.
3. Shift the selector lever to position **2** and, still holding the foot brake down, floor the accelerator briefly. Instead of revving up, the engine should run at a reduced rpm known as stall speed.
4. Read the tachometer. The stall speed should be 2000 to 2250 rpm.

If the engine is properly tuned but the stall speed is below specifications, the converter is faulty. If the stall speed is above specifications, either the clutch requires attention or the converter is not getting ATF pressure. A pressure test will show which fault is present.

3.3 Pressure Test

Install a pressure gauge as shown in Fig. 3-1. If the pressure indicated at approximately 2000 rpm is much below 52.6 psi (3.7 kg/cm^2), the ATF level is low, the pump output is low, the hoses are restricted, or there are leaks. If the pressure is above 52.6 psi (3.7 kg/cm^2), either the pressure relief valve or the pump is faulty, or the hoses to or from the converter are restricted. You should also check for restricted hoses if the seals are leaking or if the hoses are bulging.

Fig. 3-1. Pressure gauge connection on engine.

The troubleshooting chart, **Table a**, lists twelve trouble symptoms that may be encountered in a faulty Automatic Stick Shift. The numbers in bold type in the remedy column refer to the numbered headings in this section where the suggested repairs are described.

AUTOMATIC STICK SHIFT 7

Table a. Automatic Stick Shift Troubleshooting

Symptom	Possible Cause	Remedy
1. Clutch slips at full throttle, not just after shifting	a. Inadequate freeplay b. Oil or ATF (Automatic Transmission Fluid) on clutch facings c. Worn clutch facings	a. Adjust clutch linkage. See **4.1**. b. Eliminate oil/ATF leaks. Replace driven plate. See **6.2, 8.3, 8.4**. c. Replace driven plate. See **8.4**.
2. Excessive clutch slippage following gear selection	a. Hose between carburetor and valve off or leaking b. Filter on valve blocked c. Reducing valve adjusting screw in too far d. Double diaphragm defective	a. Replace hose. b. Remove and clean filter. See **5.2**. c. Adjust control valve. See **4.2**. d. Replace diaphragm. See **5.2**.
3. Clutch not disengaging properly	a. Leak in vacuum hose or vacuum tank b. Excessive freeplay c. Servo diaphragm defective d. Needle bearing in carrier plate hub defective	a. Replace hose or repair tank. See **13.1**. b. Adjust clutch linkage. See **4.1**. c. Replace servo unit. See **4.1, 8.5**. d. Replace needle bearing and seal. See **8.4**.
4. Clutch not disengaging at all	a. Solenoid circuit open b. Voltage drop in solenoid circuit c. Poor selector lever ground contact with frame d. Servo hoses kinked or collapsed e. Servo diaphragm defective	a. Check fuse. Replace if open and correct cause. If fuse is all right, clean or replace switch in selector lever. See **4.3, 12.1**. b. Check wiring. Replace relay if not delivering full voltage. See **ELECTRICAL SYSTEM**. c. Check ground connection. See **12.1**. d. Replace hoses. e. Replace servo unit. See **4.1, 8.5**.
5. Engine stalls as gear is selected	a. Hose between control valve and servo leaking b. Servo diaphragm defective	a. Replace hose. b. Replace servo unit. See **4.1, 8.5**.
6. Engine stalls, will not restart	a. Hose between carburetor and valve or between valve and vacuum tank off or leaking b. Leak in vacuum tank	a. Replace faulty hose. b. Repair tank. See **13.1**.
7. Clutch not engaging following gear selection	a. Switch in selector lever sticking b. Solenoid circuit shorted c. Control valve solenoid sticking	a. Clean contacts. Replace if necessary. See **4.3, 12.1**. b. Eliminate short or replace wire. See **ELECTRICAL SYSTEM**. c. Replace solenoid. See **5.2**.
8. Clutch grabs following gear selection	a. Oil on clutch facings b. Carrier plate distorted	a. Eliminate oil/ATF leaks, replace driven plate. See **6.2, 8.3, 8.4**. b. Machine or replace carrier plate. See **8.4**.
9. Car jerks at idle when lever is released after gear selection	a. Idle speed too high b. Control valve misadjusted	a. Adjust idle. See **FUEL SYSTEM**. b. Adjust control valve. See **4.2**.
10. Converter noisy (high-pitched hissing)	a. ATF level low b. ATF pressure low c. Converter seal leaking d. Converter housing leaking	a. Correct ATF level. See **LUBRICATION AND MAINTENANCE**. b. Remove and repair pump or replace seal on pump hub. See **ENGINE**. c. Replace seal. See **6.2**. d. Replace converter. See **6.1**.
11. Poor acceleration despite good engine output	Converter faulty	Check stall speed. If outside 2000 to 2500 rpm range, replace converter. See **3.2, 6.1**.
12. Warning lights do not work	Bulb, wire, or switch defective	Check bulb. If all right, ground wire from temperature switch. If bulb lights, switch is faulty. If not, replace wire.

8 Automatic Stick Shift

4. Service Adjustments

In some cases, simple adjustments will cure Automatic Stick Shift operating problems without the need for more extensive repairs.

4.1 Adjusting Clutch Freeplay

In time, normal clutch lining wear reduces clutch freeplay. This may eventually cause the clutch to slip. Freeplay should be checked routinely according to the schedule given in **LUBRICATION AND MAINTENANCE**.

A pair of measuring tools for checking clutch freeplay can be made from two strips of sheet steel 1 mm (about 1/16 in.) thick. Trim one strip to a width of 4 mm (about 5/32 in), the other to 6.5 mm (about 1/4 in.). Fig. 4-2 shows such a measuring tool.

To adjust:

1. Remove the vacuum hose from the servo. Then pull the servo rod out of the servo as far as possible.

2. Using the 4-mm (or about 5/32-in.) measuring strip, check the clearance indicated at **e** in Fig. 4-1.

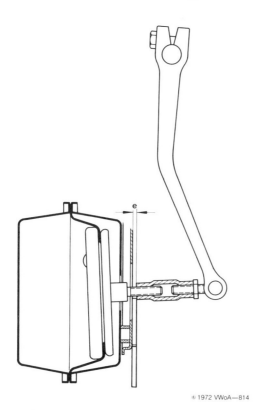

Fig. 4-1. Clearance between the servo mounting and the end of the adjusting sleeve.

3. If the clearance is greater than 4 mm (or 5/32 in.), barely loosen the adjusting sleeve's locknut.

4. Turn the adjusting sleeve five or five-and-a-half turns away from the locknut. The 6.5-mm (1/4-in.) measuring strip should then fit between the locknut and the adjusting sleeve as indicated in Fig. 4-2.

Fig. 4-2. Dimension **d** being checked with a 6.5-mm (or 1/4-in.) measuring strip.

5. Screw the locknut up against the adjusting sleeve and tighten it enough to hold the adjusting sleeve in its new position.

NOTE

If so many adjustments have already been made that the clutch operating lever is in contact with the bellhousing, further adjustment is not possible. The clutch plate is worn out and must be replaced.

6. Road test the car to check the adjustment. If you can accelerate without the clutch slipping, freeplay is adequate. However, freeplay may be excessive if it is difficult to engage reverse.

4.2 Adjusting Control Valve

The control valve is located in the engine compartment, above and to the left of the engine. The speed of clutch engagement is determined by an adjustment on the control valve. Slow clutch engagement increases slippage and consequent clutch lining wear. Conversely, after the vehicle has been in service for some time, the clutch plate contact pattern improves and may cause the clutch to grab. A road test will help you determine whether or not the control valve requires adjustment.

Driving at 30 mph (about 50 kph), move the selector lever from 2 to 1, but do not accelerate. Approximately one second after taking your hand off the selector lever, the clutch should engage fully. Minor adjustments to the speed of clutch engagements may be made according to your preference or driving habits.

Clutch Engages Too Quickly

To correct a clutch that engages too quickly, remove the protective cap from the top of the control valve. Turn the adjusting screw indicated in Fig. 4-3 clockwise one-quarter to one-half turn, then replace the protective cap.

Fig. 4-3. Reducing valve adjusting screw on control valve (arrow **1**). Vacuum line is indicated by arrow **2**.

Clutch Engages Too Slowly

To correct a clutch that engages too slowly, remove the protective cap from the top of the control valve. Turn the adjusting screw indicated in Fig. 4-3 counterclockwise one-quarter to one-half turn, then replace the protective cap.

4.3 Cleaning and Adjusting Selector Lever Contacts

Dirty or worn selector lever contacts may prevent the clutch from disengaging. Therefore, an electrical test of the selector lever contacts should be made before assuming there is trouble in the control valve. Worn contacts can be replaced. The contact clearance must, however, be adjusted carefully following cleaning or replacement.

To inspect, clean, or replace the selector lever contacts, fold the rubber selector lever boot upward to uncover the sleeve on the selector lever. Loosen the upper of the two locknuts shown in Fig. 4-4. This will permit you to unscrew the top sleeve and the lever and to take them off upward.

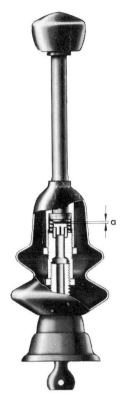

Fig. 4-4. Cutaway of selector lever. Gap between contacts is indicated as dimension **a**.

During assembly, screw the sleeve down until the contacts just touch. Then slowly unscrew it until the vertical travel of the lever is between 0.25 and 0.40 mm (.010 to .016 in.). Tighten the upper locknut. If adjustment moves the angle of the selector lever out of alignment with the car's centerline, loosen the lower locknut and turn the lever so it angles straight back. Then tighten the lower locknut and reinstall the boot.

5. Control Valve

The solenoid and air filter on the control valve can be replaced separately. In addition, rebuilding kits are available for reconditioning the entire control valve.

5.1 Removing and Installing Control Valve

The control valve is located in the engine compartment, above and to the left of the engine. Check the fuse and selector lever contacts before assuming there is trouble in the valve or its solenoid.

To remove the control valve, pull off the wires. Then loosen the vacuum hose clamps and take off the hoses. Remove the three screws that hold the control valve mounting bracket to the car. Installation is the reverse of removal.

10 Automatic Stick Shift

5.2 Disassembling and Assembling Control Valve

The control valve is shown disassembled in Fig. 5-1. When rebuilding the control valve, make certain each part is installed in its proper place.

The concave washer that holds the cover containing the reducing valve adjustment screw must be pried out with a screwdriver. All other parts are held by screws, with the exception of the air filter, which is threaded into the valve housing.

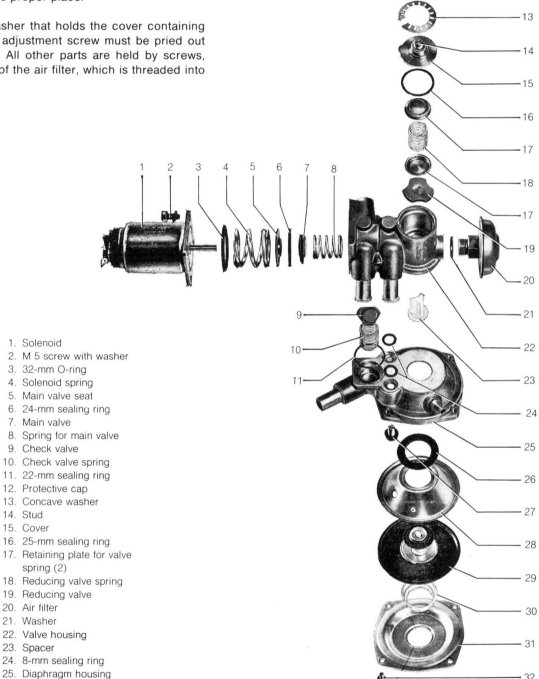

1. Solenoid
2. M 5 screw with washer
3. 32-mm O-ring
4. Solenoid spring
5. Main valve seat
6. 24-mm sealing ring
7. Main valve
8. Spring for main valve
9. Check valve
10. Check valve spring
11. 22-mm sealing ring
12. Protective cap
13. Concave washer
14. Stud
15. Cover
16. 25-mm sealing ring
17. Retaining plate for valve spring (2)
18. Reducing valve spring
19. Reducing valve
20. Air filter
21. Washer
22. Valve housing
23. Spacer
24. 8-mm sealing ring
25. Diaphragm housing
26. Seal
27. Screw with washer (2)
28. Support
29. Diaphragm
30. Spring
31. Cover
32. Screw with washer (4)

Fig. 5-1. Exploded view of control valve for Automatic Stick Shift.

To assemble control valve:

1. Check the solenoid, air filter, and all soft parts. Obtain new parts or a rebuilding kit as needed.

2. Install the check valve and sealing rings in the diaphragm housing. Then install the diaphragm housing on the valve housing.

3. Insert the spacer in the valve housing. Install the diaphragm and associated parts in the diaphragm housing. Then install the diaphragm housing cover.

4. Working through the hole in the diaphragm housing cover, loosen the locknut. Then turn the adjusting screw in the center of the diaphragm to obtain the dimensions illustrated in Fig. 5-2.

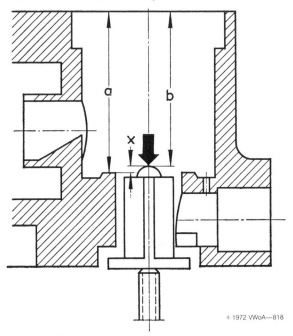

Fig. 5-2. Diaphragm screw adjustment. Dimension **b**—from dome on spacer (large arrow) to top of valve housing—should be 0.30 to 0.40 mm (.012 to .016 in.) less than dimension **a**. The difference is shown as dimension **x**.

5. Tighten the diaphragm adjusting screw locknut.

6. Be sure that the compensating port and valve seat for the reducing valve are clean. Then install the reducing valve, bottom retaining plate, valve spring, upper retaining plate, sealing ring, and cover. Secure the cover with the concave washer.

7. Install the main valve along with its springs and associated parts. Then install the O-ring and the solenoid.

After installing the control valve, check the clutch engagement speed with a road test. If necessary, adjust the control valve as described in **4.2 Adjusting Control Valve**.

6. Torque Converter

The engine must be removed as described in **ENGINE** for access to the torque converter. The torque converter will then be exposed in the bellhousing as shown in Fig. 6-1.

Fig. 6-1. Torque converter in bellhousing of transmission. After the engine is removed, the retaining strap is installed in order to keep the converter from falling out.

6.1 Removing, Checking, and Installing Torque Converter

Once the engine has been removed, the converter can be pulled off the one-way clutch support and removed from the car. Keep the opening in the converter hub covered to prevent the entry of foreign matter.

The torque converter should be checked carefully to determine whether it is suitable for further use. It is a welded assembly, however, and must be replaced only as a unit if it is faulty.

Leaks are rarely caused by cracks in the converter housing. However, if the converter is leaking, the outside of the converter and the inside of the bellhousing will be coated with ATF—just as it would if only the seal were leaking. It is therefore wise to check the converter for leaks, even though a leaking seal is usually the only source of trouble.

A considerable amount of ATF remains inside the converter even after the converter has been removed from the car. This fluid can be used to check the converter for leaks. Wipe the outside of the converter clean. Then turn it slowly, looking at the welded seams and the points where the cooling vanes and starter ring gear brackets are welded to its exterior. Seepage is most likely to appear at these points.

12 Automatic Stick Shift

Make sure that none of the welds holding the starter ring gear has broken. Inspect the converter hub for signs of scoring by the oil seal. If the scoring is deep, replace the converter. The one-way clutch inside the converter hub should permit the stator to turn in only one direction. Check its operation if acceleration is poor or if the stall speed is outside specifications.

Remove all burrs and sharp edges from the converter hub with crocus cloth. Do not use emery cloth because it will leave roughness that can damage the seal. Do not tilt or rock the converter as you install it on the transmission. This could damage the seal or the converter bushing. Instead, slowly turn the converter clockwise and counterclockwise until the turbine splines engage the splined shaft on the clutch carrier plate. Then press the torque converter straight in without allowing it to tilt. Install the retaining strap that was shown in Fig. 6-1 to prevent the converter from falling off its support before the engine is installed.

6.2 Replacing Converter Seal

In addition to the converter seal discussed here, there is also a turbine shaft ATF seal pressed into the bellhousing. Its replacement is described in **8.4 Disassembling and Assembling Clutch**.

To replace converter seal:

1. Remove the engine. Remove the converter and store it in a safe place with its hub opening covered.
2. Using a hook such as that shown in Fig. 6-2, pry the old seal off the one-way clutch support.
3. Moisten the inside of the new seal with ATF and start it onto the one-way clutch support.

Fig. 6-2. Prying off converter seal. Work at one side of the seal, then the other until it is free.

4. Using a block and rubber mallet as shown in Fig. 6-3, drive the seal into place. Make sure the seal goes on without tilting.

Fig. 6-3. Driving the converter seal onto the one-way clutch support.

5. Install a bar over the block, as shown in Fig. 6-4. Tighten the nuts alternately and uniformly until the bar contacts the bellhousing. This will ensure that the seal is seated properly.
6. Remove the bar and block from the bellhousing.
7. Install the torque converter.
8. Install the engine.

Fig. 6-4. Bar used to press the seal squarely into position on the one-way clutch support.

6.3 Replacing Converter Bushing

Before replacing the bushing, make certain that the converter itself is in serviceable condition. See **6.1 Removing, Checking, and Installing Torque Converter**. If only the bushing is worn or scored, it can be replaced. The wear limit is 36.05 mm (1.419 in.).

To replace:

1. Using an extractor and slide hammer, pull out the old bushing as shown in Fig. 6-5.

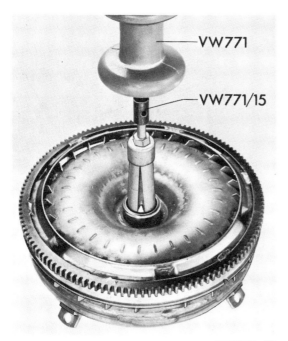

Fig. 6-5. Pulling out converter bushing.

2. Using a repair press and the proper bushing driver, press the new bushing into place in the torque converter hub.

3. Check the inside diameter of the new bushing. It should be between 35.98 and 36.03 mm (1.416 and 1.418 in.). If the diameter is less than the lower limit, the bushing will seize.

6.4 Cleaning Torque Converter

Solid particles from a damaged ATF pump, deteriorating hose, or a seized converter bushing may contaminate the ATF. When repairs are made, it is important that as much of the contaminated ATF as possible is removed from the torque converter. Otherwise, the particles may cause damage when the car is returned to service.

Make a siphon to the specifications given in Fig. 6-6. Then place the converter on a workbench as shown in Fig. 6-7 and install the siphon as illustrated. Start the siphon by blowing into the short tube. Let the converter drain overnight. The small-diameter tubing and long drain period are necessary for maximum ATF removal.

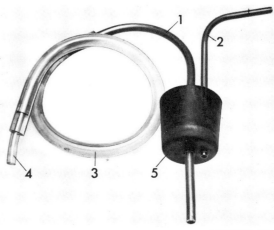

1. Steel or copper tube 3/16 × 8 in. (4 × 200 mm)
2. Steel or copper tube 1/8 × 6 in. (3 × 150 mm)
3. PVC hose 1/4 × 14 in. (6 × 350 mm)
4. PVC hose 1/8 × 1 1/4 in. (3 × 30 mm)
5. Conical rubber plug 1 1/2 in. (35 mm) diam.

Fig. 6-6. Siphon parts available at auto stores.

Fig. 6-7. Siphon and oil receptacle in position.

14 Automatic Stick Shift

7. Removing and Installing Automatic Stick Shift

The tests, repairs, and adjustments described so far do not require transmission removal. The transmission must be removed, however, for most of the procedures that follow.

To remove:

1. Disconnect the battery ground strap. If possible, place the selector lever in **1**.
2. Remove the rear seat. Then remove the inspection cover from the rear of the frame tunnel.
3. Working through the opening in the frame tunnel, remove the square head setscrew from the gearshift rod coupling. Then move the selector lever to **2** and pull the coupling off the transmission's selector lever shaft.
4. Working in the engine compartment, disconnect the accelerator cable from the carburetor. Remove all hoses, wires, and panels that must be taken off preparatory to engine removal. See **ENGINE** for details.
5. Raise the car.
6. Disconnect the ATF pressure line at the engine. Elevate the disconnected end and secure it to the chassis so that the ATF does not drain out.

   NOTE
   If the transmission or final drive is to be disassembled, drain the hypoid oil now. It is not necessary to drain the ATF.

7. Disconnect the ATF suction line at the point indicated in Fig. 7-1. Quickly seal the hose end with an M 16 × 1.5 union that has been soldered shut.

Fig. 7-1. Point where suction line from tank to pump should be separated.

8. Using a 12-point socket, remove the bolts that hold the torque converter to the engine's drive plate. The bolts are accessible through holes in the bellhousing (Fig. 7-2). Hand-turn the crankshaft to position each of the four bolts in the opening.

Fig. 7-2. TOP: Bolt access opening on carburetor engine. BOTTOM: Bolt access opening on fuel injection engine. On fuel injection engine, you must pry a cap out of the opening.

9. Using a spring clip, pinch the fuel hose closed and pull it off the line to the carburetor. Disconnect the heater flap cables, detach the heater hoses from the engine's heat exchangers, and pull the accelerator cable out of its guide.
10. Remove the engine as described in **ENGINE**.
11. Install a metal strap on one of the engine mounting studs in order to hold the torque converter in place in the bellhousing. See **6. Torque Converter**.
12. Remove the ATF hoses from the transmission. Disconnect all wires to the transmission. See Fig. 7-3.

Fig. 7-3. ATF hose banjo unions (upper arrow). One of several wire connections is at lower arrow.

Automatic Stick Shift 15

13. Apply the parking brake. Remove the two driveshaft (axles) between the transmission and the rear wheels as described in **TRANSMISSION AND REAR AXLE.**

14. Loosen the vacuum hose clamp at the servo, then pull off the vacuum hose.

15. Remove the cable and wires from the starter solenoid. Attach identifying tags to them so that you will be able to return them to their proper terminals.

16. Place a transmission jack under the transmission. Raise it until it makes firm contact.

17. On 1968 through 1972 models, remove the front transmission mounting nuts indicated in Fig. 7-4. The new mount introduced on 1973 models is shown in Fig. 7-5.

Fig. 7-4. Front transmission mount nuts (two lower arrows). Neutral safety switch wire connector is at upper arrow.

Fig. 7-5. Late-type front transmission mount. It is held by two studs on the underside of the frame crossmember.

18. Remove the two large rear transmission mount bolts. See Fig. 7-6.

Fig. 7-6. Rear transmission mount bolt in end of frame fork. This is the type introduced on 1973 cars, but earlier mountings are similar.

19. On 1973 or later cars, lower the transmission slightly, then move it to the rear before lowering it from the car. On earlier models, first move the transmission slightly to the rear and then lower it.

Installation is the reverse of removal. Torque the rear transmission mount bolts to 23.0 mkg (166 ft. lb.). Torque the front mount nuts to 3.5 mkg (25 ft. lb.) and the 12-point converter/drive plate bolts to 2.5 mkg (18 ft. lb.). Fill the transmission with hypoid oil and correct the ATF level. Make sure all ranges engage properly, then road test the car as described in **4.2 Adjusting Control Valve**. Adjust the control valve if necessary.

8. Clutch

The bellhousing must be removed from the transmission case for access to the clutch, the main drive shaft seal, and the turbine shaft seal.

8.1 Removing and Installing Bellhousing and Clutch

The hypoid oil must be drained from the transmission before beginning this disassembly procedure.

To remove:

1. Remove the engine. Remove the torque converter. Remove the transmission and put it on a repair stand.

16 Automatic Stick Shift

2. Remove the servo and its mounting bracket.

3. Loosen the clamp bolt in the clutch operating lever. Then pull the lever off the clutch release shaft.

4. Invert the transmission, then remove the oil pan from the bottom of the final drive case. Remove the two nuts indicated in Fig. 8-1.

Fig. 8-1. Two nuts inside final drive case that must be removed to free bellhousing.

5. Remove the six remaining nuts from the bellhousing mounting studs. Then separate the bellhousing, complete with clutch, from the transmission case. Fig. 8-2 shows the relationship of all the parts.

Installation is basically the reverse of removal, with the addition of the following assembly steps:

1. Lightly coat the clutch release bearing guide sleeve with molybdenum grease. Similarly lubricate the lugs of the clutch release bearing (Fig. 8-3).

Fig. 8-3. Release bearing lugs (arrows). Note O-rings around mounting studs in foreground.

2. Install new O-rings around the mounting studs.

3. Lubricate the needle bearing in the carrier plate hub with just enough multipurpose grease to coat the needles. Replace the needle bearing seal if it is cracked or worn.

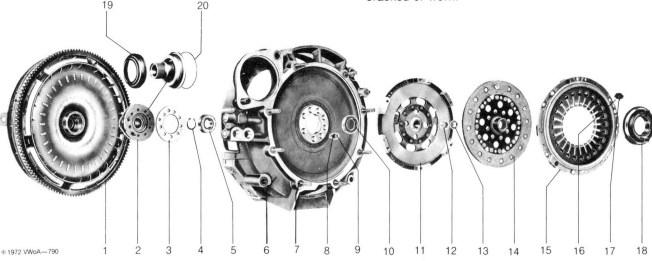

1. Torque converter
2. One-way clutch support
3. Gasket
4. Circlip for turbine shaft
5. Ball bearing
6. Bellhousing
7. O-ring for stud (2)
8. Spring washer (8)
9. Socket head screw (8)
10. Turbine shaft seal
11. Clutch carrier plate
12. Needle bearing
13. Needle bearing seal
14. Clutch plate
15. Pressure plate assembly
16. Spring washer (6)
17. Socket head screw
18. Release bearing
19. Converter seal
20. O-ring

Fig. 8-2. Exploded view of bellhousing and clutch.

8.2 Removing and Installing Clutch Release Shaft

To remove the clutch release shaft, remove the hexagon bearing sleeve bolt and its spring washer. Then push the shaft upward to force the bearing sleeve, rubber seals, and spacer bushing out of the top of the transmission case. Remove the bearing sleeve and the other parts from the shaft, then pull the shaft downward and out through the clutch housing. The slotted bushing can be pried open with a screwdriver to remove it. Press a new slotted bushing into the case. Fig. 8-4 shows the relative positions of the parts. Installation is the reverse of removal.

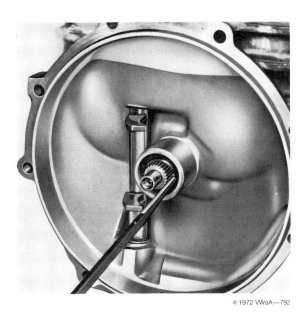

Fig. 8-5. Hook-type puller being used to pull main drive shaft seal out of clutch release bearing guide sleeve.

Lubricate the new seal with hypoid oil. Start it onto the main drive shaft with the seal lip toward the transmission case. Drive the seal into position using a sleeve-type seal driver of appropriate diameter.

8.4 Disassembling and Assembling Clutch

When the bellhousing is removed, all clutch parts except the clutch release shaft and the clutch release bearing remain with the bellhousing. The clutch release bearing usually remains on the guide sleeve. The release bearing must not be washed in solvent since this will destroy the lubrication built into it during manufacture.

To disassemble clutch:

1. Using a 12-point socket, gradually loosen all six bolts that hold the pressure plate assembly to the clutch carrier plate. Work around the pressure plate assembly so that spring tension is relieved gradually. Refer to Fig. 8-2 if necessary.

2. When the spring tension has been relieved, remove the six bolts completely and lift the pressure plate assembly off the clutch plate and clutch carrier plate.

3. Lift the clutch plate out of the clutch carrier plate.

   NOTE
   Further disassembly is unnecessary if you are replacing either the pressure plate or the clutch plate only. At this stage of disassembly, you can also replace the needle bearing in the clutch carrier plate hub if necessary.

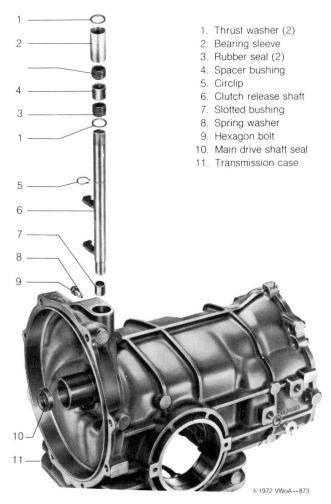

1. Thrust washer (2)
2. Bearing sleeve
3. Rubber seal (2)
4. Spacer bushing
5. Circlip
6. Clutch release shaft
7. Slotted bushing
8. Spring washer
9. Hexagon bolt
10. Main drive shaft seal
11. Transmission case

Fig. 8-4. Clutch release shaft installation.

8.3 Replacing Main Drive Shaft Seal

After removing the bellhousing and clutch, remove the main drive shaft seal inside the clutch release bearing guide sleeve as shown in Fig. 8-5. Be careful that the seal pulling hook does not damage the splines on the main drive shaft.

4. Remove the needle bearing and seal, if necessary, as shown in Fig. 8-6. Use a bushing driver of appropriate diameter to drive in the new bearing.

Fig. 8-6. Removing needle bearing and seal from clutch carrier plate. The carrier plate has been removed here, but it is not necessary to do so to replace the needle bearing.

5. Working through the round holes in the clutch carrier plate, remove the eight 6-mm socket head screws that hold the one-way clutch support tube to the bellhousing.

6. Insert a drift through the screw holes, as shown in Fig. 8-7, to push the one-way clutch support tube off the bellhousing.

Fig. 8-7. Drift being used to push one-way clutch support tube off bellhousing.

7. Removal of the one-way clutch support tube provides access to the circlip on the turbine shaft of the clutch carrier plate. Remove the circlip.

8. Knock the clutch carrier plate out of the ball bearing in the bellhousing as shown in Fig. 8-8.

Fig. 8-8. Rubber mallet being used to knock clutch carrier plate out of bellhousing.

9. Using a drift and an adapter of appropriate diameter, knock the ball bearing out of the bellhousing as shown in Fig. 8-9. This is necessary only if the bearing is noisy or damaged or if the turbine shaft must be replaced.

10. Using tools similar to those used for step 9, drive out the turbine shaft seal located in front of the ball bearing.

Fig. 8-9. Drift and adapter in position to drive out ball bearing and turbine shaft seal.

Assembly is the reverse of disassembly. It is usually best to replace the clutch plate if the original clutch plate has been in service for more than 30,000 mi. (50,000 km). Inspect the friction surfaces on the clutch carrier plate and the pressure plate assembly. Alternating bright and dull areas, with possible heat discoloration, indicate warping. Replacing such parts will help to ensure smooth clutch operation.

Check to see that the clutch release bearing contacts all the pressure plate levers evenly. If the levers are uneven, replace the pressure plate assembly. The pressure plate's diaphragm spring should be inspected for cracks. Also be sure that it is under tension. If the diaphragm spring rattles loosely when you pick up the pressure plate assembly and shake it, replace the assembly.

Use a sleeve-type seal driver, such as the one shown in Fig. 8-10, to drive in the new ball bearing and seal. Install the seal with its lip toward the torque converter.

Install a new gasket and a new O-ring under the one-way clutch support tube. Make sure that the tube and gasket are installed so that the ATF passages are open. Before installing the socket head screws for the one-way clutch support tube, coat the threads with sealer. Torque the screws to 1.5 mkg (11 ft. lb.). The 12-point bolts that hold the pressure plate assembly to the clutch carrier plate should also be torqued to 1.5 mkg (11 ft. lb.).

Fig. 8-10. Tool for driving in ball bearing and seal.

8.5 Adjusting Clutch Linkage

The clutch linkage must be given a preliminary adjustment after installing a new clutch plate or after reinstalling the clutch operating lever.

To adjust:

1. Mount the clutch servo and bracket on the transmission case.

2. Slide the clutch operating lever onto the clutch release shaft. Pull the lever away from the servo until it contacts the bellhousing, then tighten the clamp bolt slightly.

3. Using the adjustment sleeve on the servo rod, adjust the linkage to dimensions **a** and **b** as shown in Fig. 8-11. Dimension **a** is 8.5 mm ($^{11}/_{32}$ in.). Dimension **b** is 77 mm ($3^{1}/_{32}$ in.).

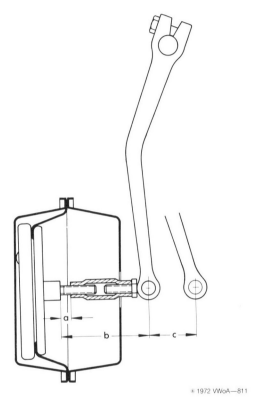

Fig. 8-11. Dimensions for clutch linkage adjustment.

4. Keeping the servo piston against the end of the vacuum chamber, loosen the clamp nut on the clutch operating lever and move the lever's clevis over the eye in the servo rod.

5. Move the clutch operating lever 40 mm ($1^{9}/_{16}$ in.) away from the servo rod eye (dimension **c** in Fig. 8-11). Then torque the clamp bolt to 3.0 mkg (22 ft. lb.).

6. Using a new plastic bushing, couple the clutch operating lever to the servo rod. Insert the clevis pin from the top and secure it at the bottom with a washer and a new cotter pin.

7. After about 300 mi. (500 km) of service, check the clutch freeplay as described in **4.1 Adjusting Clutch Freeplay**.

9. Transmission

Some parts shown in the illustrations vary slightly from those currently installed. Such differences will be pointed out whenever they alter working procedures.

9.1 Disassembling and Assembling Gearshift Housing

The gearshift housing is mounted on the front of the transmission. The parts contained in the gearshift housing are shown in Fig. 9-1.

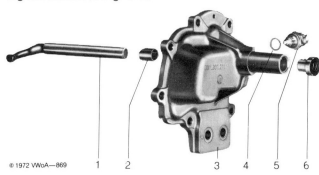

1. Inner selector lever
2. Guide bushing
3. Gearshift housing
4. Seal
5. Neutral safety switch
6. Sealing bushing

Fig. 9.1. Disassembled gearshift housing.

To disassemble:

1. Pull out the inner selector lever, then remove the sealing bushing as indicated in Fig. 9-2.

Fig. 9-2. Sealing bushing being twisted out.

2. Using a screwdriver, compress the guide bushing slot and push the bushing out of the gearshift housing.

3. Unscrew the neutral safety switch, then take off the seal.

When assembling the gearshift housing, press the new guide bushing in first. Then press in a new sealing bushing. Do not drive in the sealing bushing with a hammer or you will damage the seal. Use a new paper gasket when installing the gearshift housing on the transmission. Torque the mounting nuts to 1.5 mkg (11 ft. lb.). Torque the neutral safety switch to 2.5 mkg (18 ft. lb.).

9.2 Removing and Installing Transmission Gears in Case

It is not necessary to remove the differential prior to removing the transmission gears.

To remove:

1. Drain the hypoid oil if it was not drained previously.

2. Remove the seven mounting nuts, then remove the gearshift housing.

3. Remove the nine nuts from the studs that hold the gear carrier on the transmission case. (See Fig. 9-4 for the names and locations of the parts.)

4. Invert the transmission, then remove the oil pan from the bottom of the final drive case.

5. On 1970 and 1971 cars, remove the pinion retaining nut lockplate and the hexagonal bolt that secures it. On 1972 and later cars, the pinion retaining nut is locked by peening, as indicated in Fig. 9-3.

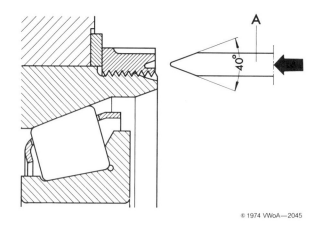

Fig. 9-3. Peening pinion retaining nut. A blunt punch, ground to a 40° angle, is driven (arrow) into a groove in the nut to expand it into a recess in the bearing. No unlocking step is required with these nuts. If a recess is ground into the bearing, peened nuts can be service-installed on earlier transmissions.

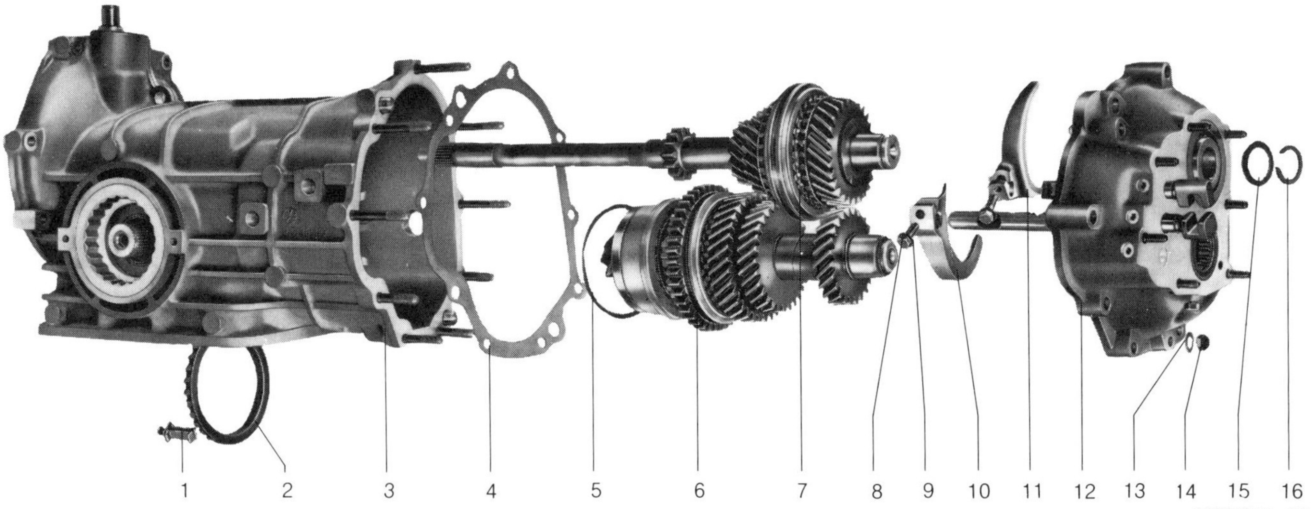

Automatic Stick Shift 21

1. Lockplate (1970-1971)
2. Pinion retaining nut
3. Transmission case
4. Gear carrier gasket
5. Shim S_3
6. Drive pinion
7. Main drive shaft
8. Clamp bolt (2)
9. Small spring washer (2)
10. Reverse/1st gear selector fork
11. 2nd/3rd gear selector fork
12. Gear carrier
13. Large spring washer (9)
14. Nut (9)
15. Dished washer
16. Lockring

Fig. 9-4. Exploded view of the gear carrier, the transmission gears, and the transmission case.

6. Using the special C-wrench shown in Fig. 9-5, unscrew the pinion retaining nut until it just touches the ring gear on the differential.

Fig. 9-5. Pinion retaining nut being unscrewed with special C-wrench.

7. Using the transmission removal tool shown in Fig. 9-6, press against the end of the pinion to force the transmission out of the case until the pinion retaining ring contacts the transmission case.

Fig. 9-6. Transmission being pressed out of case.

8. Alternately loosen the pinion retaining nut and press out the transmission until you can completely remove the pinion retaining nut.
9. Press the transmission completely out of the case.
10. Note the thickness and number of the S_3 shims installed between the flange on the double tapered-roller bearing and the end of the transmission case.
11. Clamp the gear carrier in a vise, or install the transmission in a selector fork adjusting fixture.
12. Using circlip pliers, remove the lockring from the main drive shaft as shown in Fig. 9-7.

22 Automatic Stick Shift

WARNING —
The dished washer under the lockring is under considerable tension and can cause injury if it snaps out suddenly.

Fig. 9-7. Lockring being removed with circlip pliers.

13. Loosen the clamp bolts for both selector forks. Remove only the reverse/1st gear selector fork, withdrawing its selector shaft as necessary.

14. Place the gear carrier in a suitable press jig, then press out the main drive shaft as shown in Fig. 9-8.

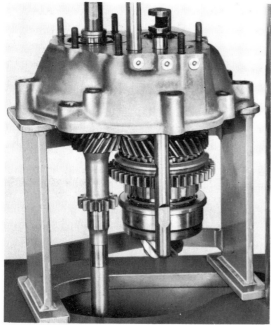

Fig. 9-8. Main drive shaft being pressed out.

NOTE —
The drive pinion will come out of its needle bearing in the gear carrier as the main drive shaft is pressed out. The 2nd/3rd gear selector fork will also slide off its selector shaft.

To install:

1. Replace any worn or damaged parts. Obtain a new dished washer if the original washer has lost its spring tension.

2. Insert the 2nd/3rd gear selector fork into the groove in the clutch gear assembly.

3. Using a special tube under the reverse gear, elevate the lower end of the main drive shaft (Fig. 9-9). Then place the main drive shaft and the drive pinion in mesh and stand them upright on a plate on the press bed.

4. Carefully place the gear carrier over the standing shafts. Position the 2nd/3rd gear selector fork so that it will receive the selector shaft properly.

5. Using a sleeve-type press tool to apply force only to the inner race of the ball bearing, press the gear carrier slowly and carefully onto the shafts as shown in Fig. 9-9.

Fig. 9-9. Gear carrier being pressed onto shafts. Note special sleeve elevating lower end of main drive shaft.

6. Raise the press tool. Then slip the dished washer and the lockring onto the main drive shaft.

7. Using the press, force the lockring and dished washer down the shaft until the lockring engages the groove on the main drive shaft.

8. Raise the press tool. Using pliers, squeeze the lockring all around until it bottoms in its groove.

9. Install and adjust the selector forks. See **9.3 Adjusting Selector Forks**.

10. Install a new paper gasket over the studs on the transmission case.

11. Install the original number of S_3 shims over the double tapered-roller bearing. Each of these shims should be of the same thickness as the original.

12. On 1971 and earlier cars, position the recess in the threaded part of the double tapered-roller bearing so that it will align with the hole for the lockplate bolt in the transmission case (or so that it will be toward the bottom on 1972 and later cars).

13. Place the pinion retaining nut in position inside the final drive case. Then insert the gears into the transmission case until the studs support the gear carrier and the pinion emerges in the final drive case. Start the pinion retaining nut onto the threads.

14. Alternately tighten the pinion retaining nut and push the gears further into the transmission case until the gear carrier touches the case.

15. Attach a torque wrench to the handle of the special C-wrench. Torque the pinion retaining nut to 22 mkg (160 ft. lb.), then loosen it. Retorque it to the same specification.

NOTE —
Since the C-wrench adds length to the torque wrench, the torque wrench reading will be only 18 mkg (130 ft. lb.) when the specified torque is attained.

16. Install the pinion retaining nut lockplate (Fig. 9-4) or lock the retaining nut by peening (Fig. 9-3).

17. Install the nine spring washers and nuts on the gear carrier mounting studs. Working diagonally, gradually torque the nuts to 2.0 mkg (14 ft. lb.).

18. Install the gearshift housing with a new paper gasket. Working diagonally, gradually torque the nuts to 1.5 mkg (11 ft. lb.).

19. Using a new gasket, install the oil pan. (Install new reinforcement plates under the bolts if the original plates are deformed.) Working diagonally, gradually torque the bolts to 1.0 mkg (7 ft. lb.).

9.3 Adjusting Selector Forks

Appliance VW 294b is required for selector fork adjustments. The S_3 shims should be in place but no paper gasket is necessary between the gear carrier and the studs, as it was with the earlier appliance VW 294. An additional 16-mm (⅝-in.) hole must be drilled in older appliances when adjusting transmissions with the parking lock.

To adjust:

1. Install the gear carrier, the drive pinion with S_3 shims, and the main drive shaft in the appliance as shown in Fig. 9-10. Tighten the pinion retaining nut lightly with the C-wrench.

Fig. 9-10. Transmission gears installed in the selector fork adjusting appliance.

2. Check to see that the selector forks are installed loosely on the selector shafts and that the clamp bolts are in place but not tightened.

3. Pull out the lower selector shaft (for the reverse/1st gear selector fork) until it engages the detent for 1st gear. Then slide the selector fork along the selector shaft in the same direction until the clutch gear slides over the teeth on the synchronizing ring and presses firmly against 1st gear.

4. Center the selector fork in the groove in the clutch gear and torque the clamp bolt to 2.5 mkg (18 ft. lb.).

5. While turning the transmission, select reverse, neutral, and 1st gear. The selector fork must not rub against the sides of the groove in any of these positions. If it does, readjust it.

6. Pull out the upper selector shaft (for the 2nd/3rd gear selector fork) until it engages the detent for 3rd gear. Then slide the selector fork along the selector shaft in the same direction until the clutch gear slides over the teeth on the synchronizing ring and presses firmly against 3rd gear.

7. Adjust the selector fork and torque the clamp screw as you did for the reverse/1st gear selector fork. While turning the transmission, select 2nd gear, neutral, and 3rd gear to see that the selector fork does not rub the sides of the groove in any of these positions.

8. Check the interlock mechanism. You should be able to move one selector shaft into a gear only when the other shaft is in neutral.

24 AUTOMATIC STICK SHIFT

9.4 Disassembling and Assembling Gear Carrier

The Automatic Stick Shift and the manual transmission use the same gear carrier. However, the Automatic Stick Shift version has no reverse relay lever. The 1970 through 1972 models have only two selector shafts rather than the three used in the manual transmission. The unused hole is sealed by two plugs. Beginning with the 1973 models, a parking lock was added to the Automatic Stick Shift. On these later transmissions, the gear carrier does have a third selector shaft for operating the parking lock. The correct procedures for replacing the ball bearing, needle bearing, and other gear carrier parts can be found in **TRANSMISSION AND REAR AXLE**.

9.5 Disassembling and Assembling Main Drive Shaft

Before proceeding, familiarize yourself with the part names given in Fig. 9-11 as they will appear frequently in the text. Some of the parts shown in the illustrations may differ slightly from those in your transmission or from those supplied as replacement parts. However, the updated parts fit the transmission perfectly and their use does not alter the repair procedures described here. Some late-type main drive shafts may have splines rather than the Woodruff key shown in the illustration. It is important that replacement parts be obtained with careful reference to the car's chassis number and to the number on the transmission. If possible, take the original parts with you to compare with those available as replacements.

To disassemble the main drive shaft, remove the thrust washer, 3rd gear, needle cage, and synchronizing ring. Then press the remaining parts off the shaft as illustrated in Fig. 9-12. Take out the Woodruff key (where fitted). Disassemble the clutch gear (if necessary).

CAUTION ——
If the main drive shaft has splines rather than a Woodruff key, there will be circlips on both sides of the clutch gear assembly and no inner race for the needle bearing. If this is the case, remove one circlip, lift off the clutch gear assembly, remove the other circlip, and then lift off the remaining parts by hand. Using a press can damage the shaft.

Fig. 9-12. Main drive shaft being pressed out of inner race for the 3rd gear needle bearing and other remaining parts.

Fig. 9-11. Exploded view of main drive shaft for Automatic Stick Shift transmission.

1. Woodruff key
2. Main drive shaft
3. 2nd gear
4. Needle cage for 2nd/3rd gear
5. Synchronizing ring for 2nd/3rd gear
6. Clutch gear assembly
7. 3rd gear
8. Needle bearing inner race
9. Thrust washer for 3rd gear
10. Dished washer
11. Lockring

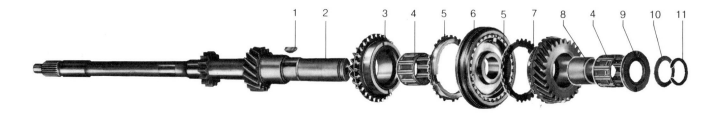

Automatic Stick Shift 25

To assemble:

1. Inspect all parts. Replace any that are worn or damaged.

   NOTE —
   Be sure to check the synchronizing rings. Modified synchronizing rings became available in 1972. These will prevent the howling noise sometimes caused by earlier-type synchronizing rings when the transmission is cold. You may wish to install these rings in earlier cars even if the original parts are not badly worn.

2. Hand-press the synchronizing rings into the gears as illustrated in Fig. 9-13. Measure clearance **a** with a feeler gauge. The wear limit is 0.60 mm (.024 in.). With new parts, dimension **a** should be between 1.00 and 1.90 mm (.039 and .075 in.).

Fig. 9-13. Clearance measurement **a**.

3. After aligning the marks indicated in Fig. 9-14, assemble the clutch gear assembly if it had been taken apart.

   NOTE —
   Late-model and replacement synchronizer hubs and operating sleeves are not marked for alignment. Mesh the teeth in various positions until a free-sliding fit is attained.

Fig. 9-14. Alignment marks (arrow).

NOTE —
The spring rings on opposite sides of the clutch gear assembly must be installed 120° apart as indicated in Fig. 9-15.

Fig. 9-15. Positions for spring rings. Make sure spring ring ends are fully hooked over lugs.

The remainder of assembly is the reverse of disassembly, with the following additional steps:

1. When installing the clutch gear assembly, the side of the operating sleeve containing the 1-mm-deep groove goes toward 3rd gear. The side of the clutch gear hub with the wide chamfer goes toward 2nd gear.

2. Heat the needle bearing inner race (where fitted) in hot oil to approximately 212°F (100°C) before you press it onto the main drive shaft as shown in Fig. 9-16.

Fig. 9-16. Needle bearing inner race being pressed onto main drive shaft. Arrow indicates 1-mm deep groove on operating sleeve.

26 Automatic Stick Shift

9.6 Disassembling and Assembling Drive Pinion

Before proceeding, familiarize yourself with the part names given in Fig. 9-17 as they appear frequently in the text. Some of the parts shown in the illustrations may differ slightly from those in your transmission or from those supplied as replacement parts.

1. Pinion
2. Shim S_3
3. Double tapered-roller bearing
4. Round nut
5. Shim for round nut
6. Clutch gear assembly
7. Synchronizing ring for 1st gear
8. 1st gear
9. Needle cage for 1st gear (needles in pairs)
10. 2nd gear
11. Large circlip
12. Spacer spring (1970-1972)
13. 3rd gear
14. Needle bearing inner race
15. Small circlip

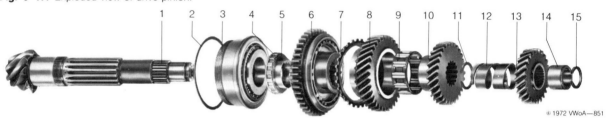

Fig. 9-17. Exploded view of drive pinion.

NOTE
Read **11. Adjusting Final Drive** before deciding to replace either the drive pinion or the double tapered-roller bearing.

To disassemble:

1. Remove the small circlip. On 1970 to 1972 transmissions, which have spacer springs, hold 3rd gear down with the repair press to keep 3rd gear from flying off.

2. If the needle bearing inner race is tight, press the drive pinion out of the inner race with 3rd gear supported.

3. Lift off the spacer spring (where fitted) and then remove the large circlip.

4. Remove 2nd gear, 1st gear, the needle cage, synchronizing ring, clutch gear assembly, and shim.

5. Mount the drive pinion in special appliance VW 293 as shown in Fig. 9-18.

Fig. 9-18. Drive pinion mounted in special appliance.

6. Unscrew the round nut by turning the drive pinion counterclockwise at the splined end.

7. Press the drive pinion out of the double tapered-roller bearing. Make sure the inner race of the bearing is supported. See Fig. 9-19.

Fig. 9-19. Drive pinion being pressed out of the double tapered-roller bearing.

8. If necessary for cleaning and inspection purposes, remove the spring rings and disassemble the clutch gear assembly.

Automatic Stick Shift 27

To assemble:

1. Inspect all parts. Replace any that are worn or damaged.

   NOTE ———
   Be sure to check the synchronizing rings. A modified synchronizing ring became available in 1972. It helps 1st gear to synchronize in both directions. You may wish to install this ring in earlier cars even if the original part is not badly worn.

2. Hand-press the synchronizing ring into the gear as illustrated in Fig. 9-20. Measure clearance **a** with a feeler gauge. With new parts, dimension **a** should be between 1.10 and 1.80 mm (.043 and .071 in.). The wear limit is 0.60 mm (.024 in.).

Fig. 9-20. Clearance measurement **a**.

3. Reassemble the clutch gear assembly (if it was taken apart earlier). See **9.5 Disassembling and Assembling Main Drive Shaft** for details.

   NOTE ———
   The operating sleeve and clutch gear are a matched pair. If either is worn, the assembly must be replaced as a unit. Also, if a gear is replaced, the meshing gear on the main drive shaft must also be replaced.

4. Heat the double tapered-roller bearing in hot oil to about 212°F (100°C) then install the bearing on the drive pinion. When it has cooled to room temperature, seat the drive pinion in the bearing using a pressure of 3 tons.

5. Using the special appliance illustrated earlier in Fig. 9-18, install the round nut. Torque it to 20.0 mkg (145 ft. lb.).

6. Using a special torque gauge, check the turning torque of the double tapered-roller bearing. The turning torque should be 6 to 21 cmkg (5.2 to 18.2 in. lb.) for a new bearing and 3 to 7 cmkg (2.6 to 6.1 in. lb.) for a bearing that has been in service for 30 mi. (50 km) or more.

   NOTE ———
   Lubricate the bearing with hypoid oil only. See **11.1 Adjusting Pinion** for complete details.

7. Using a blunt cold chisel, peen the locking shoulder of the round nut into the drive pinion splines at three places 120° apart. Make sure you do not crack or burr the locking shoulder. See Fig. 9-21.

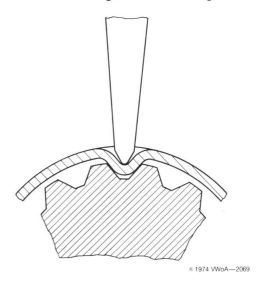

Fig. 9-21. Blunt cold chisel being used to peen round nut locking shoulder into splines.

8. Using a depth gauge, a vernier caliper, or other suitable measuring tool, find the thickness of the shim for the round nut. A shim must be selected that will adjust dimension **x** (see Fig. 9-22) to 44.40 to 44.50 mm (1.748 to 1.752 in.).

   NOTE ———
   Dimension **x** includes the shim. Keep strictly within tolerance. New shims are available in thicknesses from 0.55 mm to 0.85 mm in 0.05-mm increments.

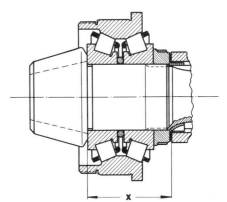

Fig. 9-22. Dimension **x**. It is measured from the end of the double tapered-roller bearing inner race to the face of the shim for the round nut that contacts the hub of the clutch gear assembly.

28 Automatic Stick Shift

9. Assemble the drive pinion up to 2nd gear. Install the large circlip, then check axial play as illustrated in Fig. 9-23. The play should be 0.10 to 0.25 mm (.004 to .010 in.) for drive pinions with a spacer spring—keeping as close as possible to the smaller figure—or 0.00 to 0.15 mm (.000 to .006 in.) for drive pinions without the spacer spring.

NOTE —
To adjust this clearance, circlips are available in thicknesses from 1.45 to 2.20 mm in increments of 0.15 mm.

Fig. 9-24. Parking lock pawl plunger **(1)**, spring **(2)**, and plug **(3)**. The pawl is installed on a third selector shaft in the transmission gear carrier.

The reverse idler gear assembly and main drive shaft needle bearing are shown in Fig. 9-25. The circlip for the reverse shaft is inside the final drive case. When driving in a new needle bearing for the main drive shaft, install the bearing's thicker side—with lettering on it—toward the transmission gears.

Fig. 9-23. Axial play for 2nd gear being measured with feeler gauge.

10. Install the spacer spring (where fitted). Then heat the needle bearing inner race to 212°F (100°C) and press it on together with 3rd gear.

11. While holding down 3rd gear against spacer spring tension (where necessary), install the small circlip.

9.7 Removing and Installing Reverse Shaft, Needle Bearings, and Parking Lock

The needle bearing for the main drive shaft is pressed into the transmission case and located by lockrings. The reverse idler gear shaft is held in the case by a circlip. On transmissions built since 1973, the parking lock pawl plunger is also in the transmission case. A plug and washer must be removed from the side of the transmission case and the spring and plunger taken out as illustrated in Fig. 9-24.

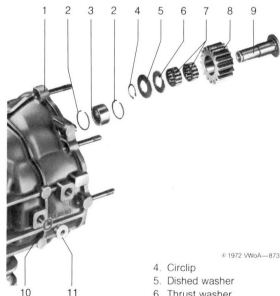

1. Transmission case
2. Lockring (2)
3. Main drive shaft needle bearing
4. Circlip
5. Dished washer
6. Thrust washer
7. Needle bearing cage (2)
8. Reverse idler gear
9. Reverse idler gear shaft
10. Washer
11. Plug

Fig. 9-25. Transmission parts installed directly in transmission case. Reverse lock pawl is not shown.

AUTOMATIC STICK SHIFT 29

10. Differential

The differential for the Automatic Stick Shift is identical to the differential for the manual transmission. However, removal and installation procedures differ because the Automatic Stick Shift transmission case has no final drive covers.

10.1 Removing and Installing Differential

Before proceeding, familiarize yourself with the part names given in Fig. 10-1 as they appear frequently in the text. On 1970 and 1971 cars, the lockplates are covered by a sealing cap and spacer (not shown) to prevent corrosion. Beginning with the 1972 models, plastic lockplates have been used for the same purpose. These should be service installed when making repairs.

NOTE
Read **11. Adjusting Final Drive** before deciding to replace the ring gear, the adjusting rings, or any part of the differential.

To remove:

1. Drain the hypoid oil if it was not drained previously.
2. Invert the transmission, then remove the oil pan from the bottom of the final drive case.
3. With a screwdriver, pierce the plastic cap in the center of each axle flange and pry off the cap.

NOTE
New caps must be used during assembly.

4. Using angled circlip pliers, remove the circlip from the hub of each axle flange.

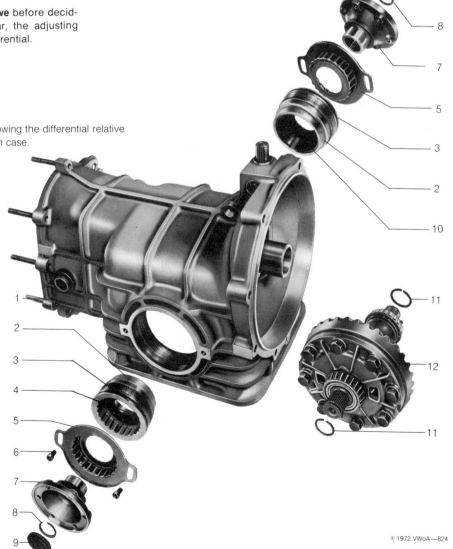

Fig. 10-1. Exploded view showing the differential relative to the transmission case.

1. Transmission case
2. Adjusting ring (2)
3. O-ring (2)
4. Oil seal for constant velocity joint flange (2)
5. Lockplate (2)
6. Phillips head screw (4)
7. Axle flange (2)
8. Circlip (2)
9. Plastic cap (2)
10. Outer race for tapered-roller bearing (2)
11. Spacer ring (2)
12. Differential with ring gear

30 Automatic Stick Shift

5. Using two levers, as shown in Fig. 10-2, pry off each axle flange.

Fig. 10-2. Axle flange being pried off differential gear splines.

6. Remove the Phillips head screws, then take off the sealing cap (where fitted) and the lockplate. Fig. 10-3 shows the latest-type plastic lockplate.

Fig. 10-3. Plastic lockplate **a**. Metal plates **b** are installed under Phillips head screws.

7. Carefully mark the position of the adjusting rings relative to the transmission case. Measure the depth to which they are screwed into the case and write it down for use during assembly.

8. With the transmission inverted, screw the adjusting rings out of the transmission case, being careful not to let the differential drop suddenly into the case.

NOTE ---

Although VW shops use a special star-shaped tool to unscrew the adjusting rings, an old lockplate with a flat steel handle attached to it is a satisfactory tool for the job.

9. Lift the differential with ring gear out of the transmission case and remove the spacer rings.

10. Press the oil seal for the axle flange out of each adjusting ring. Also remove the O-rings.

11. Inspect the outer race for the tapered-roller bearing in each adjusting ring. If it is worn or damaged, press the outer race out of the adjusting ring and press in a new one.

CAUTION ---

If you lack a suitable repair press and the special tools shown in Fig. 10-4, we suggest you leave outer race replacement to an Authorized VW dealer or other qualified and properly equipped shop.

Fig. 10-4. Special press tools being used to press out the outer race for the tapered-roller bearing. The same tools, without the lower cylindrical holder, are used to press in the new outer race. It is not necessary to heat the adjusting ring.

Automatic Stick Shift 31

To install:

1. Inspect all parts. Replace any that are worn or damaged.

2. Install new oil seals for the axle flanges in each adjusting ring.

   NOTE
   If you have made repairs that require readjustment of the tapered-roller bearings, do not install the oil seals until you have made the adjustments. (See **Table b** in **11. Adjusting Final Drive**.)

3. If the transmission gears have been removed, install them in the case and tighten the pinion retaining ring as described in **9.2 Removing and Installing Transmission Gears in Case.**

4. Install new O-rings on the adjusting rings, then coat the adjusting ring threads with molybdenum grease.

5. Screw the left-side (ring-gear side) adjusting ring into the transmission case until it is at the depth recorded and position marked during removal.

6. After lubricating the tapered-roller bearings with hypoid oil, position the differential with ring gear in the transmission case as shown in Fig. 10-5. Then screw in the right-side adjusting ring.

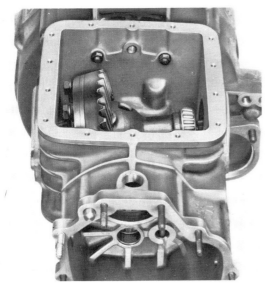

Fig. 10-5. Position of differential with ring gear in transmission case.

7. If part replacements make it necessary, adjust the ring gear depth-of-mesh and roller bearing preload. (See **Table b** in **11. Adjusting Final Drive**.)

8. If adjustment is not necessary, screw the right-side adjusting ring into the transmission case until it is at the depth and position marked during removal.

9. Coat the outer surface of the adjusting rings with anti-rust preservative, then install new plastic lockplates. Torque the Phillips head screws to 1 mkg (7 ft. lb.).

10. Install a spacer ring on one differential gear spline at a time and press it in place by installing the axle flange. Install the circlip.

    NOTE
    If necessary, use a device such as the one shown in Fig. 10-6 to draw down the axle flange. This will compress the spacer ring so that the circlip can be installed.

Fig. 10-6. Flange being drawn down. Bolt in center of bar must screw into M 10 threads in differential gear shafts.

11. Using a sleeve-type seal driver of appropriate diameter, drive in new plastic caps.

10.2 Disassembling and Assembling Differential

As stated earlier, the differential for the Automatic Stick Shift is identical to the differential for the manual transmission. Therefore, you should refer to **TRANSMISSION AND REAR AXLE** if it is necessary to disassemble the differential. The following procedures are covered there:

1. Disassembling and assembling differential includes removing and installing the ring gear.

2. Removing and installing tapered-roller bearings.

32 Automatic Stick Shift

10.3 Replacing Oil Seal for Constant Velocity Joint Flange

It is possible to replace the oil seals for the axle flanges without removing the adjusting rings. First, remove the driveshafts and constant velocity joints as described in **TRANSMISSION AND REAR AXLE**. Then remove the circlip, axle flange, lockplate, and spacer ring as described in **10.1 Removing and Installing Differential**.

Pry the old seal out of the recess in the adjusting ring. Clean the seal recess, then lubricate the seal lip with hypoid oil and drive it into place with an appropriate seal driver. Make sure the seal is seated squarely and completely in the recess. Then install the parts removed earlier.

11. Adjusting Final Drive

Mechanics without prior experience in adjusting VW final drive gears are urged to read **10. Adjusting Final Drive** in **TRANSMISSION AND REAR AXLE**. Many technical details that will not be repeated in this section are covered fully there.

The mesh of the ring gear and drive pinion is adjusted by first adjusting the pinion axially (as indicated by the horizontal arrows in Fig. 11-1). This is done by installing thicker or thinner S_3 shims between the double tapered-roller bearing and the transmission case. The ring gear's depth-of-mesh is then adjusted (as indicated by the vertical arrows in Fig. 11-1) by turning the differential adjusting rings.

If only those parts affecting ring gear depth-of-mesh have been replaced, only the ring gear needs adjusting. If only drive pinion parts have been replaced, only the pinion's S_3 thickness requires checking and adjustment.

Table b lists parts that must be adjusted after the replacement of particular final drive or transmission components.

Table b. Final Drive Replacements and Adjustments

Part replaced	Parts to be adjusted		
	Pinion	Selector forks	Ring gear
Transmission case	X	X	X
Adjusting ring			X
Differential tapered-roller bearings			X
Pinion double tapered-roller bearings	X	X	
Ring gear and pinion	X	X	X
Differential housing			X
Differential housing cover			X

Fig. 11-1 also shows the numbers that are stamped on replacement gearsets and on gearsets used in production prior to early 1972. Their explanation is as follows:

1. G358 indicates a Gleason gearset with a ratio of 8/35 teeth.

2. 369 is the matching number of the gearset. It shows that the ring gear and the drive pinion are a matched pair.

3. 26 means that **r** = 0.26 mm. This is the deviation of **R** from the master gauge length, R_o. The dimension R_o equals the length of the master gauge (58.70 mm). The dimension **R** equals the actual distance between the ring gear's centerline and the end face of the drive pinion. This dimension is selected during assembly and is the point of quietest running for the particular gearset.

The ring and drive pinion gearsets used in production since early 1972 have different markings. The position of

Fig. 11-1. Top view of ring gear and drive pinion used through early 1972. (Markings explained in text.)

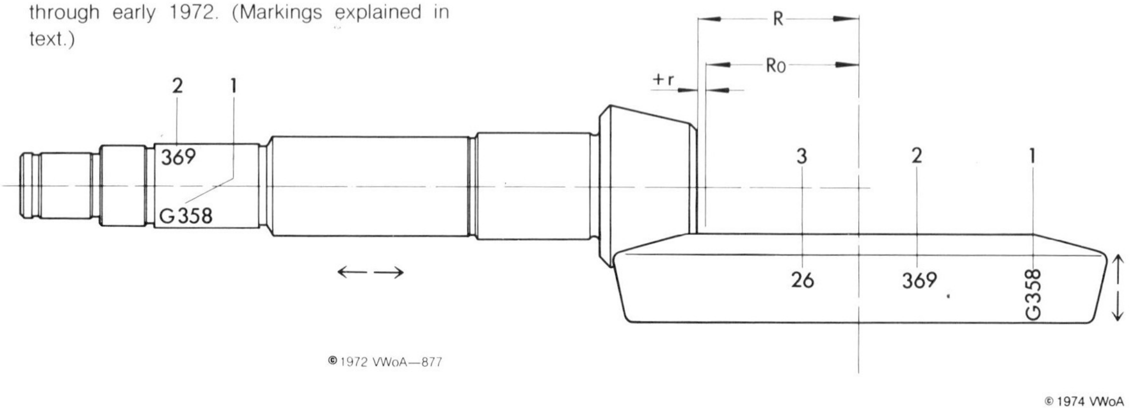

the pinion is now determined by dimension P_o as shown in Fig. 11-2. P_o is the distance from the centerline of the ring gear to the back of the pinion head. This dimension is selected during assembly and is the point of quietest running for the particular gearset. It is not marked on the gears, however.

CAUTION

Before removing the gears from the transmission for the purpose of replacing the double tapered-roller bearing, it is essential that you remove the differential so that P_o can be measured using the special measuring bar illustrated in **11.1 Adjusting Pinion.** *You must determine P_o and write it down for use in determining S_3 shim thickness during installation of transmission gears in the transmission case.*

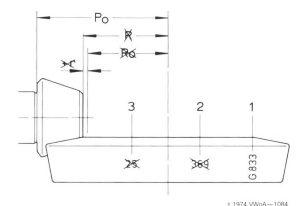

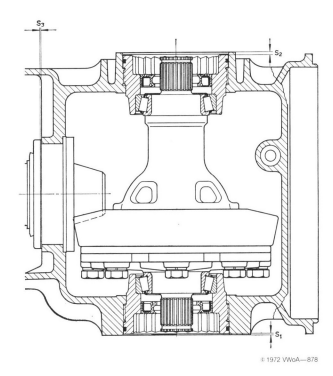

Fig. 11-3. Final drive adjustments. S_3 is determined by a shim. S_1 and S_2 are determined by the screw-in depth of the adjusting rings.

Fig. 11-2. Top view of ring and drive pinion gearset used in production since early 1972. Only manufacturer's initial and ratio are stamped on these gearsets.

NOTE

Since replacement ring and drive pinion gearsets still have the deviation r marked on them, it is not necessary to determine P_o before disassembling the transmission if the drive pinion itself is to be replaced.

11.1 Adjusting Pinion

The entire transmission gearset need not be in place while the pinion is being adjusted.

To adjust:

1. Press the drive pinion into the double tapered-roller bearing. Torque the round nut to 20.0 mkg (145 ft. lb.), but do not lock it by peening.

2. Install the drive pinion and double tapered-roller bearing without the S_3 shim. See Fig. 11-4.

In addition to the pinion adjustment dimensions determined by S_3 shim thickness, there are two adjustments that determine ring gear depth-of-mesh. These are called S_1 and S_2. They are indicated, along with pinion shim S_3, in Fig. 11-3. Unlike the manual transmission, no shims are used in the Automatic Stick Shift to establish S_1 and S_2. Adjustment is made by means of the adjusting rings only.

It is important that the bearing preload for both the drive pinion double tapered-roller bearing and the differential's two tapered-roller bearings is within specifications after final drive adjustments have been completed.

Fig. 11-4. Pinion retaining nut being torqued to about 22 mkg (160 ft. lb.). Do not lock by peening.

34 Automatic Stick Shift

3. Lubricate the double tapered-roller bearing with hypoid oil, then install a torque gauge as shown in Fig. 11-5.

Fig. 11-5. Gauge for measuring pinion turning torque.

4. Using the handle of the torque gauge, spin the pinion rapidly 15 or 20 turns in each direction. Take torque readings while you continue to spin the pinion. The turning torque should be 6 to 21 cmkg (5.2 to 18.2 in. lb.) for a new bearing and 3 to 7 cmkg (2.6 to 6.1 in. lb.) for a bearing that has been in service for 30 mi. (50 km) or more.

NOTE
Used bearings must have no visible axial play. Use only hypoid oil as a lubricant. The test results will be inaccurate if the turning torque is checked with the bearings dry or lubricated with another kind of oil.

5. Install the adjusting rings without oil seals and without the differential.

6. Screw in the right-side (the side opposite the ring gear) adjusting ring until its outer surface is flush with the transmission case.

7. Adjust the setting ring on measuring bar VW 385/1 as shown in Fig. 11-6. Dimension **a** should be approximately 70 mm (2¾ in.).

Fig. 11-6. Dimension **a** between setting ring and measuring bar center. To move setting rings, loosen the socket head clamp screw.

8. Install the other parts of the measuring bar as indicated in Fig. 11-7, then install the left-side adjusting ring to hold it in place.

NOTE
The centering disks, VW 385/2, should be lubricated lightly with hypoid oil and the left-side adjusting ring tightened until the bar can just be turned by hand. The measuring plate, VW 385/17, is held on the drive pinion face by magnets.

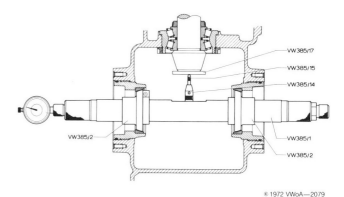

Fig. 11-7. Measuring bar in place.

9. Install a dial indicator with a 3-mm range on the measuring bar. Using the setting block VW 385/9, as shown in Fig. 11-8, zero the dial indicator. The setting block zeros the indicator at R_o (58.70 mm).

Fig. 11-8. Setting block in place so that dial indicator can be zeroed at R_o.

10. Turn the bar so that the measuring pin extension rests against the measuring plate on the pinion face.

11. Rotate the bar back and forth over center. The maximum dial indicator reading should be observed and written down. This is measurement **e**.

12. Read the deviation **r** marked on the ring gear. Using **r** and measurement **e**, determine the thickness of shim S_3 as in the following example:

Example:

1. Actual dimension fixed by applying the setting block to the bar measured reading (**e**)	58.70 mm − 0.48 mm
Actual dimension measured to VW 385/17 with no pinion shims installed (without shims)	58.22 mm
2. Design dimension R_0 + deviation **r** (marked on ring gear)	58.70 mm + 0.18
Nominal pinion dimension **R** (for quietest running) $R_0 + r$	58.88 mm
3. Nominal pinion dimension **R** − actual pinion dimension (in housing without shims)	58.88 mm − 58.22 mm
Required thickness of shims S_3	0.66 mm

CAUTION ──
Every measurement in the preceding example is imaginary. Using any for actual adjustments could cause serious damage.

NOTE ──
Table c lists the shims available as replacement parts. If the S_3 measurement you obtain exceeds a standard shim dimension, use the next thickest shim. Shims can be used singly or in combination as prescribed in **Table d**. Measure the shims carefully at several points with a micrometer and check them for burrs or damage. Use only shims that are in perfect condition.

Table c. Available S_3 Shim Sizes

Shim No.	Part No.	Thickness
1	001 311 391	0.15 mm (.0059 inch)
2	001 311 392	0.20 mm (.0079 inch)
3	001 311 393	0.30 mm (.0118 inch)
4	001 311 394	0.40 mm (.0158 inch)
5	001 311 395	0.50 mm (.0197 inch)
6	001 311 396	0.60 mm (.0236 inch)
7	001 311 397	0.70 mm (.0276 inch)
8	001 311 398	0.80 mm (.0315 inch)
9	001 311 399	0.90 mm (.0354 inch)

13. After installing the S_3 shim you have selected, recheck the measurement using the measuring bar. The reading should not deviate from the **r** dimension marked on the ring gear by more than 0.04 mm.

Table d. S_3 Shim Measurements

S_3 nominal mm (inch)	S_3 actual mm (inch)	Shim No.
0.28-0.32 (.0110-.0126)	0.30 (.0118)	3
0.33-0.37 (.0130-.0146)	0.35 (.0138)	2 + 1
0.38-0.42 (.0150-.0165)	0.40 (.0158)	4
0.43-0.47 (.0169-.0185)	0.45 (.0177)	3 + 1
0.48-0.52 (.0189-.0205)	0.50 (.0197)	5
0.53-0.57 (.0209-.0224)	0.55 (.0217)	4 + 1
0.58-0.62 (.0228-.0244)	0.60 (.0236)	6
0.63-0.67 (.0248-.0264)	0.65 (.0256)	5 + 1
0.68-0.72 (.0268-.0284)	0.70 (.0276)	7
0.73-0.77 (.0287-.0303)	0.75 (.0295)	6 + 1
0.78-0.82 (.0307-.0323)	0.80 (.0315)	8
0.83-0.87 (.0327-.0343)	0.85 (.0335)	7 + 1
0.88-0.92 (.0347-.0362)	0.90 (.0354)	9
0.93-0.97 (.0366-.0382)	0.95 (.0374)	8 + 1

NOTE ──
If you are adjusting a pinion for which **r** is not given, select a shim that will duplicate the pinion position measurement you obtained prior to disassembly.

11.2 Adjusting Ring Gear

The ring gear's depth-of-mesh is adjusted in order to bring the average backlash between the ring gear and the drive pinion within a specified range. However, prior to adjusting the backlash, the preload on the tapered-roller bearings must be adjusted to specifications. This preload must be maintained by turning the adjusting rings equally in opposite directions when backlash is finally adjusted to specifications. A tool such as that shown in Fig. 11-9 is very helpful for turning the adjusting rings.

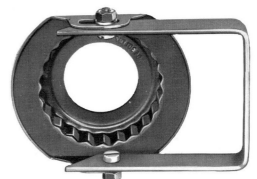

Fig. 11-9. Tool for turning adjusting rings. A similar tool can be improvised from an old metal lock ring.

36   Automatic Stick Shift

To adjust:

1. Remove the oil seals from the adjusting rings if they were not removed previously. Make sure the outer races for the tapered-roller bearings are pressed all the way into the adjusting rings.

2. Screw in the adjusting ring on the ring gear side until it is approximately 0.10 to 0.20 mm (.004 to .008 in.) below the measuring surface on the case.

3. Install the differential with ring gear. The ring gear goes on the left side of the case.

4. Screw in the other adjusting ring until the differential is supported firmly but without bearing preload.

5. Install a torque gauge on the ring gear end of the differential as shown in Fig. 11-10.

Fig. 11-10. Torque gauge in position for measuring tapered-roller bearing preload.

6. Using the handle of the torque gauge, spin the differential 15 or 20 turns in each direction while applying hypoid oil to the tapered-roller bearings.

   NOTE —
   Use only hypoid oil as a lubricant. The test results will be inaccurate if the turning torque is checked with the bearings dry or lubricated with another kind of oil.

7. While spinning the differential rapidly, slowly increase the bearing preload by screwing in the adjusting ring that is opposite the ring gear. Continue until the preload indicated by the torque gauge reaches 18 to 22 cmkg (15.6 to 19.1 in. lb.) for new bearings or 3 to 7 cmkg (2.6 to 6.1 in. lb.) for bearings that have been in service for 30 mi. (50 km) or more.

8. Measure depth S_1 and S_2 as illustrated in Fig. 11-11 and write down the readings.

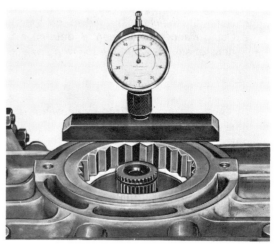

Fig. 11-11. Dial indicator with bridge being used to measure depth of adjusting ring. A depth micrometer is also suitable for this job.

9. Install the transmission gears. Torque and lock the pinion retaining nut as described in **9.2 Removing and Installing Transmission Gears in Case**.

To adjust backlash:

1. Install a measuring lever and dial indicator as illustrated in Fig. 11-12. Turn the differential until the lever contacts the gauge pin as shown.

Fig. 11-12. Measuring lever and dial indicator positioned for measuring backlash.

Automatic Stick Shift 37

2. Move the gauge in its holder to obtain a gauge preload of 1.50 mm (or .050 in.) and to position the gauge pin on a vertical line through the differential centerline. See Fig. 11-13.

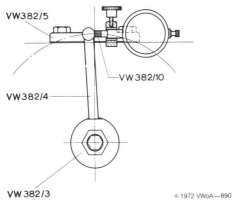

Fig. 11-13. Proper angle of measuring lever and proper position of gauge pin.

3. Clamp the drive pinion as in Fig. 11-14 to keep it from turning.

Fig. 11-14. Clamping bar bolted acrosss front of gear carrier to lock drive pinion.

4. Turn the differential with ring gear as far as it will go before it is stopped by the locked drive pinion. Zero the gauge dial with the gears in this position.

5. Turn the differential with ring gear as far as it will go in the opposite direction. Write down the reading on the gauge. This is the backlash, sometimes indicated by Sv_o.

6. Loosen the locknut on the sleeve that holds the measuring lever, loosen the clamping bar on the drive pinion, and then turn the ring gear 90°.

Tighten the locknut and clamping bar. Again measure Sv_o (backlash).

7. Repeat step 6 two more times so you will have recorded Sv_o at four points around the ring gear.

8. Determine the average Sv_o as in the following example:

Example

1st reading: Sv_o = 0.53 mm (.0208 in.)
2nd reading: Sv_o = 0.56 mm (.0220 in.)
3rd reading: Sv_o = 0.54 mm (.0212 in.)
4th reading: Sv_o = 0.52 mm (.0204 in.)
Total = 2.15 mm (.0844 in.)

Sv_o average = Total ÷ 4
= 2.15 mm (.0844 in.) ÷ 4
= 0.537 mm (.0211 in.)
= 0.54 mm (.0210 in.)

CAUTION
Every measurement in the preceding example is imaginary. Using any for actual adjustments could cause serious damage.

NOTE
If the Sv_o readings you obtain vary by more than 0.06 mm (.002 in.) from one another, there is something wrong with the ring gear or its installation.

9. Using the Sv_o average, the correction factor **w** listed for your particular gearset in the table below, and lift **h** from the same table, compute the S_1 correction factor as in the following example:

Gearset	Correction factor w	Lift h
G 358	1.00	0.20
K 835	1.10	0.22

Example (finding S_1 correction factor):

$S_1 = (Sv_o \text{ average} \times w) - h$
$S_1 = (0.54 \times 1.00) - 0.20$
$S_1 = 0.34$

10. Unscrew the adjusting ring opposite the ring gear by the distance of the S_1 correction factor. Screw in the adjusting ring on the ring gear side by exactly the same amount. Stay within a tolerance of ± 0.01 mm (.0004 in.).

11. Recheck the backlash. The Sv_o average should be 0.15 to 0.25 mm (.006 to .010 in.) and individual readings should not vary from each other by more than 0.05 mm (.002 in.).

38 Automatic Stick Shift

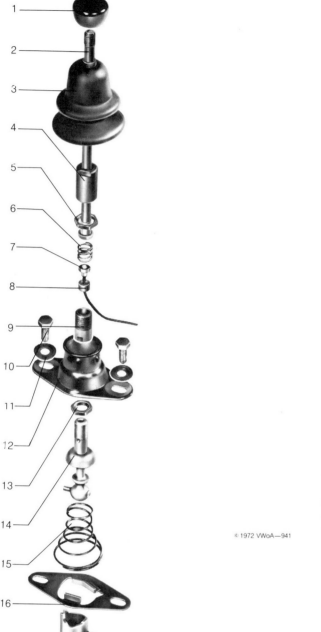

12. Coat the outer surface of the adjusting rings with anti-rust preservative. Then install new plastic lockplates. Torque the Phillips head screws to 1 mkg. (7 ft. lb.).

12. Selector Lever

The shift rod for the Automatic Stick Shift transmission is similar, but not identical, to the shift rod for the manual transmission. Also, because it contains the selector lever contacts for the automatic clutch, the selector lever itself is completely different on cars with Automatic Stick Shift. Fig. 12-1 is an exploded view of the complete selector lever assembly.

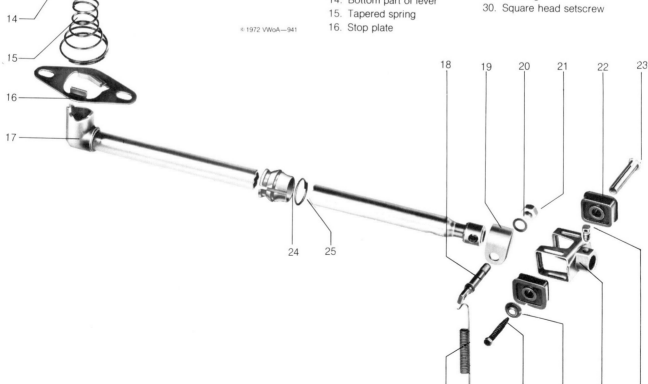

Fig. 12-1. Selector lever and shift rod dissassembled.

1. Knob
2. Upper part of lever
3. Boot
4. Shift sleeve
5. M 19 locknut
6. Contact spring
7. Contact
8. Insulating sleeve
9. Threaded sleeve
10. M 8 mounting bolt (2)
11. Spring washer (2)
12. Mounting
13. M 15 locknut
14. Bottom part of lever
15. Tapered spring
16. Stop plate
17. Shift rod
18. Screw for spring
19. Clamp for shift rod
20. Washer
21. M 8 self-locking nut
22. Rubber insert (2)
23. Clamping sleeve
24. Bushing for shift rod guide
25. Retaining ring
26. Hooked spring
27. Hexagon head sheet metal screw
28. Locking cap
29. Housing
30. Square head setscrew

Automatic Stick Shift

The selector lever can be removed as a unit for subsequent disassembly at the workbench. It must be removed as a unit if you need to remove the shift rod from inside the car's frame tunnel. Although shift rod removal and installation follow procedures similar to those given in **TRANSMISSION AND REAR AXLE**, there are several additional parts that must be dealt with that are not used on cars with manual transmissions.

Electrical contact replacements are available for installation in the selector lever. Their adjustment is covered in **4.3 Cleaning and Adjusting Selector Lever Contacts**. It is possible to clean or replace the contacts without removing the selector lever mounting and the bottom part of the selector lever from the car.

12.1 Removing and Installing Selector Lever and Shift Rod

If only the selector lever is to be removed, adjusted, and installed, carry out only the procedures that apply to it. Usually, the shift rod is removed only if it is rattling or binding inside the frame tunnel.

To remove

1. Lift out the rear seat. Remove the floor mats to obtain access to the M 8 mounting bolts and the selector lever ground wire.
2. Separate the selector lever ground wire at the in-line connector indicated in Fig. 12-2.

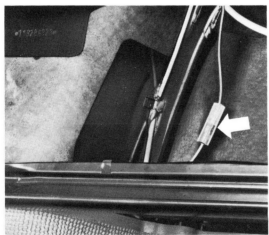

Fig. 12-2. Selector lever ground wire connector. Connector is located under rear seat, on left-hand side of frame tunnel. Disconnect it by carefully pulling the wires apart.

3. Remove the two M 8 mounting bolts, then lift the selector lever off the frame tunnel.

NOTE

Press downward on the lever while loosening the bolts so that the tapered spring does not force the mounting up at an angle.

4. Take off the tapered spring and the stop plate.
5. Working in the area under the rear seat, remove the inspection cover from the top of the frame fork.
6. Remove the hexagon head screw, the hooked spring, and the screw for the spring indicated in Fig. 12-3.

Fig. 12-3. Hexagon head screw is at **1**, hooked spring at **2**, and screw for spring at **3**.

7. Remove the cover plates from both the frame head and the front body apron. On Super Beetles, remove the deformation element in front of the frame head. Working with a pair of pliers through the selector lever opening, push the shift rod out through the front of the car.

To install:

1. Inspect the bushing for the shift rod guide shown in Fig. 12-4. Replace the bushing if it is cracked, worn, or loose.

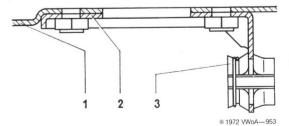

Fig. 12-4. Bushing **3** for shift rod guide **2**. Frame tunnel is at **1**.

40 Automatic Stick Shift

NOTE —
If the guide sleeve must be replaced, first install the retaining ring on the guide sleeve, and then press the sleeve into the bracket. Grease the sleeve following installation. Be sure the slot is toward the left side of the car.

2. Lightly grease the shift rod along its entire length.

3. Insert the shift rod at the front of the car and guide it through the bushing. Then push it to the rear and attach it to the transmission using the hexagon head screw, the hooked spring, and the screw for the spring.

4. Install the body apron cover plate and the frame head cover plate and gasket. On Super Beetles, also install the deformation element.

5. Install the stop plate, the tapered spring, and the selector lever on the frame tunnel. Then move the selector lever to **L**.

6. Loosen the M 15 locknut. Carefully align the lever in the L position so that it inclines directly backward and does not tilt to either side. Tighten the M 15 locknut.

7. Loosen the M 8 mounting bolts slightly. Then, holding the lever to keep it vertical, press the stop plate to the left until it contacts the lever as indicated in Fig. 12-5.

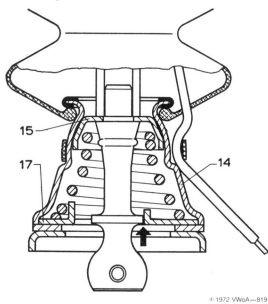

Fig. 12-5. Stop plate **(17)** contacting lever (arrow). Mounting is at **14**, bottom part of lever at **15**.

8. Using downward pressure, hold the lever, the mounting, and the stop plate in this position. Torque the M 8 mounting bolts to 2.0 mkg (14 ft. lb.).

9. Route the selector lever ground wire along the frame tunnel and under the rear seat, then join it at the in-line connector.

10. If necessary, adjust the selector lever contacts as described in **4.3 Cleaning and Adjusting Selector Lever Contacts.**

11. Bind the selector lever ground wire to the threaded sleeve and to the mounting as indicated in Fig. 12-6.

Fig. 12-6. Electrical tape binding the ground wire.

12. Carefully pull down the rubber boot and install it over the flange on the gearshift lever's mounting bracket.

13. Tanks

Cars with Automatic Stick Shift have two tanks that are related to the transmission. One is the vacuum tank and the other is the ATF (automatic transmission fluid) tank. Because of the redesigned bumper mountings on 1974 Beetles and Super Beetles, the tanks for these models have been modified. Since the new tanks cannot be service installed on earlier cars, the earlier type tanks are being kept available as replacement parts.

13.1 Removing and Installing Vacuum Tank

The vacuum tank is mounted under the left rear fender on Beetles and Super Beetles. On Karmann Ghia models it is in the engine compartment, at the left.

To remove:

1. Loosen the clamps on the vacuum hoses, then pull the hoses off the tank.

AUTOMATIC STICK SHIFT 41

2. Remove the tank by taking out the bolts indicated in Fig. 13-1.

Fig. 13-1. Retaining strap bolt (lower arrow) and mounting bracket bolt (upper arrow).

3. Remove the bolts that hold on the vacuum tank mounting bracket on the car.

Installation is the reverse of removal.

13.2 Removing and Installing ATF Tank

The ATF tank is mounted under the right rear fender on Beetles and Super Beetles. On Karmann Ghia models it is in the engine compartment, at the right.

To remove:

1. Except on cars with fuel injection, disconnect the ATF return line from the ATF filler as indicated in Fig. 13-2.

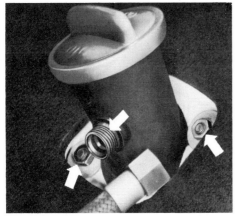

Fig. 13-2. ATF hose removed from connection (center arrow). Outer two arrows indicate nuts that hold trim plate and rubber seal on 1970 through 1974 cars.

2. On 1970 through 1974 cars, remove the nuts indicated in Fig. 13-2 in order to free the filler neck trim plate and rubber seal. On later models, loosen the large filler hose from the ATF tank neck.

3. Disconnect the pump suction line from the ATF tank, draining off the ATF as the union is loosened. See Fig. 13-3.

Fig. 13-3. Union for ATF suction line (lower arrow) and retaining strap bolt (upper arrow).

4. Remove the retaining strap bolt indicated in Fig. 13-3, then take off the ATF tank.

5. Remove the bolts that hold the ATF tank mounting bracket on the car.

Installation is the reverse of removal. Make sure the tank has been filled before starting the engine. After the engine has run for a few minutes, check and correct the ATF level as described in **LUBRICATION AND MAINTENANCE**.

14. AUTOMATIC STICK SHIFT TECHNICAL DATA

I. Ratios

	No. of teeth	Ratio
1st gear	35/17	2.06
2nd gear	29/23	1.26
3rd gear	24/27	0.89
Reverse	43/14	3.07
Converter: maximum torque multiplication	—	2.1
Final drive	8/33	4.125

42 Automatic Stick Shift

Rear Suspension and Rear Wheel Alignment Data

The rear suspension is the same on cars with either Automatic Stick Shift or manual transmissions. For rear suspension repair precedures, rear wheel alignment procedures, and technical data, consult **TRANSMISSION AND REAR AXLE**.

II. Tolerances, Wear Limits, and Settings

Designation	Automatic Stick Shift	
	New part mm (inch)	Wear limit mm (inch)
A. Gears, drive pinion		
1. 2nd gear .... end play	0.10-0.25 (.004-.010) keep to lower limit	—
2. 3rd gear .... end play	0.10-0.25 (.004-.010) keep to lower limit	—
3. Synchromesh units clearance **a** between coupling teeth and synchronizer ring 1st gear .... clearance	1.10-1.80 (.043-.071)	0.60 (.024)
2nd/3rd gears .... clearance	1.10-1.90 (.043-.075)	0.60 (.024)
4. Shift fork/operating sleeves for 1st and 2nd/3rd gears .. end play	0.10-0.30 (.004-.012)	—
sleeves with wider slots .... end play	0.50-0.85 (.020-.033)	—
5. Preload of pinion tapered-roller bearing Turning torque .... new	6-21 cmkg (5.2-18.2 in. lb.)	—
Turning torque .... used (30 mi [48 km] or more)	3-7 cmkg (2.6-6.1 in. lb.)	—
B. Gearbox and gearshift housing		
1. Preload of adjusting rings on tapered-roller bearings Turning torque .... new	18-22 cmkg (15.6-19.1 in. lb.)	—
Turning torque .... used (30 mi. [48 km] or more)	3-7 cmkg (2.6-6.1 in. lb.)	—
2. Shift rod shift pressure	6.5 kg (14 lb.)	—
3. Gearshift housing bushings .... inside diameter	15.03-15.05 (.592-.593)	15.25 (.600)
4. Inner shift lever .... diameter	14.96-15.00 (.589-.591)	14.75 (.581)
5. Starter bushing .... inside diameter	12.55-12.57 (.494-.495)	12.65 (.498)
6. Starter shaft/bushing .... radial clearance	0.09-0.14 (.0035-.006)	0.25 (.010)

III. Tightening Torques

Location	Designation	mkg	ft. lb.
Automatic Stick Shift			
Clutch housing	temperature switch	2.5	18
Transmission case	selector switch	2.5	18
Gearshift housing	neutral safety switch	2.5	18
Converter/drive plate	socket head screw	2.5	18
Gears/transmission case	retaining nut	22.0	160
Lockplate/retaining ring	tapping screw	1.0	7
Gear carrier/transmission case	nut	2.0	14
Lockplate/adjusting ring	fillister head screw	1.0	7
Gearshift housing/transmission case	nut	1.5	11
Converter housing/transmission case	nut	2.0	14
Cover/transmission case	fillister head screw	1.0	7
Bonded rubber mounting/converter housing	nut	2.0	14
Transmission mounting/bonded rubber mounting	nut	2.0	14
Selector shaft/fork	bolt	2.5	18
Bearing lock bolt	bolt	1.0	7
Pinion	round nut	20.0	145
Ring gear	bolt	4.5	32
Gearshift housing/bonded rubber mounting	nut	3.5	25
One-way clutch support tube/converter	socket head screw	1.5	11
Clutch/clutch carrier plate	socket head screw	1.5	11
Oil feed line/transmission	union	3.5	25
Oil return line/transmission	union	3.5	25
Oil suction line/tank	union	3.5	25
Clamp bolt/clutch lever	bolt	3.0	22
Joint to flange	socket head screw	3.5	25

© 1974 VWoA

Section 10

LUBRICATION AND MAINTENANCE

Contents

Introduction 3	Checking Distributor 10
1. Test Equipment 4	Checking and Adjusting Engine Idle 11
2. Lubricants 4	Checking Ignition Timing 11
Oil Viscosities 5	Checking Exhaust System 11
Greases 5	Servicing Air Cleaner 11
3. VW Maintenance Schedules for 1975 and Later Cars 5	Servicing Exhaust Gas Recirculation Filter 12
3.1 Oil Change Service 5	Replacing Activated Charcoal Filter 12
3.2 Brake and Valve Adjustment 5	Checking Clutch Pedal Freeplay (Manual Transmission only) 12
3.3 Scheduled Vehicle and Emission Control Maintenance 5	Checking Freeplay Clearance at Servo Rod (Automatic Stick Shift only) 12
3.4 Scheduled Vehicle and Emission Control Checks 6	5.2 Manual Transmission and Automatic Stick Shift 13
4. VW Maintenance Schedules for 1970 through 1974 Cars 7	Checking Hypoid Oil and Cleaning Magnetic Plugs 13
4.1 Oil Change Service 7	Changing Hypoid Oil 13
4.2 Scheduled Vehicle and Emission Control Checks 7	Checking Constant Velocity Joint Boots . 13
Test Drive 8	Checking Constant Velocity Joint Screws 13
4.3 VW Maintenance 8	Checking and Correcting ATF Level (Automatic Stick Shift only) 13
5. Lubrication and Maintenance Operations 9	Lubricating Rear Wheel Bearings (except 1973 and later models) 14
5.1 Engine and Clutch (including fuel system) 9	Checking Shock Absorbers 14
Changing Engine Oil and Cleaning Oil Strainer 9	Checking Transmission and Final Drive for Leaks 14
Checking and Adjusting Generator or Alternator Belt 9	Checking Pan Screw Torque (Automatic Stick Shift only) 14
Checking Compression 10	5.3 Front Axle and Steering 14
Servicing and Replacing Spark Plugs .. 10	Lubricating Front End 14
Checking and Adjusting Valve Clearance 10	Checking Steering Play 15
	Checking Ball Joint and Tie Rod Seals . 15
	Checking Ball Joint Play 15
	Lubricating Front Wheel Bearings (except 1973 and later models) 15

©1978 VWoA

2 Lubrication and Maintenance

	Checking Wheel Alignment	15
	Checking Shock Absorbers	16
5.4	Brakes and Wheels	16
	Checking and Changing Brake Fluid	16
	Checking Brake Pedal Travel and Freeplay	16
	Checking Brake Linings	16
	Checking Parking Brake Adjustment	16
	Checking Brake System for Leaks	16
	Checking Brake Lines and Hoses	16
	Checking Brake Light Operation	16
	Checking Brake Light Warning Switch	16
	Checking Tire Condition and Tire Pressures	16
5.5	Electrical System	17
	Checking Lights	17
	Checking Windshield Wipers and Washers	17
	Checking and Filling Battery	17
5.6	Body and Frame	17
	Lubricating Door and Hood Hinges	17

	Lubricating Locks	18
6. Basic Car Care		18
6.1	Care of Car Finish	18
	Washing	18
	Waxing	18
	Polishing	18
	Washing Chassis	18
	Special Cleaning	18
6.2	Care of Interior	18
	Cloth Upholstery and Carpet	18
	Leatherette Upholstery and Trim	18
6.3	Tires and Accessories	18
	Accessories	19

TABLES

a.	Engine Oil Viscosities	5
b.	Conventional Tire Inflation Pressures (bias ply)	17
c.	Radial Tire Inflation Pressures (radial ply)	17

Lubrication
And Maintenance

The service life of your car depends on the kind of maintenance it receives. The procedures described in this section of the Manual include all periodic checks and maintenance steps necessary for long and reliable operation as well as instructions for basic car care. Washing your VW and cleaning its interior need not be done at specific intervals of course, but doing so will keep your VW looking good.

The Owner's Manual and the VW Maintenance Record — or Owner's Manual and Warranty & Maintenance booklet — originally supplied with the car contain the analysis and maintenance schedules that apply to your VW. Following these schedules will ensure safe and dependable operation. Because several model years are covered in this Manual, some of the procedures described may not apply to your car. If you are in doubt, always take the Owner's Manual, the Maintenance Record, or the Warranty & Maintenance booklet as your guide.

Some maintenance procedures, such as oil change service, require no special tools and can be carried out by almost all car owners, regardless of their mechanical experience. However, certain other analysis and maintenance operations require tools and equipment specifically designed for those operations. Wheel alignment checks, ignition timing, generator testing, and idle speed adjustments are a few examples. If you lack the skills, tools, or a suitable workshop for performing any of the service steps described, we suggest you leave this work to an Authorized VW Dealer or other qualified shop. We especially urge you to consult your Authorized VW Dealer before attempting any repairs on a car still covered by the new-car warranty.

4 Lubrication and Maintenance

1. Test Equipment

The test equipment shown in this Manual is the kind of equipment that is often found in VW workshops. The methods described for using these pieces of equipment apply only to the equipment that is shown or described. Many car owners and mechanics have other kinds of test equipment. Their equipment should be connected to the vehicle in the manner that is recommended by the test equipment's manufacturer.

Expensive and complicated equipment is not necessary for routine maintenance. However, it is necessary to use a stroboscopic timing light when you check the engine's ignition system. Static timing is not suitable to the kinds of distributors used on the cars covered by this Manual. Additional information on timing lights and their applications can be found in **ENGINE**.

Many of the cars covered by this Manual are equipped with a system of sensors and test wiring that terminates in a central control socket located in the engine compartment. This socket is designed to receive a plug from the cable of the VW Computer Analysis system. The socket should not be used in conjunction with other kinds of test equipment.

CAUTION—
Never connect any device other than the test plug of the VW Computer Analysis system to the test network central socket in the engine compartment. Incorrect equipment may damage the plug connections, test sensors, or the vehicle components containing them.

2. Lubricants

Because of recent improvements in the quality of commercially available lubricating oils, completely new oil recommendations for VW cars became effective in 1975. These new oil recommendations, given below, should be applied to all the VWs covered by this Manual.

The superseded oil recommendations, given in the Owner's Manuals supplied with 1970 through 1974 VWs, can still be used in servicing 1970 through 1974 models. You will find, however, that there are many advantages to be gained in adopting the new recommendations. Only the new oil recommendations, given in this Manual, should be applied to 1975 and later VWs.

The lubricants used in your VW have a vital influence on its operation. Use only name brand oils labeled "For Service SD" (or "For Service SE" or both) in the engine. Oils used in 1975 and later cars must be labeled "For Service API/SE". Automatic transmission fluid (ATF) must be labeled DEXRON®. The hypoid oil used both in the manual transmission and in the Automatic Stick Shift must meet specification MIL-L 2105 API/GL4. No additives should be used in the engine oil, hypoid oil, or ATF. Experience has shown that name-brand lubricants of the correct specification and viscosity meet all operating needs of VW engines and transmissions.

Oil viscosity must be suitable to climatic conditions. Viscosity is a term used to describe how readily a liquid flows. High viscosity oils seem thicker and pour more slowly at room temperature than do low viscosity (thinner) oils. When heated, however, oil loses some of its viscosity. A high viscosity oil heated to 93°C (200°F) may pour as readily as a low viscosity oil does at room temperature. If an oil has too low a viscosity, it will not maintain an adequate lubricating film between moving parts. A thin, low viscosity oil may maintain this film at low temperatures but become so much thinner after it has warmed up that it leaves the engine parts unprotected.

It might seem that a high viscosity oil is all that is necessary to properly lubricate an engine. Unfortunately, this is not true. If a high viscosity oil is used during cold weather, it will become so thick and resistant to flow that it cannot properly circulate and reach the parts of the engine requiring lubrication. A thick, high viscosity oil will also become so gummy in cold weather that the starter cannot turn the engine fast enough to start it. The proper viscosity oil will remain fluid enough after the engine has cooled to permit easy starting, yet after the engine has reached operating temperature will retain sufficient viscosity to maintain an adequate lubricating film.

Single-grade engine oils, such as SAE 30, were formerly recommended for use in VW engines because of the unreliable quality of the then available multi-grade engine oils. The new high standard for engine oils that conform to the API (American Petroleum Institute) service rating API/SE has made such multi-grade oils suitable for use in the VW engine. Car owners will find that these high-quality multi-grade oils offer many advantages in convenience, performance, and economy—even in older model VWs.

For example, a single-grade oil may have to be discarded after a short period of service owing to the early arrival of winter temperatures. A multi-grade oil, suitable for both summer and winter temperatures, can be left in the engine until the normal oil change mileage has been reached. This feature of multi-grade oils can save the expense of oil changes necessitated by climatic conditions.

Also to be considered is that some oil grades and ratings, as originally recommended for 1970 through 1974 models, may eventually become unavailable. As old stocks of motor oil become depleted, new containers bearing the identifying marks of the new rating system will

LUBRICATION AND MAINTENANCE

replace those bearing the identifying marks of older systems. By becoming familiar with the new VW oil recommendations, you can be sure that the oil you buy is the correct kind for your car. Inferior lubricants, no matter how attractively priced, are not a good investment; using the wrong oil will greatly shorten the service life of your VW.

Oil Viscosities

The viscosity grade of oil is designated by an SAE (Society of Automotive Engineers) standard number. An oil designated SAE 40 has a higher viscosity (greater resistance to flow) than an oil designated SAE 30. Multi-grade oils have an extended viscosity range and can be used in place of a number of single-grade oils. For example, an SAE 10W-30 oil is suitable for use within a range of temperatures that would require three different single-grade oils in order to cover it (SAE 10W, SAE 20W/20, and SAE 30).

Table a lists the proper oil viscosity for VW engines under specific climatic conditions. The SAE viscosity number of the oil should be selected for the lowest anticipated temperature at which engine starting will be required, and not for the temperature at the time of the oil change. Because the temperature ranges of the different oil grades overlap, brief variations in outside temperatures are no cause for alarm.

Table a. Engine Oil Viscosities

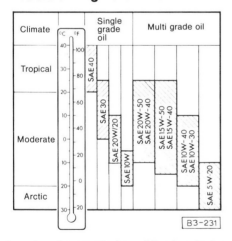

The viscosity of transmission oil is also designated by SAE numbers. Use SAE 80W or SAE 80W/90 for general year-round service in both the Automatic Stick Shift transmission and in the manual transmission. ATF is not graded for viscosity and requires no seasonal change when used in the Automatic Stick Shift transmission's torque converter. However, in arctic areas where temperatures remain consistently below −25°C (−13°F), you can also use ATF in place of hypoid oil in the manual transmission and in the gearcase of the Automatic Stick Shift transmission. The ATF should be replaced promptly with hypoid oil if summer temperatures exceed −25°C (−13°F).

Greases

Two types of grease are used for lubrication of chassis and driveline parts. Multipurpose grease (lithium grease) has a wider temperature tolerance range than ordinary grease and should be used for most lubrication purposes. The additives in multipurpose grease give it increased pressure resistance and anticorrosion capabilities. It is suitable for use both in plain bearings and in roller bearings. Molybdenum grease is lithium grease with a friction-reducing molybdenum disulfide additive. It is used in the constant velocity joints of the rear axle. Use dry stick lubricant on the hood locks and on the sliding surfaces of the door striker plates. The battery terminals should be coated either with silicone spray or petroleum jelly.

3. VW Maintenance Schedules for 1975 and Later Cars

The VW Maintenance schedules given here should be followed carefully even if your car is not being serviced by your Authorized VW Dealer.

3.1 Oil Change Service

Change the engine oil every 5000 mi. (8000 km) on 1975 and 1976 cars or 7500 mi. (12,000 km) on 1977 and later cars. Change the oil every six months if you drive less than the specified distance during that time. Under severe driving conditions, you may have to change the oil more frequently than scheduled. An oil change includes draining the old oil, cleaning the oil strainer, and refilling the crankcase with new oil of the correct grade.

3.2 Brake and Valve Adjustment

The brakes and the engine valves of 1975 and 1976 cars should be checked and, if necessary, adjusted after 10,000 mi. (16,000 km) of driving. On 1977 and later cars, perform these services at 7500 mi. (12,000 km). Repeat these services on 1975 and later cars at each additional 15,000 mi. (24,000 km) thereafter. Brake adjustment includes (1) checking the brake system for damage and leaks, (2) checking the fluid level and, if necessary, adding brake fluid, and (3) adjusting both the foot brakes and the parking brake. The valve clearance adjustment specified at these mileage intervals is recommended for maximum engine life but is not necessary for maintaining the VW Emission Control System Warranty.

3.3 Scheduled Vehicle and Emission Control Maintenance

To keep your Emission Control System Warranty in effect, all maintenance services affecting vehicle emission control have to be performed every 15,000 mi. (24,000 km) or once a year—if you drive less than 15,000 mi. (24,000 km) within 12 months. For additional information,

6 Lubrication and Maintenance

please refer to the Emission Control System brochure supplied with the car. The following emission and vehicle maintenance schedule applies to 1975 and later VWs.

1. Engine: Change oil, clean oil strainer.
2. Valves: Adjust clearance.
3. Spark plugs: Replace.
4. Ignition distributor: Replace ignition points. Adjust dwell angle and timing. Visually check ignition wires, distributor cap, and distributor rotor. Replace damaged or deteriorating parts.
5. Compression: Check.
6. V-belt: Check tension and condition; replace worn-out belts.
7. Fuel filter: Replace.
8. EGR (exhaust gas recirculation) system: Check condition and operation visually.
9. Crankcase ventilation hoses: Check visually.
10. Exhaust system: Check for damage.
11. Fuel tank and EEC (evaporative emission control) system: Visually check fuel filler cap to see that it seals correctly; visually check charcoal filter canister and hose connections.
12. Clutch: Check and, if necessary, adjust freeplay.
13. Transmission: Check oil level and, if necessary, add oil.
14. Automatic Stick Shift: Check and, if necessary, correct ATF (automatic transmission fluid) level.
15. Brakes: Check and, if necessary, adjust.
16. Door hinges: Lubricate.
17. Front axle: Lubricate (necessary only on the VW 111 Sedan (Beetle for 1976) which has a torsion bar front axle).

NOTE —
Road test the car after you have completed the above maintenance work. During the road test you should check brake operation and observe the efficiency of the heating and ventilation systems. After the road test, check and, if necessary, adjust the engine idle speed and exhaust CO level. Check to see that the cylinder head covers are not leaking oil.

In addition every 30,000 mi. (48,000 km):

1. Air cleaner: Replace filter element.
2. Vacuum passages in throttle valve housing: Clean (where applicable).

In addition every two years:

1. Brakes: Replace brake fluid.
2. Brake warning light switches: Check operation.

3.4 Scheduled Vehicle and Emission Control Checks

A physical checkup is extremely important for determining the amount of additional maintenance your car may need for continuing peak performance. The checkup should include the items listed below.

Engine

1. Compression: Check.
2. Ignition system and timing: Check.
3. Exhaust system: Check for damage.

Rear axle and transmission

4. Driveshafts: Check boots for leaks.

Front axle and steering

5. Front axle: Check dust seals on ball joints and dust seals on the tie rod ends. Check the tie rods.
6. Ball joints: Check play (VW 111 Sedan (Beetle for 1976) only).
7. Steering: Check play.
8. Steering gearbox: Check boots for leaks.
9. Front wheels: Check camber and toe.

Brakes, wheels and tires

10. Brake system: Check for damage and leaks.
11. Brake linings: Check thickness.
12. Brake fluid: Check level and, if necessary, add fluid.
13. Tires, including spare wheel: Check for wear and damage; check and, if necessary, correct pressures.

Electrical system

14. Starting current: Check with electronic equipment.

©1978 VWoA

Lubrication and Maintenance

15. Charging system: Check with electronic equipment.
16. Lights: Check all bulbs and switches for correct operation.
17. Headlights: Check adjustment.
18. Windshield wipers: Check operation.
19. Windshield washer: Check operation and, if necessary, add fluid.
20. Battery: Check electrolyte level; check voltage under load.
21. Safety belt warning light, buzzer and ignition/steering lock: Check.
22. Heater lever spot light: Check.
23. Interior lights and instrument lights: Check.
24. Warning lights for alternator and oil pressure: Check.

4. VW Maintenance Schedules for 1970 through 1974 Cars

The schedules given here list all the analysis and maintenance tasks specified for your VW. They should be followed carefully even if your car is not being serviced by your Authorized VW Dealer.

4.1 Oil Change Service

The engine in your VW requires little oil. But for long engine life, this oil should be changed every 3000 mi. (5000 km). An oil change at an Authorized VW Dealer includes the services listed below:

1. Engine: Change the oil, clean the oil strainer, check for leaks.

   NOTE
   When changing the oil, most mechanics also check the crankcase breather rubber valve on cars built prior to January 1972.

2. Battery: Check the electrolyte. If necessary, add distilled water. Clean and grease the terminals.
3. Windshield washer: Check the fluid.

4.2 Scheduled Vehicle and Emission Control Checks

(every 6000 mi. or 10,000 km)

A periodic checkup of your VW is extremely important in order to determine the amount of additional maintenance your car may need to ensure continued peak performance. The following VW schedule applies to the 1970 through 1974 VWs covered in this Manual.

©1978 VWoA

Engine and clutch (including fuel system):
1. V belt: Check the tension and condition.
2. Ignition system: Check the dwell and the timing with electronic equipment.
3. Throttle positioner: Check for proper functioning (where applicable).
4. Compression: Check.
5. Exhaust system: Check for damage.
6. Manual transmission: Check the clutch pedal freeplay.
7. Automatic Stick Shift: Check the freeplay clearance at the servo rod.
8. Engine: Check the engine oil level.

Transmission and rear axle:

Driveshafts: Check the boots for leaks.

NOTE
When checking the driveshaft boots, mechanics usually check the transmission for leaks, look for loose constant velocity joint screws, and inspect the shock absorbers.

Front axle and steering:
1. Cars with torsion bar front axles: Check the dust seals and the ball joint plug (where applicable) for a proper fit.
2. Cars with torsion bar front axles: Check for excessive ball joint play.
3. Steering: Check the play. Check the tie rods and the dust seals on the tie rod ends.

NOTE
When checking the front axle and steering, the shock absorbers and steering damper can be inspected and the steering gear checked for leaks.

Brakes, wheels, and tires:
1. Brake system: Check for damage and leaks.
2. Brake pedal: Check the pedal freeplay and the pedal travel (brake adjustment).
3. Parking brake: Check the adjustment.
4. Brake fluid: Check the level.
5. Brake linings or pads: Check thicknesses.
6. Tires (including spare): Check for wear and damage. Check and correct the pressure.

Electrical system:
1. Starting system: Check with electronic equipment.
2. Charging system: Check with electronic equipment.

8 LUBRICATION AND MAINTENANCE

3. Check the operation of the headlights, high beam indicator light, parking lights, side marker lights, license plate light, emergency flasher, stop lights, taillights, back-up lights, turn signals, horn, rear window defogger, and brake warning light.
4. Headlights: Check the adjustment.
5. Windshield wipers: Check the operation.
6. Windshield washer: Check the operation and the fluid level.
7. Battery: Check the electrolyte level. Check the voltage under load.

Test Drive

NOTE ——
If maintenance or repair is to follow diagnosis, delay the test drive until the completion of all repairs.

1. Check the braking, clutch, steering, heating, ventilation system (including fresh air fan), and overall vehicle performance.
2. Check the interior lights, the instrument lights, and—on 1973 and 1974 cars—the heater lever spot light.
3. Check the ignition/steering lock and the buzzer alarm.
4. Check the safety belt warning light and the buzzer alarm.
5. Check the warning lights for the generator and for oil pressure.
6. Check the operation of the Automatic Stick Shift transmission (where applicable). Check the ATF level following the test drive.

4.3 VW Maintenance

The following jobs can be performed either at the same time as the preceding diagnosis, after the diagnosis, or after VW Computer Analysis by an Authorized VW Dealer.

Every 6000 mi. (10,000 km):

1. Engine: Change the oil. Clean the oil strainer.
2. Valves: Check and adjust the clearance.
3. Ignition distributor: Check and adjust the dwell angle and the timing.
4. Engine idle: Check and, if necessary, adjust.
5. Door hinges and door checks: Lubricate.

NOTE ——
If necessary, lubricate the door and hood locks at this time.

6. Manual transmission: Check the oil level. Add oil if necessary.
7. Automatic Stick Shift: Check the oil level. Add oil if necessary. Check the torque of the pan bolts.
8. Front axle: Lubricate (1970 through 1972 cars with torsion bar front axles).
9. Test drive as described earlier. Then check the cylinder head covers for leaks.

In addition every 12,000 mi. (20,000 km):

1. Ignition distributor: Replace the breaker points. Then adjust the dwell angle and the timing.
2. Ignition system: Visually check the distributor cap and rotor.
3. Spark plugs: Replace.
4. Activated charcoal filter: Check visually.

In addition every 18,000 mi. (30,000 km):

1. Air cleaner: Clean and refill the lower part with oil. Or, where applicable, replace the paper filter element. Check the intake air preheating flaps.
2. Front axle: Lubricate (1973 and later cars with torsion bar front axles).

In addition every 24,000 mi. (40,000 km):

1. Exhaust recirculation filter (Automatic Stick Shift only): Clean on 1972 California cars, replace on later models (at least every two years).
2. Exhaust recirculation valve (Automatic Stick Shift only): Test.
3. Ignition wires, distributor cap, and distributor rotor: Check and, if necessary, replace.

In addition every 30,000 mi. (50,000 km):

1. Clean and repack the front wheel bearings (1970 through 1972 cars).
2. Clean and repack the rear wheel bearings (1970 through 1972 cars).

In addition every 48,000 mi. (80,000 km):

Activated charcoal filter: Replace (1970 cars sold in California and all 1971 and later models).

Every two years:

1. Brakes: Replace the brake fluid.
2. Brake warning light switch: Check functioning.

© 1976 VWoA

5. Lubrication and Maintenance Operations

The maintenance schedules given earlier can be used as a checklist when you are servicing your car. For your additional help, all the necessary specifications and instructions for performing the more complex checks and maintenance tasks follow.

5.1 Engine and Clutch
(including fuel system and fuel injection)

In addition to the steps already listed, you should change engine oil at 600 mi. (1000 km) following engine rebuilding. Doing so gets rid of any metallic particles that have accumulated in the oil during break-in.

Changing Engine Oil and Cleaning Oil Strainer

The oil should be drained from the engine while it is hot. Drain the oil by placing a pan of at least one gallon capacity beneath the engine. Loosen all six cap nuts on the oil strainer cover. After removing five of the nuts, pry the oil strainer cover loose. When most of the oil has drained, remove the oil strainer for cleaning as shown in Fig. 5-1.

The oil strainer should be cleaned with every oil change. To prevent oil leaks, use new gaskets and copper washers when installing the strainer. Wipe the area around the strainer following its installation. Doing so will aid in the early detection of leaks.

Fill the engine with 5.3 U.S. pints (2.5 liters, 4.4 Imperial pints) of oil labeled "For Service SD" (or "For Service SE", or both). On 1975 and later cars, use only oil marked "For Service API/SE". Consult **2. Lubricants** to determine the proper viscosity. After filling the crankcase, check the oil level with the dipstick. Pull the dipstick out and wipe it dry. Insert it all the way and pull it out again. The oil level should be at the upper mark.

After running the engine, look closely at the oil strainer cover to see that no oil is leaking past the gaskets. Replace the gaskets if you find leaks. On 1970 through early 1972 cars, check the condition of the rubber valve over the lower end of the crankcase breather tube. Replace the valve if it is torn or if its slot does not remain closed.

Checking and Adjusting Generator Or Alternator Belt

To check the generator or alternator belt tension, press down on the belt halfway between the generator and the crankshaft pulleys. The belt should deflect about 10 mm (3/8 in.). If deflection is significantly greater or smaller than this amount, remove the nut from the center of the generator or alternator pulley. Wedge a screwdriver between one of the notches in the rear of the pulley and one of the generator's through bolts to keep the pulley from turning. Separate the pulley halves. To tighten the belt, remove shim washers as necessary from between the pulley halves; to loosen the belt, add shim washers. Store leftover shim washers between the pulley outer half and the hat-shaped special washer (Fig. 5-2).

1. Gasket
2. Oil strainer
3. Gasket
4. Cover plate
5. Cap nut with washer
6. Oil drain plug with washer

Fig. 5-1. Oil strainer removal. Oil drain plug with washer (**6**) was discontinued beginning with the 1973 models.

Fig. 5-2. Belt tension adjustment. Transfer shim washers from one side of the pulley outer half to the other as indicated by the double arrow.

10 LUBRICATION AND MAINTENANCE

Checking Compression

To check the cylinder compression, remove all four spark plugs. Then install a compression testing gauge in one of the spark plug holes. Crank the engine with the starter for a few moments while pressing the accelerator to the floor. Note the gauge reading and repeat the procedure with the other cylinders. Compression pressure for cars built before 1972 should be 114 to 142 psi (8.0 to 10.0 kg/cm^2). For 1972 and later cars, compression should be 107 to 135 psi (7.5 to 9.5 kg/cm^2). If the pressure is below 100 psi (7.0 kg/cm^2) on 1970 and 1971 cars or below 85 psi (6.0 kg/cm^2) on 1972 and later cars, the valves may need to be ground, piston rings replaced, or new cylinders and pistons installed.

Before assuming the problem is in the valves, however, check the valve adjustment to make sure that there is at least a small amount of clearance between the rocker arms and the valve stems. If there is no measurable gap, loosen the adjusters enough to obtain clearance and repeat the compression test. If pressure is still low, the valves need grinding.

Servicing and Replacing Spark Plugs

The spark plugs, which were removed for the compression test, should not be reinstalled until after the valves have been adjusted. This will make it easier to turn the engine by hand. On 1970 through 1974 models, check the gap and the condition of the spark plug electrodes each 6000 mi. (10,000 km). Install new spark plugs each 12,000 mi. (20,000 km). The fuel injection engine introduced in 1975 uses long-life spark plugs. You need only replace such spark plugs each 15,000 mi. (24,000 km).

Adjust the gaps of new spark plugs before you install the plugs in the engine. The correct gap for 1970 through 1974 engines is 0.60 mm (.024 in.). The correct gap for fuel injection engines is 0.70 mm (.028 in.).

> **CAUTION** —
> Adjust the spark plug gap by bending the side electrode only. Attempting to bend the center electrode will break the insulator nose, ruining the spark plug.

The following or similar spark plugs are suitable for use:

Carburetor Engines, Normal Service: Champion L-88-A; Bosch W 145 T1; Beru 145/14.

Carburetor Engines, High-speed Service: Champion L-85 or L-86; Bosch W 175 T1; Beru 175/14.

Fuel Injection Engines, Normal Service: Champion L-288; Bosch W 145 M1; Beru 145/14.

Fouled plugs can be cleaned with a sharp stick, a plastic rod, or in a sandblast machine. Never sandblast an oil-fouled plug until after the oily deposit is removed with solvent. Do not clean plugs with a wire brush, which can leave electrically conductive "pencil marks" on the insulator nose. After cleaning or inspecting used spark plugs, file the electrodes to the square profiles of new spark plug electrodes before you regap and install the spark plugs.

Checking and Adjusting Valve Clearance

Valve adjustments should be checked while the engine is completely cold (oil temperature no more than 50°C [122°F]) and the cylinder is in position to fire. Clearance between the rocker arm and valve stem should be 0.15 mm (.006 in.). Complete step-by-step instructions for checking and adjusting the valve clearance can be found in **ENGINE**.

Checking Distributor

Before you check the distributor breaker point dwell (or gap) or adjust the ignition timing, there are several service checks that should be made. Begin by removing the distributor cap. Wipe it clean inside and out. Check to see that the carbon contact in the center of the cap is not worn away. Carefully inspect the cap and the rotor for carbon tracks. Use a magnifying glass if necessary.

Clean away grease, dust, or other foreign matter that may have accumulated inside the distributor. Check to see that the breaker points do not have excessive pitting or buildup. If a feeler gauge is used to check breaker point gap, first clean the gauge. Otherwise, oil or other foreign matter that may be adhering to it could contaminate the point contacts. Lubricate the distributor as shown in Fig. 5-3.

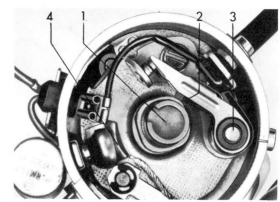

Fig. 5-3. Distributor lubrication. Apply a few drops of engine oil to the felt wick (**1**) and one drop to the pivot (**3**). After cleaning away old lubricant, put a small amount of multipurpose grease on the cam and rubbing block (**2**) and on the advance plate guide ball (**4**).

Complete information on replacing and adjusting the breaker points, testing the rotor electrically, and checking the high voltage cables can be found in **ENGINE**. The section also contains details for disassembling the distributor and checking the spark advance mechanisms.

Checking and Adjusting Engine Idle

For accuracy, adjust the idle speed after the road test. On 1970 through 1974 cars with Automatic Stick Shift, set the idle to 900 to 1000 rpm. For 1970 through 1974 cars with manual transmissions, set the idle to 800 to 900 rpm. Fuel injection cars with Automatic Stick Shift should idle at 850 to 1000 rpm. Fuel injection cars with manual transmissions should idle at 800 to 950 rpm. To adjust the idle, use only the procedures given in **FUEL SYSTEM** or in **FUEL INJECTION**.

Checking Ignition Timing

The breaker point dwell (or point gap) and the idle speed must be correct before you check or adjust the timing. Set the breaker point dwell to 44° to 50° for new points or 42° to 58° for used points. Alternatively, you can use a feeler gauge to set the point gap to 0.40 to 0.50 mm (.016 to .020 in.). By turning the bypass screw (see **FUEL SYSTEM** or **FUEL INJECTION**), adjust the idle of carburetor engines to 800 to 900 rpm. Adjust fuel injection engines with manual transmissions to 800 to 950 rpm; adjust fuel injection engines with Automatic Stick Shift to 850 to 1000 rpm. Never adjust the idle rpm by turning the screw in the throttle valve lever.

On cars that have a single vacuum hose attached to the distributor, disconnect and plug the hose. On other engines, with two hoses, leave both hoses attached. Adjust the timing, if necessary, so that, when illuminated by a stroboscopic timing light, the V-notch in the crankshaft pulley (the left V-notch on 1970 engines) is aligned with the central vertical seam in the crankcase. More detailed procedures, including pictures of the timing marks, can be found in **ENGINE**.

Checking Exhaust System

Momentarily cover the exhaust pipe openings with pads of rags while the engine is running. Hissing sounds coming from under the car signal leakage. Darker or lighter areas around joints in the system may indicate escaping gases.

Check the heater flaps for free movement. Lubricate the pivots with dry stick lubricant. If the flaps are difficult to move, treat them with penetrating oil.

Servicing Air Cleaner

To check the oil bath air cleaner, release the clips that hold the top part of the air cleaner. Do so by pulling their wire bails upward. Disconnect any hoses attached to the top part of the air cleaner, carefully noting their position for future installation. Remove the top part of the air cleaner, but do not lay it down with the filter up. Doing so allows contaminated oil to drain into the upper part of the unit.

Inspect the lower part of the air cleaner. If no more than 8 mm (5/16 in.) of oil remain above the sludge layer in the bottom of the air cleaner, the air cleaner must be removed, cleaned, and filled with fresh oil.

When you remove an oil bath air cleaner, hold it level to prevent spillage. Removal requires that the various hoses be disconnected from the air cleaner and that the clamp holding the air cleaner on the carburetor be loosened. Note carefully the location of each hose since accidentally interchanging them will adversely affect the operation of the engine.

On 1970 cars, the cable for the intake air preheating flap must be disconnected from the air cleaner. Begin by removing the spring clip that holds the cable eye on the warm air control flap arm. Unhook the cable eye from the arm. Then loosen the screw in the cable retainer. Pull the cable out of the retainer but leave it attached to the engine. During installation, adjust the cable as described in **FUEL SYSTEM**.

A solvent may be used for final cleaning after the sludge has been wiped out of the air cleaner's lower part. Fill the lower part of the air cleaner either to the baffle (Fig. 5-4) or to the mark, using fresh SAE 30 engine oil. SAE 10 W oil should be used in subarctic climates. Use the same viscosity oil all year round.

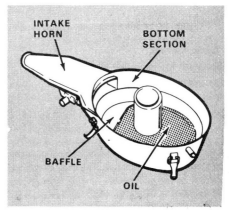

Fig. 5-4. Air cleaner filling level.

The top part of the air cleaner does not normally require cleaning. However, if the bottom part of the filter has become so dirty that the air inlet holes are partially blocked, you should clean them with a scraper. Never use solvent to wash the upper part of the air cleaner with-

12 LUBRICATION AND MAINTENANCE

out allowing the part to dry completely before reinstallation. Otherwise, dripping solvent will contaminate the oil in the bottom part.

Installation is the reverse of removal. However, when you service and install any type of air cleaner, check the intake air preheating system as described in **FUEL SYSTEM**. Be sure that the warm air control flaps move freely.

NOTE —
On 1971 and later Beetles and Super Beetles, hook the green vacuum hose behind the top release clip as shown in Fig. 5-5. This will keep the hose from contacting the moving generator or alternator belt.

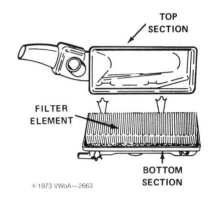

Fig. 5-6. Filter element positioned in bottom section of air cleaner.

Servicing Exhaust Gas Recirculation Filter

The exhaust gas recirculation (EGR) system on cars with Automatic Stick Shift sold in California during the 1972 model year has a cyclone filter that must be removed for cleaning each 24,000 mi. (40,000 km). The filter comes out of its housing by removing two bolts and the housing cover. Beginning with the 1973 model year, all U.S. cars with Automatic Stick Shift are equipped with the EGR system, but with a different filter. Rather than being disassembled for cleaning, the new filter is replaced after each 24,000 mi. (40,000 km) or two years, whichever comes first. The new-type filter installs readily in place of the cyclone filter to simplify servicing. There is no filter in the EGR system used on fuel injection engines.

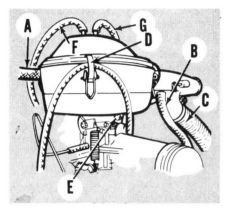

A. Hose from charcoal filter
B. Crankcase ventilation hose
C. Preheated intake air hose
D. Release clip supporting green vacuum hose
E. Air cleaner clamp screw
F. Vacuum hose from carburetor
G. Vacuum hose to intake air valve

Fig. 5-5. Hoses connected to late-type air cleaner.

The paper element air cleaner introduced in 1973 does not need to be removed from the engine when the filter is cleaned or replaced. To remove the filter element, pry off the four release clips. Then lift the top section of the air cleaner. Remove the filter element and shake off its accumulation of dirt. Tap the filter element—dirty side down—on a hard surface. Install a new filter element each 18,000 mi. (30,000 km) on carburetor engines or each 30,000 mi. (48,000 km) on fuel injection engines.

CAUTION —
Do not clean the filter element with solvents or gasoline or saturate it with oil. These practices will destroy its effectiveness.

With the top section of the air cleaner lifted, as illustrated in Fig. 5-6, install the filter element in the position shown. Replace the original filter element if it is damaged, clogged with dust, or if it cannot be cleaned by shaking.

Replacing Activated Charcoal Filter

Except on 1975 and later models, the activated charcoal filtering canister for the evaporative emission control system must be replaced after each 48,000 mi. (80,000 km) or two years, whichever comes first. The filter is located under the right-hand rear fender on all Beetle models and at the right-hand side of the engine compartment on Karmann Ghias. To remove it for replacement, detach the hoses and unscrew the mounting bracket that holds the canister.

Checking Clutch Pedal Freeplay
(manual transmission only)

It should be possible to depress the clutch pedal for a distance of 10 to 20 mm (⅜ to ¾ in.) before feeling resistance. If the freeplay is outside this range, adjust it as described in **ENGINE**.

Checking Freeplay Clearance at Servo Rod
(Automatic Stick Shift only)

To check and adjust the freeplay clearance at the servo rod on Automatic Stick Shift transmissions, follow the procedure given in **AUTOMATIC STICK SHIFT**. Instructions for making a clearance-measuring gauge are also given in that section.

5.2 Manual Transmission and Automatic Stick Shift

Both the manual transmission and the Automatic Stick Shift transmission must be kept properly filled with hypoid gear oil. Two types of gear oil are commonly available. The first is high pressure transmission oil, commonly designated HP for high pressure. Hypoid transmission oil suitable for VW cars is designated EP for extreme pressure. The hypoid oil used should meet specification MIL-L-2105 API/GL4. Hypoid oils meeting this specification have a sulphur-phosphorus additive base that creates a protective coating on gear surfaces.

Checking Hypoid Oil and Cleaning Magnetic Plugs

The hypoid oil level in both manual and Automatic Stick Shift transmissions should be kept at the lower edge of the filler hole in the side of the transmission case. The level is correct if hypoid oil just barely drips from the hole when the plug is removed. The level may be considered satisfactory if you can feel it with your fingertip just below the edge of the filler hole. Hypoid oil is added, if necessary, through the oil filler hole (Fig. 5-7).

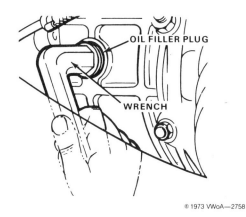

Fig. 5-7. Oil filler plug being removed from oil filler hole. A 17-mm hexagonal wrench is required.

The drain plug in the bottom of the manual transmission contains a magnet that traps metallic particles as they settle in the oil. The accumulation can be cleaned from the magnet periodically by removing the plug. Have a cork of appropriate size ready to plunge into the transmission drain hole as soon as the magnetic plug is removed. Very little hypoid oil will be lost if you leave the filler plug installed so that air cannot enter readily. After cleaning the magnetic plug and reinstalling it, correct the hypoid oil level to make up for any hypoid oil that was lost.

Changing Hypoid Oil

It is unnecessary to change the hypoid oil unless it has become contaminated, a temperature change has made it necessary to use an oil of a different viscosity or if the transmission has been rebuilt. The hypoid oil should always be changed 600 mi. (1000 km) after rebuilding the transmission.

The oil will drain faster if it is warm. On manual transmissions, remove both the filler plug and the drain plug. On Automatic Stick Shift transmissions, remove the oil pan. Install the drain plug after drips have slowed to one every 20 seconds. Use a new gasket when you reinstall the oil pan on Automatic Stick Shift transmissions. Torque the pan screws to 1.0 mkg (7 ft. lb.).

Refill the transmission with hypoid oil of the proper specification and viscosity (see **2. Lubricants**). While refilling, pause occasionally to give the oil time to flow into the final drive portion of the case. If you attempt to fill the transmission too rapidly, the oil may overflow after only 2 or 3 pints have been put in and give the impression that the case is already full. The manual transmission requires 5.3 U.S. pints (2.5 liters or 4.4 Imperial pints). The Automatic Stick Shift transmission requires 6.3 U.S. pints (3.0 liters or 5.4 Imperial pints).

Checking Constant Velocity Joint Boots

The rubber boots over the constant velocity joints should not be cracked or torn. Instructions for removing the driveshafts and replacing the boots can be found in **TRANSMISSION AND REAR AXLE**.

Checking Constant Velocity Joint Screws

The socket head screws that hold the constant velocity joints to the flanges should be torqued to 3.5 mkg (25 ft. lb.).

Checking and Correcting ATF Level
(Automatic Stick Shift only)

The dipstick for the ATF tank is inside the ATF filler at the right-hand side of the engine. The dipstick is attached to the ATF filler cap. To check the ATF level, remove the cap and withdraw the dipstick. The engine should not be running. Wipe the dipstick. Then reinsert and withdraw it again. The fluid level should be between the two marks and should never fall below the lower mark. If necessary, add ATF to the tank. Check for leaks if the level is seriously low.

The ATF does not have to be changed unless it has been contaminated by material from a damaged pump, a damaged torque converter bearing, or a deteriorating hose. In such cases, the torque converter and the ATF tank should both be removed, drained, and cleaned as described in **AUTOMATIC STICK SHIFT**.

14   Lubrication and Maintenance

Lubricating Rear Wheel Bearings
(except 1973 and later models)

After 30,000 mi. (50,000 km) of service, the rear wheel bearings should be removed, cleaned, and packed with fresh multipurpose grease. The rear wheel driveshafts and constant velocity joints must be removed first. The joints should be checked at this time to see that they work smoothly and do not feel gritty. If the joints feel stiff or gritty, disassemble and clean them and then pack them with molybdenum grease as described in **TRANSMISSION AND REAR AXLE**.

Remove the brake drums, bearing covers, and brake backing plates. Drive or press the rear wheel shafts out of the rear wheel bearings.

NOTE
There is one roller bearing (outer) and one ball bearing (inner) at each rear wheel.

Withdraw the ball bearings and the roller bearings from the bearing housing in the rear suspension diagonal arm. This and the preceding operations are described in **TRANSMISSION AND REAR AXLE**.

Carefully clean the bearings in solvent and dry them with compressed air. Clean the bearing housing and other related parts. Pack the bearing races with multipurpose grease. A pressure bearing lubricator is best for this, but the bearings can be packed by hand. Work the grease in between the balls or rollers, then turn the bearing slightly and repeat the process until the bearing is completely filled with grease. Reassemble the parts, replacing the oil seal in the bearing cover so that no grease can reach the brake linings.

Checking Shock Absorbers

If the shock absorbers seem to lack control during the road test or if you can see that the shock absorbers are damaged, check them and, if necessary, replace or repair them.

The best way to check a shock absorber is to remove it from the car and to extend and compress it by hand. Damping action should be uniform and extend throughout the entire range of travel. If possible, compare it with a new shock absorber. Faulty shock absorbers should be replaced. Leakage around the rod does not demand that the shock absorber be replaced unless its action has been impaired. Shock absorbers can be replaced individually as long as the new shock absorber is the same type as the old one.

Checking Transmission and Final Drive for Leaks

A leak from the transmission (or final drive) should be considered serious if there is wet, fresh oil present at all times. Such leaks should always be corrected by installing new gaskets or seals and never by tightening screws or bolts above their specified torque. The best time to check for leaks is just after a road test.

Checking Pan Screw Torque
(Automatic Stick Shift only)

Each 6000 mi. (10,000 km) torque the Automatic Stick Shift oil pan screws to 1 mkg (7 ft. lb.).

CAUTION
Never tighten pan screws over 1 mkg (7 ft. lb.) in an attempt to cure a leaking gasket. Overtightening will deform the pan and make it impossible to get a good seal. Always install a new gasket to correct leaks.

5.3 Front Axle and Steering

Most of the maintenance steps involving the front axle can be performed with the car on a lift. It is therefore convenient to do these jobs while you drain the engine oil or service the transmission.

Lubricating Front End

The torsion bar front axle has four grease fittings (Fig. 5-8). On 1970 through 1972 models, the front axle should be greased every 6000 mi. (10,000 km); on 1973 and 1974 models, the front axle should be greased every 18,000 mi. (30,000 km). On 1975 and later cars, grease the axle every 15,000 mi. (24,000 km).

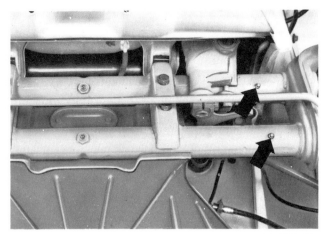

Fig. 5-8. Front axle grease fittings. The wheels must be off the ground while the axle is being greased.

Wipe the grease fittings clean before pumping grease into them. Pump until fresh grease appears in the joint between the torsion arm and axle beam. Afterward, wipe all surplus grease from the fittings and the suspension parts.

Lubrication and Maintenance

Check the oil in the steering gearbox by removing the access plate and prying out the plastic plug in the steering gearbox cover. The oil level in the steering gearbox should be at the bottom edge of the hole. Add transmission oil, if necessary, and install a new plastic plug.

Checking Steering Play

On 1975 and later cars with rack and pinion steering, there should be no steering play. See **STRUT FRONT SUSPENSION**. On other models, which have worm and roller steering, there should be no more than 25 mm (1 in.) of play measured at the rim of the steering wheel when the steering is centered and the front wheels are on the ground. If play exceeds this, adjust the steering as described in **FRONT AXLE** or in **STRUT FRONT SUSPENSION**.

Checking Ball Joint and Tie Rod Seals

Check the rubber seals on the suspension ball joints and tie rod ends to make certain that none are torn or cracked. To replace seals, use the procedures given in **FRONT AXLE**. Also check that the plastic plugs in the tops of the 1970 through early 1973 suspension ball joints have not been lost. Procedures for lubricating the ball joints are given in **FRONT AXLE**, but lubrication is required only if broken seals have allowed dirt to enter or if the joints have become stiff.

Checking Ball Joint Play

To check the play in the upper and lower ball joints for the front suspension, use the procedures described in **FRONT AXLE**. Play specifications are also given there.

Lubricating Front Wheel Bearings

(except 1973 and later models)

After 30,000 mi. (50,000 km) of service, the front wheel bearings should be removed, cleaned, and packed with fresh multipurpose grease.

CAUTION —
Use only multipurpose (lithium) grease to lubricate the wheel bearings. Other greases will not maintain adequate lubrication and may lead to bearing failure.

Loosen the socket head clamping screw and remove the clamp nut from the stub axle. Then, by hand, pull the brake drum off the stub axle. Exercise care to avoid dropping the thrust washer and outer wheel bearing race. The outer bearing race can be lifted out of the hub. Before the inner bearing race can be removed, however, it is necessary to pry the oil seal out of the back of the hub. Install a new oil seal during assembly.

Clean the hub cavity with solvent and inspect the condition of the bearing races that are inside the hub. These steel cups should not be roughened, worn irregularly, or blued by heat. If they are, replace them as described in **FRONT AXLE**. Wash the roller bearing races in solvent and dry them carefully with compressed air.

CAUTION —
Do not use solvents such as gasoline that remove every trace of lubrication. Also, do not spin the races with a blast of compressed air. Without adequate lubrication, the bearings can be damaged by movement.

If a pressure bearing lubricator is not available, get a palmful of multipurpose grease and thrust the edges of the roller bearing races into it, continuing around the bearing until it is completely filled. Coat the races inside the hub with the same lubricant. Do not pack large quantities of grease inside the hub; just coat the interior lightly to prevent corrosion. Clean the stub axle, inspecting its bearing seating surfaces to see that they are not roughened or blued by heat. Then apply a corrosion-preventing layer of multipurpose grease. Install the bearings and a new oil seal in the hub. Clean every trace of grease from the braking surface of the brake drum. Then install the drum on the stub axle.

The wheel bearings must then be adjusted. Install a dial indicator as shown in Fig. 5-9. The brake drum should be turned slightly and axial play measured at several different positions. The readings should not vary greatly. Their average should fall within a range of 0.03 to 0.12 mm (.001 to .005 in.). If necessary, bring axial play within this range by turning the clamp nut. See **FRONT AXLE** for additional instructions.

Fig. 5-9. Checking front wheel bearing axial play.

Checking Wheel Alignment

Check the camber and toe-in of the front wheels. Camber for cars with torsion bar front axles should be 0° 30′ ± 20′; camber for cars with strut front suspension should

16 Lubrication and Maintenance

be 1° + 20' or − 40'. Toe-in for cars with torsion bar front axles should be 30° ± 15', or 5' ± 15' with approximately 11 kg (24 lb.) of weight applied above the wheel. Toe-in for cars with strut front suspension should be 30° ± 15', or 10' ± 15' with approximately 11 kg (24 lb.) of weight applied above the wheel. If necessary, consult the detailed procedures given in **FRONT AXLE** and in **STRUT FRONT SUSPENSION**.

Checking Shock Absorbers

The instructions given for checking shock absorbers in **5.2 Manual Transmission and Automatic Stick Shift** also apply to the shock absorbers fitted to the front axle.

5.4 Brakes and Wheels

Properly servicing the brake system at the prescribed intervals is vital to the safe operation of your vehicle. Carrying out the following procedures will ensure full brake system efficiency.

Checking and Changing Brake Fluid

The fluid level in the brake fluid reservoir should be maintained at 15 to 20 mm (about ¾ in.) below the top of the reservoir (just above the seam near the top of the reservoir). Brake fluid added to the system must meet SAE recommendation J 1703 and conform to Motor Vehicle Safety Standard 116. The brake fluid should be changed every two years. Detailed procedures for this job are in **BRAKES AND WHEELS**. Also test the warning light switch at this time.

Checking Brake Pedal Travel and Freeplay

It should be possible to depress the brake pedal for a distance of 5 to 7 mm (³⁄₁₆ to ⁹⁄₃₂ in.) before encountering resistance. If you find that the freeplay is outside this range, adjust the pushrod clearance as described in **BRAKES AND WHEELS**.

If the pedal must be depressed an excessive distance to obtain braking action, the drum brakes require adjustment. For access to the adjusters, remove the rubber plugs from the brake backing plates. Consult the adjustment procedures given in **BRAKES AND WHEELS**.

Checking Brake Linings

On 1970 through 1974 models, brake lining wear is to be checked at the 6000-mi. (10,000-km) maintenance interval. On 1975 and later cars, brake lining wear is checked as part of the 15,000-mi. (24,000-km) VW Computer Analysis test sequence. Disc brake pads should not be worn to a remaining thickness of less than 2.00 mm (.080 in.). Drum brake linings should have at least 2.5 mm (³⁄₃₂ in.) of friction material left above the brake shoe. Lining thickness can be checked through holes in the backing plate when the rubber plugs have been removed.

Checking Parking Brake Adjustment

You should not be able to raise the parking brake lever more than five clicks without obtaining noticeable braking action. Make this check only after the rear wheel brakes have been adjusted. If the lever requires more than five clicks to obtain braking, adjust the parking brake as described in **BRAKES AND WHEELS**.

Checking Brake System for Leaks

If the fluid level in the brake fluid reservoir is noticeably low, carefully inspect the brake system for leaks. Check the lines, hoses, unions, and bleeder valves. If there is no sign of leakage, remove the brake drums and lift the rubber boots at the ends of each wheel cylinder. If fluid leaks out, the wheel cylinder is faulty and must be rebuilt or replaced as described in **BRAKES AND WHEELS**. On Karmann Ghias, if there is no leakage at the rear brakes, make a similar check of the disc brake calipers by lifting the dust seals away from the pistons. If there is leakage past the pistons, rebuild the caliper as described in **BRAKES AND WHEELS**.

Checking Brake Lines and Hoses

The brake lines must not be deeply pitted by corrosion, flattened due to bending, or dented by flying stones. These conditions may soon lead to leaks or may already be interfering with uniform braking action. The brake hoses must not be soft, swollen, or coated with grease and oil. They must not be abraded from rubbing against moving suspension parts or cut due to road hazards. Replace any brake hose that has breaks in its fabric outer cover even if bulges are not yet apparent. Details can be found in **BRAKES AND WHEELS**.

Checking Brake Light Operation

The rear brake lights should come on as soon as the brake pedal encounters resistance when depressed. If they do not come on, test the system as described in **BRAKES AND WHEELS**.

Checking Brake Light Warning Switch

Check the brake warning light system every two years (as described in **BRAKES AND WHEELS**) while changing the brake fluid.

Checking Tire Condition and Tire Pressures

Inspect the tires for cuts, separated plies, and abnormal wear. Replace any tires on which the tread wear indicators are showing. The causes of abnormal tire wear are described in **BRAKES AND WHEELS**. Keep 42 psi (3.0 kg/cm^2) of air in the spare tire. Inflate the tires on the road wheels to the following specifications.

Table b. Conventional Tire Inflation Pressures (bias ply)

	With 1-2 persons		Fully loaded	
	Front	Rear	Front	Rear
Beetle, Ghia, to Dec. 1972; 1970 Convertible	16 psi	24 psi	17 psi	26 psi
Super Beetle, Convertible from 1971 up to Dec. 1972	16 psi	27 psi	18 psi	27 psi
All models, from Jan. 1973	18 psi	29 psi	18 psi	29 psi

Table c. Radial Tire Inflation Pressures (radial ply)

	With 1-2 persons		Fully loaded	
	Front	Rear	Front	Rear
Beetle, Ghia to Dec. 1972; 1970 Convertible	18 psi	27 psi	18 psi	27 psi
Super Beetle, Convertible from 1971 up to Dec. 1972	18 psi	27 psi	18 psi	27 psi
All models from Jan. 1973	18 psi	29 psi	18 psi	29 psi

NOTE —
Increase the pressures in conventional bias ply tires by 3 psi for prolonged high speeds but never exceed the maximum tire inflation pressure designated on the tire sidewall.

5.5 Electrical System

Headlight adjustment can be checked by following the instructions given in **ELECTRICAL SYSTEM**. However, note that in some states there are legal specifications for headlight aim that may differ from those recommended by VW. In some areas it is also a legal requirement that headlights be adjusted only by state-authorized shops.

Checking Lights

VW Computer Analysis makes it possible to check most lights on the car in a few moments. Without the VW Computer Analysis system, each light and light system must be switched on and checked visually. Make certain that both the generator and oil pressure warning lights in the instrument panel light when the ignition is turned on and the engine is not running.

Check the operation of all electrical switches, including those in the door frames. Make certain that the turn signal and emergency flasher switches each produce the appropriate lighting action. Be sure to check the neutral safety switch on cars with Automatic Stick Shift transmissions. On 1970 through 1972 cars, the engine should start only with the selector lever in neutral. The 1973 and later models should also start in **P** (park). Also check the headlight dimmer switch. This switch is combined with the turn signal lever. Pulling the lever toward the steering wheel should actuate a relay that changes the headlights from high to low beam or from low to high beam.

Checking Windshield Wipers and Washers

On 1970 and 1971 models, the windshield wiper control switch is located on the dashboard. On 1972 and later cars, the wiper switch is a lever on the right-hand side of the steering column. Make certain that the wiper blades are in good condition. A troubleshooting chart for wiper defects can be found in **ELECTRICAL SYSTEM**.

The windshield washers work by air pressure from the spare tire. Keep the reservoir filled with water or a cleaning solution such as VW's Windshield Washer Anti-Freeze and Solvent. Mix as directed on the container. The reservoir holds 3½ U.S. pints (1.7 liters or 3 Imperial pints). Spare tire pressure should be kept at 42 psi (3.0 kg/cm^2).

Checking and Filling Battery

The battery is under the rear seat cushion on the right, as seen by a passenger riding in the car. If the battery terminals are corroded, correct the condition as described in **ELECTRICAL SYSTEM**.

Check the specific gravity of the electrolyte with a hydrometer. A reading of 1.250 to 1.280 is good. A reading of 1.225 to 1.250 is fair. Anything below 1.225 is poor. The electrolyte level can be checked by VW Computer Analysis only on cars with the central diagnosis socket. Even if a battery with an electrolyte level sensor has been service-installed in an earlier model, its electrolyte level must still be checked visually. Unscrew each filler plug, then bring the electrolyte level up to the indicators with distilled water. Never overfill the battery.

5.6 Body and Frame

Though there are only a few routine maintenance steps for the body of your car, they are important steps.

Lubricating Door and Hood Hinges

Pry the plastic plug out of the top of each passenger door hinge. Then pump in multipurpose grease with a grease gun that has a tapered nozzle. Press the plug back in and wipe away excess grease. Use SAE 30 engine oil on all hood and lid hinge pins, again being sure to wipe away excess oil.

18 Lubrication and Maintenance

Lubricating Locks

Lubricate the rubbing surfaces of all latches and striker plates with dry stick lubricant. Remove the plastic plug from the edge of each passenger door (located in the top of the recessed area in the door that contains the lock). Then squirt a few drops of SAE 30 engine oil into the hole to lubricate the lock mechanism. Lubricate door lock cylinders with graphite. Dip the key into graphite powder and turn it in the lock a few times.

6. Basic Car Care

The following brief guide will help you keep your VW looking as good as it runs.

6.1 Care of Car Finish

The longer dirt is left on the paint, the greater the risk of damaging the glossy finish, either by scratching or simply due to the chemical effect dirt particles have on the painted surface.

Washing

Never wash the car in direct sunlight. Beads of water not only leave spots when dried rapidly by the sun's heat, they also act as tiny magnifying glasses that burn spots into the finish. Use plenty of water, a car-wash soap such as VW's Car Wash and Wax, and a soft sponge.

Begin by spraying water over the dry car to remove all loose dirt. Then apply lukewarm soapy water. Rinse the car after sponging off the soapy water, using plenty of clear water under as little pressure as possible. Wipe the car dry with a chamois or soft terrycloth towels to prevent water-spotting.

Waxing

For a long-lasting, protective, and glossy wax finish after the car has been washed and dried, apply a hard wax, such as VW's Classic Car Wax. Waxing is not needed after every washing, and a more effortless shine can be obtained by using a car-wash liquid that contains wax. You can tell when waxing is required by looking at the finish while it is wet. If the water coats the paint in smooth sheets instead of forming beads that roll off, waxing is in order.

Polishing

Use paint polish, such as VW's Paint Polish, only if the finish assumes a dull look after long service. You can use polish on the car's brightwork to remove tar spots and tarnish, but afterwards apply a coat of wax to protect the clean plating.

Washing Chassis

The best time to wash the underside of the car is just after it has been driven in the rain. Spray the chassis with a powerful jet of water to remove dirt and deicing salt that may have accumulated there.

Special Cleaning

Tar spots can be removed with tar remover. Never use gasoline, kerosene, nail polish remover, or other unsuitable solvents. Insect spots also respond to tar remover. A bit of baking soda dissolved in the wash water will facilitate their removal. You can also use this method to remove spotting from tree sap.

To clean debris from the windshield wiper blades, remove the blades periodically and scrub them with a hard bristle brush and alcohol or a strong detergent solution. The windows can be cleaned with a sponge and warm water and then dried with a chamois or soft towel. If you use commercial window washing preparations, make certain they are not damaging to automotive finishes.

6.2 Care of Interior

The rubber weatherstrips around windows and doors must be kept pliable if they are to remain effective. Either spray these parts with silicone spray, or coat them with talcum powder. Petroleum products are harmful to rubber and should never be used.

Cloth Upholstery and Carpet

Clean the cloth interiors with a vacuum cleaner or whisk broom. Dirt spots can usually be removed with lukewarm soapy water. Use spot remover for grease and oil spots. Do not pour the liquid directly onto the upholstery, but dampen a clean cloth and rub carefully, starting at the edge of the spot and working inward.

Leatherette Upholstery and Trim

Either use VW's All Purpose Cleaner or a dry foam cleaner. Grease or paint spots can be removed by wiping with a cloth soaked in VW's All Purpose Cleaner. Use the same cleaner, applied with a soft cloth or brush, on the headliner and side trim panels.

6.3 Tires and Accessories

Never use tar remover, gasoline, or any other petroleum-based substance for cleaning tires. Such liquids damage rubber. Rubber paints, commonly sold as tire dressing, are largely cosmetic. White sidewalls can be cleaned with VW's All Purpose Cleaner.

© 1974 VWoA

LUBRICATION AND MAINTENANCE

Accessories

Most chrome-plated accessories can be polished and waxed along with the rest of the car's bright trim. Radio antennas should be lubricated only if hardened grease and collected dirt are interfering with raising or lowering the antennas. Do not use abrasive polish or cleaners on aluminum trim or accessories. They will destory the mirror-like shine of anodized surfaces.

The safety belts in your VW should be kept clean. If cleaning is necessary, wash them with a mild soap solution without removing them from the car. Do not bleach or dye safety belts or use any other cleaning agents. They may weaken the webbing.

Carefully check the condition of the webbing while you are cleaning the belts or the interior of the car. Frayed or damaged belts should be replaced.

Section 11

FUEL INJECTION

Contents

Introduction 2	Testing and Adjusting Microswitch (1977 and later California cars) 17
1. General Description 3	4.8 Testing and Replacing Temperature Sensors 17
Fuel Circuit 3	4.9 Testing and Replacing Injectors and Series Resistance Block 18
Air System 4	
Electronic Controls 4	4.10 Testing and Replacing Fuel Pump and Fuel Filter 20
Emission Controls 4	
2. Maintenance 4	4.11 Making Tests at Control Unit Plug and Replacing Control Unit 21
3. Troubleshooting Fuel Injection 5	**5. Adjusting Fuel Injection** 22
Troubleshooting Procedures 6	5.1 Adjusting Idle Speed 22
4. Testing and Repairing **Fuel Injection System** 7	Checking and Replacing Decel Valve (manual transmission only) 23
4.1 Testing and Replacing Double Relay 7	5.2 Adjusting Idle Mixture 24
4.2 Testing and Replacing Cold-start Valve 9	**6. Intake Air System** 25
4.3 Testing and Replacing Intake Air Sensor 10	6.1 Removing and Installing Air Cleaner 25
	Checking Thermostatic Valve 26
4.4 Testing and Replacing Thermo-time Switch 13	6.2 Removing and Installing Throttle Valve Housing 26
4.5 Testing and Replacing Pressure Regulator 13	6.3 Removing and Installing Intake Air Distributor 26
4.6 Testing and Replacing Auxiliary Air Regulator 14	Adjusting Accelerator Cable 27
	Removing and Installing Manifold Tubes 27
4.7 Testing and Replacing Throttle Valve Switch or Microswitch 16	
Testing and Replacing Throttle Valve Switch (except 1977 and later California cars) 16	**7. Testing and Replacing EGR** **(Exhaust Gas Recirculation) Valve** 27

1978 VWoA

Fuel Injection

The electronic fuel injection system covered in this section of the Manual is used only on 1975 and later models. Also covered here are the emission controls used on 1975 and later VWs. The fuel tank, the evaporative emission control, and the fuel gauge sending unit are covered in **FUEL SYSTEM.** These components have remained virtually unchanged from those used on VWs with carburetors. Whether equipped with carburetors or fuel injection, all VWs covered by this Manual are designed to operate on regular (91 octane) gasoline. Lead free gasoline is required for fuel injection cars that have catalytic converters.

The electronic fuel injection system includes an electric fuel pump, fuel lines, an air cleaner, and a complex system for mixing fuel and air in precisely controlled proportions and for delivering the mixture to the cylinders of the engine. An electronic control unit, commonly called the computer or "black box", monitors various engine operating factors, including the volume of air flowing into the engine. From this data, the control unit computes the proper amount of fuel to be delivered for every driving condition. This AFC (air flow controlled) fuel injection system also provides the exacting mixture control necessary for reduced exhaust emissions.

Unlike a carburetor, fuel injection does not depend on the velocity of the incoming air to draw fuel into the engine. Instead, fuel is injected into the airstream under pressure in the areas directly ahead of the intake valves. The injection pulses are timed to the firing of the engine, ensuring that fully vaporized fuel reaches the combustion chambers. The amount of fuel delivered is primarily determined by the length of the injection pulse. At closed throttle the mixture is made leaner through a reduction of the fuel pressure.

Electronic testing requires only a test light and either an ohmeter or a VOM (a common electrical testing instrument that can be used to measure accurately either ohms resistance or voltage). You should, however, have a thorough basic knowledge of electricity and be familiar with reading wiring diagrams before you attempt such tests. If you lack the skills, instruments, or other equipment necessary for testing the electronic fuel injection system, we suggest you leave such tests or repairs to an Authorized VW Dealer or other qualified and properly equipped shop. We especially urge you to consult your Authorized VW Dealer before attempting repairs on a car still covered by the new-car warranty.

FUEL INJECTION 3

1. General Description

Fig. 1-1 is a schematic view of the AFC (air flow controlled) electronic fuel injection system that was introduced on the 1975 Type 1 Volkswagen. To help you understand this diagram, a brief explanation is given here of the function of each of the system's components. A detailed description of each component can be found under the heading for the particular component; for example, **4.3 Testing and Replacing Intake Air Sensor**.

For the sake of description, it is convenient to divide the fuel injection system into three subsystems. These subsystems are the fuel circuit, the air system, and the electronic controls.

Fuel Circuit

The electric fuel pump draws fuel from the tank through the filter and pumps the fuel into the pressure line to the ring main. The ring main consists of the fuel hoses that supply the four injectors and the hose that supplies the cold-start valve. In order to limit the maximum pressure in the ring main to a predetermined level, the pressure regulator allows surplus fuel from the pump to flow back to the fuel tank. (Both the pump pressure line and the return line are enclosed in the car's frame tunnel.) The pressure regulator is not adjustable, its output being controlled by a calibrated spring and by engine vacuum.

Fig. 1-1. Schematic view of AFC fuel injection.

1. Fuel filter
2. Fuel pump
3. Pressure regulator
4. Cold-start valve
5. Injector
6. Auxiliary air regulator
7. Intake air sensor
8. Throttle valve housing
9. Intake air distributor
10. Temperature sensor I (1976 and later models)
11. Thermo-time switch
12. Potentiometer with fuel pump switch
13. Throttle valve switch or microswitch (except 1978 California cars)
14. Resistor
15. Temperature Sensor II
16. Control unit
17. Ignition distributor

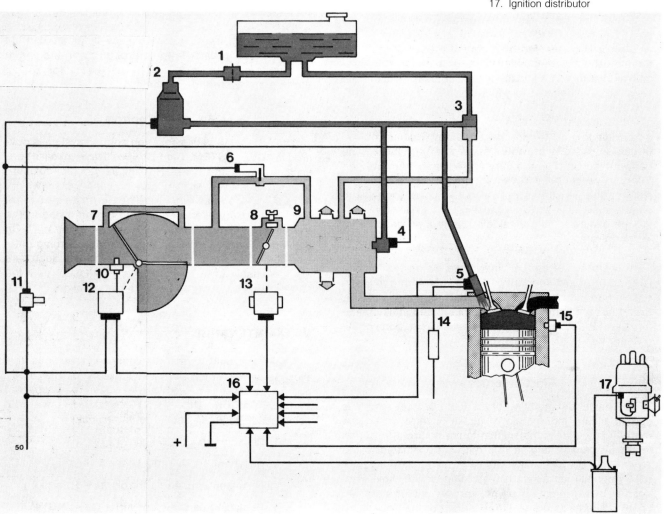

4 Fuel Injection

Each injector consists of a jet orifice with a solenoid-operated needle valve. All four injectors are triggered simultaneously at every other opening of the ignition distributor's breaker points. The length of the electrical impulse that holds the injector needle valves open is determined by the control unit.

The cold-start valve, sometimes called the fifth injector, injects fuel into the intake air distributor only while the starter is being operated. However, if the thermo-time switch senses that the air surrounding the engine is warm, the cold-start valve does not operate. The thermo-time switch also cuts off current to the cold-start valve if the engine is cranked for a longer than normal period of time in order to get the car started in cold weather. The purpose of the time limit on the cold-start valve is to prevent flooding.

Air System

The components of the air system are located atop the engine. All air enters the engine through an air cleaner (not shown in Fig. 1-1). From the air cleaner, the air enters the intake air sensor. The intake air sensor consists of a two-vane stator flap that operates a potentiometer and a fuel pump switch. As installed on the car, the intake air sensor is connected by a curved rubber duct to the throttle valve housing which, in turn, is bolted to the intake air distributor. A bypass screw, located in the throttle valve housing, is used to adjust the engine's idle speed.

When the engine is cold, the auxiliary air regulator admits additional air to the intake air distributor at closed throttle in order to provide a faster idle until the engine has warmed up. The auxiliary air regulator consists of a rotary valve operated by an electrically heated thermostatic spring. The spring begins to heat up as soon as the engine is started, gradually closing the valve and returning the engine to its normal idle rpm at a predetermined warm-up rate.

As installed on the car, the intake air distributor is a large, boxy, aluminum alloy casting that is integral with the alternator support. The two outlets from the intake air distributor are joined by hoses to bifurcated intake manifold tubes that conduct intake air to the intake ports of the cylinder heads.

Electronic Controls

The basic fuel metering of the fuel injection system is dependent on the volume of intake air and on engine rpm. Air flow data is transmitted to the control unit by the intake air sensor; engine rpm data is transmitted to the control unit by the breaker points of the ignition distributor. The intake air sensor's stator flap, which moves according to the volume of air being drawn into the engine, constantly adjusts the electrical output of the potentiometer—thereby transmitting to the control unit an electrical signal that varies proportionally to air flow.

In addition to the intake air sensor and the ignition distributor, which are the two main sources of data for the control unit, there are three other electronic controls that signal the control unit—when special operating conditions demand it.

The throttle valve switch (the microswitch on 1977 California cars) provides the control unit with a signal for full load enrichment whenever the accelerator pedal is fully depressed. This function was discontinued on the 1978 models by deleting the microswitch on California cars and by leaving vacant two terminals (3 and 18) of the throttle valve switches on cars sold elsewhere in the U.S. Except on 1977 and later California models, the throttle valve switch also controls the operation of the EGR (exhaust gas recirculation) valve. The throttle valve shuts off the EGR valve both at full load and at idle. (1977 and later California EGR valves are operated mechanically.)

Temperature sensor I, located inside the intake air sensor of 1976 and later models, senses intake air temperature and incorporates this data in the output signal of the potentiometer. (1975 cars have no temperature sensor I.) Temperature sensor II, located in the cylinder head, provides the control unit with the engine temperature information that is necessary for correct starting and warm-up enrichment.

Emission Controls

The EEC (evaporative emission control) system is covered in **FUEL SYSTEM**. The PCV (positive crankcase ventilation) system and the EGR (exhaust gas recirculation) system are integral with the fuel injection system which, in itself, functions as a highly effective emission control system (the Controlled Combustion System listed in the VW emission control booklet supplied with the car). On some fuel injection VWs, an Emission Afterburning catalytic converter is built into the exhaust system. Such cars are identifiable by a fuel filler that will accept only the small pump nozzles used for lead-free gasoline.

2. Maintenance

There are only a few maintenance operations that must be carried out at a specified mileage or after a certain period of service. These are listed below and covered in **LUBRICATION AND MAINTENANCE** or under the listed headings in this section of the Manual.

1. Servicing the air cleaner
2. Replacing the fuel filter (see **4.10**)
3. Checking emission controls (see **5.1** and **7.**)
4. Replacing catalytic converter (see **ENGINE**).

FUEL INJECTION 5

3. Troubleshooting Fuel Injection

Two things are important prerequisites to successful troubleshooting of the fuel injection system. First, you should be thoroughly familiar with the function of the system as a whole. See **1. General Description.** Before making any electrical tests or fuel circuit tests, make certain that no hoses in the air system are leaking and that there are no leaks in the air system components themselves. Further information about the function of each injection system component will be found under the heading for testing and replacing that component.

The second prerequisite to successful troubleshooting is your ability to read the wiring diagram given in Fig. 3-1. If you need help in understanding any of the symbols given in this diagram, please refer to the **Wiring** topic in **ELECTRICAL SYSTEM.** There you will find explanations of all the various symbols. Note especially that the two diodes shown in the double relay in Fig. 3-1 permit current to flow only in the direction indicated by the arrow-like shape of the diode symbol. The part of the wiring diagram that is below the broken line supplements Wiring Diagrams G and J given in **ELECTRICAL SYSTEM.**

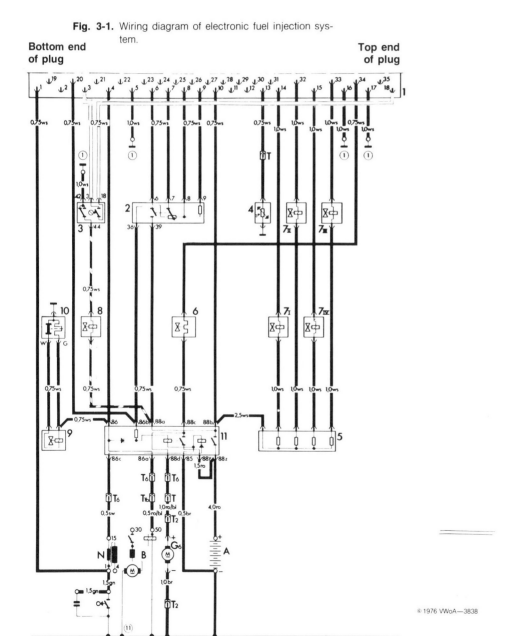

Fig. 3-1. Wiring diagram of electronic fuel injection system.

1—Electronic control unit plug
2—Intake air sensor
3—Throttle valve switch/microswitch (except 1978 California cars)
4—Temperature sensor II
5—Series resistance block
6—Auxiliary air regulator
7—Injectors
8—EGR valve (except 1977 and late California cars)
9—Cold-start valve
10—Thermo-time switch
11—Double relay
N. Ignition coil
G₆. Fuel pump
A. Battery
B. Starter
T. Wire connector
T₁ᵦ. Wire connector
T₂. Wire connector, double
T₆. Wire connector, 6-point
① Ground connector near alternator
⑪ Ground connector, headlights

Color codes
ws = white
sw = black
br = brown
ro = red
gn = green
ro/bl = red/blue

——— Circuit not used on 1977 and later California cars

– – – Circuit used both with throttle valve switch and with microswitch, discontinued on 1978 cars.

6 Fuel Injection

Before you begin troubleshooting the fuel injection system, make sure that the engine trouble is not caused by something other than a fuel system problem. As with carburetors, no fuel injection test or adjustment should be made until you are confident that the engine has adequate compression and that the ignition system is not faulty. Because low charging system output can also upset the operation of the fuel injection system, make sure that the voltage in the electrical system is between 12.5 and 14.5 volts.

Erratic fuel metering can be caused by bouncing or arcing ignition distributor breaker points. The fuel injection system control unit, or computer, is programmed to trigger an injection pulse at every other opening of the ignition distributor's breaker points. If the breaker points are bouncing, owing to incorrect dwell or a weak spring, extra injection pulses will be triggered. If the points are arcing, the control unit may fail to sense the opening of the points. Breaker point dwell adjustment, point inspection, and ignition timing are described in **ENGINE**.

Misfiring is not a typical sympton of a faulty fuel injection system. If you encounter misfiring—or if the car starts hard, fails to start, or has inadequate power—check for carbon tracking at the distributor rotor, distributor cap (Fig. 3-2), and the coil (Fig. 3-3). Replace faulty components. Also check the resistance of the spark plug connectors and the distributor rotor as described in **ENGINE**. Neither rotor nor connector resistance should exceed 10,000 ohms.

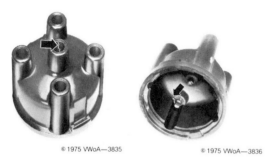

Fig. 3-2. Typical carbon tracks in distributor cap.

Fig. 3-3. Typical carbon track on coil tower.

Finally, do not overlook incorrect valve clearances, faulty spark plugs, a dirty air cleaner, or a restricted exhaust system as possible causes of engine trouble. Because the fuel metering is highly dependent on the measurement of intake air volume, any air entering the engine that does not pass through the intake air sensor will cause the engine to operate incorrectly. Typically, it will idle slowly or stall at idle and give symptoms of lean operation at driving speeds. Look for a loose-fitting oil filler cap, leaking or disconnected ventilation hoses, a leaking cylinder head cover gasket, or leaking pushrod tubes on the engine.

Troubleshooting Procedures

The Bosch AFC (air flow controlled) fuel injection system has been designed so that you can make all the necessary electrical tests using only a test lamp and an ohmmeter. In addition to these two electrical instruments, you will need a fuel pressure gauge to test the fuel pump and the pressure regulator.

The double relay, which is located adjacent to the control unit on the vehicle, controls the supply of electrical current to every electrical component of the fuel injection system. Consequently, if current fails to reach the fuel pump or any of the injection system components in the engine compartment, your next step should be to test the double relay.

For example, if the cold-start valve fails to operate during the cranking of a cold engine in cold weather, you would first use a test lamp to determine whether current is reaching the cold-start valve during cranking. If there is no current, you should test the double relay terminals which supply current to the cold-start valve. First, you would check for current to double relay terminal 86a while the starter is being operated. If there is no current, the wire to starter terminal 50 is faulty. If there is current, repeat the test at terminal 86 of the double relay. If there is no current, the double relay is faulty; if there is current, you know that the wire from terminal 86 to the cold-start valve is faulty.

Even though the wiring diagram in Fig. 3-1 shows that many components receive current from the control unit, you should keep in mind that the control unit must first receive this current from the double relay. Therefore, test the double relay thoroughly before you assume that there is trouble in the control unit. Experience has shown that faulty double relays are encountered many times more often than faulty control units.

No attempt should be made to test the electronic control unit itself. Applying test current to its terminals or bridging the terminals could ruin the control unit. The control unit should be judged faulty only if, after careful testing, all other components of the fuel injection system are proved to be in good working order.

Fuel Injection 7

Before you begin testing the fuel injection system components, select mentally those components that are most likely to be causing the trouble. If the engine is running rich, begin by checking the components of the fuel circuit, then proceed to the intake air sensor and temperature sensor II. If the engine is running lean, concentrate your initial troubleshooting on the components related to the air system—with special attention to the intake air sensor and the auxiliary air regulator.

Once you have (1) determined that there is no engine fault outside the fuel injection system; (2) thoroughly checked the air system for leaking or disconnected hoses; and (3) determined which components are most likely to be causing the trouble, go to the appropriate component topics under **4. Testing and Repairing Fuel Injection System.** There you will find a description of each component's function, the method of testing the component, and instructions for replacing faulty components.

4. Testing and Repairing Fuel Injection System

The order in which the components are discussed under the following headings is not intended to indicate which components should be tested first. The sequence in which you test the components should be based upon your own analysis of the operating problem, an understanding of the system as a whole, and the wiring diagram that was given in **3. Troubleshooting Fuel Injection.**

> **CAUTION**
> If you lack the skills or equipment necessary for testing and repairing the fuel injection system, we suggest you leave such work to an Authorized VW Dealer or other qualified shop. We especially urge you to consult your Authorized VW Dealer before attempting tests or repairs on a car still covered by the new-car warranty.

4.1 Testing and Replacing Double Relay

The double relay is a small device, approximately 20 mm by 20 mm and about 45 mm in length (3/4 in. by 3/4 in. and about 2 in. in length). It has eleven numbered terminals. The number of each terminal is printed beside the terminal on the relay. Inside the double relay are two electromagnetic switches (relays), two diodes, and a resistor. These components and their connections were shown earlier in Fig. 3-1.

The double relay is located under a composition board cover in the right-hand side of the rear luggage compartment, behind the rear seat backrest. After you have lifted off the cover, you will see the double relay on a bracket adjacent to the electronic control unit (Fig. 4-1). Two white plastic multiple connector plugs connect the double relay to the fuel injection system wiring harness.

Fig. 4-1. Location of double relay (arrow).

The double relay controls the supply of battery voltage to all parts of the fuel injection system—including the control unit. When the engine is stopped, the double relay breaks all the circuits between the fuel injection system and the car's electrical system. The basic sequence of double relay operation is as follows:

1. When the starter is operated, current from terminal 50 on the starter solenoid supplies positive (+) current to terminal 86a of the double relay, causing one of the relays—the fuel pump relay—to close and send battery voltage from terminal 30 on the starter to the electric fuel pump, via double relay terminals 88y and 88d.

2. Because the ignition is turned on when the starter is operating, ignition primary terminal 15 delivers positive (+) current to double relay terminal 86c, thereby closing the power relay that sends positive (+) current from the battery to all electrical parts of the fuel injection system, with the exception of the fuel pump, the cold-start valve, and the auxiliary air regulator—all of which receive positive (+) current from the pump relay. The power relay remains closed as long as the ignition is on and will not open again until the ignition is turned off.

3. When the engine has started, the driver moves the key from the start (3) position to the running (2) position. This removes the supply of positive (+) current that was reaching double relay terminal 86a from terminal 50 on the starter—the current that had originally closed the pump relay. However, the pump relay now receives current from the pump switch in the intake air sensor, via terminal 86b and the resistor of the double relay. Whether or not the pump operates depends on whether or not the

8 Fuel Injection

pump switch in the intake air sensor opens or closes the pump relay.

4. When the driver stops the engine by turning off the ignition, ignition current no longer reaches double relay terminal 86c. The power relay opens, breaking all positive (+) connections between the fuel injection system and the car's electrical system.

The two diodes in the double relay isolate the current that reaches the relay while the engine is running from the current sources that closed the relays during starting. Failure of either diode will prevent its associated relay from closing during starting, causing a malfunction of the fuel injection system. In testing the double relay, make the following two tests before making any other test. In doing so you will check (1) that the double relay is getting current from the car's electrical system and (2) that the diodes are intact and that the relays are functioning correctly.

To test current supply:

1. Connect one test probe of a test lamp to terminal 85 of the double relay (which is grounded to the negative post of the battery).

2. Apply the second test probe to terminal 88y, as shown in Fig. 4-2. Repeat the test by applying the second test probe to terminal 88z. The test lamp should light during both repetitions of the test. If not, positive (+) battery current is not reaching the terminal to which the second probe is connected.

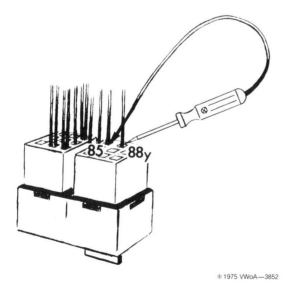

Fig. 4-2. Test for positive (+) battery current with ignition switch turned off.

3. With the ignition key turned to the running (2) position, apply the second test probe to terminal 86c. If the test lamp fails to light, positive (+) battery current is not reaching terminal 86c from the ignition system.

4. While running the starter, again apply the second test probe to terminal 86c. The test lamp should light, indicating positive (+) battery current from the ignition system.

5. Again running the starter, apply the second test probe to terminal 86a. The test lamp should light, indicating that positive (+) battery current is reaching the double relay from terminal 50 on the starter solenoid.

NOTE

If positive (+) battery current reaches double relay terminal 86a whether the starter is running or not, the wires to terminals 30 and 50 of the starter solenoid have accidentally been reversed. Correct the wire positions. If positive (+) battery current is reaching the double relay correctly, continue with the next test procedure.

To test relay function:

1. Connect one test probe of a test lamp to terminal 85 of the double relay (which is grounded to the negative post of the battery).

2. While running the starter, apply the second test probe to terminal 88d, as shown in Fig. 4-3. If the lamp does not light, the pump relay is not closing and the double relay should be replaced.

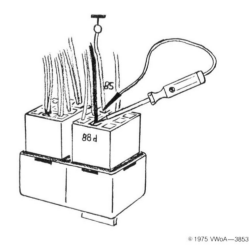

Fig. 4-3. Test for positive (+) battery voltage to fuel pump during starting.

3. With the ignition key turned to the running (2) position, apply the second test probe to double relay terminal 88b. If no positive (+) battery current reaches terminal 88b, the power relay is not closing and the double relay should be replaced.

FUEL INJECTION 9

If the double relay, and the wiring that supplies current to the double relay, pass the two tests that have just been described, you can continue by making tests that will determine whether positive (+) battery current is reaching the other relay terminals at the correct times and under the correct conditions. Further tests should be prompted by the failure of the double relay or the control unit to deliver current to some other component of the fuel injection system. Your tests should be based on a study of the wiring diagram given earlier in **3. Troubleshooting Fuel Injection**.

To replace double relay:

1. Disconnect the battery ground strap.
2. Remove the screw that holds the double relay to the car body. Then disconnect the double relay from the two multiple connector plugs.
3. Install the multiple connector plugs on the new double relay, mount the relay on the car body, then reconnect the battery ground strap.

4.2 Testing and Replacing Cold-start Valve

The cold-start valve, sometimes called the fifth injector, sprays fuel into the intake air distributor only during the first few seconds of starter operation and only when the engine and the surrounding air are cold. The valve's operation is controlled by the thermo-time switch and by current from terminal 50 of the starter solenoid—current that reaches the valve by way of the double relay. The cold-start valve is located atop the intake air distributor, as indicated in Fig. 4-4.

Fig. 4-4. Location of cold-start valve on engine.

If the cold-start valve fails to inject fuel during the cranking of a cold engine, it will be difficult or impossible to start the engine. If the cold-start valve is leaky, the engine may flood during starting—especially if the engine is hot.

To test cold-start valve:

1. Carefully disconnect the electrical plug from the cold-start valve. Put tape over the plug terminals to prevent accidental electrical sparks.
2. Remove the two screws that hold the cold-start valve to the intake air distributor. Remove the valve from the intake air distributor, but leave the ring main fuel hose connected to the valve.
3. Disconnect the ignition primary wire from terminal 1 of the ignition coil so that the engine will not start. Tape the terminal of the wire to prevent accidental electrical sparks.
4. Hold the cold-start valve as shown in Fig. 4-5 while someone operates the starter.

WARNING —
Do not smoke or work near heaters or other fire hazards. Have a fire extinguisher handy.

5. If the cold-start valve leaks fuel while the starter is operated (owing to normal pressure from the fuel pump), replace the cold-start valve.

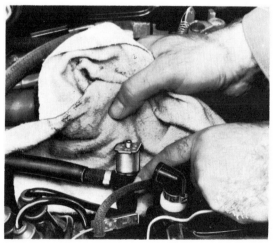

Fig. 4-5. Cold-start valve being checked for leakage. Use a cloth as shown in order to catch any fuel.

6. If the valve did not leak, reconnect the electrical plug to the cold-start valve. Then disconnect the electrical plug from the thermo-time switch and,

10 Fuel Injection

using a jumper wire, ground the **W** terminal of the disconnected plug. The thermo-time switch plug is indicated in Fig. 4-6.

Fig. 4-6. Plug on thermo-time switch (being pointed to by finger).

7. Hold the cold-start valve as shown previously in Fig. 4-5 while someone briefly operates the starter.

WARNING —
Do not smoke or work near heaters or other fire hazards. Have a fire extinguisher handy.

8. Fuel should be expelled from the cold-start valve when the starter is operated. If so, the valve is working correctly and can be reinstalled on the intake air distributor. Test the thermo-time switch if you suspect that the cold-start valve is failing to operate when the engine is cold.

9. If no fuel is expelled from the cold-start valve during starter operation, disconnect the electrical plug from the cold-start valve. Then connect one lead of a test lamp to the plug terminal that receives current from terminal 86 of the double relay. See the wiring diagram given previously in **3. Troubleshooting Fuel Injection.** Connect the other test lead to ground on some clean, unpainted metal part of the engine.

10. Operate the starter. If the test lamp does not light, either the wiring to the double relay is faulty, the double relay itself is faulty or has poor connections, or the wire between terminal 50 on the starter solenoid and double relay terminal 86a is faulty or disconnected. Also, you may be testing the wrong terminal of the cold-start valve plug.

NOTE —
If the test lamp lights whether the starter is operating or not, someone has misconnected the wires at the starter solenoid, installing the wire that belongs on terminal 50 on terminal 30 instead.

11. If battery current is reaching the cold-start valve, but the valve does not operate, the valve is faulty and must be replaced. You can also use an ohmmeter to test for continuity between the two terminals on the cold-start valve. If there is no continuity, the valve is definitely faulty.

To replace cold-start valve:

1. Disconnect the electrical plug from the cold-start valve. Loosen the hose clamp, then disconnect the ring main fuel hose from the cold-start system.

WARNING —
Disconnect the battery ground strap. Do not smoke or work near heaters or other fire hazards. Have a fire extinguisher handy.

2. Remove the two screws that hold the cold-start valve to the intake air distributor. Then remove the cold-start valve and its gaskets.

3. Install the new cold-start valve and a new gasket. Then reconnect the electrical plug, the fuel hose, and the battery ground strap.

4.3 Testing and Replacing Intake Air Sensor

The intake air sensor (Fig. 4-7) is a mechanical device that operates a potentiometer and the fuel pump switch. The sensor's mechanical operation depends upon the free movement of the spring-loaded stator flap inside the sensor's main air passage.

Fig. 4-7. Intake air sensor location.

FUEL INJECTION 11

Though usually prevented by the back pressure valve (see Fig. 4-11 given later), backfiring can cause the stator flap of the intake air sensor to jam closed. Dirt or physical impact can cause the stator flap to bind or jam partway open. Therefore, check the stator flap for free movement before you make any electrical tests. To check flap movement, release the four spring clips that hold the air cleaner together. Then separate the two halves of the air cleaner and remove the pleated paper filter element. If necessary, consult **6. Intake Air System.**

Insert a wooden dowel or other suitable probe at the point indicated in Fig. 4-8. Using the probe, move the stator flap throughout its entire range. If the flap is merely stuck closed, owing to a backfire, it will usually operate correctly after it is once pressed loose. If the flap binds at any other point, remove the intake air sensor, as described later under this heading, and then carefully clean away any foreign matter that might cause the flap to bind.

CAUTION —
To prevent damage to the connector plug and wiring, first remove the rubber boot from the plug and slide the boot down onto the wiring harness. Then, grasping the connector plug (not the wires), pull the plug off the intake air sensor terminals.

Connect an ohmmeter to intake air sensor terminals 36 and 39. (The numbers are marked on the sensor.) Hand-operate the stator flap. With the flap closed, there should be no electrical continuity between sensor terminals 36 and 39; when the flap is slightly opened, there should be continuity. If there is no continuity, replace the intake air sensor.

To check the potentiometer, connect an ohmmeter to terminals 6 and 9 of the intake air sensor (Fig. 4-9). The resistance should be 200 to 400 ohms—except on 1976 models, which have a resistance of 100 to 300 ohms. (The 200- to 400-ohm unit is usually installed on 1976 cars as a replacement part.) Repeat the test at terminals 7 and 8. The resistance should be 120 to 200 ohms on 1975 cars, 80 to 200 ohms on 1976 cars, or 100 to 500 ohms on 1977 and 1978 cars—as well as on 1975 and 1976 cars that have been equipped with the current-type intake air sensor. If the resistance is incorrect for either test, replace the intake air sensor.

NOTE —
Temperature sensor I, installed on 1976 models, is built into the intake air sensor. This temperature sensor is checked when you measure the resistance between intake air sensor terminals 6 and 9.

Photo by Alex Walordy

Fig. 4-8. Air intake opening in body of air cleaner. Carefully insert probe until it contacts stator flap.

© 1975 VWoA—3825

Fig. 4-9. Ohmmeter test probes applied to potentiometer terminals on intake air sensor.

If necessary, you can disassemble the intake air sensor by taking out the screws and carefully prying the upper side of the housing out of the main part of the housing. This cover part is held in by sealer. If the flap cannot be made to move freely, install a new intake air sensor. In assembling an intake air sensor, be sure to use a suitable sealer at the periphery of the cover and on the screws. Do not wash the sensor with solvent, which could damage the potentiometer.

If the fuel pressure test described in **4.10 Testing and Replacing Fuel Pump and Fuel Filter** shows that the pump is not operating after the engine has started—or if the engine refuses to start—test the fuel pump switch in the intake air sensor. To test the switch, remove the connector plug from the terminals of the intake air sensor.

To replace intake air sensor:

1. Release the four clips that hold the cover and the body of the air cleaner together. Remove the cover and the pleated paper air filter.

12 Fuel Injection

2. Carefully pull back the rubber boot that covers the multiple connector plug on the intake air sensor. Slide the boot down the wiring harness. Then, grasping the connector plug (not the wires), carefully pull the plug off the terminals of the intake air sensor.

3. Loosen the hose clamp screw indicated in Fig. 4-10. Then disconnect the rubber elbow duct from the intake air sensor.

4. Remove the two nuts indicated in Fig. 4-10. Then remove the body of the air cleaner and the intake air sensor from the car as a unit.

5. Remove the four bolts that hold the body of the air cleaner to the intake air sensor. Then remove the air cleaner body from the intake air sensor.

Installation is the reverse of removal. If you install a new intake air sensor, or if you have disassembled the old sensor for cleaning or inspection, you should carry out the adjustments described in **5. Adjusting Fuel Injection** after installation is complete. If you disassemble the intake air sensor, use sealer on the screws and around the periphery of the cover during assembly. The plastic cover can be pried off the potentiometer which, in some cases, may permit you to correct minor mechanical faults. However, the electrical components cannot be replaced or repaired and, if there is an electrical fault in the potentiometer, the entire intake air sensor must be replaced as a unit. Fig. 4-11 is a cutaway view of the intake air sensor.

Photo by Alex Walordy

Fig. 4-10. Hose clamp screw (left arrow) and air cleaner mounting nuts (right arrows).

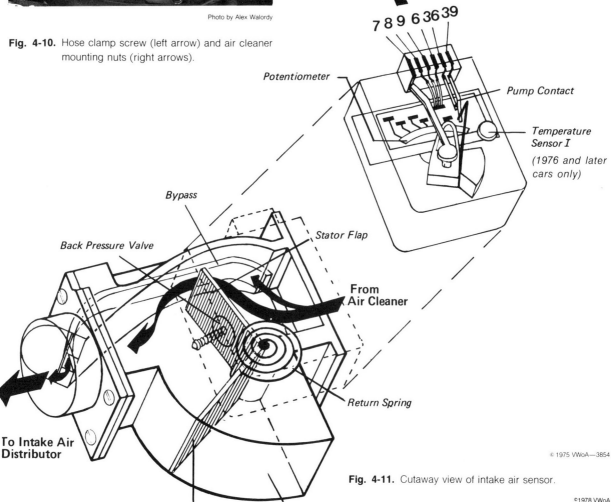

Fig. 4-11. Cutaway view of intake air sensor.

FUEL INJECTION 13

4.4 Testing and Replacing Thermo-time Switch

The thermo-time switch (Fig. 4-12) controls the operation of the cold-start valve. If the temperature of the air around the engine is below the temperature marked on the thermo-time switch, the switch permits operation of the cold-start valve. Above that temperature, the thermo-time switch breaks the circuit to the cold-start valve. In addition to air temperature, an electrical heating element influences the operation of the thermo-time switch. This heating element starts to warm up when the starter begins to operate. After the number of seconds marked on the thermo-time switch, the heating element will have raised the temperature of the thermo-time switch above the marked temperature and the thermo-time switch will cut off the activating current to the cold-start valve.

Fig. 4-12. Location of thermo-time switch on engine.

To test thermo-time switch:

1. Make sure that the temperature of the thermo-time switch is below the temperature marked on its housing. If necessary, cool the thermo-time switch with ice.

2. Carefully disconnect the electrical plug from the cold-start valve. Disconnect the ignition primary wire from terminal 1 of the ignition coil in order to prevent the engine from starting. Position the terminal of the disconnected wire so that it will not accidentally contact any metal part of the engine.

3. Attach the leads of a test lamp to the terminals of the disconnected plug. Operate the starter without interruption for at least twelve seconds. The test lamp should first light brightly and then become noticeably dimmer—or go out—at the time marked on the thermo-time switch—eleven seconds, for example. If the lamp lights but does not dim or go out, replace the thermo-time switch.

4. If the test lamp does not light at all, attach one of the test lamp leads to ground on some clean, unpainted metal part of the engine. The other lead should remain connected to one of the plug terminals.

5. Operate the starter. If the test lamp does not light, repeat the test with the test lamp connected to the other terminal of the disconnected plug. If the lamp lights in neither test, the wiring to the double relay is faulty, the double relay itself is faulty or has poor connections, or the wire between terminal 50 on the starter solenoid and double relay terminal 86a is faulty or disconnected.

   NOTE
   If the test lamp lights whether the starter is operating or not, someone has misconnected the wires at the starter solenoid, installing the wire that belongs on terminal 50 on terminal 30 instead.

6. If battery current is reaching the plug, but the test lamp did not light when the starter was operated in step 3, the thermo-time switch is faulty and should be replaced.

To replace thermo-time switch:

1. Disconnect the electrical plug from the thermo-time switch.

2. Using a socket wrench, unscrew the thermo-time switch from the mounting bracket.

3. Install a new thermo-time switch in the mounting bracket. Then reconnect the electrical plug to the new thermo-time switch.

4.5 Testing and Replacing Pressure Regulator

When the fuel pump is operating, fuel from the fuel pump enters the pressure regulator and exerts pressure against a spring loaded diaphragm. Fuel pressure in excess of about 35 psi (2.5 kg/cm^2) causes the diaphragm to deflect against spring tension, thereby uncovering a fuel outlet that returns excess fuel to the tank.

In addition to fuel pressure, engine vacuum can act upon the diaphragm, cooperating with the fuel pressure in moving the diaphragm against spring tension. When

14 Fuel Injection

there is high vacuum available to the pressure regulator from the intake air distributor, a fuel pressure in excess of approximately 28 psi (2.0 kg/cm^2) will move the diaphragm and permit excess fuel to return to the tank.

The pressure regulator is located on the engine's front cover plate, ahead of the fan housing on the right-hand side of the engine. You can find it by following the vacuum line from the intake air distributor. The pressure regulator cannot be adjusted or repaired and must, therefore, be replaced if it is faulty.

To test pressure regulator:

1. If the engine is cold and the air temperature is 50°F (10°C) or less, start the engine and allow it to warm up.

2. With the engine turned off, disconnect the ring main fuel hose from the cold-start valve. (If necessary, see **4.2 Testing and Replacing Cold-start Valve**.)

   WARNING —
   Gasoline may be expelled when you disconnect the hose. Do not smoke or work near heaters or other fire hazards. Have a fire extinguisher handy.

3. Connect a fuel pressure gauge to the disconnected hose as shown in Fig. 4-13. Disconnect the vacuum hose from the pressure regulator and plug the hose.

Fig. 4-13. Fuel pressure gauge installed on ring main fuel hose of cold-start valve.

4. Start the engine and observe the pressure gauge. The gauge should read approximately 35 psi (2.5 kg/cm^2). (33 to 37 psi (2.35 to 2.65 kg/cm^2) is considered acceptable by experienced mechanics.) If the pressure is too high or too low, replace the pressure regulator.

5. If the fuel pressure is correct with the vacuum hose disconnected, repeat the test with the vacuum hose connected to the pressure regulator. The gauge should read approximately 28 psi (2.0 kg/cm^2). (26 to 30 psi (1.85 to 2.15 kg/cm^2) is considered acceptable by experienced mechanics.) If the pressure is too high or too low, replace the pressure regulator.

   NOTE —
   Correct pressure is most important at idle, where fuel pressure has the greatest influence on exhaust emissions. However, if the engine seems to lack power, check to see whether the pressure drops considerably below 35 psi (2.5 kg/cm^2) at full throttle. If so, look for a kinked or restricted fuel line, a blocked fuel filter, dirt in the fuel tank, or inadequate pump output.

To replace pressure regulator:

1. Disconnect the vacuum hose from the pressure regulator. Disconnect and quickly plug the pressure regulator's two fuel hoses.

   WARNING —
   Disconnect the battery ground strap. Do not smoke or work near heaters or other fire hazards. Have a fire extinguisher handy.

2. Working under the car, remove the ring nut that holds the pressure regulator to the engine's front cover plate. The nut is threaded onto the regulator's fuel outlet tube.

3. Remove the pressure regulator. Then install the new pressure regulator on the engine's front cover plate.

4. Reconnect the hoses. The fuel pressure line from the pump should be connected to the side pipe of the pressure regulator.

5. Reconnect the battery ground strap. Because the pressure regulator has an important effect on exhaust emissions, you should test the new regulator to make sure that it maintains the correct fuel pressures.

4.6 Testing and Replacing Auxiliary Air Regulator

Fig. 4-14 is a schematic view of the auxiliary air regulator. When open, a rotary valve in the auxiliary air regulator provides additional air—and consequently additional fuel—during engine warm up. The auxiliary air regulator's function in the air system can be seen in Fig. 1-1 in **1. General Description.** On the car, the auxiliary air regula-

FUEL INJECTION 15

tor is located between the crankshaft pulley and the alternator pulley.

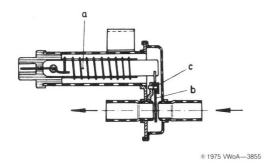

a. Bimetallic strip   b. Rotary valve   c. Pivot

Fig. 4-14. Schematic view of auxiliary air regulator. Arrows indicate air flow through open valve.

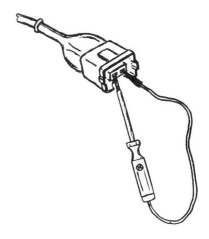

Fig. 4-15. Test light being used to check for current at auxiliary air regulator plug while engine is running.

WARNING —
To avoid serious injury, be careful of the moving V-belt while you make tests with the engine running

When the engine is started, current from the double relay begins to warm the heating element that is wound around the auxiliary air regulator's bimetallic strip. Heat—whether from the heating element or from hot air surrounding the engine—causes the bimetallic strip to deflect gradually, closing the rotary valve and cutting off the additional air.

To test the auxiliary air regulator, disconnect its fabric-covered intake air hose from the large rubber elbow duct that joins the intake air sensor with the throttle valve housing. Then start the engine (the engine should be cold). There should be vacuum at the disconnected hose. If you cover the hose, the engine should slow down. After several minutes of operation, the vacuum should diminish and, eventually, disappear almost completely.

If the auxiliary air regulator does not close after a reasonable period of engine operation (in cold weather, considerable time may be required), stop the engine. Disconnect the electrical plug from the auxiliary air regulator. Using an ohmmeter, measure the resistance between the two terminals on the auxiliary air regulator. The reading should be 30 ohms. If the ohmmeter reads infinity, or if the resistance is considerably less than 30 ohms, replace the auxiliary air regulator.

If the auxiliary air regulator resistance is in the correct range, use a test light as shown in Fig. 4-15 to determine whether battery voltage is reaching the heating element while the engine is running. If not, you should make further tests at the double relay. Especially check that someone has not accidentally exchanged the positions of the wires that go to terminals 30 and 50 on the starter solenoid. See the wiring diagram given earlier in **3. Troubleshooting Fuel Injection.**

To replace auxiliary air regulator:

1. Disconnect both air hoses from the auxiliary air regulator.

2. Disconnect the electrical plug from the auxiliary air regulator. Then remove the two screws that hold the auxiliary air regulator to its mounting bracket. One of these screws is indicated in Fig. 4-16.

3. Remove the old auxiliary air regulator. Then install the new auxiliary air regulator using a reverse of the removal procedure.

Fig. 4-16. Arrow indicating one of two screws that hold auxiliary air regulator to mounting bracket.

16 Fuel Injection

4.7 Testing and Replacing Throttle Valve Switch or Microswitch

The throttle valve switch (Fig. 4-17) is a black, rectangular device on the right side of the throttle valve housing, just to the left of the alternator. In addition to signaling the control unit to provide full load enrichment, the throttle valve switch controls the EGR (exhaust gas recirculation) valve's operation. Beginning with the 1977 models, cars sold in California have no throttle valve switch; the EGR valve is mechanical and a microswitch handles the full load enrichment function. (1977 only).

Fig. 4-17. Location of throttle valve switch (arrow).

Testing and Replacing Throttle Valve Switch
(except 1977 and later California cars)

Fig. 4-18 shows the inner parts of the throttle valve switch. At closed throttle, the control terminal (18) is connected to terminal 2 (42 in the wiring diagram given in **3. Troubleshooting Fuel Injection**). In this position, the EGR valve does not recirculate exhaust gases. At (or near) full throttle, the control terminal (18) is connected to terminal 3, again cutting off exhaust gas recirculation, but also signaling the control unit to provide full load enrichment. Exhaust gas recirculation occurs only in the mid-range of throttle valve operation.

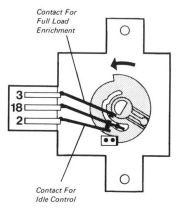

Fig. 4-18. Schematic view of throttle valve switch.

To test throttle valve switch:

1. With the engine stopped, carefully disconnect the electrical plug from the throttle valve switch.

   NOTE
   For access to the throttle valve switch and its plug, loosen the hose clamp, then remove the large rubber elbow from the top of the throttle valve housing.

2. With the throttle valve closed, take an ohmmeter reading between switch terminals 18 and 2 (or 42). If there is no electrical continuity, adjust the switch as described in step 5 of the replacement procedure. If there is still no continuity, replace the switch.

3. If there is continuity between terminals 18 and 2 (or 42) at closed throttle, gradually open the throttle. Electrical continuity should be broken at about 11° of throttle valve shaft rotation. If continuity does not break, replace the throttle valve switch.

4. If the tests for terminals 18 and 2 (or 42) were satisfactory, reconnect the ohmmeter to switch terminals 18 and 3. At closed throttle there should be no electrical continuity but, when the throttle valve is opened about 75° or more, there should be continuity. If not, replace the throttle valve switch.

To replace throttle valve switch:

1. Carefully disconnect the electrical plug from the throttle valve switch.

   NOTE
   For access to the throttle valve switch and its plug, loosen the hose clamp, then remove the large rubber elbow from the top of the throttle valve housing.

2. Using an offset screwdriver, remove the two screws that hold the throttle valve switch to the throttle valve housing.

3. Pull the switch toward the alternator so that the switch slides off the end of the throttle valve shaft.

4. Loosely install the new switch. To avoid damaging the switch, do not press it against the throttle valve housing until you are sure that the throttle valve shaft is correctly engaged. It may be necessary to rotate the switch slightly in order to engage the cam on the shaft correctly.

5. To adjust the switch, hold the throttle valve in its closed position. Turn the switch body in the direction of open throttle until a light resistance is felt. Then tighten the mounting screws.

Fuel Injection 17

Testing and Adjusting Microswitch
(1977 California cars)

The microswitch controls full load/full throttle enrichment. If the car fails to reach top speed and has poor full-throttle acceleration, carry out the checks given in the procedure that follows.

To check full load enrichment:

1. Connect an ohmmeter to the microswitch terminals with the throttle valve closed (Fig. 4-19), disconnecting the wires beforehand. The ohmmeter should indicate no continuity (infinite ohms). Hand-operate the microswitch by raising up the roller. The ohmmeter should indicate continuity (zero ohms). Replace the switch if it fails to operate correctly in either position.

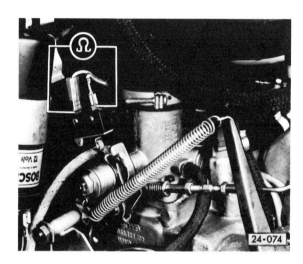

Fig. 4-19. Microswitch being tested with ohmmeter (ohmmeter indicated schematically). Though wires are shown connected to the switch, they should be disconnected when making an actual test.

2. Allow the roller to return to its rest position. Then have someone press the accelerator pedal fully to the floor. Just before full throttle is reached, the ohmmeter should indicate continuity (zero ohms). If not, adjust the switch as described in the next three steps.

3. To adjust the microswitch, slightly loosen the switch's mounting screws.

4. Have someone hold down the accelerator. Adjust the switch position until the ohmmeter reading changes from infinite ohms to zero ohms. Then tighten the mounting screws.

5. Recheck the adjustment by having someone press down and release the accelerator pedal several times while you watch the changes in the ohmmeter readings.

6. If the switch worked correctly during the test you made in step 2, yet full-throttle performance is poor, test the switch with the engine running. To do this, reconnect the wires and then allow the engine to idle until it is fully warmed up (oil temperature 50° to 70°C (122° to 158°F)).

7. With the engine idling at 800 to 950 rpm, hand-operate the microswitch by raising up on the roller. The idle speed should increase by about 100 rpm.

8. If the idle speeds up, the poor full-throttle performance is not being caused by faulty full load enrichment. If the idle fails to speed up, test the wires that connect the microswitch to terminals 3 and 18 of the control unit. Replace or repair faulty wires.

9. If the wires are not faulty, test the control unit by temporarily installing another control unit that is known to be in good condition.

4.8 Testing and Replacing Temperature Sensors

Temperature sensor I, used on 1976 and later models, is built into the intake air sensor. This temperature sensor is tested as part of the potentiometer test described in **4.3 Testing and Replacing Intake Air Sensor.** Temperature sensor I measures the temperature of the air that enters the engine. This data is incorporated into the potentiometer's data and sent to the control unit.

Temperature sensor II is screwed into the rear of the left-hand cylinder head, above the combustion chamber of the No. 4 cylinder. The sensor is concealed by the rubber seal indicated in Fig. 4-20. Temperature sensor II provides the control unit with the engine temperature information that is needed for correct starting and warm-up enrichment.

Fig. 4-20. Location (arrow) of temperature sensor II.

18 Fuel Injection

To test temperature sensor II, disconnect its lead wire. Then connect an ohmmeter as shown in Fig. 4-21. At 68°F (20°C), the resistance should be approximately 2500 ohms. If the resistance is considerably higher or lower, replace temperature sensor II.

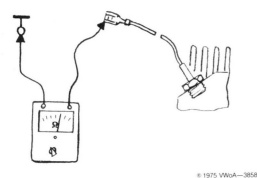

Fig. 4-21. Ohmmeter being used to test temperature sensor II.

To replace temperature sensor II:

1. Remove the left-hand air duct hose that connects the engine fan housing with the heat exchanger of the exhaust system. (The duct is already removed in Fig. 4-18, given previously.)

2. Disconnect temperature sensor II's lead wire. Then carefully pry the rubber seal out of the cylinder cover plate and slide the seal up and off the wire as shown in Fig. 4-22.

3. Using a deep socket wrench (so that the wire can be folded into the top of the socket), remove the temperature sensor.

4. Install the new sensor, using a reversal of the removal procedure. Torque the sensor to 1.5 mkg (11 ft. lb.). Make certain that the rubber seal fits tightly in the cylinder cover plate in order to prevent the loss of cooling air.

Fig. 4-22. Rubber seal (arrow) being slid up wire. Sensor can be seen at bottom of wire.

4.9 Testing and Replacing Injectors and Series Resistance Block

A constant supply of positive (+) battery current is delivered to the four injectors. This supply of current comes from terminal 88b of the double relay and is routed through four individual resistors that are contained in the series resistance block. See the wiring diagram given previously in **3. Troubleshooting Fuel Injection.** The purpose of the series resistance is to keep a stable voltage available to the injectors.

The control unit supplies negative (− or ground) current to all four injectors simultaneously at every other opening of the ignition distributor's breaker points. Each time the circuit through the injectors is completed by the application of ground, the solenoids of the injectors open the injector needle valves, allowing fuel from the ring main to spray into the engine's intake ports.

In the event of severe engine flooding (the trouble may be intermittent, becoming a major problem in damp weather), check the injectors for internal grounding. If one injector winding is grounded, owing to a breakdown of the insulation inside the injector, it will also ground the windings of the other injectors via the connections with the control unit. As a result all four injectors will open and discharge fuel continuously for as long as the ignition is turned on.

To test an injector for internal ground, disconnect the electrical plug. Apply one test probe of an ohmmeter to the injector case and the other test probe to first one and then the other of the two electrical terminals on the injector. If there is electrical continuity between either terminal and the injector case, the injector is faulty and should be replaced.

NOTE —
It is not necessary to remove the injectors from the engine in order to test them.

If none of the injectors is faulty, test the wiring by applying one ohmmeter test probe to a clean, unpainted metal part of the car and the other test probe to each of the terminals inside the four injector plugs. With the ignition turned off, there should be no electrical continuity between the injector wires and the car's chassis. Replace the wiring harness if any of the wires is grounded.

When the engine fails to run or fails to run properly, always troubleshoot the ignition system thoroughly before assuming that the problem is in the fuel injection. If the engine fails to start even though the fuel pump is functioning, it is sometimes best if you determine whether fuel is being injected before you proceed to more detailed troubleshooting.

Fuel Injection 19

To determine whether or not fuel is being injected, remove the injectors from the engine, leaving the wires and the fuel hoses attached. To keep the engine from starting during the test, disconnect the high tension cable from coil terminal 4.

> *CAUTION —*
>
> *Vaporized gasoline will be expelled during the following test. Do not disconnect the high tension cable from the distributor instead of the coil, as doing this could cause electrical sparks that may ignite gasoline being discharged from the injectors. Do not work near heaters or other fire hazards. Have a fire extinguisher handy.*

Run the starter briefly and see whether fuel is discharged from the injectors. If not, the fuel injection system is faulty. If one or more injectors fail to discharge fuel, disconnect the electrical plug(s) from the malfunctioning injector(s). Using an ohmmeter, test between the two terminals of each injector that failed to function. If there is no continuity, the solenoid winding of the injector is open-circuited and the injector must be replaced. If the injectors are not faulty, use a test light as shown in Fig. 4-23 to determine whether positive (+) battery current is reaching the series resistance block from terminal 88b of the double relay when the ignition is on.

Fig. 4-23. Series resistance block connector plug being tested for presence of current from double relay terminal 88b.

Once you have determined that positive (+) current is reaching the series resistance block, you can continue testing to determine whether any resistor is faulty. If no current is reaching the series resistance block, test the double relay and the wiring that connects double relay terminal 88b with the series resistance block. See the wiring diagram given in **3. Troubleshooting Fuel Injection.** You can test for voltage at the injector plugs as shown in Fig. 4-24.

Fig. 4-24. Test lamp being used to determine whether current is reaching injectors. Run starter during test. Light should flicker.

To replace a faulty series resistance block, disconnect its connector plugs from the wiring harness. Remove the two screws that hold the series resistance block cover to the car body, then remove the block. Installation is the reverse of removal.

In addition to failing to inject fuel, an injector can become faulty owing to leakage. To check for leakage, wipe clean the ends of the removed injectors (fuel hoses and wires still attached). Then, without operating the starter, turn on the ignition. If the injector jets become wet with fuel but lose no more than two drops of fuel per minute, you should test the pressure regulator in order to determine whether the pressure is excessive. If an injector jet loses more than two or three drops of fuel per minute, the injector itself is faulty and must be replaced.

> **NOTE —**
>
> White exhaust smoke after the engine has warmed up usually indicates leaking injectors or a leaking cold-start valve. High fuel consumption, starting trouble, and oil dilution are more likely to result from continuous operation of the cold-start valve than from leaking injectors. Continuous cold-start valve operation can be caused by accidentally reversing the wires at terminals 30 and 50 of the starter solenoid during engine or starter installation.

To replace injector(s):

1. Remove the large air duct hose(s) that connect(s) the engine's fan housing with the heat exchangers of the exhaust system.

2. Carefully disconnect the electrical plug from the injector(s) that must be removed.

3. Using a 10-mm socket wrench, remove the single bolt that holds the injector retainer plate to the intake manifold.

20 Fuel Injection

4. Being careful not to let the tips of the injectors rub or bump against the cylinder cover plate, withdraw the injector(s) from the intake manifold. Remove the injector seals and the retainer plate.

NOTE —
The inner seals sometimes stick inside the injector seat instead of coming out along with the injectors. Be sure to remove the seals. Whether you intend to install new injectors or to reinstall the original injectors, the nozzles must be clean and uncorroded. You can carefully twist off the plastic caps to check them.

5. Loosen the hose clamps, then remove the injector(s) from the ring main fuel hose(s). Two removed injectors are shown in Fig. 4-25.

WARNING —
Fuel may be expelled. Do not smoke or work near heaters or other fire hazards. Have a fire extinguisher handy.

CAUTION —
Do not drop the injectors or handle them carelessly. Doing so may damage the needle tips that project from the injector jets.

Fig. 4-25. Injectors removed from engine.

Installation is the reverse of removal. Use new injector seals during installation (there should be a small seal around the tip of each injector and a larger seal around the barrel of the injector). Torque the retainer bolt to 60 cmkg (52 in. lb.).

4.10 Testing and Replacing Fuel Pump and Fuel Filter

Fuel injection cars have an electric fuel pump. The pump and the replaceable in-line fuel filter are located beneath the front, right-hand part of the car, near the right front wheel (Fig. 4-26). The operation of the fuel pump is controlled by the pump relay in the double relay. During starting, the pump relay is operated by current from the starter solenoid. With the engine running, the pump relay is operated by current from the intake air sensor.

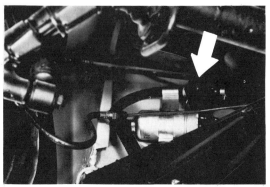

Fig. 4-26. Fuel pump and filter. Filter is at arrow; pump is silvery object behind and to the left of filter.

To test pump:

1. With the engine turned off, disconnect the ring main fuel hose from the cold-start valve.

    WARNING —
    Gasoline may be expelled when you disconnect the hose. Do not smoke or work near heaters or other fire hazards. Have a fire extinguisher handy.

2. Connect a fuel pressure gauge to the disconnected hose as shown in Fig. 4-27.

Fig. 4-27. Fuel pressure gauge correctly installed.

FUEL INJECTION

3. Operate the starter. The pressure gauge should indicate a pump pressure of approximately 35 psi (2.5 kg/cm²).

4. If the pump supplied pressure during the test but the pressure was far outside the normal range, test the pressure regulator as described in **4.5 Testing and Replacing Pressure Regulator.** If the pump failed to supply pressure, attach a test lamp to the fuel pump connector as shown in Fig. 4-28.

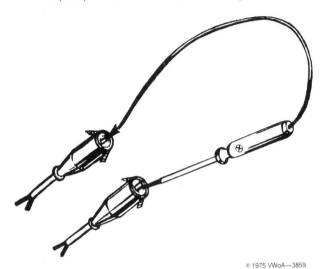

Fig. 4-28. Test lamp attached to fuel pump connector.

5. Operate the starter. If the test lamp lights, the fuel pump is faulty and should be replaced. If the test lamp does not light up, test the double relay as described in **4.1 Testing and Replacing Double Relay.** Also test the wiring using as a guide the wiring diagram given in **3. Troubleshooting Fuel Injection.**

If, after replacing the pump, there is still no fuel pressure, check for the following possible causes:

1. Fuel filter blocked
2. Rust or other foreign matter in fuel tank
3. Defective pressure regulator
4. Wires for fuel pump or fuel pump connections interchanged
5. Hoses interchanged at pump or at pressure regulator.

To replace fuel pump:

1. Raise the car on a hoist or support it on jack stands. If you are working near floor level, it will be helpful to remove the right front road wheel.

2. Disconnect the battery ground strap. Then disconnect the connector plugs from the pump.

3. Loosen the hose clamps. Then remove and quickly plug the fuel hoses that are attached to the pump.

WARNING ——
Fuel will be expelled when you disconnect the hoses. Do not smoke or work near heaters or other fire hazards. Have a fire extinguisher handy.

4. Take out the screws, then remove the pump from the car.

Installation is the reverse of removal. The fuel hose from the tank, which contains the filter, should be connected to the rounded end of the pump. The output line that goes to the pressure regulator should be connected to the flat end of the pump which also receives the connector plugs. Use new hose clamps and make sure that the hoses are tight fitting and in good condition.

To replace fuel filter:

1. Apply pinch clamps to squeeze shut the hoses from the old filter.
2. Loosen the hose clamps, then detach the hoses from the old filter.
3. Using new hose clamps, install a new filter; the arrow indicating flow direction should point toward the pump.
4. Remove the pinch clamps and check for leaks.

4.11 Making Tests at Control Unit Plug and Replacing Control Unit

If, in troubleshooting the fuel injection, you need to test the wiring that connects some component of the fuel injection system with the control unit, you can apply an ohmmeter or a test lamp to the terminals of the disconnected control unit plug. Disconnect the battery ground strap before you disconnect the plug from the electronic control unit. Then press back the spring catch that is at the rear of the control unit; this catch retains the rear end of the plug. Carefully pull out the rear end of the control unit plug in order to free it from the terminals of the control unit. If necessary, use a screwdriver to raise the plug gently.

If for any reason you must install a replacement control unit, make sure that the new control unit has the same part number as the original control unit. Several changes have been made since the first fuel injection models were introduced at the beginning of the 1975 model year. For example, the 1977 fuel injection system has an electronic control unit, Part No. 043 906 021 D,

22 Fuel Injection

which is different from any used on the 1975 and 1976 models. This new control unit makes possible the service installation of a high altitude corrector on vehicles that are regularly operated at altitudes greater than 4000 feet above sea level.

Under no circumstances should you attempt to make tests at the terminals of the control unit itself. Fig. 4-29 shows a typical test being made at the control unit plug. If necessary, you can reconnect the battery ground strap after the plug has been completely detached from the control unit.

Fig. 4-30. Control unit (left) and double relay (arrow) after removal of composition board cover.

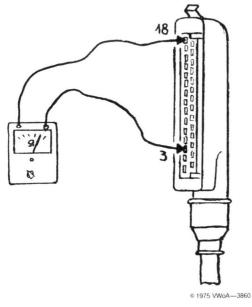

Fig. 4-29. Ohmmeter being used at control unit plug to test wiring to throttle valve switch.

In disconnecting or reconnecting the plug, use the procedure described in conjunction with control unit replacement. Be especially careful not to damage the plug connectors either during electrical testing or when you press the plug onto the terminals of the control unit.

Do not judge the control unit faulty until you have (1) measured the engine's compression and found it within specifications, (2) thoroughly checked the ignition system with special emphasis given to breaker point condition and correct dwell, and (3) tested all fuel injection system components and wiring. If no other fuel injection system fault is found, and the engine still does not run right, replace the control unit.

To replace control unit:

1. Working in the right-hand side of the luggage compartment, behind the rear seat backrest, lift off the composition board cover that houses the control unit and the double relay. See Fig. 4-30.

2. Press back the spring catch that is at the rear of the control unit, then pull out the rear end of the control unit plug—prying it up with a screwdriver if necessary.

3. Fully remove the plug and position it so that none of its terminals can come into contact with anything that could either damage the system electrically or damage the plug physically.

4. Remove the mounting strap for the control unit. Then remove the control unit from the car.

Installation is the reverse of removal. Be especially careful, when installing the plug, to engage all the control unit terminals squarely. Forcing the plug onto the control unit may bend or break a misaligned terminal. If you encounter difficulty, remove the plug and check whether any of the terminals is misaligned or bent before you again attempt to press the plug into position.

5. Adjusting Fuel Injection

During routine maintenance, you should adjust the idle speed only. Do not attempt to adjust the idle mixture, as described in **5.2 Adjusting Idle Mixture,** without a good quality infrared exhaust gas analyzer such as the Sun® EPA 75 Performance Analyzer.

5.1 Adjusting Idle Speed

Before you attempt to adjust the idle speed, it is imperative that you adjust the valves and the ignition timing to specifications as described in **ENGINE**. Also, for accurate idle speed adjustments, the engine should be thoroughly warmed up by a test drive so that the oil will be hot and the auxiliary air regulator will be fully closed.

Fuel Injection 23

On cars that have a throttle valve adjustment screw, you should never turn the screw to adjust the idle speed. Doing this will upset the operation of the ignition distributor's spark advance. The basic throttle valve adjustment is set at the factory and should not be changed. If the screw (Fig. 5-1) has been accidentally turned, you can correct the throttle valve adjustment as follows. First, turn the screw out until there is a gap between the tip of the screw and the stop plate of the throttle valve arm. Then turn in the screw until it just barely touches the stop plate. From this position, turn the screw in an additional ¼ to ½ turn.

Fig. 5-2. Idle speed being adjusted.

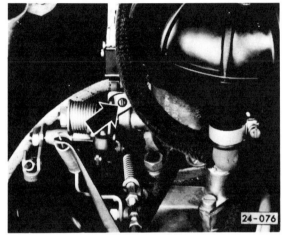

Fig. 5-1. Throttle valve adjustment screw (arrow). Never turn this screw to adjust idle speed.

To adjust idle speed:

1. When you return from the test drive, stop the engine. Then connect the red lead of a precision dwell meter/tachometer to ignition coil terminal No. 1; connect the black lead to ground on some clean, unpainted metal part of the engine. Adjust the instrument to measure rpm for a four cylinder engine.

2. Make sure that the tachometer leads are not near the V-belt and its pulleys. Then start the engine, speed it up, and allow it to return to idle.

3. The idle speed for cars with manual transmissions should be 800 to 950 rpm; the idle speed for cars with Automatic Stick Shift should be 850 to 1000 rpm. If the idle speed is too fast or too slow, adjust it.

4. To adjust the idle speed, use a screwdriver to turn the bypass screw in the throttle valve housing as shown in Fig. 5-2. Turning the screw clockwise slows the idle; turning the screw counterclockwise increases the idle speed. This adjustment does not alter exhaust emissions so long as the idle rpm is within the prescribed limits.

If turning the bypass screw fully out fails to increase the idle speed noticeably—or if the idle speed was exceptionally slow prior to adjustment—check for air leaks into the crankcase or around the intake manifold connecting hoses. Check especially that the oil filler cap is on tight and that no hoses are disconnected from the intake air distributor. If no air leak is found, check the EGR (exhaust gas recirculation) system as described in **7. Testing and Replacing EGR (Exhaust Gas Recirculation) Valve.**

If turning the bypass screw makes little change in the idle speed, it is likely that someone has tampered with the factory setting of the throttle valve adjustment screw. Check and, if necessary, correct the adjustment as described earlier under this heading.

If the bypass screw must be turned fully in, or nearly so, in order to obtain a slow enough idle—or if the idle speed was exceptionally fast prior to adjustment—check the auxiliary air regulator as described in **4.6 Testing and Replacing Auxiliary Air Regulator.** If the auxiliary air regulator is neither open nor faulty, check the pressure regulator as described in **4.5 Testing and Replacing Pressure Regulator.** Also inspect the pressure regulator's vacuum hose for leaks. If you find no other faults on a car with a manual transmission, check the deceleration air enrichment valve (commonly called the decel valve) as described under the following heading.

Checking and Replacing Decel Valve
(manual transmission only)

The deceleration air enrichment valve (commonly called the decel valve) is located on the car's left hood hinge mounting. Its purpose is to prevent the emission of excessive unburned hydrocarbons during closed throttle deceleration. If the decel valve is faulty (fails to close), the idle speed will be too fast. To check whether the decel valve is responsible for the fast idle, pinch closed the de-

24 Fuel Injection

cel valve's large, fabric-covered hose—not the vacuum hose indicated in Fig. 5-3. If the speed drops, the decel valve is faulty and must be replaced.

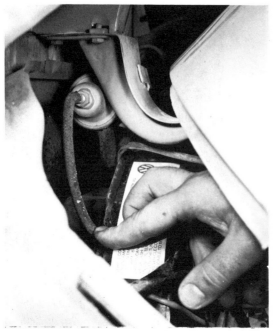

Fig. 5-3. Decel valve location. Vacuum hose is indicated by finger. Note location of emission decal on left side of fan housing.

To replace decel valve:

1. Stop the engine.
2. Disconnect the hoses from the decel valve.
3. Remove the screws that hold the decel valve to the car, then remove the decel valve.

Installation is the reverse of removal. The small vacuum hose should be connected to the fitting on the domed end of the new decel valve.

5.2 Adjusting Idle Mixture

It is unnecessary to adjust the idle mixture during a tune-up or routine maintenance unless exhaust emissions are outside the prescribed range. The idle mixture must be adjusted if you have replaced or repaired the intake air sensor or after extensive engine rebuilding. Professional mechanics should check their state laws to determine whether a shop without state authorization can make adjustments that affect exhaust emissions.

Do not attempt to adjust the idle mixture without a good quality infrared exhaust gas analyzer such as the Sun® EPA 75 Performance Analyzer. The instrument that you use for making idle mixture adjustments should measure parts per million of hydrocarbons and CO percentages.

To adjust idle mixture:

1. Adjust the idle speed as described in **5.1 Adjusting Idle Speed.**
2. Connect the exhaust gas analyzer to its power source and turn it on. Insert the probe into the car's tailpipe or into the port provided ahead of the catalytic converter on cars that are so equipped.
3. Compare the exhaust analyzer readings with the prescribed ranges for CO emissions and, where applicable, for hydrocarbon emissions. The applicable specifications are listed on the engine's emissions decal or are prescribed by the state government. (The emissions decal is on top the fan housing on the left side, as shown previously in Fig. 5-2.) If the decal is missing, check to see that the CO volume percentage at idle is between 0.2% and 2.0%.
4. If the CO level is outside the prescribed range, carefully pry out the plastic plug that covers the idle mixture adjusting screw. Then choose an offset screwdriver, or a short screwdriver that will fit the recessed screw, as shown in Fig. 5-4.

WARNING —
Be careful of the moving V-belt and pulleys. Serious injury will result if your fingers or clothing are caught in the moving machinery.

Fig. 5-4. Idle mixture screw in intake air sensor being adjusted. Pry out plastic plug for access to screw.

Fuel Injection 25

5. Turn the idle mixture adjusting screw one way or the other until emissions, as indicated by the exhaust gas analyzer, are within the correct range.

   NOTE
   Tuning the idle mixture adjusting screw clockwise (screwing it in) makes the mixture richer.

6. If necessary, readjust the idle speed to the prescribed range by turning the bypass screw.

7. If both the idle speed and the idle mixture are correct, replace the plastic plug that covers the mixture adjusting screw. Then disconnect the exhaust gas analyzer.

If the CO level cannot be adjusted to the prescribed range, check the following possible causes:

1. Fuel pressure too high owing to a faulty pressure regulator
2. Temperature sensor I in the cylinder head having too high a resistance (this can be caused by corrosion between the threads in the cylinder head and the threads of the sensor)
3. Leaking injector(s) or cold-start valve
4. Air flow restricted by a dirty air filter
5. Valve clearances incorrect, or CO level influenced by the re-evaporation of fuel that has condensed in the crankcase (especially in cold weather)
6. Poor combustion owing to worn-out spark plugs or faulty ignition cables.

6. Intake Air System

The intake air system includes the air cleaner, the throttle valve housing, the intake air distributor, and the intake manifold tubes. The intake air sensor, which is also part of the intake air system, is covered separately in **4.3 Testing and Replacing Intake Air Sensor**.

6.1 Removing and Installing Air Cleaner

Use the procedure given here if you intend to remove the air cleaner only. If you must remove the intake air sensor, remove the air cleaner body and the intake air sensor as a unit. See **4.3 Testing and Replacing Intake Air Sensor.**

To remove air cleaner:

1. Open the hood. Release the four clips indicated in Fig. 6-1.

2. Remove the cover from the air cleaner, carefully disengaging the cover from the warm air duct indicated in Fig. 6-1.

   CAUTION
   Be careful not to tear or damage the duct if it is stuck in the cover.

Fig. 6-1. Air cleaner cover removal. ATF dipstick is at **A**, clips are at **B**, and warm air duct is at **C**.

3. Remove the filter element from the body of the air cleaner (Fig. 6-2).

Fig. 6-2. Cover and filter element removed from air cleaner body.

4. Remove the four bolts that hold the air cleaner body to the intake air sensor; remove the two nuts that hold the air cleaner body to the hood's right-hand hinge mounting. See Fig. 6-3. Remove the body of the air cleaner.

Fig. 6-3. Nuts that hold air cleaner body to hinge mount (top arrows) and one of four bolts (lower arrow) that holds air cleaner body to intake air sensor.

Installation is the reverse of removal. Replace the filter element if it is damaged or clogged by dirt. Replace the warm air duct if it is damaged.

Checking Thermostatic Valve

The thermostatic valve, located in the cover part of the air cleaner, can be checked by removing the intake air duct from the cover. The flap of the thermostatic valve should be in position to close off cool air from the intake air duct when the thermostatic bulb is at room temperature. With the engine fully warmed up in warm weather, the flap should close off the supply of warm air from the warm air duct and open the path for cool air from the intake air duct. If the flap is not functioning correctly, replace the thermostatic valve assembly.

6.2 Removing and Installing Throttle Valve Housing

The throttle valve housing contains the throttle valve, which is cable-operated by the accelerator pedal. The housing also contains the bypass screw used to adjust idle speed. Except on 1977 and later California models, there is an inlet port for the EGR (exhaust gas recirculation) system in the throttle valve housing. On the later fuel injection engines, there is a throttle valve adjustment screw that prevents the throttle valve from contacting the housing bore, which could result in undesirable wear.

To remove throttle valve housing:

1. Remove the intake air sensor and the air cleaner body as a unit. See **4.3 Testing and Replacing Intake Air Sensor.** Then loosen the hose clamp and detach the large rubber elbow from the top of the throttle valve housing.

2. Disconnect the throttle return spring from the bracket on the EGR valve body. Except on 1977 and later California cars, remove the EGR inlet pipe and remove the EGR valve. See **7. Testing and Replacing EGR (Exhaust Gas Recirculation) Valve.** On 1977 and later California cars, detach the operating rod from the EGR valve.

3. Pry the E-clip off the clevis pin that holds the throttle linkage ball link to the arm on the throttle valve shaft. Remove the clevis pin, then detach the upper end of the ball link from the arm.

4. Disconnect the vacuum hoses from the throttle valve housing. If necessary, mark their positions for correct installation.

5. Remove the two nuts that hold the throttle valve housing to the intake air distributor. Then remove the throttle valve housing and its gasket.

Installation is the reverse of removal. Use a new gasket and torque the throttle valve housing nuts to 2.0 mkg (14 ft. lb.). Where applicable, torque the EGR system nuts to 1.0 mkg (7 ft. lb.). If it is necessary to clean the EGR ports in the throttle valve housing, be careful not to burr or deform the housing.

6.3 Removing and Installing Intake Air Distributor

The intake air distributor is an integral part of the alternator support. Normally, you will need to remove the intake air distributor only when you are preparing to disassemble the crankcase with the engine removed from the car. If you must remove the intake air distributor with the engine installed, you must partially remove the fan housing as described in **ENGINE** in order to raise the alternator off the intake air distributor.

To remove:

1. If not previously removed, remove the intake air sensor and the air cleaner as described in **4.3 Testing and Replacing Intake Air Sensor.** Remove the large rubber elbow from the top of the throttle valve housing.

2. Disconnect the throttle return spring from the bracket on the EGR valve. Remove the EGR inlet pipe. Remove the EGR valve. See **7. Testing and Replacing EGR (Exhaust Gas Recirculation) Valve.** (On 1977 and later California cars, you can leave

the EGR valve attached to the intake air distributor.)

3. Loosen the screw in the accelerator cable pivot pin of the throttle linkage. Then pull the cable out of the pivot pin and remove the pivot pin so that it will not become lost.

4. Remove the alternator and the fan housing as a unit. See **ENGINE**.

5. Disconnect all hoses from the intake air distributor. Remove the auxiliary air regulator together with its mounting bracket. See **4.6 Testing and Replacing Auxiliary Air Regulator**.

6. Remove the nuts that hold the oil filler pipe to the intake air distributor. Then remove the filler pipe and its gasket.

7. If the cylinder heads and intake manifold tubes are installed, slide the two manifold hoses off the intake air distributor and onto the intake manifold tubes.

8. Remove the four nuts that hold the intake air distributor to the crankcase. Then lift the intake air distributor off the crankcase and remove it toward the rear of the engine.

Installation is the reverse of removal. Use new gaskets at all points. Torque the nuts that hold the intake air distributor to the crankcase 2.0 mkg (14 ft. lb.). Torque the nuts that hold the oil filler pipe to the intake air distributor to 1.5 mkg (11 ft. lb.). Adjust the accelerator cable as described under the next heading.

Adjusting Accelerator Cable

To adjust the accelerator cable, loosen the setscrew that holds the cable end in the pivot pin of the throttle linkage. Have someone fully depress the accelerator pedal so that it is resting against its stop on the floorboard. Open the throttle by hand, sliding the pivot pin down onto the cable end. When the throttle valve is fully open, back it off by the smallest possible amount. Then tighten the setscrew in order to lock the cable end in the pivot pin of the throttle linkage.

The purpose of your adjustment should be to have the throttle valve as far open as possible when the accelerator pedal is against its stop. However, the accelerator pedal travel must be limited by the accelerator pedal stop and not by the tension on the cable that would result from the throttle valve contacting its stop.

Removing and Installing Manifold Tubes

To remove the manifold tubes (engine installed), remove the duct hoses that connect the fan housing to the heat exchangers of the exhaust system. Remove the injectors as described in **4.9 Testing and Replacing Injectors and Series Resistance Block**. Remove the two nuts that hold each intake manifold tube to the cylinder head. Then pull each intake manifold tube, its gasket, and its connecting hose off the cylinder head and the intake air distributor.

Installation is the reverse of removal. Use new gaskets. Replace loose or leaking connecting hoses. Torque the nuts that hold the manifold tubes to the cylinder heads to 2.0 mkg (14 ft. lb.).

7. TESTING AND REPLACING EGR (EXHAUST GAS RECIRCULATION) VALVE

On 1977 and later California models, the EGR valve is operated mechanically by a rod attached to the throttle valve lever (Fig. 7-1). As on other fuel injection cars covered by this Manual, no exhaust gases are recirculated at idle or at full throttle.

Fig. 7-1. 1977 and later California EGR linkage. Arrow indicates hex-with-pin that is turned in adjusting the EGR valve operation.

To adjust mechanical EGR valve operation:

1. Start the engine and allow it to warm up thoroughly. Make sure that the idle speed is correct, between 800 and 950 rpm.

2. Loosen both locknuts for the hex-with-pin adjusting rod, which was shown previously in Fig. 7-1. Then, by turning the hex-with-pin, shorten the length of the rod until the idle speed suddenly drops (EGR valve opens).

3. From this position, turn the hex-with-pin in the opposite direction by 1½ turns (the pin on the hex can be used for orientation).

4. Without moving the adjustment, tighten both locknuts.

28 Fuel Injection

To check the operation of a mechanical EGR valve, remove the E-clip that holds the operating rod to the EGR valve. With the engine idling, hand-open the EGR valve. The engine should slow down or stall, indicating that exhaust gases are being recirculated. If not, the EGR valve or the EGR pipe is clogged.

To check electro/vacuum EGR valve (except 1977 and later California cars):

1. With the engine idling, pull the plug (Fig. 7-2) off the EGR valve's vacuum unit. The engine should slow down noticeably or stall, indicating that exhaust gases are being recirculated.

2. If there is no change when the plug is pulled off the EGR valve at idle, stop the engine. Then turn the ignition key to the running (2) position without restarting the engine.

3. Connect a test lamp as shown in Fig. 7-2. Then operate the throttle valve by hand, moving the throttle valve shaft from its idle position into the mid-speed range.

Fig. 7-2. Test lamp being used to check whether battery current is reaching EGR valve plug. Arrow indicates installed location of plug on vacuum unit.

4. If the test lamp does not light at idle and at full throttle, test the throttle valve switch as described in **4.7 Testing and Replacing Throttle Valve Switch or Microswitch.** If neither the throttle valve switch nor the wire to the EGR valve is faulty, test the double relay as described in **4.1 Testing and Replacing Double Relay.** Make your tests with the help of the wiring diagram given in **3. Troubleshooting Fuel Injection.**

5. If the test lamp goes off when the throttle is moved off the idle position, and lights at idle and at or near full throttle, the EGR valve is faulty and should be replaced.

To replace EGR valve and inlet pipe:

1. To remove the inlet pipe only, remove the two nuts at each end of the pipe (Fig. 7-3). Then remove the pipe and its gaskets. Installation is the reverse of removal. Use new gaskets and torque the nuts to 1.0 mkg (7 ft. lb.).

   NOTE ——

   The procedure for removing the inlet pipe is the same on mechanically operated EGR valves. Removal and installation of the EGR feed pipe, which is under the car, is described in **ENGINE.**

Fig. 7-3. Nuts (arrows) that hold EGR inlet pipe.

2. If you must remove an electro/vacuum EGR valve, first remove the inlet pipe as described in step 1. Then disconnect the vacuum hose and the electrical plug from the EGR valve's vacuum diaphragm.

3. On both kinds of EGR valve, unhook the throttle return spring from the bracket on the EGR valve. Remove the two nuts (electro/vacuum valve) or the two socket-head screws (mechanical valve) that hold the EGR valve to the pipe or to the intake air distributor. Then remove the EGR valve and the gasket.

Installation is the reverse of removal. Use new gaskets at all points. Torque the nuts or screws to 1.0 mkg (7 ft. lb.).